Modern Approaches in Solid Earth Sciences

Volume 19

Background and motivation

Earth Sciences are going through an interesting phase as the traditional disciplinary boundaries are collapsing. Disciplines or sub-disciplines that have been traditionally separated in the past have started interacting more closely, and some new fields have emerged at their interfaces. Disciplinary boundaries between geology, geophysics and geochemistry have become more transparent during the last ten years. Geodesy has developed close interactions with geophysics and geology (tectonics). Specialized research fields, which have been important in development of fundamental expertise, are being interfaced in solving common problems.

In Earth Sciences the term System Earth and, correspondingly, Earth System Science have become overall common denominators. Of this full System Earth, Solid Earth Sciences – predominantly addressing the Inner Earth - constitute a major component, whereas others focus on the Oceans, the Atmosphere, and their interaction. This integrated nature in Solid Earth Sciences can be recognized clearly in the field of Geodynamics. The broad research field of Geodynamics builds on contributions from a wide variety of Earth Science disciplines, encompassing geophysics, geology, geochemistry, and geodesy. Continuing theoretical and numerical advances in seismological methods, new developments in computational science, inverse modelling, and space geodetic methods directed to solid Earth problems, new analytical and experimental methods in geochemistry, geology and materials science have contributed to the investigation of challenging problems in geodynamics. Among these problems are the high-resolution 3D structure and composition of the Earth's interior, the thermal evolution of the Earth on a planetary scale, mantle convection, deformation and dynamics of the lithosphere (including orogeny and basin formation), and landscape evolution through tectonic and surface processes. A characteristic aspect of geodynamic processes is the wide range of spatial and temporal scales involved. An integrated approach to the investigation of geodynamic problems is required to link these scales by incorporating their interactions. Scope and aims of the new series

The book series "Modern Approaches in Solid Earth Sciences" provides an integrated publication outlet for innovative and interdisciplinary approaches to problems and processes in Solid Earth Sciences, including Geodynamics.

It acknowledges the fact that traditionally separate disciplines or sub-disciplines have started interacting more closely, and some new fields have emerged at their interfaces. Disciplinary boundaries between geology, geophysics and geochemistry have become more transparent during the last ten years. Geodesy has developed close interactions with geophysics and geology (tectonics). Specialized research fields (seismic tomography, double difference techniques etc), which have been important in development of fundamental expertise, are being interfaced in solving common problems.

Accepted for inclusion in Scopus.

Prospective authors and/or editors should consult one of the Series Editors or the Springer Contact for more details. Any comments or suggestions for future volumes are welcomed.

More information about this series at http://www.springer.com/series/7377

Ochir Gerel • Franco Pirajno •
Bayaraa Batkhishig • Jaroslav Dostal
Editors

Mineral Resources of Mongolia

Springer

Editors
Ochir Gerel
Geoscience Center
Mongolian University of Science
and Technology
Ulaanbatar, Mongolia

Franco Pirajno
Centre for Exploration Targeting
The University of Western Australia
Crawley, WA, Australia

Bayaraa Batkhishig
Department of Geology and Hydrogeology
Mongolian University of Science
and Technology
Ulaanbaatar, Mongolia

Jaroslav Dostal
Department of Geology
Saint Mary's University
Halifax, NS, Canada

Responsible Series Editor: F. Pirajno

ISSN 1876-1682 ISSN 1876-1690 (electronic)
Modern Approaches in Solid Earth Sciences
ISBN 978-981-15-5942-6 ISBN 978-981-15-5943-3 (eBook)
https://doi.org/10.1007/978-981-15-5943-3

This Springer imprint is published by the registered company Springer Nature Singapore Pte Ltd.
The registered company address is: 152 Beach Road, #21-01/04 Gateway East, Singapore 189721, Singapore

Preface

Mongolia occupies the central part of the giant Central Asian Orogenic Belt—an accretionary orogen with long-time evolution since the early Palaeozoic to Mesozoic. The orogenic belt formed by accretion of Cambrian, Ordovician and Devonian-Carboniferous arcs, back-arcs and accretionary wedges contains world-class copper porphyry, and orogenic gold deposits originated in subduction zones and a number of important deposits related to the post-orogenic and intraplate settings.

The mineral deposit studies in Mongolia began in the middle of the last century by Russian and Mongolian geologists during systematic geological mapping and prospecting work on a scale of 1:1,000,000 and 1:200,000, as well as the research work of the joint Mongolian-Russian expedition of the Mongolian and Russian Academies of Sciences in the second half of the last century. The results of these detailed and comprehensive studies were summarized in a book "Geology of the Mongolian People's Republic, Volume III: Mineral Deposits", published in 1977. In parallel with this, several monographs by Russian and Mongolian scientists on deposits of copper, tungsten, gold, phosphates, coal, uranium, fluorite and others were published, and a database of mineral deposits and occurrences in Mongolia was completed, resulting in the publication of a "Distribution Map of Mineral Deposits and Occurrences in Mongolia at the scale of 1:1,000,000" developed by Mongolian Geological Information Center in 2002 in Mongolian and English languages. Data of Geological Information Center were summarized in the book "Guide to the Geology and Mineral Resources of Mongolia", published in English by Geological Exploration, Consulting and Service (GCS) Co. Ltd. A series of eight books, including "Metallic Mineral Deposits", "Non-metallic Mineral Deposits" and "Fuel Minerals", was published in 2010, and the second edition in 2012 in Mongolian language supported by the Erdenet Mine Corporation. Important data are available in the USGS professional paper 1765 (Nokleberg ed., 2010), which provides a comprehensive synthesis of the regional geology, tectonics and metallogenesis of Northeast Asia, including Mongolia.

The first chapter introduces readers to the geology and metallogeny of Mongolia. The next 12 chapters contain descriptions of 12 commodities of metallic, e.g. copper, gold, rare metals, rare earths, lead and zinc, iron group, silver and platinum group, non-metallic, namely fluorite and phosphates, and energy and fuel, namely uranium and coal, mineral deposits and occurrences. Every chapter includes tectonic position, classification, major metallogenic belts, host rocks and examples of major mineral deposits that are important for Mongolian economy based on existing data and new data on mineral commodities.

The main copper reserves are the porphyry Cu-Mo and Cu-Au deposits. The Oyu Tolgoi has reserves of 42 Mt of Cu and 1850 tonnes of Au. Other types of mineral systems include copper skarn, massive sulphide and sedimentary rock-hosted Cu in the Devonian and Silurian sequences. Mongolia has large proven coal reserves of Carboniferous, Permian, Jurassic and Cretaceous age. The Tavan Tolgoi deposit is a unique deposit of cooking coal of about 5 billion tonnes of coal reserves. Jurassic coal and Cretaceous deposits of coal and oil shales are of high economic potential. There are more than 600 fluorite deposits and occurrences of hydrothermal and epithermal origin associated mainly with Mesozoic plutonic and volcanic rocks. Gold occurs in placers and in deposits from Neoproterozoic to late Mesozoic age of hydrothermal, mesothermal and epithermal origin. According to the Mongolian Geological Information Center, uranium resources account for 1.39 Mt in volcanogenic and sedimentary deposits. In Mongolia, traditionally, tungsten and tin were mined in the last century from placers, greisen, stockworks, and quartz vein type deposits associated with late Paleozoic and Mesozoic granites. Rare earth mineral deposits and occurrences are associated with carbonatites and alkaline granites and syenites and some of them are of economic interest.

This book was compiled by authors from the Mongolian University of Science and Technology, Mongolian National University, and Ministry of Mining and Heavy Industry of Mongolia.

Ulaanbaatar, Mongolia Ochir Gerel
Crawley, WA, Australia Franco Pirajno
Ulaanbaatar, Mongolia Bayaraa Batkhishig
Halifax, NS, Canada Jaroslav Dostal

Acknowledgements

We are very grateful to Springer for the opportunity to publish this book. We would like to thank Mike Porter and Porter Publishing for permission and high-resolution figures in Copper chapter; thanks to professor Reimar Seltmann to give permission to use CERCAMS Book series "Geodynamics and Metallogeny of Mongolia with a Special Emphasis on Copper and Gold Deposits" by Reimar Seltmann, Ochir Gerel and Douglas Kirwin (eds); Elsevier and Springer publishing, Russian Nauka, Nedra and VSEGEI printing for permission to reuse data and figures from published book; special thanks to the Erdenet Mining Corporation (Mr. A. Undrakhtamir) for permission to reuse eight books series of the "Geology and Mineral Resources of Mongolia" J. Byamba (Ed); gratefully thanks Professors L.V. Agafonov, Yu. B. Mironov, J. Lkhamsuren, G. Ukhnaa, D. Altangerel and Dr. G. Jamsrandorj for permission to use figures from their published papers and books.

Also our authors thank to Imants Kavalieris for detailed comments, review, as well as providing important ideas in terms of geodynamic interpretations; Peter Lehmann provided generous, constructive, in-depth reviews, which improved the text for Chap. 12 (Uranium). Thanks to the Bayan Airag Exploration Company's exploration group, who provided information (unpublished Exploration results) to Chap. 2 (Copper).

We are grateful to Dr. J. Undaraya, A. Battushig, A. Nomuulin, B. Manzshir and A. Bolor-Erdene for their help to produce figures in the book chapters.

Contents

About the Editors

Ochir Gerel is Professor, Director of the Geoscience Center, and Consulting Geologist. She obtained her bachelor's and master's degrees from Charles University (Prague), PhD from Institute of Earth Crust, and ScD from Vinogradov Institute of Geochemistry, both of the Russian Academy of Sciences. O. Gerel was for 30 years head of the Department of Geology at the Mongolian University of Science and Technology. She is the author of more than 350 scientific publications, including scientific reports, textbooks, and books, and leader of many international projects. She is part of the editorial board of many international journals. Dr. Gerel is Past Vice President of the International Union of Geological Sciences, Executive Committee member of other professional societies, and Adjunct Professor at the Institute of Mineral resources, CAGS, Beijing. She is an Honored Scientist of Mongolia and State Prize Winner. Her scientific interests include petrology, geochemistry, mineral resources, and metallogeny.

Franco Pirajno received his PhD from University Federico II, Naples, Italy. He has considerable experience in tectonics, ore deposit geology, and mineral exploration in many parts of the world. He worked for the Anglo American Corporation of South Africa Ltd, as Exploration Geologist, following a spell as a postdoctoral research scientist at the Vesuvius Volcano Observatory, Exploration Manager of the Anglo American Corporation's activities in the South West Pacific, and the Chair of Exploration Geology at Rhodes University, Grahamstown, South Africa. In 1993 he joined the Geological Survey of Western Australia till October 2015. Currently, Franco is Adjunct Professor at the University of Western Australia, Institute of Mineral Resources at the CAGS, Beijing, Honorary Professor at China University of Geosciences, Honorary Research Fellow at CECAMS Natural History Museum, London, and Editor-in-Chief of *Ore Geology Reviews*. Franco is the author of 5 books. He was nominated as a top 1% Highly Cited Researcher in 2019.

Bayaraa Batkhishig is Associate Professor at the Department of Geology and Hydrogeology of the Mongolian University of Science and Technology since 2005. She earned a BSc degree (1996) and MSc degree (2000) in Geology from the Mongolian Technical University at Ulaanbaatar and obtained her PhD in Environmental Geochemistry in 2005 from the Tohoku University, Japan, with a thesis on the magmatic-hydrothermal system of the Shuteen Cu-Au Mineralized Complex (South Gobi, Mongolia). Dr. Batkhishig worked as a geologist in national projects and surveys and has been a Researcher at the Geoscience Center of the Mongolian Technical University. She is a member of the Society of Economic Geologists, the Geochemical Society, and the International Association on the Genesis of Ore Deposits (IAGOD). Her areas of expertise include economic geology, petrology, and geochemistry. She is Chief Editor of the open access journal *Mongolian Geoscientist.*

Jaroslav Dostal, PhD (McMaster University, Canada) is Professor Emeritus at Saint Mary's University (Halifax, Nova Scotia, Canada) where he has taught since 1975. He published over 300 peer-reviewed papers in international geology journals. At Saint Mary's, Dr. Dostal established a regional geochemical center. His research has been recognized by scientific awards including the Distinguished Scientist Award of the Atlantic Geoscience Society (Gesner Medal), the Career Achievement Award of the Volcanology and Igneous Petrology Division of the Geological Association of Canada, and the Hawley Medal, Mineralogical Association of Canada. Currently he is director of a resource company Ucore Rare Metals Ltd. His interests are geochemistry, mineral resources, igneous petrology, and geodynamics.

Chapter 1
Geology and Metallogeny of Mongolia

Ochir Gerel

1.1 Introduction

Mongolia occupies a central part of the giant Central Asian Orogenic Belt (CAOB) or Central Asian Foldbelt (Zonenshain et al. 1990; Mossakovsky et al. 1993), Altaids (Şengör et al. 1993) located between the Siberian Craton in the north, and Tarim and Sino-Korean cratons in the south. Different tectonic models were invoked to explain the origin of CAOB. The CAOB was formed by accretion of Cambrian, Ordovician, and Devonian-Carboniferous arcs, back-arcs, and accretionary wedges (Badarch et al. 2002; Windley et al. 2007; Tomurtogoo 2005) and Mesozoic-Cenozoic cover. Another view suggests that the CAOB comprises a collage of microcontinents and oceanic arcs that collided with one another and eventually accreted to the Siberian, Tarim, and Northern China cratons (Zonenshain et al. 1990, Mossakovsky et al. 1993, Dobretsov et al. 1995; Kröner et al. 2007; Safonova et al. 2011). This view corresponds to models of the continent and arc-continent collisions. In general, this interpretation looks similar to the present Southern-western Pacific style of accretion and may be considered as a major mechanism of large-scale orogen building. The CAOB according to Safonova (2016) is a Pacific-type belt consisting of numerous occurrences of accretionary complexes, intra-oceanic arcs, and oceanic plate stratigraphical units, and MORB-OIB-derived blueschist belts. Şengör et al. (1993), on the other hand, proposed a single intra-oceanic arc model for the evolution of the CAOB. This model regards the CAOB as an originally single and long chain of the Paleozoic intra-oceanic island arc systems and back-arc basins that were produced by continuous subduction-accretion processes (Şengör et al. 1993; Şengör and Natal'in 1996). Yakubchuk (2004) proposed that the Altaids are an

O. Gerel (✉)
Geoscience Center, School of Geology and Mining, Mongolian University of Science and Technology, Ulaanbaatar, Mongolia
e-mail: gerel@must.edu.mn

O. Gerel et al. (eds.), *Mineral Resources of Mongolia*, Modern Approaches in Solid Earth Sciences 19, https://doi.org/10.1007/978-981-15-5943-3_1

orogenic collage of Neoproterozoic-Paleozoic rocks consisting of three oroclinally bent Neoproterozoic-Early Paleozoic magmatic arcs, separated by sutures of their former back-arc basins, which were stitched by new generations of overlapping magmatic arcs. In addition, the Neoproterozoic to Cenozoic mantle plume-related magmatic rocks were superimposed on the accreted fragments. All these assemblages host important, many world-class, Late Proterozoic to Early Mesozoic gold, copper-molybdenum, lead-zinc, and other deposits. The Nd-Sr isotopic systematics of granitoid rocks in Central Asia (Jahn 2000) imply that more than 50% of the crust is juvenile, i.e., derived through melting of depleted mantle sources and thus implying exceptionally high crustal growth (Jahn 2004, Safonova 2009, Xiao et al. 2010). Kröner (2015) thought that the continental growth in the CAOB has been overestimated because this model includes a vast amount of Mesozoic granitoid rocks that are not related to the orogenic evolution of the CAOB but are now mainly explained as a result of mantle plume activity (Kovalenko et al. 2002). The Mongolian part of the CAOB is an orogenic structure comprising Neoproterozoic, Paleozoic, and Mesozoic subduction-accretionary structures and numerous Proterozoic microcontinents, mainly of the Gondwana series (Zavkhan, Tuva-Mongolia, Central Mongolia, and Southern Gobi). Subduction-accretionary structures are favorable for a different type of mineral deposits.

1.2 Geology of Mongolia

Historically, the territory of Mongolia is divided into two parts or domains: Northern and Southern. The Northern domain is mainly comprised of structures of Neoproterozoic and Early Paleozoic (Caledonian), whereas the Southern domain has Late Paleozoic (Hercynian) both bordered by a series of faults known as Main Mongolian Lineament. The Northern domain contains Proterozoic and lower Paleozoic rocks, including a cratonic fragment of Precambrian rocks, and a series of back- and fore-arc sequences, Proterozoic to lower Paleozoic intrusions, accretionary wedge sequences, and Late Neoproterozoic (Vendian) to Early Cambrian ophiolites. The Southern domain is largely Mid to Late Paleozoic, composed of Devonian to Carboniferous island arc volcanic rocks, but also includes sporadic Ordovician and Silurian volcanics, as well as Ordovician to Carboniferous sedimentary rocks, and is intruded by voluminous Permian-Carboniferous granitoids in the south (Badarch 2005). The tectonic setting of Mongolia is characterized by a complex block and mosaic structure, determined by a combination of faults, deformations, and extensive fractures. Major fold systems from north to south were established as follows: Northern Mongolian, Mongol Altai, Khangai-Khentii, Central Mongolian, Southern Gobi, and Inner Mongolian fold systems (Marinov et al. 1973; Tomurtogoo 2002; Yanshin 1974). The north and south domains are succeeded by Permo-Triassic magmatic arcs. In the Northern domain, the Mongol-Okhotsk suture marks the similar aged, progressive west to east, scissor-like closure of the Mongol-Okhotsk Sea that separated the main, previously accreted Central Asian Orogenic Belt from the Siberian craton, resulting in the formation of the Mongolian Orocline. This

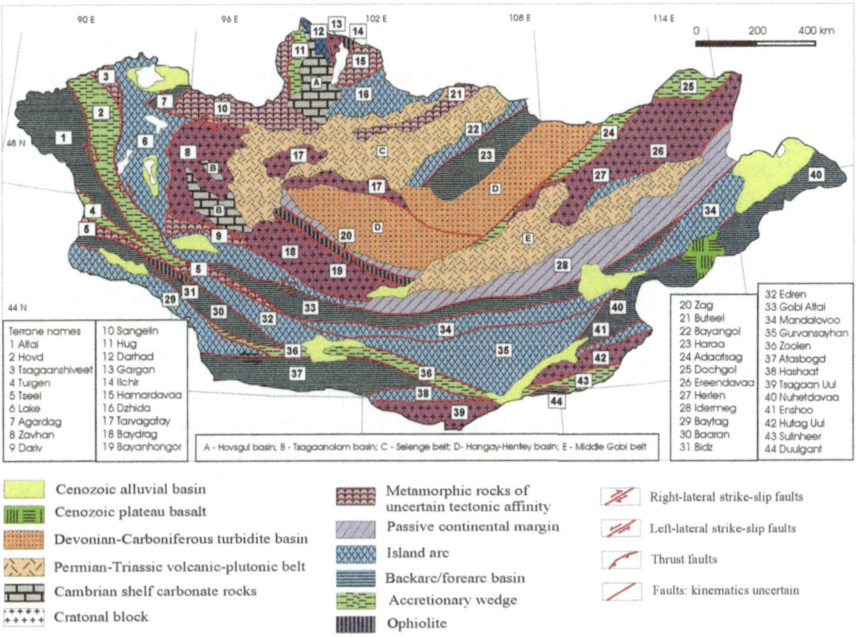

Fig. 1.1 Terrane tectonic map (Badarch et al. 2002)

closure was accompanied by the formation of the Permo-Triassic Orkhon-Selenge magmatic arc with extensive magmatism. During the Late Paleozoic, eastern and southern Mongolia underwent a period of basin-and-range style extension, accompanied by bimodal, basalt-peralkaline granite-comendite magmatism (Kovalenko et al. 1995) in a mature continental setting. The amalgamation of continental blocks and magmatic arcs within the main Central Asian Orogenic Belt was largely completed by the end of the Paleozoic-Early Mesozoic (Zonenshain et al. 1990; Şengör et al. 1993) accompanied by widespread uplift and associated thrusting which unroofed the magmatic arcs. During this period, Early Mesozoic continental sediments were deposited in thrust-controlled foreland basins. Major regional structures related to this period in western and central Mongolia include the SE-trending complex zone of sedimentary basins. Extensive intracontinental rifting and subsidence took place in southeastern Mongolia during the Late Jurassic to Early Cretaceous, with associated uplift and formation of syn-rift basins. Deposition of alluvial plain and eolian red beds continued into the Late Cretaceous. This extension was followed by Cenozoic transpressional tectonic events related to the Himalayan collision.

Badarch et al. (2002) proposed a "Terrane tectonic map of Mongolia" distinguishing 44 Paleozoic terranes: cratonal, island arc, active continental margin, accretionary wedge, passive continental margin, ophiolite, metamorphic, and overlap complexes (Fig. 1.1). A similar map with 35 terranes was proposed by Tomurtogoo (2002).

Cratonic terranes are composed of Proterozoic metamorphic complexes and Neoproterozoic metasedimentary and volcanic rocks. Badarch et al. (2002) distinguished seven cratonal blocks. The recognized 11 island arc terranes consist of ophiolites and tholeiitic to calc-alkaline volcanic and volcaniclastic rocks intruded by diorite and granodiorite plutons. The back-arc-fore-arc basin terranes contain lower Paleozoic volcaniclastic and sedimentary succession and minor slivers and melanges of ultramafic and volcanic rocks. Eight terranes are distinguished. Accretionary wedge terranes are narrow linear belts of highly deformed metamorphosed rocks containing melanges, thrust sheets, slivers of serpentinite, gabbro, fragments of ophiolitic rocks, and high-pressure schists. Five accretionary wedge terranes are distinguished. Three ophiolitic terranes contain ophiolitic rocks and melanges. Six metamorphic terranes contain Proterozoic and Paleozoic metamorphic complexes. Passive continental margin terranes comprise Neoproterozoic-lower Paleozoic shelf carbonate-quartzite sequences and deep marine sediments overlain by Devonian-Carboniferous and Permian volcanic and sedimentary rocks (Badarch et al. 2002). The main stages of geodynamic development and their metallogeny are discussed below.

1.2.1 Precambrian Microcontinents

Precambrian blocks or cratonic terranes (Badarch et al. 2002) are composed of Proterozoic metamorphic complexes and Neoproterozoic metasedimentary and metavolcanic rocks. The Proterozoic metamorphic complexes represent a crystalline basement. Badarch et al. (2002) described seven separate terranes, five cratonic terranes (Baidrag, Zavkhan, Tarvagatai, Gargan, Ereendavaa) in Northern and two (Khutag Uul, Tsagaan Uul) in Southern Mongolia (Fig. 1.1). Dergunov et al. (2001) distinguished three microcontinents: Tuva-Mongolia, Zavkhan, and Central Mongolia. Obtained ages confirm the existence of the Neoarchean-Neoproterozoic basement in the region. The Baidrag terrane (18, Fig. 1.1) composed of the Baidrag and Bumbuger metamorphic complexes contains Archaean tonalitic gneiss (U $\pm$ Pb zircon age of 2650 $\pm$ 30 Ma), granulite and amphibolite, and minor quartzite (Kozakov et al. 1997, 2014). The Bumbuger Complex comprises schist, gneiss, marble, quartzite, and charnockite (U-Pb zircon ages of 2364 $\pm$ 6), intruded by granite and granodiorite dikes (U-Pb zircon age of 1854 $\pm$ 5 Ma) (Kotov et al. 1995). Paleoproterozoic age of the Baidrag cratonic terrane is confirmed by the K-Ar phlogopite isotopic age of skarn (1.9 Ga). Thus, the U-Pb isochron and Pb-Pb zircon isochron ages range from 2.65 to 2.8 Ga for tonalite gneiss in the Baidrag metamorphic complex and 2.4 Ga for charnockite in the Bumbuger intrusive complex. Neoproterozoic ultramafic and mafic rocks, gabbro-norite, gabbro, gabbrodiorite, and gabbro-amphibolite, occur in cratons as small bodies and granodiorite and granite as larger plutons of 1120–650 Ma in Western and Northern Mongolia. In Southern Mongolia, gneissic biotite-muscovite granite in Khutag Uul cratonal terrane is dated by the U-Pb zircon method at 952 $\pm$ 8 Ma.

1.2.2 Neoproterozoic-Early Paleozoic

In Neoproterozoic-Early Paleozoic, Northern Mongolia was dominated by an island arc, whereas Southern Mongolia was an oceanic environment. In Neoproterozoic-Silurian (1000–410 Ma), passive continental margin of the Paleoasian ocean was formed along the Northern Asian craton followed by the formation of Neoproterozoic, Ediacaran (Vendian)-Cambrian, and Early Paleozoic magmatic arcs in subduction zones and associated metallogenic belts in the eastern part of the Paleoasian ocean. At the end of the Early Paleozoic, the eastern part of the Paleoasian ocean closed as a result of subduction and terrane accretion toward the Northern Asia craton (the largest Precambrian continental fragment) and cratonal margin (Parfenov et al. 2010).

The Neoproterozoic-Early Cambrian period is characterized by ophiolite associations formed mainly in the back-arc and fore-arc environment and island arc basalt-andesite-rhyolite series rocks. In Northern Mongolia, the eldest Shishged ultramafic massif of ophiolite complex composed of serpentinized dunite, harzburgite, and lherzolite is dated at 800 Ma (Kuzmichev et al. 2005). The Baidrag metamorphic complex is composed of mafic crystalline schist and amphibolite after basaltic andesite.

Ediacaran (Vendian)-Early Cambrian tonalite-granodiorite-granite suites formed in an island arc environment in the Lake (Nuur), Bayankhongor, Zed (Dzida), and Kherlen zones with the formation of island arc and back-arc ophiolite (Khain et al. 1995, 2003; Windely 1995). The isotopic ages of ophiolites vary between 573 and 498 Ma in Western Mongolia. The Lake zone, Khantaishir, and Bayankhongor gabbro Sm-Nd age is 522–569 Ma, and island arc basalt and plagiogranite have εNd (T) from +9.6 to +8 close to depleted mantle. Gabbro-norite of layered intrusion was dated at 695–531 Ma, and Darvi gabbro at 457 $\pm$ 40 Ma by the Sm-Nd method. Some gabbro and plagiogranite of the Bayankhongor ophiolite complex have higher values of εNd (T) from +11.8 to +11.5, while Lake zone gabbro and Darvi plagiogranite have εNd(T) of +7.0 $\pm$ 6.6 and +5.6 $\pm$ 5.4, respectively, showing that during Ediacaran through Cambrian, the source for the ophiolites was heterogeneously depleted mantle. Island arc, fore-arc, and back-arc depressions and microcontinents have collided before 490 Ma that has been confirmed by amphibolite to granulite facies metamorphism in island arc and microcontinents (Kovalenko et al. 1995; Windely 1995; Jahn et al. 2000). The ages of collisional and post-collisional tonalite-granodiorite-granite series are in the Zed belt 506 $\pm$ 2 Ma, in the Lake zone range from 495 $\pm$ 5 to 441 $\pm$ 5 Ma, and in the Darvi ridge are 457 $\pm$ 3 and 451 $\pm$ 6 Ma. Their εNd (T) is ranging from 7.7 to 5.2 and model ages from 0.74 to 0.61 Ga. Large batholith-like granitoid bodies common in Western and Central Mongolia were formed in a subduction environment and have a typical I type geochemical signature. The Lake zone is a giant island arc system (Kozakov et al. 2002) that consists of the ophiolite sequences of serpentinized peridotites, gabbros, pillow lavas, and bedded cherts (U-Pb zircon age of 570 Ma), island arc volcanics and siliceous-terrigenous complexes, and arc-related magmatic rocks (Kovach et al.

2011). Basalt-andesite-dacite island arc volcanic rocks of 545 Ma (Kovalenko et al. 2004) and sedimentary sequence are intruded by arc-related layered gabbro and quartz diorite-tonalite trondhjemite-granodiorite plutons dated between 530 and 480 Ma (Rudnev et al. 2009; Yarmolyuk and Kuzmin 2011; Janoušek et al. 2018).

1.2.3 Middle Paleozoic

In the Early and Middle Paleozoic, the accretionary-collision-related continental margin was destroyed, and new oceanic basins opened (Parfenov et al. 2010). Middle Paleozoic intrusive rocks are known from Mongolian Altai and Northern and Southern Mongolia. In Mongolian Altai, the Khovd and Altai complexes of gabbrodiorite-diorite-tonalite-plagiogranite and granite series dated at 426–413 Ma and in Southern Mongolia dated at 433 $\pm$ 12 Ma (Yarmolyuk et al. 2008b) occur. In Western Mongolia abundant subduction-related granitoid previously dated as Late Silurian Khovd and collisional Altai complexes were dated at 365 $\pm$ 10 Ma of Pb-Pb method (Demin 1990) and 356–382 Ma (Mustafa 2007) and by U-Pb method at 280–308 Ma (Žaček et al. 2016). Altai complex granites previously described as Late Devonian are dated even by the U-Pb method at 244 $\pm$ 1 Ma and 225–230 Ma (Demin 1990).

The Mongolian Altai system with prolonged calc-alkaline magmatism from Late Silurian to Carboniferous includes the Kharkhiraa accretionary wedge composed of low-grade ultramafics, metabasalts, and greenschist complex (Tomurtogoo 2002). The Khovd and Mongolian Altai turbidites are composed of sandstone and siltstones with gabbro and melanged dismembered ophiolite complex. The low-grade rocks are represented by terrigenous and volcanogenic rocks. The dominant metamorphic rocks include high-grade paragneiss, granite-gneiss, and schist with subordinate amphibolite. These high-grade rocks are interpreted to be high-grade metamorphic equivalents of the Ordovician sedimentary sequence (Jiang et al. 2011). Abundant granitoid rocks of arc type dated as Silurian-Devonian intruded the Ordovician gneiss and low-grade rocks (Yuan et al. 2007). Ordovician gabbro-diorite-plagiogranite calc-alkaline and tholeiitic series, low K, and metaluminous granitoid were formed in the active continental margin.

In Late Silurian to Early Devonian, the Mongol-Okhotsk intracontinental ocean opened (440 Ma). The volcano-sedimentary sequences of Late Silurian-Early Devonian age formed in troughs overlaid the old Kherlen-North Gobi microcontinent and granitoid intrusion of the same age. In the period of Middle Devonian to Early Carboniferous, the structural reconstruction of the Mongol-Okhotsk Ocean and opening of the Inner Mongolian oceanic basin with the formation of Sulinkheer ophiolite complex resulted in subduction-related magmatism. In Southern Mongolia silica-carbonate-volcanogenic sequence was deposited, and ophiolite remnants are found within Late Paleozoic sequences. Subduction-related volcanic rocks and granitoid have arc affinity with porphyry-style mineralization (Oyu Tolgoi and Tsagaan Suvarga deposits).

Granitoid rocks in Northern Mongolia were generated in the Paleozoic paleoocean active continental margin between 440–360 and 340–250 Ma. Granitoids have εNd (T) from +4.3 to +2.7 and +3.8 to +0.9 accordingly and model ages of 0.92–0.76 and 0.96–0.77 Ga.

1.2.4 Late Paleozoic

In the Late Carboniferous, during subduction, the continental margin arcs were formed, e.g., Khangai and Orkhon-Selenge. The Khangai arc was formed during the subduction of the Northern part of the Mongol-Okhotsk Ocean plate under the Siberian craton margin and previously accreted terranes. The Orkhon-Selenge arc (Permian to Jurassic) intrudes the Tuva-Mongolia superterrane. The arcs are related to the subduction of the Late Paleozoic and Early Mesozoic Mongol-Okhotsk Ocean plate beneath the Northern Asian craton and margin. In the Carboniferous, the back-arc basins of the Mongol-Okhotsk oceanic crust accreted to the Siberian craton, and in Southern Mongolia nappe-folded structures formed.

In Southern Mongolia, the ophiolite sequence started to form in an island arc (Ruzhentsev et al. 1985; Ruzhentsev and Pospelov 1992; Dergunov et al. 2001). Late Paleozoic continental crust originated from a transformation of oceanic crust to juvenile island arc and continental crust. Ophiolites have εNd from +7.9 to +6.9. Kovalenko et al. (2004) proposed that it was a relatively enriched mantle. Granitoid rocks of 300–260 Ma and 125 Ma and felsic volcanic rocks at εNd (T) +6.1 and +3.3 show Neoproterozoic-Cambrian model ages of 0.8–0.5 Ga.

Late Paleozoic continental magmatism is widespread in Central Mongolia. In Late Paleozoic, it was an active continental margin of Euroasia, and elongated volcanic-plutonic belts of calc-alkaline andesite, basaltic andesite, granodiorite-granite-leucogranite, and alkaline monzonite-syenite-quartz syenite series were generated. Many researchers recognized the Late Devonian-Early Carboniferous stage. Early Carboniferous granitoid rocks are known from Southern Mongolia, and Late Carboniferous granitoids formed in Mongolian Altai, Khangai-Khentii, and Northern and Southern Mongolia.

Large batholiths and plutons are present in Mongolia. Early Carboniferous continental arc-related granites are composed of biotite tonalite, quartz diorite, granodiorite, and granite. In Northern Mongolia, alkaline nepheline syenite and syenite with K-Ar age of 342–256 Ma (Yashina 1982) were formed in a rift or back-arc environment.

In Southern Mongolia, Late Devonian-Early Carboniferous and Carboniferous granitoids formed subduction-related volcanic-plutonic belts with calc-alkaline I-type granitoids, andesite series volcanic rocks, and alkaline monzodiorite-monzonite series rocks. Porphyritic quartz-monzodiorite is the main causative intrusion for the Oyu Tolgoi porphyry copper deposits dated at 373 $\pm$ 1 to 366.7 $\pm$ 2.6 Ma (Wainwright et al. 2011), and volcanic andesite in this area has Rb-Sr age of 333.8 $\pm$ 19.4 (Amar-Amgalan 2004). The Shuteen granitoid complex hosting the same name porphyry Cu-Au deposit has Rb-Sr age of 321.4 ± 9.6 (Batkhishig and Iizumi 2001) with adakitic geochemistry. All these porphyry Cu-Au and associated granitoid complexes are related to volcanic arc granites.

Permian volcanic and granitic rocks are the most abundant in Mongolia developing volcanic-plutonic complexes with volcanogenic and volcanogenic-sedimentary rocks. Many researchers classified the Permian period together with Early Triassic (208–240 Ma) (Kovalenko et al. 2006). Based on Kovalenko et al. (2006), granitoid rocks are mainly related to riftogenesis in Early Permian (270 Ma) in Gobi-Altai and Late Permian (265–249 Ma) in Northern Mongolian rift systems. The Khangai Batholith has a similar age (270–250 Ma) (Kovalenko et al. 2006). The Khangai Batholith that occupies an area of 120,000 km^2 intruded Devonian-Carboniferous turbidites and Precambrian and Neoproterozoic-Cambrian rocks consisting of subduction-related calc-alkaline I-type granitoid related to an active continental margin. The eNd(t) values in the Khangai Batholith granitoids vary from −0.1 to −1.3 showing depleted source containing approximately 80% of juvenile mantle-derived component (Jahn 2004). The U-Pb zircon ages of the Khangai Batholith granitoid vary from 261 ± 3 Ma to 241.3 ± 1.5 (Orolmaa et al. 2008; Yarmolyuk et al. 2008a, b; Yarmolyuk et al. 2019). The Khangai Batholith is surrounded by the Northern Mongolian and Gobi-Altai belts on its periphery (Yarmolyuk and Kovalenko 1991; Yarmolyuk et al. 2008a). These belts comprise a bimodal volcanic series (basalts, comendites, pantellerites, trachydacites, and trachyrhyolites) and alkaline intrusions. Whole-rock Rb-Sr isochron age of comendites and trachydacites located in the Northern Mongolian belt is 264 ± 4 Ma (Yarmolyuk et al. 2008b). In this stage the Paleozoic arc accreted to the Siberian continent and resulted in the intrusion of calc-alkaline granitic rocks and small- and medium-sized alkaline plutons. Permian granitoids are widespread in Central, Eastern, and Southern Mongolia. In Southern Mongolia, one of the world's largest Khanbogd alkaline granite plutons of 295 ± 5.3 and 267 ± 7 Ma Rb-Sr age (Amar-Amgalan 2004) was formed. In the Orkhon-Selenge, the Selenge complex shows Permian-Early Triassic age of 290–230 Ma.

1.2.5 Mesozoic

Mesozoic rocks are spread mainly in Eastern Mongolia. Two stages are distinguished: Early Mesozoic, e.g., Triassic-Early Jurassic (230–175 or 250–170 Ma), and Late Mesozoic, e.g., Mid-Late Jurassic-Early Cretaceous (175–135 or 171–100 Ma). Mesozoic magmatism is associated with the Mongol-Okhotsk belt that extends from Khangai to the Pacific Ocean for more than 3000 km (Zonenshain et al. 1990). The Mongol-Okhotsk Ocean as an embayment of the Paleo-Pacific existed in the Late Paleozoic-Early Mesozoic between the Siberian and Mongolia-Northern China continents (Zonenshain et al. 1990; Parfenov et al. 2001). The formation age of the Mongol-Okhotsk Ocean and related complexes is still not clear. According to Zonenshain et al. (1990), the Mongol-Okhotsk Ocean was a gulf of the Paleo-Pacific, which was separated after the joining of the Mongolian (Central-Mongolian) continental block and the Siberian continent approximately in the Khangai area in Early Carboniferous-Earliest Permian. Parfenov et al. (2001, 2003) and Bussien et al. (2011) considered that the Mongol-Okhotsk Ocean opened in the Late Ordovician-Early Silurian as a result of large-scale displacement. After Şengör et al. (1993), the Mongol-Okhotsk Ocean (Khangai-Khentii Ocean) was opened in Ediacaran-Cambrian time and separated the Siberian craton and the Tuva-Mongol massif. There is no consensus about the closure time of the Mongol-Okhotsk Ocean. After Maruyama et al. (1997), the Mongol-Okhotsk Ocean closed in the Triassic, after Zonenshain et al. (1990) in the Triassic-Late Jurassic whereas after Zorin (1999) and Parfenov et al. (2001) in the Early to Middle Jurassic. Most researchers agree that the eastern part of the Mongol-Okhotsk Ocean closed later, i.e., in the Late Jurassic-Early Cretaceous (e.g., Şengör and Natal'in 1996; Kravchinsky et al. 2002; Cogne et al. 2002; Yakubchuk and Edwards 1999). Relics of this large ocean remain as fragments of metamorphosed sediments and volcanics within the Mongol-Okhotsk belt/suture which fromed during its closure. The large Khentii Batholith and surrounding volcanic-plutonic belts established during subduction of the Mongol-Okhotsk Ocean beneath the continental margin of the Siberian continent (Parfenov et al. 2001; Tomurtogoo 2005). Yarmolyuk and Kovalenko (1991), Kovalenko et al. (2004), Kuzmin et al. (2010), and Yarmolyuk and Kuzmin (2011) proposed that the Khentii Batholith and volcanic-plutonic belts could have formed along an active continental margin above a mantle plume. Wickham et al. (1995) and Reichow et al (2010) are proposed that the same granitoid bathoiths and volcano-plutonic belts are related to an anorogenic intracontinental setting that had no connection to the evolution of the Mongol-Okhotsk Ocean. The Khentii Batholith of 225–195 Ma (Yarmolyuk and Kovalenko 2003) is a Mongolian part of large Khentii-Daurian Batholith. It includes more than 40 plutons composed of diorite-granodiorite and granite-leucogranites characterized by a zonal structure. The batholith is surrounded by rift zones with basaltic, bimodal, and peralkaline-granite magmatism. The granitoids of the Khentii Batholith have $\varepsilon Nd(T) = -2$ to +2. Granitoids of the batholith are similar in geochemistry and have a composition close to that of the continental crust, which points to the crustal source of anatectic magmas during the batholith formation (Yarmolyuk and Kovalenko 2003). The juvenile continental crust was composed of the Early Paleozoic mantle component and old continent (Kovalenko et al. 2006). Abundant granitoid rock of the Khentei Batholth range in age from 230 to 180 Ma and preserved arc-related geochemical signature (Gerel 2012).

Thereby, the Mongol-Okhotsk tectonic collage of Devonian to Late Jurassic, accreted in the Late Paleozoic through Early Mesozoic, consists mainly of the Permian-Jurassic Orkhon-Selenge and the Late Carboniferous-Early Permian Khangai continental margin arcs. These arcs are composed of continental margin igneous overlap assemblages and continental margin turbidite terranes and overlap the Southern margin of the Northern Asian craton and cratonic margin and previously accreted terranes. This tectonic collage is formed during the long-lived closure of the Mongol-Okhotsk Ocean with oblique subduction of terranes beneath the Southern margin of the Northern Asian craton and previously accreted terranes.

Starting from the Middle-Late Triassic, all tectonic processes appeared in a within-plate environment (Dergunov et al. 2001) with the development of rift structures and post-sedimentary basins with resources of hydrocarbons as well as structures with metamorphic core complexes (Badarch et al. 2002). The Mongol-Okhotsk and Sulinkheer belts (Kovalenko et al. 2004) originated during the closure of the Mongol-Okhotsk Ocean and Sulinkheer belt by Northern Asia China-Korean continent collision. In Mongol-Okhotsk belt, ophiolite age is 320 Ma (Tomurtogoo et al. 2005) and in Sulinkheer is 253 Ma. Basalts from the Adaatsag ophiolite have εNd (T) = +9.8 to +6.2 Mesozoic granites from Jargalant depression have εNd (T) +0.8 and model age of 0.33 Ga juvenile island arc crust. Intraplate magmatism has enriched mantle source EMI, EMII, also DM and HIMU (Kovalenko et al. 2004).

Continental clastic sediments were deposited in alluvial, deluvial, and lake environments. Lower-Middle Triassic sequence consisting of conglomerate and sandstones in the upper part is dominated by fine-grained shale with coal layers. The Upper Jurassic sequence contains alkaline volcanic rocks and associated REE deposits. Cretaceous sequences developed mainly in Eastern and Southern Mongolia are composed of rift-related bimodal volcanics and sedimentary rocks. Sedimentary rocks are rich in fossils and lignite deposits.

1.2.6 Cenozoic

Volcanic activity that started from Late Jurassic to Early Cretaceous was dominated by tholeiitic basalts and rhyolites in Central Mongolia and was formed mainly in a rift environment. Early Paleogene, Late Neogene-Pleistocene, and Holocene volcanic belts that formed in Northern, Central, and Southern Mongolia are interpreted to be a result of a mantle plume or continental rift and are represented by alkali K basalts, Na basalts, tholeiite, and calc-alkaline. Alkali basalts contain mantle xenoliths and megacrysts. Cenozoic terrigenous sediments are widespread in Western, Central, and Southern Mongolia.

1.3 Metallogeny and Mineral Deposits

During systematic mapping and prospecting work done in the past century, many energy and mineral deposits and occurrences were discovered: oil, coal, iron, copper, molybdenum, tungsten, tin, gold, silver, fluorite, phosphorite, salts, precious stones, and building materials (Marinov et al. 1977; Lkhamsuren 2012). Presently, more than 500 deposits and 6000 mineral occurrences of 80 different mineral commodities are known (Lkhamsuren et al. 2002). Three large metallogenic provinces have been identified in Mongolia: Altai-Sayan or North Mongolian with W, Mo, Cu, Au, and

Fe; Mongol-Transbaikalian or Eastern Mongolian with Cu, Mo, W, Zn, fluorite, Au, and Fe; and South Mongolian with Cu, Pb-Zn, Au, Mo, W, and REE (Marinov et al. 1977; Rodionov et al. 2004). From Neoproterozoic through Phanerozoic, subduction and accretion processes occurred, resulting in the formation of island and continental magmatic arcs with calc-alkaline andesitic magmatism and granitoids. In this setting the main types of economic metallic mineral deposits are porphyry Cu-Mo-Au; orogenic Au and Au placers; post-orogenic stockwork and vein complex of Sn, W, Mo, F, Be, Li, Ta, Nb, and Sn placers; REE carbonatite and alkaline intrusion-related REE-Nb-Zr; and sedimentary uranium, coal, and oil, all related to rift environment.

The main types can be grouped as follows: (1) copper-molybdenum and copper-gold porphyry; (2) greisen, vein, and stockwork of Sn, W, Mo, F, Be, and associated Sn placers; (3) skarn of Sn, W, Mo, Zn, and Cu; (4) vein gold and gold placers; (5) alkaline rock-related REE-Zr-Nb and Ta; (6) sedimentary deposits of phosphate and uranium; (7) fluorspar; and (8) coal and oil deposits.

1.3.1 Neoproterozoic-Early Cambrian Metallogeny

Neoproterozoic-Early Cambrian metallogenic belts are interpreted as pre-accretionary (Dejidmaa and Badarch 2005), related to an ancient island arc, continental margin arc, or passive continental margin. These metallogenic belts include Fe, Mn, and Au occurrences and large deposits of sedimentary phosphorite in Northern Mongolia. The metallogenic belt contains different deposits and occurrences associated with ultramafic rocks (podiform Cr, serpentine-hosted asbestos, and talc-carbonate) and mafic rocks (gabbroic Ni-Cu- Ti-magnetite) and Cu-Zn massive sulfide occurrences and occurrences related to calc-alkaline intrusions with Au-Cu-Fe skarn and gabbroic Cu-Ni with PGE (Fig. 1.2). The Ediacaran Early Cambrian Khuvsgul metallogenic belt is hosted in the Paleozoic passive continental margin terrane (Nokleberg 2010; Dejidmaa and Badarch 2005). This belt contains sedimentary phosphate deposits and occurrences and sedimentary Mn and sedimentary Fe-V occurrences. Sedimentary phosphate deposits are related to Ediacaran-Early Cambrian lower siliceous dolomite member, and sedimentary Fe and Mn and Fe-F occurrences are mainly above the productive phosphate deposition.

About 30 phosphate deposits and occurrences are known at the Khuvsgul phosphate basin. This basin which occupies 30,000 km^2 area trends north-south and is 300 km long and ranges from a few tens to 120 km wide (Ilyin 2004). In the Early Paleozoic calc-alkaline series and endogenic mineralization like Fe and Cu skarn, massive sulfide and polymetallic mineralization associated with ophiolite complexes were generated. The magmatic deposits are interpreted as having formed in the Late Neoproterozoic through the Early Cambrian Zed-Lake island arc. The sediment-hosted deposits are formed during seafloor spreading volcanism and related mafic-ultramafic magmatism. Middle Cambrian and Late Ordovician cratonal blocks

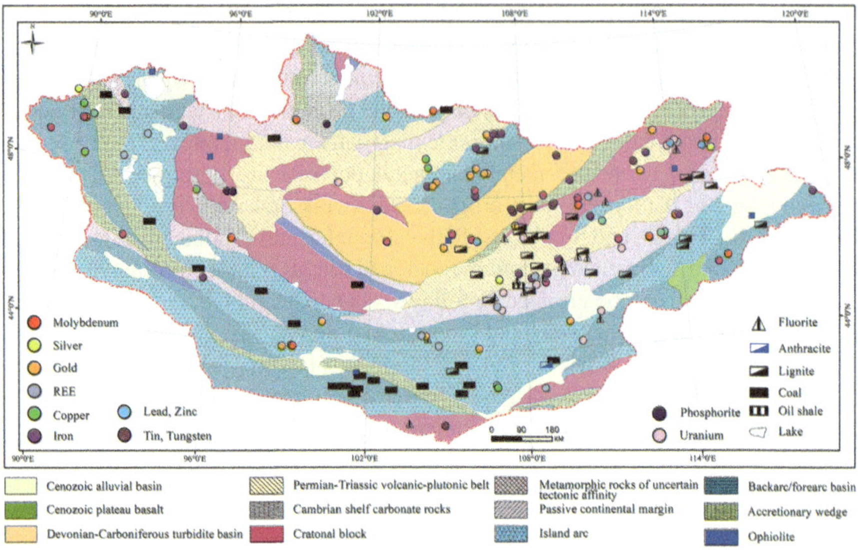

Fig. 1.2 Major mineral deposits in Mongolia (Terrane subdivision after Badarch et al. (2002))

contain Fe, Fe-Ti and Pb-Zn, and BIF metamorphic graphite, muscovite pegmatite associated with high metamorphic rocks, and bauxite.

1.3.2 *Middle Paleozoic Metallogeny*

Metallogenic belts related to island arcs contain a wide variety of island arc magmatism-related deposits and occurrences of volcanogenic Cu-Zn massive sulfide, volcanogenic Zn-Pb-Cu massive sulfide, volcanogenic-sedimentary Mn, volcanogenic-sedimentary Fe, barite vein, volcanic-hosted metasomatite, polymetallic (Pb, Zn $\pm$ Cu, Ba, Ag, Au) volcanic-hosted metasomatite, porphyry Cu-Mo ($\pm$Au, Ag), porphyry Cu ($\pm$Au), porphyry Cu-Au, and granitoid-related Au vein. The isotopic ages of the deposits or hosting units range from Devonian through Early Carboniferous (440–396 Ma). Several Cambrian-Silurian (540–410 Ma) metallogenic belts in Western and Central Mongolia possess geologic units favorable for major granitoid-hosted or related deposits, including the Bayankhongor belt in Central Mongolia (with Au in shear zone and quartz vein, granitoid-related Au vein, Cu-Ag vein, Cu skarn deposits), the Khovd belt in Mongolian Altai (with granitoid-related Au vein, Au skarn, and Cu skarn deposits), and Gobi-Altai with volcanogenic-sedimentary Fe and Mn occurrences. In Middle Paleozoic subalkaline, alkaline, and leucocratic magmatism occur, and these complexes are related to accretion (Dejidmaa and Badarch 2005). In western Mongolia Nb-Zr-REE deposits are associated with alkali granites and syenites of Devonian-Early Carboniferous magmatism, while in Northern Mongolia Nb-Zr-REE and Ta-Nb-REE are associated with nepheline syenites. In Southern Mongolia world-class porphyry Cu-Au deposit of Oyu Tolgoi and Tsagaan Suvarga, and also granitoid-related Au, polymetallic Pb-Zn + Cu Ag, and Au occurrences are known.

1.3.3 Late Paleozoic Metallogeny

The Late Paleozoic is characterized by accretion-rifting processes with an increase in the role of rare metals (REE, Zr, Nb, W, Sn with Mo, Pb-Zn, Au). Also, chromitite and Fe skarns formed. The majority of the deposits are associated with active continental margins with calc-alkaline and subalkaline volcanic rocks and intraplate riftogenic bimodal volcanics and alkaline rocks.

Late Carboniferous-Middle Triassic contains large belts, e.g., Central Mongolian belt with Fe-Zn skarn, Sn, Zn-Pb (±Ag, Cu) skarn, Au vein, and porphyry Mo deposits and occurrences, in Northern Mongolia the Orkhon-Selenge metallogenic belt with porphyry Cu-Mo deposits (e.g., Erdenet, and number of occurrences), and in Southern Mongolia the Kharmagtai-Khunguut-Oyut belt with Cu-Mo ± Au, Ag and Au-Ag epithermal veins.

In Southern Mongolia, the Middle Carboniferous to Early Permian metallogenic belt is related to granitoids of the Carboniferous Mandakh intrusive complex of the South Mongolian volcanic-plutonic belt. This Kharmagtai-Khunguut-Oyut belt extends from the southwest to the northeast for 450 km and ranges from 30 km to 60 km wide. Yakovlev (Marinov et al. 1977) first defined the Mandakh Cu district, and later Sotnikov et al. (1984) defined the South Mongolian porphyry Cu (±Au) metallogenic belt. The Mandakh intrusive complex consists of monzodiorite, granodiorite, granite, and deposit-hosting diorite porphyry and granodiorite porphyry stocks and dikes. The complex is comagmatic with andesite, dacite, and rhyolite volcanic rock of the Dushiin Ovoo Formation. Geological and isotopic age data indicate that plutons in the eastern belt are Late Carboniferous (Sotnikov et al. 1984), whereas in the western belt they are Late Carboniferous to Early Permian.

Permian Central Mongolian metallogenic belt occurs around the Khangai Batholith forming a sickle shape in Central Mongolia and hosting Fe-Zn and Sn; Zn-Pb (±Ag, Cu), W ± Mo ± Be, and Cu (±Fe, Au, Ag, Mo) skarns; porphyry Cu-Mo (±Au, Ag) and Mo (±W, Bi); granitoid-related Au vein and Cu-Ag vein; W-Mo greisen, stockwork, and quartz veins; and basaltic native Cu deposits and occurrences (Dejidmaa et al. 1996). The belt is situated along the Selenga transform continental margin arc along the Northern margin of the Mongol-Okhotsk Ocean (Nokleberg 2010). This belt is overlapping the Late Carboniferous-Late Permian continental margin arc that was tectonically linked to a subduction zone on the margin of the Mongol-Okhotsk Ocean. The skarn deposits are related to subalkaline granitoids. Vein and stockwork Mo occurrences in the central part of the belt are interpreted as having formed during the intrusion of Permian subalkaline leucogranite stocks. Various porphyry Mo deposits and occurrences are associated with Late Carboniferous or Permian granodiorite and monzonite porphyry stocks. Granitoid-related vein and stockwork Cu occurrences are more extensive in the northern and central parts of the belt. Various basalt native Cu occurrences are closely related to Permian basalts.

In Southern Mongolia, the Late Paleozoic metallogenic belts contain porphyry Cu-Mo (±Au, Ag), porphyry Au, granitoid-related Au vein, and Au-Ag epithermal vein deposits.

The Orkhon-Selenge assemblage overprints parts of the Late Archean and Paleoproterozoic Baidrag cratonal, Ediacaran through Middle and Late Cambrian Lake island arc and Neoproterozoic through Early Cambrian Idermeg passive marginal terranes. The Orkhon-Selenge metallogenic belt (Permian-Early Triassic) is situated along the Selenga transform continental margin arc along the Northern margin of the Mongol-Okhotsk Ocean. The transform margin consisted of oblique subduction of oceanic crust of the Mongol-Okhotsk Ocean under the Southern margin of the Siberian continent. The Late Permian through Early Jurassic plutonic rocks of the Orhon-Selenge metallogenic belt are part of the mainly Permian Selenga sedimentary-volcanic plutonic belt (Tomurtogoo et al. 1999). Remnants of this ocean are preserved in a narrow band that extends 3000 km from central Mongolia to the Okhotsk Sea. The major deposits are Erdenetiin Ovoo, Central, and Oyut; Shand Cu-Mo deposits; and the Zuiliin gol Cu-Mo occurrence (Gavrilova et al. 1984; Sotnikov et al. 1985a, b; Dejidmaa and Naito 1989, Lamb and Cox 1998, Gerel and Munkhtsengel 2005, Kavalieris et al. 2017).

1.3.4 Mesozoic Metallogeny

In the Mesozoic, the lithophile character of the metallogeny is still preserved, and Sn, W, Mo, Au. Pb-Zn, Ta, Nb, Be, and Au deposits occur in the active continental margin and collisional belts. In the Late Mesozoic, riftogenic structures formed, and the calc-alkaline character of the volcanic and intrusive rocks changed to alkaline with complex fluorite, REE, Sr, Be, Ta, Au, Pb-Zn, and fluorite mineralization.

In Central Khentii, Sn-W and W-Mo-Be greisen, stockwork, and quartz vein, REE-Li pegmatite, Ta-Li ongonite, Ta-Nb-REE alkaline metasomatite, peralkaline granitoid-related Nb-Zr-REE, and W ± Mo ± Be skarn deposits and occurrences are hosted in a Late Triassic and Early Jurassic granite belt that forms the Khentii megadome that is 600 km long, as much as 200 to 220 km wide, and trends north-east.

Late Triassic-Early Jurassic is characterized by granitoid-related mineral deposits in Central Mongolia: Sn-W greisen, stockworks, and quartz veins; REE-Li pegmatite; Ta-Li ongonite; Ta-Nb-REE alkaline metasomatite; peralkaline granitoid-related Nb-Zr-REE; W-Mo-Be greisen, stockworks, and quartz veins; and W ± Mo ± Be skarn deposits and occurrences. The host granites have K-Ar isotopic age of 190.5 ± 4.7 Ma and an Rb-Sr isotopic age of 225 to 188 Ma and are composed of two- or three-stage plutons formed in post-orogenic setting (Kovalenko and Koval

1984). There are several metallogenic belts with porphyry Cu; granitoid-related Au; Au in shear zone and quartz vein; Fe-Zn skarn, Cu skarn, Zn-Pb skarn, and Sn-skarn; Sn-W greisen, stockworks, and quartz veins; W skarn; Ta-Nb-REE alkaline metasomatite; and REE carbonatite.

In Mongol Altai, W-Mo-Be greisen, stockworks, and quartz vein deposits are related to small bodies of Early Jurassic age leucogranite that intrudes the Altai and Khovd terranes and is controlled by deep Khovd fault. In this region various REE deposits are related to Middle Devonian collisional, Carboniferous post-collisional, and Permian and Early Jurassic late-stage and post-orogenic granitoid (Dandar 2012). For the Mesozoic Mongol Altai metallogenic belt, the W-Mo-Be deposits and occurrences are associated with Early Jurassic granite. This belt is interpreted as having formed during Mesozoic continental intraplate rifting that resulted from the impinging of a mantle plume.

In Northern Khentii, the Middle Triassic-Middle Jurassic granitoid-related Au vein and Au in shear zone and quartz vein deposits are associated with small stocks and dikes in the margin of calc-alkaline granitoid of the Khentii Batholith along deep faults. These deposits are interpreted also as being of orogenic type. Early Mesozoic intrusive stocks consist of simple gabbro, and (or) multiphase plutons are composed of gabbro, diorite, and granite and single granite plutons with abundant gabbro schlieren. The K-Ar isotopic age of the granitoids ranges from 235 to 166 Ma (Koval 1998). Large Au-bearing conglomerate and placer Au deposits and occurrences have Late Cretaceous age (Tcherbakov and Dejidmaa 1984; Dejidmaa 2012). The metallogenic belt is interpreted as having formed during the generation of collisional granitoids in the final closure of the Mongol-Okhotsk at Mesozoic (Late Jurassic-Early Cretaceous). East Mongolian metallogenic belt contains polymetallic metasomatic carbonate and volcanic-hosted Zn-Pb (Ag, Cu, W) skarn; Au skarn; W-Mo-Be greisen, stockworks, and quartz veins; porphyry Mo (±W, Bi) (W, Sn, Bi); granitoid-related Au vein; carbonate-hosted As-Au metasomatite; Au-Ag epithermal vein; Sn-W greisen, stockwork, and quartz vein; fluorspar vein; and volcanic-hosted U deposits. Volcanic-hosted U deposits are associated with Late Jurassic-Early Cretaceous bimodal volcanic rocks and characterized by F-Mo-U and U-F apatite mineralization. The East Mongolian belt is affected by subduction zones, passive continental margin, and island arc terranes. Au deposits and occurrences are subdivided into two age groups. Many deposits are related to diorite, granodiorite, monzonite, and granite that occur in hypabyssal stocks and have K-Ar isotopic ages of 190–180 and 175–165 Ma. Other deposits are closely related to a Late Jurassic through Late Cretaceous basalt and rhyolite bimodal sequence. Granitoid-related vein and replacement Au-Te occurrences are closely related to microsyenite, lamprophyre, and diabase dikes with K-Ar isotopic ages of 220–190 Ma (Mironov 2006). Porphyry Mo-dominated Cu-Mo (±Au, Ag) (W, Au, Ag) occurrence and Au-Ag epithermal vein occurrences are located along the Main Mongolian Lineament. Major W-Mo deposits are related to Late Jurassic leucogranite and granite porphyry stocks in North-East Mongolia.

The major sediment-hosted U deposits occur in Mesozoic basins in southeastern Mongolia. The uranium mineralization is accumulated in the medium- to coarse-grained sandy, lacustrine sediments.

1.4 Concluding Remarks

Mongolia as a central part of the giant Central Asian Orogenic Belt situated between Siberian North Chinese and Tarim cratons is an accreted collage composed of Precambrian microcontinents (basement), Neoproterozoic-Early Cambrian ophiolitic units, Paleozoic arcs and back-arcs and accretionary wedges, and Mesozoic-Cenozoic cover. Mongolia is subdivided into two domains Northern and Southern bordered by faults known as a Main Mongolian Lineament.

The Northern domain contains cratonic fragments of Precambrian rocks and a series of back- and fore-arc sequences, Proterozoic to lower Paleozoic intrusions, and accretionary wedge sequences that include Late Neoproterozoic (Vendian) to Early Cambrian ophiolites. The Southern domain is largely a Mid to Late Paleozoic composed of Devonian to Carboniferous island arc volcanic rocks but also includes sporadic Ordovician and Silurian volcanic and Ordovician to Carboniferous sedimentary rocks and is extensively intruded by voluminous Permo-Carboniferous granitoids. Both North and South domains are succeeded by Permo-Triassic magmatic arcs. During the Late Paleozoic, western and southern Mongolia underwent a period of basin-and-range style extension, accompanied by bimodal, basalt-peralkaline granite-comendite magmatism in a mature continental setting. The amalgamation of continental blocks and magmatic arcs within the main Central Asian Orogenic Belt was largely completed by the end of the Paleozoic-Early Mesozoic accompanied by widespread uplift and associated thrusting which unroofed the magmatic arcs. This extension was followed by Cenozoic transpressional tectonic events related to the Himalayan collision. The position of Mongolia and accretionary structures is favorable for world-class deposits. From Neoproterozoic through Phanerozoic subduction and accretion, in island and continental magmatic arc environment, calc-alkaline andesitic magmatism and granitoids were generated, and the main types of economic metallic mineral deposits are porphyry Cu-Mo-Au; orogenic Au and Au placers; post-orogenic stockwork and vein complex of Sn, W, Mo, F, Be, Li, Ta, Nb, and Sn placers; REE carbonatite and alkaline intrusion-related REE-Nb-Zr formed in rift environment; and sedimentary uranium, coal, and oil.

References

Amar-Amgalan S (2004) Petrographical and geochemical studies on the Khanbogd alkaline pluton. Shimane University, Japan, Master thesis

Badarch G (2005) Tectonics of South Mongolia. In: Seltmann R, Gerel O, Kirwin DJ (eds) Geodynamics and Metallogeny of Mongolia with a special emphasis on copper and gold deposits. SEG-IAGOD Field Trip, 14–16 August 2005, 8th Biennial SGA Meeting, IAGOD Guidebook Series, CERCAMS/NHM, London, vol 11, pp 119–129

Badarch G, Cunningham WD, Windley BF (2002) A new terrane subdivision for Mongolia: implications for the Phanerozoic crustal growth of Central Asia. J Asian Earth Sci 21:87–110

Batkhishig B, Iizumi S (2001) Petrographical, petrochemical and geochronological study of the carboniferous Shuteen complex in South Mongolia. Geology 2:135–145

Bussien D, Gombojav N, Winkler W, Quadt A (2011) The Mongol-Okhotsk Belt in Mongolia – an appraisal of the geodynamic development by the study of sandstone provenance and detrital zircons. Tectonophysics 510:132–150

Cogne JP, Kravchinsky VA, Halim N, Hankard F (2002) Late Jurassic – early cretaceous closure of the Mongol-Okhotsk Ocean demonstrated by new Mesozoic palaeomagnetic results from the trans-Baikal area (SE Siberia). Geophys J Int 63(2):813–832

Dandar S (2012) Wolfram, tsagaan tugalgany ord ilrel (tungsten, tin deposits and occurrences). In: Lkhamsuren J (ed) Metallic mineral deposits. Soyombo Printing, Ulaanbaatar, pp 120–192

Dejidmaa G (2012) Gold deposits. In: Lkhamsuren J (ed) Metallic mineral deposits. VI Soyombo Printing, Ulaanbaatar, pp 215–264

Dejidmaa G, Badarch G (2005) Summary of pre-accretionary and accretionary metallogenic belts of Mongolia. In: Seltmann R, Gerel O, Kirwin D (eds) Geodynamics and Metallogeny of Mongolia with a special emphasis on copper and gold deposits, CERCAMS/NHM, London, pp 25–31

Dejidmaa G, Gantomor B, Gundsambuu Ts et al. (1996) Mongol ornii metallogenii zurag masshtab 1: 1,000,000 (Metallogenic map of Mongolia at the scale of 1,000,000: Geologic Information Center, Ulaanbaatar, Mongolia, Open-File Report 5023

Dejidmaa G, Naito K (1989) Previous studies on the Erdenetiin Ovoo porphyry copper-molybdenum deposit, Mongolia. Bull Geol Surv Jpn 49:299–308

Demin A (1990) Razrabotka skhem stratigrafii, magmatizma i voprosov metallogeni dlya s"yemochnykh i poiskovykh rabot masshtaba 1:50000 v Mongol'skom Altaye za 1987–90 g. (Development of stratigraphy, magmatism, and metallogeny schemes for mapping and prospecting on a scale of 1: 50,000 in the Mongolian Altai for 1987–90) Open-file report

Dergunov AB, Kovalenko VI, Ruzhentsev SV, Yarmolyuk VV (2001) Tectonics, Magmatism, and Metallogeny of Mongolia. Routledge, Taylor and Francis Group, London, New York

Dobretsov NL, Berzin NA, Buslov MM (1995) Opening and tectonic evolution of the Paleo-Asian Ocean. Int Geol Rev 37:335–360

Gavrilova SP, Maximyk IE, Orolmaa D (1984) Features of magmatism and composition of ore of Erdenetiin Ovoo copper-molybdenum deposit, in Endogenic ore-formations of Mongolia: Nauka, Moscow, 101–105

Gerel O (ed) (2012) Intrusive chuluulag (Intrusive rocks)Volume III, Second ed.. Soyombo Printing, Ulaanbaatar, p 460

Gerel O, Munkhtsengel B (2005) Erdenetiin Ovoo porphyry copper-molybdenum deposit in northern Mongolia. In: Seltmann R, Gerel O, Kirwin D (eds) Geodynamics and metallogeny of Mongolia with special emphasis on copper and gold deposits, London, pp 85–103

Ilyin AV (2004) The Khubsugul phosphate-bearing basin: new data and concepts. Lithol Miner Resour 39(5):454–467

Jahn BM (2004) The central Asian Orogenic Belt and growth of the continental crust in the Phanerozoic. Trans R Soc Edinb Earth Sci 91:181–193

Jahn BM, Wu F, Chen B (2000) Granitoids of the central Asian Orogenic Belt and continental growth in the Phanerozoic. Geol Soc Am Spec Pap 350:181–193

Janoušek V, Jiang Y, Buriánek D, Schulmann K, Hanžl P, Soejono I, Kröner A, Battushig A, Erban V, Lexa O, Turbat G, Košler J (2018) Cambrian–Ordovician magmatism of the Ikh-Mongol arc system exemplified by the Khantaishir magmatic complex (Lake Zone, south–Central Mongolia). Gondwana Res 54(2):122–149. https://doi.org/10.1016/j.gr.2017.10.003

Jiang YD, Sun M, Zhao GC, Yuan C, Xiao WJ, Xia XP, Long XP, Wu FY (2011) Precambrian detrital zircons in the early Paleozoic Chinese Altai: their provenance and implications for the crustal growth of Central Asia. Precambrian Res 189:140–154

Kavalieris I, Khashgerel D, Morgan LE, Undrakhtamir A, Borohul A (2017) Characteristics and 40Ar/39Ar geochronology of the Erdenet Cu-Mo deposit, Mongolia. Econ Geol 112:1033–1053

Khain EV, Bibikova EV, Sal'ikova EB, Kröner A, Gibsher AS, Didenko AN, Degtyarev KE, Fedotova AA (2003) The Palaeo-Asian Ocean in the Neoproterozoic and early Palaeozoic: new geochronologic data and palaeotectonic reconstructions. Precambrian Res 122:329–358

Khain EV, Neimark LA, Amelin Yu V (1995) Caledonskiy etap remobilizatsii dokembriiskogo fundamenta Grganskogo bloka, Vostochnye Sayany (the Caledonian stage of remobilization of the Precambrian basement of the Gargan block), the Eastern Sayan (isotopic geochronological data). Dokl Akad Nauk 342:776–780

Kotov AB, Kozakov IK, Bibikova EV, Sal'nikova EB, Kirnozova TI, Kovach VP (1995) Prodolzhitel'nost regional'no metamorphicheskych epizodov v oblastyach politsiklicheskych endogennych protsessov—U-Pb geochrnologiya (duration of regional metamorphic episodes in areas of polycyclic endogenic processes—a U–Pb geochronological study). Petrology 3:567–575

Kovach VP, Yarmolyuk VV, Kovalenko VI, Kozlovsky AM, Kotov AB, Terent'eva LB (2011) Composition, sources, and mechanisms of formation of the continental crust of the Lake Zone of the Central Asian Caledonides. II. Geochemical and Nd isotope data. Petrology 19:399–425

Koval PV (1998) Regional'ny geochemicheskiy analiz granitoidov (Regional geochemical analysis of granitoids). Siberian Branch RAS SPC UIGGM, Novosibirsk 489

Kovalenko VI, Koval PV (1984) Endogennye redkozemel'nye I redkometal'nye rudnye formatsii Mongolii (Endogenic rare earth element and rare metal ore formations in Mongolia. In: Endogenic ore formations of Mongolia 38:50–75

Kovalenko VI, Yarmolyuk VV, Bogatikov OA (1995) Magmatism, geodynamics and metallogeny of Central Asia. MIKO, p 272

Kovalenko VI, Yarmolyuk VV, Kovach VP, Kotov AB, Kozakov IK, Sal'nikova EB, Larin AM (2004) Isotope provinces, mechanisms of generation and sources of the continental crust in the central Asian mobile belt: geological and isotopic evidence. J Asian Earth Sci 23(5):605–627

Kovalenko VI, Yarmolyuk VV, Kovach VP et al. (2006) Tipy magm i ikh istochniki v istorii Zemli (Types of magmas and their sources in the history of the Earth) Part II, M, IGEM RAN, 275

Kovalenko VI, Yarmolyuk VV, Vladykin NV, Ivanov VG, Kovach VP, Kozlovsky A, Kostitsyn YA, Kotov AB, Sal'nikova EB (2002) Epochs of formation, geodynamic setting, and sources of rare-metal magmatism in Central Asia. Petrology 10(3):199–221

Kozakov IK, Kotov AB, Kovach VP, Sal'nikova EB (1997) Crustal growth in the geologic evolution of the Baidarik Block, Central Mongolia: evidence from Sm-Nd isotopic systematics. Petrology 5(3):201–207

Kozakov IK, Kovach VP, Bibikova EV et al (2014) Late Riphean episode in the formation of crystalline rock complexes in the Dzabkhan microcontinent: geological, geochronologic, and Nd isotopic-geochemical data. Petrology 22(5):480–506

Kozakov IK, Sal'nikova EB, Khain EV, Kovach VP, Berezhnaya NG, Yakovleva SZ, Plotkina YV (2002) Early Caledonian crystalline rocks of the Lake zone in Mongolia: formation history and tectonic settings as deduced from U–Pb and Sm–Nd datings. Geotectonics 36:156–166

Kravchinsky VA, Cogné JP, Harbert WP, Kuzmin MI (2002) Evolution of the Mongol-Okhotsk Ocean as constrained by new palaeomagnetic data from the Mongol-Okhotsk suture zone, Siberia. Geophys J Int 148:34–57

Kröner A (2015) The Central Asian Orogenic Belt—present knowledge and comparison with the SW Pacific. In: Central Asian Orogenic Belt: geology, evolution, tectonics, and models. Borntraeger Science Publishers, Stuttgart, pp 1–6

Kröner A, Windley BF, Badarch G, Tomurtogoo O, Hegner E, Jahn BM, Gruschka S, Khain EV, Demoux A, Wingate MTD (2007) Accretionary growth and crust formation in the Central Asian

Orogenic Belt and comparison with the Arabian-Nubian shield, 4-D Framework of Continental Crust, 200:181–209
Kuzmichev A, Kröner A, Hegner E, Dunyi L, Yusheng W (2005) The Shishkhid ophiolite, Northern Mongolia: a key to the reconstruction of a Neoproterozoic island-arc system in central Asia. Precambrian Res 138:125–150
Kuzmin MI, Yarmolyuk VV, Kravchinsky VA (2010) Phanerozoic hot spot traces and paleogeographic reconstructions of the Siberian continent based on interaction with the African large low shear velocity province. Earth Sci Rev 102:29–59
Lamb MA, Cox D (1998) New 40Ar/39Ar age data and implications for porphyry copper deposits of Mongolia. Econ Geol 93:524–529
Lkhamsuren J (ed) (2012) Metal ashigt maltmal (metallic mineral resources). Soyombo Printing, Ulaanbaatar, p 362
Lkhamsuren J, Dejidmaa G, Gerel O, Dandar S, Batjargal Sh, Bold-Erdene B, Batbold D, Begzsuren B, Bat-Erdene B (eds) (2002) Distribution Map of mineral deposits and occurrences in Mongolia. Scale 1:1000000, Mineral Resources Authority of Mongolia, Geologic Information Center, Ulaanbaatar
Marinov NA, Khasin RA, Khurtz C (eds) (1977) Geologiya Mongol'skoi Narodnoi Respubliky: poleznye iskopaemye (Geology of the People's Republic of Mongolia, vol. III: Mineral deposits). Nedra, Moscow, p 703
Marinov NA, Zonenshain LP, Blagonravov VA (eds) (1973) Geologiya Mongol'skoi Narodnoi Respubliky: magmatism, metamorphism, tectonika (Geology of the Mongolian People's Republic) 2. Magmatism, metamorphism, tectonics. Nedra, Moscow, p 782
Maruyama S, Isozaki Y, Kimura G, Terabayashi M (1997) Paleogeographic maps of the Japanese Islands: plate tectonic synthesis from 750 ma to the present. Island Arc 6:121–142
Mironov YB (2006) Uranium of Mongolia. Center for Russian and Central Eur Asian Mineral Studies (CERCAMS), London, 230
Mossakovsky AA, Ruzhentsev SV, Samygin SG, Kheraskova TN (1993) Tsentral'no-Aziatskiy skladchaty poyas: geodynamicheskaya evolyutsiya I istoriya fromirovaniya (the Central Asia foldbelt: geodynamic evolution and formation history). Geotektonika 27:445–473
Mustafa S (2007) Geology and metallogeny of the central part of Mongol Altaids (Hovd) final report
Nokleberg WJ (ed.) (2010) Metallogenesis and Tectonics of Northeast Asia. USGS Professional paper 1765
Orolmaa D, Erdenesaikhan G, Borisenko AS et al (2008) Permian–Triassic granitoid magmatism and metallogeny of the Hangay (Central Mongolia). Russ Geol Geophys 49(7):534–544
Parfenov LM, Berzin NA, Badarch G et al. (2010) Tectonic and Metallogenic Model for Northeast Asia. Chapter 9. Metallogenesis and Tectonics of Northeast Asia (ed. Nokleberg) U.S. Geological Survey Professional Paper 1765
Parfenov LM, Berzin NA, Khanchuk AI, Badrach G, Belichenko VG, Bulgatov AN, Dril' SI, Kirillova GL, Kuz'min MI, Nokleberg WJ, Prokop'ev AV, Timofeev VF, Tomurtogoo O, Yab H (2003) Model formirovaniya orogennych poyasov Central'noi I Severo-Vostochnoi Azii (a model for the formation of orogenic belts in central and Northeast Asia). Tikhookeanskaya Geologiya 22(6):7–41
Parfenov LM, Popeko LI, Tomurtogoo O (2001) Problems of tectonics of the Mongol-Okhotsk orogenic belt. Geology of the Pacific Ocean 16(5):797–830
Reichow MK, Litvinovsky BA, Parrish RR, Saunders AD (2010) Multi-stage emplacement of alkaline and peralkaline syenite-granite suites in the Mongolian–Transbaikalian Belt, Russia: evidence from U–Pb geochronology and whole-rock geochemistry. Chem Geol 273:120–135
Rodionov SM, Obolensky AA, Dejidmaa G, Gerel O, Hwang DH, Miller RJ, Nokleberg WJ, Ogasawara M, Smelov AP, Yan H, Seminskii ZV (2004) Descriptions of metallogenic belts, methodology, and definitions for Northern Eastern Asia mineral deposit location and metallogenic belt maps: U.S. Geological Survey Open-File Report 2004–1252, CD-ROM, explanatory text, 442

Rudnev SN, Izokh AE, Kovach VP, Shelepaev RA, Terent'eva LB (2009) Age, composition, sources, and geodynamic environments of the origin of granitoids in the northern part of the Ozernaya zone, western Mongolia: growth mechanisms of the Paleozoic continental crust. Petrology 17:439–475

Ruzhentsev SV, Badarch G, Voznesenskakaya TA (1985) Tectonics of the Trans-Altai zone of Mongolia (Gurvansaikhan and Dzolen ranges). Geotectonika 19:276–284

Ruzhentsev SV, Pospelov II (1992) The south Mongolian Variscan fold system. Geotectonics 30:383–395

Safonova I (2016) Juvenile versus recycled crust in the Central Asian Orogenic Belt: implications from ocean plate stratigraphy, blueschist belts and intra-oceanic arcs. Gondwana Res 47:6–27. https://doi.org/10.1016/j.gr.2016.09.003

Safonova I, Seltmann R, Kröner A, Gladkochub D, Schulmann K, Xiao W, Komiya T, Sun M (2011) A new concept of continental construction in the Central Asian Orogenic Belt (compared to actualistic examples from the Western Pacific). Episodes 34:186–194

Safonova IY (2009) Intraplate magmatism and oceanic plate stratigraphy of the Paleo-Asian and Paleo-Pacific Oceans from 600 to 140 Ma. Ore Geol Rev 35:137–154

Şengör AMC, Natal'in BA, Burtman VS (1993) Evolution of the Altaid tectonic collage and Paleozoic crustal growth in Eurasia. Nature 364:299–307

Şengör AMC, Natal'in BA (1996) Paleotectonics of Asia: fragments of a synthesis. In: The tectonic evolution of Asia. Cambridge University Press, London, pp 486–640

Sotnikov VI, Berzina AP, Bold D (1984) Regularity of distribution of Cu and Mo of Mongolian People's Republic. In: Endogenic ore-bearing formations of Mongolia. Nauka, Moscow, pp 89–101

Sotnikov VI, Berzina AP, Jamsran M (1985a) Mednye rudnye formatsii Mongol'skoi Nrodnoi Respubliky (copper ore formations of Mongolian People's Republic). Nauka, Novosibirsk, p 225

Sotnikov VI, Berzina AP, Jamsran M, Shabalovskii AE, Garamjav D, Bold D (1985b) Metallogeny of Mongolian People's Republic (copper, molybdenum): Novosibirsk, United Institute of Geology and Geophysics, U.S.S.R. Academy of Sciences, 40

Tcherbakov YG, Dejidmaa G (1984) Zolotorudnye formatsii Mongolii (gold-bearing ore formations of Mongolia). In: Endogenic ore-bearing formations of Mongolia. Nauka, Moscow, pp 42–50

Tomurtogoo O (ed.) (2002) Tectonic map of Mongolia, scale 1: 1,000,000 (with brief explanatory nots, 23 p) Ulaanbaatar, Geological Information Center, 15 sheets

Tomurtogoo O (2005) Tectonics and structural evolution of Mongolia. In: Seltmann R, Gerel O, Kirwin D (eds) Geodynamics and Metallogeny of Mongolia with a special emphasis on copper and gold deposits, London, CERCAMS/NHM, pp 5–13

Tomurtogoo O, Badarch G, Orolmaa D, Byamba J (1999) Terranes and accretionary tectonics of Mongolia. Mongolian Geosci 14:5–10

Tomurtogoo O, Windley BF, Kroner A, Badach G, Liu DY (2005) Zircon age and occurrence of the Adaatsag ophiolite and Muron shear zone, Central Mongolia: constraints on the evolution of the Mongol–Okhotsk Ocean, suture and orogen. J Geol Soc 162:125–134

Wainwright AJ, Tosdal RM, Wooden JL, Mazdab FK, Friedman RM (2011) U-Pb (zircon) and geochemical constraints on the age, origin, and evolution of Paleozoic arc magmas in the Oyu Tolgoi porphyry Cu-Au district, southern Mongolia. Gondwana Res 19:764–787

Wickham SM, Litvinovsky BA, Zanvilevich AN, Bindeman IN (1995) Geochemical evolution of Phanerozoic magmatism in Transbaikalia, East-Asia–a key constraint on the origin of K-rich silicic magmas and the process of cratonization. J Geophys Res Solid Earth 100 (B8):15641–15654

Windely BF (1995) The evolving continents. Geol Mag 133(6):776–777

Windley BF, Alexeiev D, Xiao W, Kröner A, Badarch G (2007) Tectonic models for accretion of the Central Asian Orogenic Belt. J Geol Soc Lond 164:31–47

Xiao WJ, Huang BC, Han CM, Sun S, Li JL (2010) A review of the western part of the Altaids: a key to understanding the architecture of accretionary orogens. Gondwana Res 18:253–273
Yakubchuk A (2004) Architecture and mineral deposit settings of the Altaid orogenic collage: a revised model. J Asian Earth Sci 23:761–779
Yakubchuk AS, Edwards AC (1999) Auriferous Palaeozoic accretionary terranes within the Mongol-Okhotsk suture zone, Russian Far East. In: Weber G (Ed.) Proceedings Pacrim '99. Australasian Institute of Mining and Metallurgy, Publications Series 4/99:347–358
Yanshin AL (ed.) (1974) Tectonika Mongol'dkoi Narodnoi Respubliky (tectonics of Mongolian People's republic) Nauka, Moscow, 283
Yarmolyuk VV, Kovalenko VI (1991) Rift-originated magmatism of active continental margins and its ore potential. Nauka, Moscow
Yarmolyuk VV, Kovalenko VI (2003) Batholiths and geodynamics of their formation in the Central Asian fold belt. Geol Geofiz 44(12):1305–1320
Yarmolyuk VV, Kovalenko VI, Kozakov IK, Sal'nikova EB, Bibikova EV, Kovach VP, Kozlovsky AM, Kotov AB, Lebedev VI, Eenjin G, Fugzan MM (2008a) The age of the Khangai batholith and the problem of batholith formation in Central Asia. Dokl Earth Sci 423(8):1223–1228
Yarmolyuk VV, Kovalenko VI, Sal'nikova EB, Kovach VP, Kozlovsky AM, Kozlovsky AB, Kotov AB, Lebedev VI (2008b) Geochronology of igneous rocks and formation of the late Paleozoic South Mongolian active margin of the Siberian Continent. Stratigr Geol Correl 16:162–181
Yarmolyuk VV, Kozlovskya AM, Travin AV, Kirnozova TI, Fugzan MM, Kozakov IK, Plotkina YV, Eenjin G, Oyunchimeg T, Sviridov OE (2019) Duration and geodynamic nature of giant central Asian batholiths: geological and geochronological studies of the Khangai batholith. Stratigr Geol Correl 27(1):73–94
Yarmolyuk VV, Kuzmin MI (2011) Rifting and silicic Large Igneous Provinces of the Late Paleozoic—Early Mesozoic in the Central Asia. Large Igneous Provinces Commission. http://www.largeigneousprovinces.org/11dec
Yashina RM (1982) Shchelochnoy magmatizm skladchato-glybovykh oblastey (alkaline magmatism of folded-blocky areas). Nauka, Moscow, p 274
Yuan C, Sun M, Xiao WJ, Li XH, Chen HL, Lin SF, Xia XP, Long XP (2007) Accretionary orogenesis of the Chinese Altai: insights from Paleozoic granitoids. Chem Geol 242:22–39
Žaček V. et al. (2016) Geological mapping 1: 50 000 and assessment of economic potential of selected region in western Mongolia (Mongol Altai 50, MA-50)
Zonenshain LP, Kuzmin MI, Natapov LM (1990) Geology of the USSR: a plate tectonic synthesis. In: Page BT (ed) Geodynamics series, geodynamic series, 21. American Geophysical Union, Washington, DC, p 242
Zorin YA (1999) Geodynamics of the western part of the Mongolia–Okhotsk collisional belt, Trans-Baikal region (Russia) and Mongolia. Tectonophysics 306:33–56

Part I
Metallic Mineral Resources

Chapter 2
Copper Deposits

Ochir Gerel, Bayaraa Batkhishig, Baatar Munkhtsengel, Baasandorj Altanzul, and Dorjyunden Altankhuyag

2.1 Introduction

Copper in Mongolia is the major mining industry and source of income for the country. The copper-molybdenum deposit Erdenet (1.54 billion tonnes estimated ore reserves) has operated since 1978 as a joint project between the governments of Mongolia and Russia. The mine produced a copper concentrate with molybdenum. Copper from Erdenet mine was Mongolia's largest export until 2010, when it was bypassed by coal. Mongolia is ranked 12th in the world for copper reserves. In the early 2000s, copper sent abroad accounted for 30% of Mongolia's export earnings. In 2001, Canadian-based Ivanhoe Mines discovered the gold-copper porphyry deposit Oyu Tolgoi (438 Mt. averaging 0.52% copper and 0.25 ppm gold). Recently, there are only two companies that produce copper in Mongolia: the Erdenet Mining Corporation, a Mongolian state-owned enterprise, and the Oyu Tolgoi mine, a joint venture between Rio Tinto Group, Turquoise Hill Resources, and the Government of Mongolia.

2.2 Copper Deposit Types and Metallogeny

Sotnikov et al. (1984) recognized three major metallogenic belts: North, Central, and South Mongolian where Cu mineralization mainly of porphyry type is controlled by a northwest-trending structural zone. These belts are spatially coinciding with

O. Gerel (✉) · B. Batkhishig · B. Munkhtsengel · B. Altanzul · D. Altankhuyag
Geoscience Center, School of Geology and Mining, Mongolian University of Science and Technology, Ulaanbaatar, Mongolia
e-mail: gerel@must.edu.mn; bkhishig@must.edu.mn; tsengel@must.edu.mn; altanzul@must.edu.mn; altankhuyag@mmhi.gov.mn

O. Gerel et al. (eds.), *Mineral Resources of Mongolia*, Modern Approaches in Solid Earth Sciences 19, https://doi.org/10.1007/978-981-15-5943-3_2

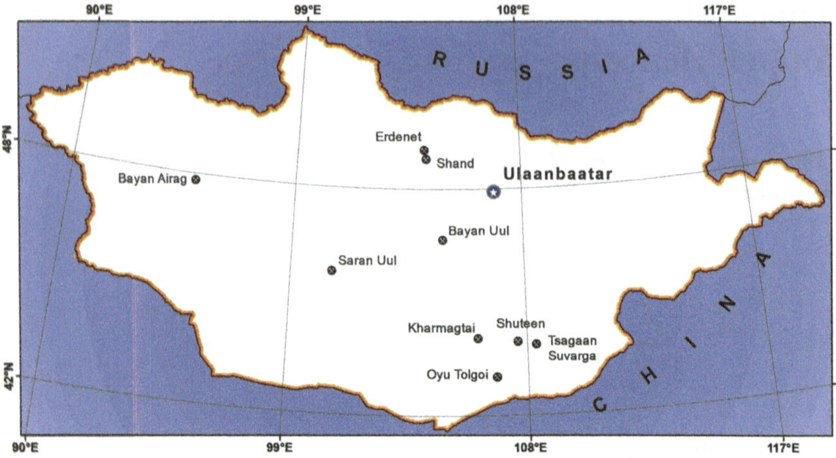

Fig. 2.1 Distribution of largest Cu deposits in Mongolia

overlapping volcanic belts and have a latitudinal orientation. The North Mongolian metallogenic belt includes Erdenet and Shand porphyry Cu-Mo deposits and numerous occurrences, Central Mongolian belt includes the Bayan Uul porphyry Cu-Mo deposit, and South Mongolian belt includes the Tsagaan Suvarga deposit (Sotnikov et al. 1984). According to Yakubchuk (Yakubchuk 2004; Yakubchuk et al. 2005, 2012) based on the Altaid tectonic model by Şengör and Natal'in (1996), the Erdenet deposit belongs to the Late Paleozoic Orkhon-Selenge belt, whereas the Oyu Tolgoi deposit is included in late Paleozoic arc, and Oyu Tolgoi is a part of the mid-late Paleozoic Kazak-Mongol arc. Dejidmaa and Badarch (2005) classified the late Devonian-early Carboniferous Tsagaan Suvarga metallogenic belt with Tsagaan Suvarga and Oyu Tolgoi cluster deposits and the Carboniferous Kharmagtai-Khunguit-Oyut metallogenic belt with the Kharmagtai deposit. The Orkhon-Selenge metallogenic belt overlaps Tuva-Mongol and Khangai-Khentii superterranes along the Central Mongolian orocline and composed of Permian-Triassic volcanic and intrusive rocks. Late Devonian and Carboniferous deposits are hosted in island arc and continental arc (Oyu Tolgoi, Tsagaan Suvarga, Shuteen, and Kharmagtai) and could be related to two individual arcs: late Devonian-early Carboniferous with Oyu Tolgoi and Tsagaan Suvarga and Carboniferous Kharmagtai and Shuteen.

The geological information accounts for about 35 Cu deposits and a large number of occurrences of different genetic types, but economic is the subduction-related porphyry deposit type. Copper deposits and occurrences in Mongolia include (1) porphyry Cu-Mo and Cu-Au; (2) Cu skarn; (3) volcanogenic Cu-Zn massive sulfide and Zn-Pb-Cu; (4) basalt-hosted native Cu; and (5) sediment-hosted Cu types (Dejidmaa et al. 2005; Nokleberg 2010). Position of major mineral deposit is shown in Fig. 2.1.

2.2.1 Porphyry Deposit Type

Porphyry Cu (±Au) deposits consist of stockwork veinlets and veins of chalcopyrite, bornite, and magnetite in porphyry intrusions and coeval volcanic rocks. The host intrusive rocks vary in composition from tonalite and monzogranite to syenite and monzonite. Granitoids are calc-alkaline I type, with high K or shoshonitic, and metaluminous and show typical subduction-related geochemistry, enrichment in LILE and depletion in HFSE and HREE, with high Sr, high Sr/Y and La/Yb ratios, and low Y content (Gerel et al. 2013). Coeval volcanic rocks consist of dacite and andesite lavas and tuffs. Ore minerals are chalcopyrite and bornite with associated magnetite, pyrite, rare native gold, electrum, sylvanite, and hessite, and rare PGE minerals; gangue minerals are quartz, K-feldspar, biotite, sericite, and chlorite, and rare actinolite, anhydrite, calcite, and clay minerals. Alteration consists of (1) an inner zone of quartz, biotite, rare K-feldspar, chlorite, actinolite, and anhydrite; (2) an outer zone of propylitic minerals; and (3) late-stage quartz-pyrite-white mica-clay minerals that overprint early feldspar alteration. The deposits are cylindrically or bell-shaped and are centered on a volcanic intrusive center. Highest-grade ore commonly occurs at the level where stock divides into branches. The depositional environment consisted of subduction-related continental margin or island arcs with porphyry stocks, dykes, and large-scale breccia intruding coeval volcanic rocks of a nearby volcanic center and adjacent passive-continental-margin sedimentary rocks. Granitoids hosting the deposits generally intruded during the waning stage of a volcanic cycle. Example of this mineral-deposit type is the Oyu Tolgoi cluster deposits, Kharmagtai and Shuteen in South Mongolia.

Porphyry Cu-Mo (±Au, Ag) deposits consist of stockwork veinlets and veins of quartz, chalcopyrite, and molybdenite in or near porphyritic intrusions. The host igneous rocks are predominantly tonalite to monzogranite plutons occurring mainly in stocks that intrude granitic, volcanic, or sedimentary rocks. Granitoids are calc-alkaline, metaluminous I-type, medium to high K, enriched in LILE and depleted in HFSE. Breccia pipes and dykes are common. Ore minerals are chalcopyrite, molybdenite, pyrite, sphalerite, Ag-rich galena, and gold; alteration minerals are quartz, K-feldspar, sericite, and biotite or chlorite. Anhydrite occurs in the deeper levels of the deposits. Most deposits exhibit varying degrees of hypogene alteration, including sodic, potassic, and phyllic alteration. The earlier mineralization stage commonly starts with microclinization, followed by the metasomatic deposition of molybdenite and later overprinting by sericite alteration with subsequent deposition of Cu sulfides, sometimes in open-space filling veins and veinlets. Alteration zones, from inner to outward, are sodic-calcic, potassic, phyllic, and argillic to propylitic. The depositional environment consisted of shallow porphyry intrusions that were contemporaneous with abundant dykes, faults, and breccia pipes associated with andesite stratovolcanoes in the backarc zones of subduction-related continental margin or mature island arcs. Examples of this mineral-deposit type are Erdenetiin Ovoo (Erdenet), Tsagaan Suvarga, Saran Uul and Bayan Uul.

2.2.2 Cu (±Fe, Au, Ag, Mo) Skarn Copper Deposit Type

Skarn-type copper deposits consist of chalcopyrite, magnetite, and pyrrhotite in calc-silicate skarn that replaces carbonates along intrusive contacts with plutons ranging in composition from quartz diorite to granite and from diorite to syenite. Zn-Pb-rich skarn generally occurs farther from the intrusion, whereas Cu- and Au-rich skarn generally occurs closer. Major ore minerals are pyrite, hematite, galena, molybdenite, sphalerite, and scheelite. Mineralization is multistage. The deposit type is commonly associated with porphyry mineralization with a higher grade of gold and shows close spatial relation with Cu-Mo porphyry mineralization system. The depositional environment consisted mainly of calcareous sedimentary sequences intruded by felsic to intermediate-composition granitic plutons that form parts of continental arcs. Examples of this mineral-deposit type are some occurrences in western, central, and southwestern Mongolia.

2.2.3 Volcanogenic Cu-Zn Massive Sulfide Deposit Type

Volcanogenic massive sulfide deposits consist of massive to disseminated Zn-Cu and Zn-Pb sulfide minerals hosted in island-arc volcanic belts. The volcanic rocks comprise bimodal rhyolite and basalt, andesite, dacite, and rhyolite with subordinate felsic rocks. Ore-controlling structures are volcanotectonic depressions, calderas, domes, and synvolcanic faults. The most widespread are lens-shaped deposits that are concordant with host rocks. Less abundant deposits are funnel-shaped or T-shaped deposits. Multilevel deposits are typical. Ore minerals are mainly pyrite, chalcopyrite, and sphalerite with minor galena, tennantite, tetrahedrite, and bornite; gangue minerals are quartz, sericite, chlorite, and carbonate. The volcanic and sedimentary host rocks are widely altered. The root zones consist of sericite-quartz metasomatite grading upward and outward into quartz-sericite-chlorite and quartz-carbonate-sericite-chlorite with albite and epidote zones. Silica, epidote, and hematite alterations are widespread above the deposits. Sulfides are zoned with Cu and Zn enrichment from footwall to hanging-wall and from core to periphery. Most minerals are massive but locally may be banded or brecciated or may occur in stringers on the flanks of the deposits. The formation of thick gossan in weathering zones of deposits is typical. The depositional environment consisted of an ensimatic island arc constructed on oceanic crust containing differentiated basalt and other volcanic rocks. The deposit type is a variant of Cyprus Cu-Zn massive sulfide deposits. Bayan Airag (Golden Hill) deposit is an example of this type, and similar occurrences are known in western Mongolia.

2.2.4 *Basaltic Cu Deposit Type*

Native copper is hosted in volcanic rocks generally interbedded with red sandstone, conglomerate, and siltstone and commonly associated with sediment-hosted Cu deposits. The basalt is generally potassic or alkaline including shoshonite and trachybasalt. Major ore minerals are native copper, chalcocite, bornite, chalcopyrite, and native silver, both in the matrix and as amygdules in the porous roofs of basalt flows and in veinlets within the basalts. Minerals occur as disseminations, stringers, lenses, and irregular patchy accumulations. Wall rocks are altered mainly to epidote, calcite, chlorite, and zeolite. The depositional environment consisted of continental, rift-related, flood-basalt sequences, and continental-margin and island arcs. In Mongolia, this type was originally named as the Cu-zeolite formation by Yakovlev (Marinov et al. 1977) and as the native Cu formation (Sotnikov et al. 1985). This deposit type occurrence is known in two regions of Mongolia: (1) the Orkhon-Selenge volcanic belt where Cu is hosted in andesite, ambygdaloidal basalt, and basaltic tuff of Permian-early Triassic age; and (2) the Khan Khukhii mountain range with Cu hosted in amygdaloidal basalt and andesite porphyry in lower to middle Devonian volcanic and sedimentary rocks. Copper mineralization is associated with quartz, epidote, carbonate, and chlorite.

2.2.5 *Sediment-Hosted Cu Deposit Type*

This sediment-hosted Cu deposit type first was defined by Yakovlev (Marinov et al. 1977) as a Cu-bearing sandstone formation and subsequently by Sotnikov et al. (1985) who named the deposit type as Cu-bearing sandstone. Significant sediment-hosted Cu occurrences are distributed in the Mongol Altai and in Kharkhiraa terranes and are hosted in upper Ordovician, in middle to upper Devonian and Lower Jurassic sedimentary rocks. The occurrences consist of sandstone, rare siltstone lenses, and tabular bodies with visible chalcocite, malachite, and azurite. Covellite and chalcopyrite also are identified by optical microscopy. The thickness of the ore bodies ranges from 5.0 m to 50.0 m, and the length ranges up to 1.5 km. The grade of Cu is from 0.01% to 1.47%, and Ag can be up to 20.0 g/t. A few Cu occurrences of sediment-hosted Cu deposits are present in western Mongolia (Obolensky et al. 1989). Here are about 25 occurrences known in Mongolian Altai.

2.3 Description of Selected Deposits

2.3.1 Oyu Tolgoi Cluster Deposits

Introduction. The Oyu Tolgoi porphyry copper-gold-molybdenum cluster of deposits are located in the South Gobi Desert 550 km south of the capital, Ulaanbaatar and 80 km north of the Chinese border. The Oyu Tolgoi includes seven porphyry Cu-Au-Mo deposits (South Oyu, South West Oyu, Central Oyu, Hugo Dummett North and South, and Heruga and Heruga North), which define a narrow, linear, 12-km-long, continuously mineralized trend (Fig. 2.2), containing in excess of 438 Mt. of Cu and 1850 tonnes of Au (Perelló et al. 2001; Kirwin et al. 2005). The Ulaan Khud South, Ulaan Khud North, and Javkhlant prospects are all within the Oyu Tolgoi trend, respectively located ~7.5 and 11 km to the NNE of Hugo North and ~4.5 km SW of Heruga. There is a distinct variation in deposit characteristics along the trend of the mineralized corridor, with a high sulfidation phase partially telescoped onto the underlying porphyry systems in the northern half of the trend, from the Hugo Dummett North to Central deposits. These variations are summarized in Table 2.1 (after Kirwin et al. 2005; Khashgerel et al. 2009; Crane and Kavalieris 2012; Porter 2016).

Geology. All deposits lie within a 600 × 250 km Gurvansaikhan island arc terrane composed of dismembered ophiolite mélanges, Ordovician to Silurian greenschist facies sandstones, argillite, chert and volcaniclastic rocks, late Silurian to mid-Devonian radiolarian chert, tholeiitic pillow basalt and andesitic tuff, overlain by late Devonian to lower Carboniferous basaltic to dacitic volcano-sedimentary rocks, chert, sedimentary rocks, and intermediate to felsic volcanic rocks (Tomurtogoo 1999). All of these rock units are intruded by late Devonian granitoid and by Permian-Carboniferous diorite, monzodiorite, granite, granodiorite, and syenite bodies, ranging in size from dykes to batholiths that are tens of kilometers across (Lamb and Badarch 1997; Badarch et al. 2002; Badarch 2005). The upper parts of the succession consist of mafic to intermediate volcanic units evolved from juvenile Devonian to more mature Carboniferous calc-alkaline compositions, reflecting progressive marine arc maturity and thickening, while felsic rock formations are dominantly high-K calc-alkaline, regardless of age (Crane and Kavalieris 2012).

Mineralization at Oyu Tolgoi is associated with multiple quartz monzodiorite intrusions of late Devonian age (~372 to 370 Ma) intruding Devonian arc-related, basaltic lavas, and volcaniclastic rocks, unconformably overlain by late Devonian (~370 Ma) basaltic to dacitic pyroclastic and volcanic-sedimentary rocks. These quartz monzodiorite intrusions range from early-mineralization porphyritic dykes to larger, linear, syn-, late- and post-mineralization dykes and stocks. Ore was deposited within syn-mineral quartz monzodiorites but is dominantly hosted by augite basalts. Following ore deposition, an allochthonous plate of older Devonian (or pre-Devonian) rocks was overthrust and a post-ore biotite granodiorite intruded at ~365 Ma (Fig. 2.3). Mineralization is characterized by varying, telescoped stages

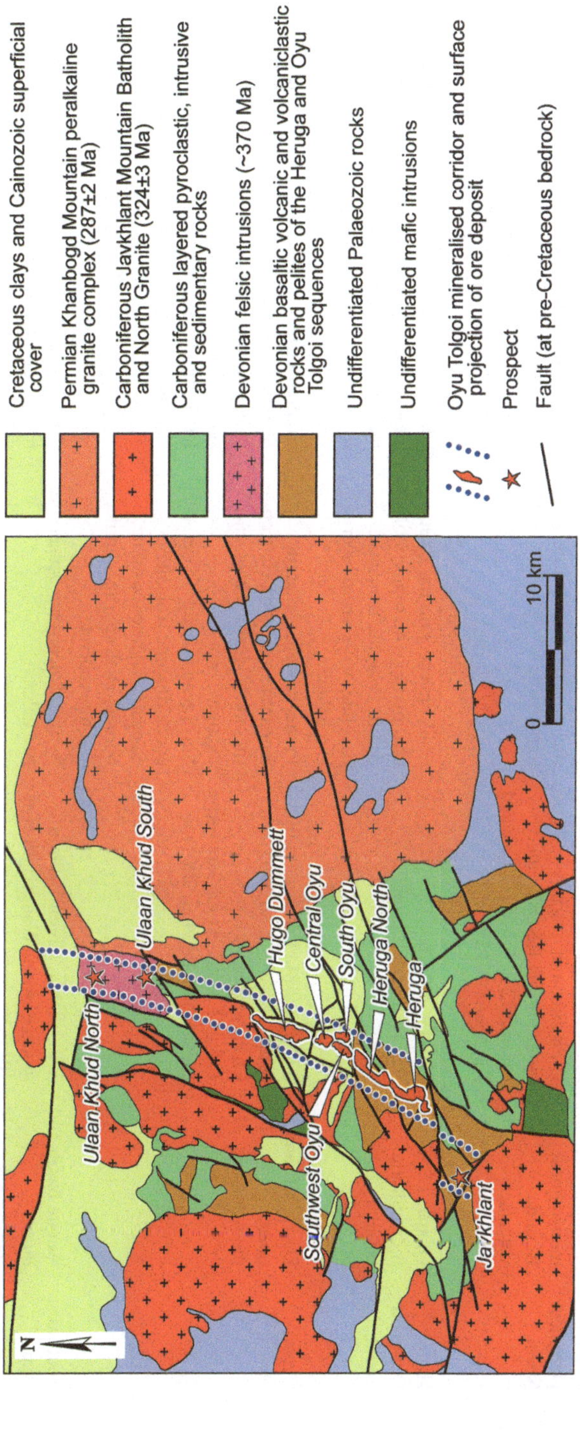

Fig. 2.2 Geological setting of Oyu Tolgoi cluster deposits (after Porter (2016), and published with permission of the PGC Publication)

Table 2.1 Summary of mineral deposits of the Oyu Tolgoi cluster (data from Kirwin et al. 2005; Crane and Kavalieris 2012; Khashgerel et al. 2009; Porter 2016, and published with permission of the PGC Publication)

Deposit name	Mineralization hosting rocks	Age, Ma	Mineralogy	Alteration	Ore reserves and grade Cu (%), Au(g/t)
Hugo Dummett North	Quartz monzodiorite Basaltic volcanics	361.4 ± 2 Ma U-Pb zircon in quartz monzodiorite Re-Os in molybdenite 372 ± 1.2 Ma (Crane and Kavalieris 2012)	Bornite, chalcocite, chalcopyrite, molybdenite, enargite, tetrahedrite	Chlorite-illite advanced and intermediate argillic, quartz-sericite/illite-(muscovite), chlorite-muscovite/illite-hematite/siderite-(biotite-magnetite), K-silicate	499 Mt. @ 1.65% Cu, 0.35 g/t Au, 3.39 g/t Ag
Hugo Dummett South	Quartz monzodiorite Basaltic volcanics	370.0 ± 1.2 Ma Re-Os in molybdenite	Chalcopyrite, bornite, chalcocite, pyrite, with minor molybdenite, enargite, tennantite, covellite, with rare sphalerite and galena	Chlorite-illite, pyrophilite-kaolinite, advanced and intermediate argillic, sericite/illite-(muscovite), chlorite-muscovite/illite-hematite	Low grade Average 0.2% Cu No significant rich zones
Central Oyu	Dacite tuff and quartz-monzodiorite	373 ± 1.2 Ma Re/Os in molybdenite (Kirwin et al. 2005) Chalcosite blanket 93–117 ± 1 Ma K-Ar for supergene alunite (Perelló et al. 2001)	Pyrite, covellite, enargite, tennantite, cubanite, chalcopyrite, molybdenite, chalcocite/digenite Supergene blanket: Pyrite, chalcocite/digenite, colusite, enargite, tenorite, covellite, bornite, chalcopyrite, cuprite, and molybdenite	Biotite-magnetite Sericite, advanced argillic (topaz, quartz, zunyite, diaspore, alunite, illite, andalusite, late kaolinite, and dickite), intermediate argillic	438 Mt. @ 0.52% Cu, 0.25 g/t Au Very low grade with average 0.1% Cu
Southwest Oyu	Augite basalt and quartz monzodiorite	372 ± 1 Ma; zircon U-Pb in monzodiorite, (Wainwright et al. 2011b)	Pyrite, chalcopyrite, sphalerite, and galena	Quartz-sericite albite, with minor fluorite and rare tourmaline Sodic-calcic with chlorite-biotite, hematite-magnetite, propylitic alteration	High grade core >1 g/t Au, low grade > 0.3% Cu 1.0 to 1.2% Cu

South Oyu	Quartz monzodiorite, basalt, dacitic tuff and breccia	374 ± 3 Ma; SHRIMP zircon U-Pb (Wainwright et al. 2011b)	Chalcopyrite, bornite, molybdenite, magnetite, covellite and primary chalcocite	Chlorite, biotite, hematite-magnetite, weak sericite, and pink albite, tourmaline, fluorite, advanced argillic alteration	High grade zone >0.8% Cu
Heruga and Heruga North	Basaltic volcanics and quartz monzodiorite	365.3 ± 1.5 Ma (Crane and Kavalieris 2012)	Chalcopyrite, pyrite, bornite	Sodic-calcic, propylitic, biotite-magnetite, hematite-chlorite-sericite, quartz-sericite-pyrite, tourmaline-sericite	0.87% Cu, 0.06 g/t Au, 12 ppm Mo
Ulaan Khud South and North	Basaltic volcanics and quartz monzodiorite	361.4 ± 3.71 SHRIMP U-Pb zircon (Crane and Kavalieris 2012)	Chalcopyrite, molybdenite, pyrite, sphalerite, galena	Sericite, and lesser secondary biotite and magnetite alteration	Low grade ranges from 0.2 to 1.0%, 0.3% Cu
Javkhlant prospect	Granodiorite	324 ± 3 Ma (Wainwright 2008)	Chalcopyrite, pyrite	Advanced argillic higher temperature assemblage of corundum, andalusite, muscovite, pyrophyllite, kaolinite and distinctive blue lazulite	0.96% Cu, 0.01 g/t Au from 1422 m

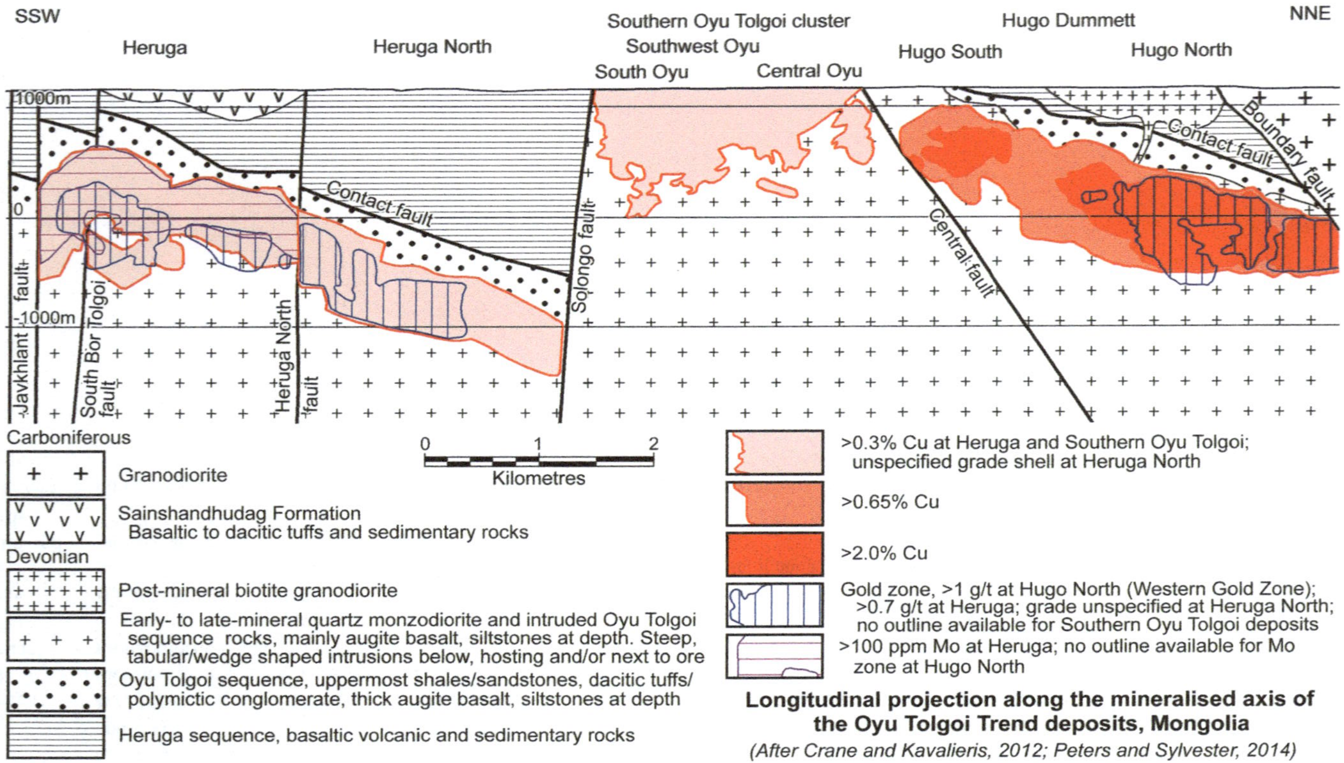

Fig. 2.3 Longitudinal projection along the mineralized axis of the Oyu Tolgoi trend de-posits, Mongolia, showing the outline of the ore deposits, mineral zonation and generalized geological relationships (Crane and Kavalieris 2012; Peters and Sylvester 2014; Porter 2016, and published with permission of the PGC Publishing)

of intrusion and alteration (Porter 2016). Early A-type quartz veins (Gustafson and Hunt 1975) were followed by Cu-Au mineralization associated with potassic alteration, mainly K-feldspar in quartz-monzodiorite and biotite-magnetite in basaltic hosts. Downward, late-magmatic hydrothermal fluid resulted in intense quartz-sericite retrograde alteration in the upper parts of the main syn-mineral intrusions, and an equivalent chlorite-muscovite/illite-hematite assemblage in basaltic host rocks. Uplift, facilitated by syn-mineralization longitudinal faulting, brought sections of the porphyry deposit to shallower depths, to be overprinted and upgraded by late stage, shallower, advanced argillic alteration and high sulfidation mineralization. Key controls on the location, size, and grade of the deposits include (1) a long-lived, narrow faulted corridor; (2) multiple pulses of overlapping intrusion within the same structure; and (3) enclosing reactive, mafic dominated wall rocks, focusing ore (Crane and Kavalieris 2012).

Alteration. The main styles of alteration are summarized below (Crane and Kavalieris 2012; Porter 2016).

The sodic-calcic alteration, represented by an early assemblage of actinolite-magnetite-albite-apatite-titanite and green biotite, is common in the southern deposits within augite basalt and generally precedes biotite-magnetite alteration.

Biotite-magnetite alteration is characteristic of gold-rich chalcopyrite mineralization in the southern deposits. Brown biotite partially replaces actinolite-altered augite phenocrysts, while secondary magnetite occurs as pervasive alteration or in micro-veinlets.

K-feldspar is generally restricted to quartz-monzodiorite, rarely in basaltic wall rocks obliterating the original rock texture. Pink K-feldspar rims, or completely replaces, plagioclase phenocrysts occur as selvages to some quartz veins, while in strongly altered rocks, the groundmass is completely replaced by coarser K-feldspar.

Quartz-sericite (muscovite) alteration is found in quartz monzodiorite, at the Central Oyu and Hugo Dummett deposits, where it persists to depths of up to 1000 m. The original quartz-monzodiorite texture is totally replaced forming a pale amorphous rock, but it is much less intense in augite basalt producing chlorite-muscovite/illite-hematite association and secondary quartz is less abundant.

Tourmaline-sericite is found in the southern deposits, occurring as late fine-grained tourmaline in the sericite alteration zone, forming large rosettes of tourmaline, commonly nucleated on pyrite and large crystals of pink-white albite.

Intermediate argillic alteration occurs as a yellow-brown to a greenish assemblage of chlorite-muscovite-illite-siderite specularite, with minor pyrophyllite-kaolinite (after magnetite-biotite) in augite basalt wall rocks at Hugo Dummett South.

Advanced argillic alteration is dominantly composed of residual quartz and pyrophyllite, with lesser corundum, diaspore, K-alunite, aluminium-phosphate-sulfate minerals, zunyite, topaz, dickite, kaolinite, anhydrite, gypsum, and relict andalusite. It is developed over a strike length of ~5.9 km from Hugo Dummett North to the Central deposit.

The sericite-muscovite and the early advanced argillic alteration formed by a late magmatic-hydrothermal fluid without a significant meteoric water component (Khashgerel et al. 2006, 2009). Stable isotope studies (Khashgerel et al. 2006,

2008, 2009) indicate that later alunite formed at moderate temperatures (~270 °C) from condensation of magmatic vapor that mixed with up to 25% meteoric water, while dickite formed at low temperatures (~150 °C) during the ingress of further meteoric water into the advanced argillic zone.

Propylitic alteration is only observed in the western parts of the Southwest and Heruga deposits and is characterized by pervasive alteration and veining of epidote, magnetite and hematite, veins of semi-massive pyrite, and albite alteration.

Mineralization. Mineral deposits distributed over ~12 km of a more than 25 km long, NNE trending structural and mineralized corridor, with gaps between the deposits of <1 to 2 km. Where completely preserved, economic mineralization persists over a vertical interval of ~1 km. These deposits appear to represent five porphyry centers, grouped into the Hugo Dummett (Hugo North and Hugo South), Central-Southern Oyu Tolgoi cluster, and the Heruga North and Heruga deposits (Crane and Kavalieris 2012).

The Ulaan Khud South, Ulaan Khud North, and Javkhlant prospects are all within the Oyu Tolgoi trend, respectively located ~7.5 and 11 km to the NNE of Hugo North, and ~ 4.5 km SW of Heruga (Fig. 2.2).

Hugo Dummett deposits lie within a NNE trending, post-ore, structural high, bounded by faults. Although the latest movement on these bounding faults displaces the Carboniferous post-mineral strata, they may represent the shallow expression of a longer-lived, deposit-controlling structural zone. Deposit represents a high sulfidation system that has been telescoped onto porphyry-style mineralization formed at an earlier stage in the evolution of the hydrothermal center. In Hugo North, bornite and hypogene chalcocite are prominent, although chalcopyrite becomes the dominant sulfide at deeper levels. These are overprinted by a high sulfidation assemblage of enargite, tennantite-tetrahedrite, and covellite. The Hugo South bornite dominated highest-grade core, with 3–5% Cu, progressively grading outward through chalcopyrite in augite basalt, passing into a lower grade fringe of pyrite-chalcopyrite ± enargite, tennantite, bornite, chalcocite in an advanced argillic-altered dacitic tuff. High-grade copper mineralization extends over a distance of more than 4.5 km in two NNE trending connected segments, Hugo South, and Hugo North (Fig. 2.3). In Hugo South deposit the gold and copper-rich zone is hosted in augite basalt and quartz monzodiorite and in the Hugo South the gold-poor mineralization is hosted in dacitic tuffs and augite basalt. Mineralization in the two deposits is dominantly bornite, chalcocite, and chalcopyrite, with subordinate pyrite, enargite, and tetrahedrite-tennantite. The sulfides are directly related to the associated alteration assemblage. The outward zonation from high- to low-grade Cu ore corresponds to the progression from bornite, chalcosite to chalcopyrite (±tetrahedrite-tennantite) to pyrite (±enargite) (Oyunchimeg 2008).

The Southern Oyu Tolgoi cluster of deposits includes the main Southwest, South, and Central Oyu deposits and a number of smaller, fault-bounded zones (Fig. 2.3). These zones form contiguous sectors of mineralization representing multiple mineralizing centers, each with distinct styles of mineralization, alteration and host rock lithology. The boundaries between the individual deposits and zones are defined by the major faults. These faults resulted in differential movement of the deposits and variations in the erosional histories of adjacent zones, depending on the depth to which a zone has been down-faulted or uplifted relative to its neighbor. The high sulfidation system at South and South-West Oyu is present only in the roots and mid-levels of the porphyry system, although part of the high sulfidation system from South-West Oyu is preserved. A supergene chalcocite blanket was developed over the Central deposit and protected by the overlying Cretaceous sequence. At the Southwest Oyu deposit, the syn-mineralized quartz monzodiorite is only found as a ~250 m diameter by 700 m high, pipe-like body in the core of the deposit. However, late-mineralized quartz monzodiorite bodies are more extensive in the vicinity of all of these deposits (Peters and Sylvester 2014). This group of deposits is characterized by a generally higher Au (g/t) to Cu (%) ratio ranging from 0.8 to 3:1 compared to the northern deposits. Chalcopyrite is the principal copper sulfide, with lesser pyrite, bornite, and molybdenite. In the southern deposits of Southwest Oyu, South Oyu, Heruga North, and Heruga; the remaining mineralization is dominantly hosted by biotite-magnetite altered augite basalt of Oyu Tolgoi sequence unit, overprinted by hematite-chlorite-sericite alteration.

The Heruga deposits are separated from the Southern Oyu Tolgoi cluster by the subvertical, E to ENE-striking, Solongo fault (Fig. 2.3). A series of other ENE striking faults are prominently reflected on both magnetic and satellite images. Geological mapping shows a 500 m apparent dextral displacement of dykes and stratigraphic contacts across each of these faults. Two of these, the Heruga North and (or South Sparrow) faults, offset the Heruga trend.

Genesis. Oyu Tolgoi porphyry Cu-Au-Mo cluster deposits are the largest in the world. The key controls on the location, tonnage, and grade of these deposits include: (1) a long-lived, narrow structural corridor, (2) repeated pulses of overlapping intrusion and mineralization, concentrated along the same narrow structural corridor, and (3) enclosing reactive wall rocks that restricted the lateral dispersion of mineralization and focused ore deposition (Porter 2016). The Oyu Tolgoi augite basalt was deposited as a juvenile marine calc-alkaline mafic volcanic complex in an intra-oceanic arc (Wainwright et al. 2011a, b; Crane and Kavalieris 2012; Dolgopolova et al. 2013). This arc formed from a north dipping oceanic slab subducting below oceanic crust that was overlain by fore-arc and distal ocean floor sediments derived from the lower- to mid-Paleozoic magmatic arcs to the north. The Oyu Tolgoi sequence siltstones underlying the augite basalts would represent the uppermost units of this sedimentary pile. The Heruga sequence possibly represents a very similar succession, deposited elsewhere in the terrane, with the two sequences being subsequently brought together as a result of arc migration and collision (Fig. 2.3). The Gurvansaikhan terrane, which hosts the Oyu Tolgoi mineralized system, during the subduction and associated compression has been uplifted, exposed and partially eroded the upper augite basalts. Subsequently, by ~370 Ma, extensive magmatic activity from the large, deep parental magma chamber produced an andesitic volcanism with a quartz diorite core that was eroded and reworked into a shallow water environment. The resultant volcano-sedimentary rocks formed the coarse, poorly sorted, polymictic volcanic conglomerate and lesser sandstone that unconformably overlies the augite basalt. Further compression and uplift resulted in

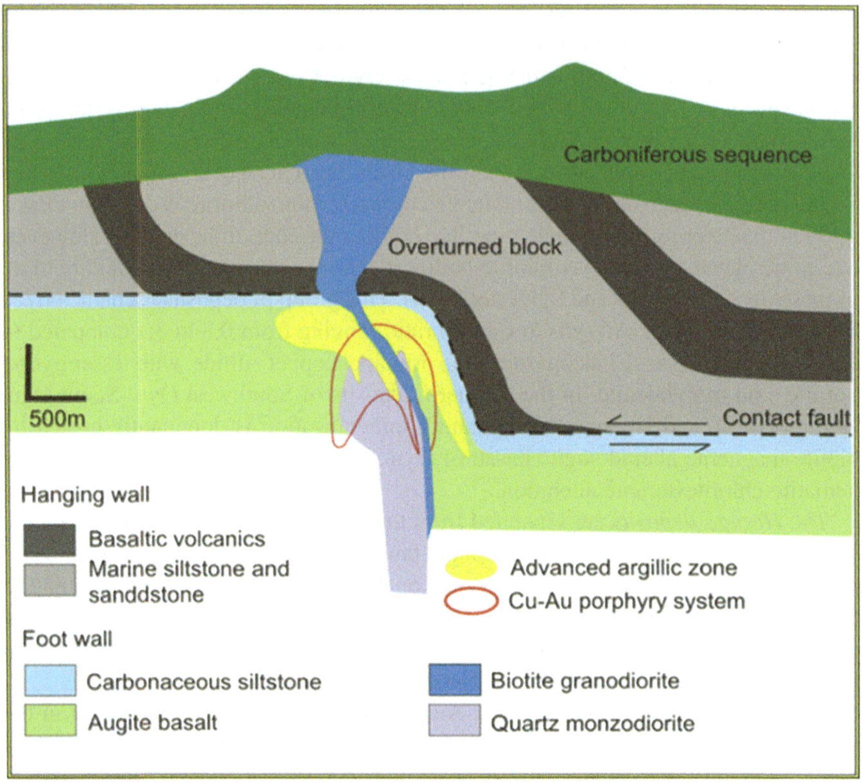

Fig. 2.4 Simplified model of Oyu Tolgoi deposit (unpublished, courtesy of David Crane)

subaerial conditions and the development of thick ignimbritic dacitic ash flow, and the block and ash tuffs erupted from the adjacent volcanic edifice. The parental andesitic magma differentiated and produced quartz-monzodiorite intrusion ponded as a shallow, >12 km long, elongate upper magma chamber underlying the interval that was to host the Oyu Tolgoi deposits cluster, possibly from as early as ~372 Ma. The Oyu Tolgoi mineralizing system was initiated immediately before or during the deposition of the dacitic volcanics, when a series of narrow quartz-monzodiorite porphyry dykes emanated from the top of the upper magma chamber, and stopped upward into the zone that was to host the Oyu Tolgoi deposits (Fig. 2.4).

The NNE trending corridor of structures, active from the late Devonian to late Mesozoic, have had a marked influence on the location, ore deposition, enhancement of ore grade, and exposure of the Oyu Tolgoi mineralized system.

The bulk of the ore is hosted by the augite basalts. The dominant mafic mineralogy of this lithology is strongly reactive and buffered mineralizing fluids to readily precipitate metals from hydrothermal fluids. Therefore, the mineralizing fluids and volatiles could not carry metals for a significant distance laterally or upward and precipitated them close to the source. Repeated pulses of overlapping intra-mineral intrusive activity and associated prograde hydrothermal fluids and volatiles, focused within the narrow structural corridor, produced a progressive increase in the quantity and grade of precipitated copper-gold. This was enhanced by retrograde to high sulfidation fluids, telescoped onto the earlier phases through cooling in the upper parts of the system, and subsequently by access of meteoric waters during uplift (Crane and Kavalieris 2012; Porter 2016).

Mineralization is associated with varying, telescoped stages of intrusion and alteration. The earliest alteration is a pre-ore sodic-calcic phase, characterized by an actinolite-magnetite-albite-apatite assemblage, most likely largely caused by fluids produced from the quartz-monzodiorite as a result of interaction with, and assimilation of, the contrasting augite basalt wall rock, and is not directly related to mineralization.

Early intense A-type quartz (Gustafson and Hunt 1975) veining accompanied quartz monzodiorite dykes, ahead of the main intrusive body. The latter was characterized by a distal propylitic halo and a core of Cu-Au mineralization associated potassic alteration, mainly K feldspar in quartz-monzodiorite and biotite-magnetite in basaltic hosts. Downward reflux of cooled, late magmatic- hydrothermal fluids, resulted in intense quartz-sericite (phyllic) retrograde alteration in the upper parts of the main intrusion, and an equivalent chlorite-muscovite/illite-hematite assemblage in basaltic wall rocks. This zone contains elevated Mo mineralization at Hugo North (>75 ppm Mo) and Heruga (>100 ppm Mo) but is relatively depleted in gold.

2.3.2 Tsagaan Suvarga Deposit

Introduction. Tsagaan Suvarga porphyry Cu-Mo deposit (255 Mt. at 0.55% Cu, 0.02% Mo) is located in South Gobi 650 km SE of Ulaanbaatar (Fig. 2.5). The deposit was explored during 1977–1982, and the reserves of 240 Mt. at 0.53% Cu and 0.19% Mo were calculated. Later MAK Corporation made a detailed exploration and ore reserves increased up to 225 Mt. (at 0.55% Cu, 0.02% Mo) of primary sulfide ore equivalent to 1.6 Mt. of oxide ore and 66, 000 tonnes of Mo.

Geology. The deposit is situated in the Gurvansaikhan island arc terrane within NE trending tectonic structure with number of mineralized prospects (Sotnikov et al. 1985) and associated with the Late Devonian Tsagaan Suvarga intrusion (Fig. 2.5). The age of mineralization at the Tsagaan Suvarga was determined by Re-Os dating to be 370.6 ± 1.2 Ma (Watanabe and Stein 2000) and 370.0 ± 5.9 Ma (Hou et al. 2010). Orolmaa and Tungalag (2015) reported a younger age on zircon U-Pb of 369 ± 1.2 and 365 ± 3.6 Ma. Sericite alteration has been dated by $^{40}Ar/^{39}Ar$ at 364.9 ± 7 Ma (Lamb and Cox 1998).

Host rocks. The Tsagaan Suvarga intrusion composed of quartz monzodiorite, monzonite, quartz syenite, syenite, and alkali granite is hosted in Carboniferous andesitic and dacitic volcanic and sedimentary rocks, covered by Mesozoic and

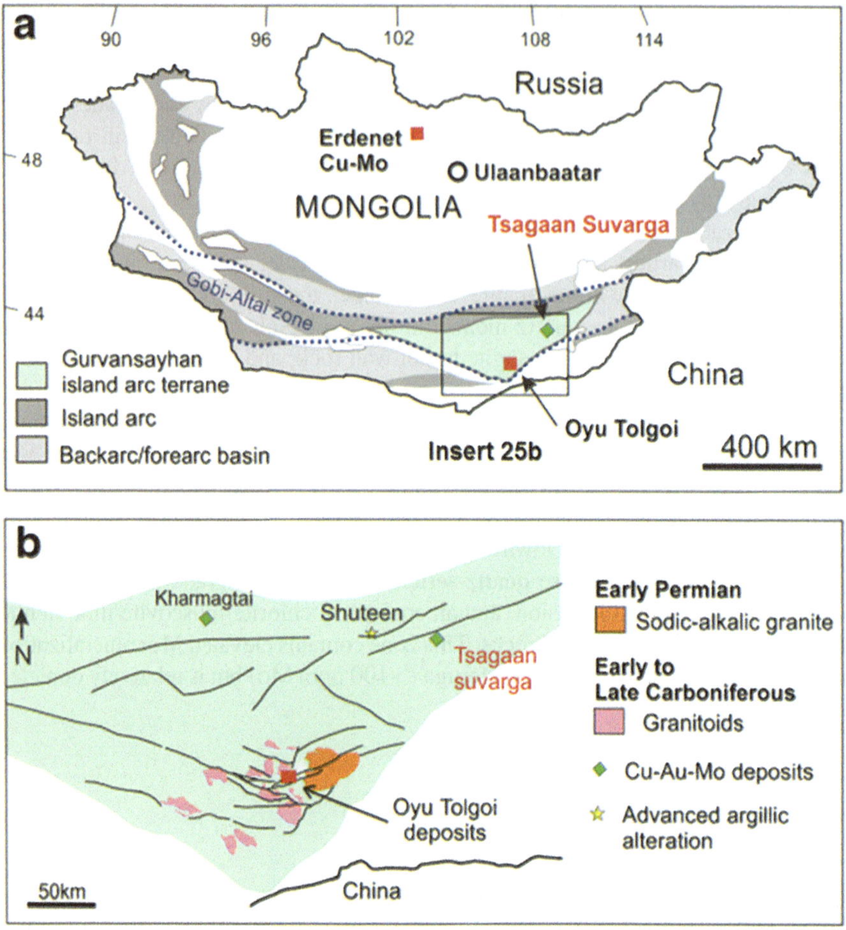

Fig. 2.5 Position of Tsagaan Suvarga deposit (Tungalag et al. 2018)

Cenozoic sedimentary sequence. A zone of brown plagioclase phyric trachyrhyolite dykes dated at 313.0 ± 5.8 Ma (Lamb and Cox 1998) cut the Tsagaan Suvarga intrusion and Carboniferous volcanics. The Tsagaan Suvarga ore mineralization is deposited in porphyritic body on the NW side of the Tsagaan Suvarga intrusion. This porphyritic body is composed of medium-grained equigranular amphibole-biotite quartz monzodiorite to quartz monzonite. The Tsagaan Suvarga intrusion is characterized by B-type quartz veins (Gustafson and Hunt 1975) and miarolitic cavities in quartz monzonite (Tungalag et al. 2018). In cross-section, the Tsagaan Suvarga deposit dips at 45° NW and forms a tabular body about 300 m wide and extends down dip for over 1 km (Fig. 2.6). Drilling shows that the ore body is fault bounded. The hanging wall comprises relatively unaltered Carboniferous andesite-dacite ash flow tuff about 100 m thick, which dips at a similar angle as the ore body. The base

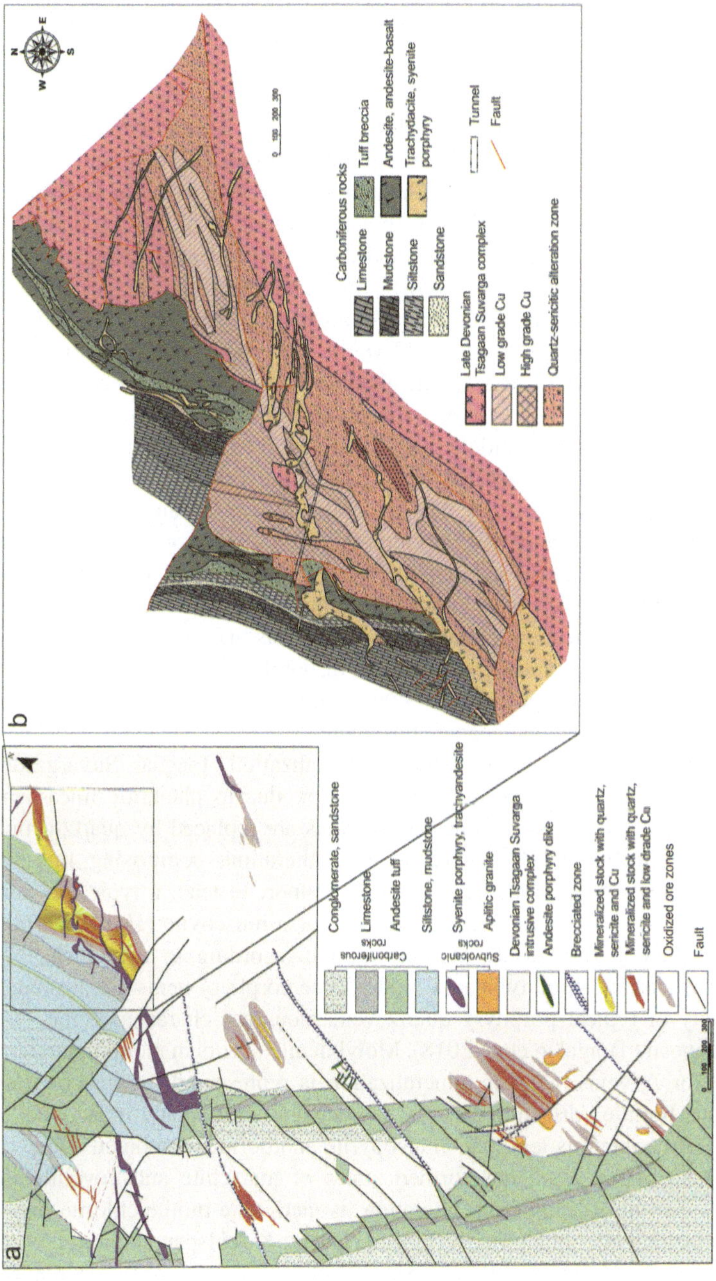

Fig. 2.6 Geological map of Tsagaan Suvarga deposit (Tungalag 2014)

of the ore body is marked by a fault (Fig. 2.6). The andesite-dacite ash flow tuff is overlain by fine siltstones and impure limestone, with marine fossils.

Amphibole-biotite quartz monzonite is composed of 40% plagioclase, 10% altered amphibole and biotite, 30% K-feldspar, and 10% interstitial quartz, as well as accessory, magnetite, titanite, zircon, and apatite. Quartz monzonite and quartz monzodiorite are high K calc-alkaline series of I-type, with a typical subduction related signature, e.g., enrichment in LILE (K, Rb, Th, U) and depletion in HFSE (Ta, Nb). The REE curved pattern shows enrichment in LREE and depletion in mid HREE. Such pattern, which lacks of Eu anomaly, was described for Erdenet (Gerel 1990; Kavalieris et al. 2017), Oyu Tolgoi (Kavalieris et al. 2017), and Tsagaan Suvarga (Tungalag et al. 2018) porphyry deposits and reflects hornblende fractionation (Loucks 2014) and is characteristic for fertile porphyry magma (Lang and Titley 1998). Whole-rock analyses and Sr-Nd isotopes, $^{87}Sr/^{86}Sr = 0.7027–0.7038$ (n = 12) and εNd = +4.26 to +2.77 (n = 12), show that the granitoids are subduction-related I-type, high K-calc-alkaline to shoshonitic series and derived from a mantle source. They exhibit fractionated light rare earth elements, without depleted Eu and depleted middle heavy rare earth elements and Y, typical of oxidized, fertile porphyry magmatic rocks (Fang et al. 2007)

Alteration and mineralization. The mineralized zone is fault bounded by Carboniferous rocks above and relatively unmineralized quartz monzonite below and forms a tabular body 1.8 km long trending NE, 300 m thick, and dipping at about 45° to the NW. It has been drilled to about 300 m depth, and the down dip extension remains unexplored. Structural interpretation suggests that the currently defined ore body is offset by thrust faulting. Two stages of porphyry alteration and mineralization are identified (Tungalag et al. 2009; Tungalag et al. 2018). Early potassic alteration is overprinted by coarse muscovite-quartz on NE-trending structures, with associated chalcopyrite and bornite mineralization. Tsagaan Suvarga deposit hosted intrusion is characterized by a red color, due to phengitic mica-hematite alteration of feldspars. Ferromagnesian minerals are replaced by quartz-actinolite-biotite-chlorite-magnetite. Pervasive sericitic alteration, comprising fine-grained muscovite replacing plagioclase, is relatively minor. Instead, a typical feature for the Tsagaan Suvarga is mm to dm zones of coarse muscovite (100–500 μm) with associated chalcopyrite-bornite mineralization. According to the Gustafson and Hunt (1975) classification, typical A- and B-type porphyry veins are present, but a high density of typical porphyry quartz veins does not characterize the Tsagaan Suvarga deposit (Tungalag et al. 2018). Molybdenite occurs in monomineralic veins (1–5 mm) or A veins. Copper mineralization is represented by chalcopyrite and subordinate bornite, disseminated and associated with quartz-muscovite veins. Pyrite (vol.%) content is less than chalcopyrite and bornite combined. Deep oxidation to about 50 m depth has formed zones of malachite and covellite in late fractures. The most important alteration is actinolite-biotite-chlorite-magnetite replacing hornblende and primary biotite. Quartz-K-feldspar alteration is minor. Late albite replaces primary K-feldspar and enhances sodic rims on plagioclase crystals. Quartz-muscovite (or sericitic alteration) overprints actinolite-biotite and porphyry-type quartz veins. Field observations and petrographic studies suggest that the bulk of the chalcopyrite-bornite mineralization at the Tsagaan Suvarga formed together with coarse muscovite alteration (Tungalag et al. 2018). Fluid inclusions temperature in quartz veins is 277–340 °C with an average of 333 °C (Tungalag 2014).

2.3.3 *Kharmagtai Cu-Au Deposit*

The Kharmagtai deposit consists of multiple co-genetic gold-rich and tourmaline breccia pipes occurring within the Carboniferous Kharmagtai igneous complex. The Kharmagtai igneous complex consists of a series of intrusive bodies ranging between diorite through monzodiorite, quartz monzodiorite to monzonite and granodiorite compositions. This cluster of intrusions is interpreted to have resulted from a very large fractionating magma chamber at depth. The extent of this magma chamber can be seen in the recent gravity survey result. According to the geophysical survey, the Kharmagtai lease has been covered by 25 m spaced ground magnetics and 50–100 m spaced gravity surveys. Magnetic data are useful at Kharmagtai as most of the known porphyry mineralization is related to or proximal to magnetic anomalies. Magnetic data are also useful to help map major structures. Recently a gravity survey was conducted aimed at defining the margins of the Kharmagtai intrusive complex. New magnetics and gravity inversions are being run to allow more efficient targeting and interpretation of structures and intrusive boundaries at depth.

Most of the copper-gold mineralization is associated with multiple porphyritic intrusive stocks of diorite to monzodiorite. Three styles of copper-gold mineralization are identified at the deposit, namely, porphyry style stockworks, tourmaline breccia style, and epithermal gold associated with base metal veins. Porphyry stockwork forms the majority of the mineralization. Copper and gold are identified in the chalcopyrite and bornite sulfides and are associated with porphyry style veining. Tourmaline breccia mineralization consists of chalcopyrite-tourmaline-pyrite as infill to hydrothermal breccia, while epithermal style mineralization is hosted by carbonate-pyrite-chalcopyrite-sphalerite-galena veins controlled by late north-west trending faults.

A JORC 2012 compliant resource was released for the three previously discovered porphyry deposits at Kharmagtai in April 2015. There are three deposits with resources at Kharmagtai, Stockwork Hill, White Hill, and Copper Hill (Fig. 2.7). The resource estimation report shows that combined indicated and inferred re-sources are 157 Mt., grading at 0.32% Cu, 0.28 g/t Au and by metal, 0.49% CuEq (Open resource n.d., Xanadu mines http://www.Xanadumines.com/site/proejcts/kharamgtai/resources).

Stockwork Hill consists of two high-grade ores, the northern and southern stockwork zones with a larger body of high grade tourmaline breccia at depth (Fig. 2.8). White Hill consists of a much larger, moderate-grade zone of stockwork mineralization 800 m south of Stockwork Hill. Copper Hill is a smaller, high-grade zone of stockwork mineralization offset at depth by a low angle fault. Work is

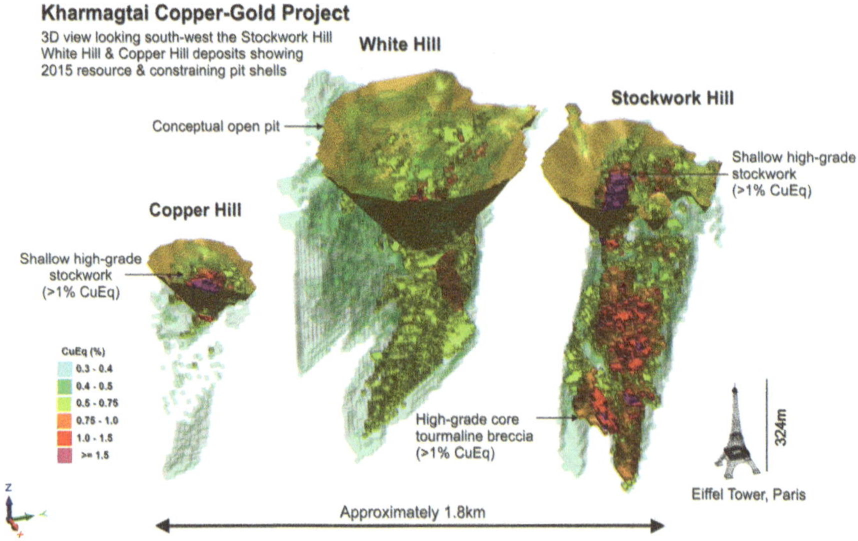

Fig. 2.7 Section through the Stockwork Hill, White Hill, and Copper Hill (from open resource of Xanadu mines)

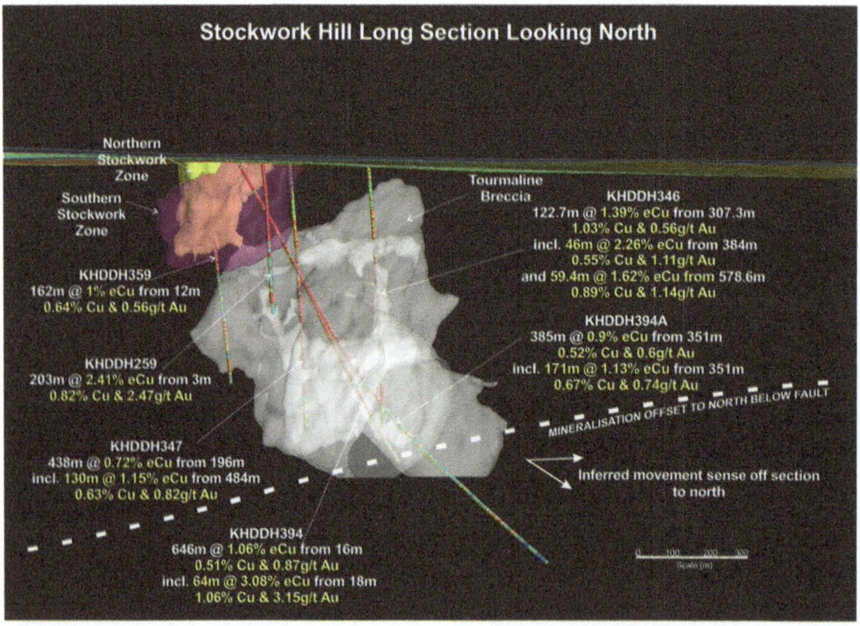

Fig. 2.8 Long section through the Stockwork Hill deposit showing the two outcropping stockwork zones and high-grade tourmaline breccia (from open resource of Xanadu mines)

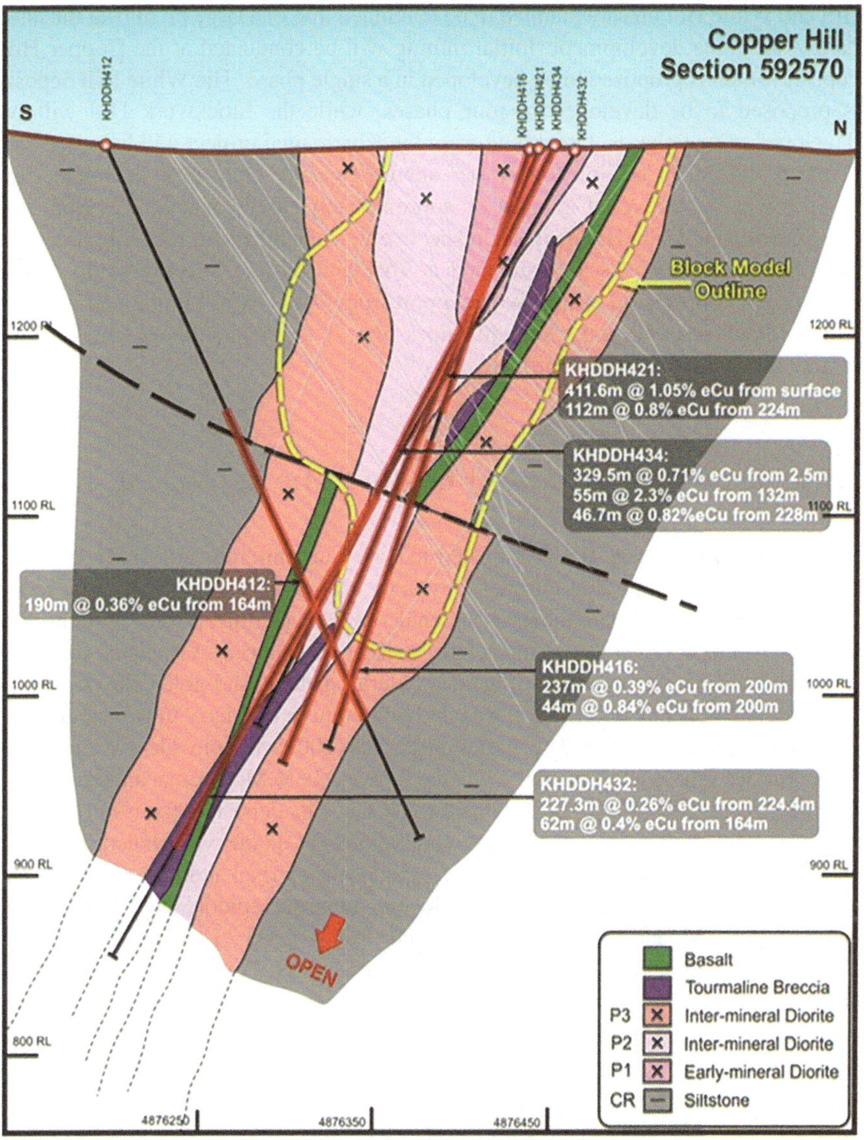

Fig. 2.9 Cross-section through the high grade Copper Hill deposit (from open resource of Xanadu mines)

underway to determine the direction of offset and the location of this offset mineralization, as the grades at Copper Hill make this an attractive target (Fig. 2.9).

Conventional open-cut mining method, involving drill-blast-load-haul operations, will be employed at Stockwork Hill, Copper Hill, and White Hill deposits. The initial plan involves mining the three deposits separately, while the Stockwork

Hill and White Hill pits are planned to be combined into one large pit during the later part of the mine development. Initial mining will be conducted at the Copper Hill deposit, which is proposed to be developed in a single phase. The White Hill deposit is proposed to be developed in four phases, while the Stockwork Hill will be developed in two phases. Ore mined from the Kharmagtai project will be processed at the proposed 20 million tonnes per annum capacity plant near the mine site. Copper will be recovered at rates of approximately 90.9% and 85.7% from the Copper Hill and White Hill deposits, respectively, while the respective gold recovery rates at the two deposits are estimated at 76% and 69.1%. Scoping study of the Kharmagtai copper-gold project with support from Oyu Tolgoi Mining and Xanadu Mines (Open resource n.d., Xanadu mines http://www.xanadumines.com/irm/content/kharmagtai.aspx?RID=368).

2.3.4 Kharmagtai 2 Porphyry Cu-Mo (±Au, Ag) Deposit

This deposit (Nokleberg 2010) is hosted in Late Carboniferous and Early Permian diorite and granodiorite that intrudes Devonain tuff, andesite, and tuffaceous sandstone and siltstone. The ore minerals are chalcopyrite, covellite, bornite, and molybdenite. Oxidation minerals are malachite, azurite, and cuprite. Associated minerals are pyrite and magnetite and peripheral sphalerite, galena, and gold. The deposit is related to subvolcanic bodies of diorite and granodiorite porphyry in two stocks and bodies of explosive breccia. The bodies range from 200 to 400 m wide and are 900 m long. Surface grades are 0.05–0.4% Cu and 0.003–0.03% Mo over an area of 400 by 900 m. A zone 100 by 300 m contains >0.3% Cu. The deposit extends at least to a depth of 250 m and is defined by stockwork veinlets of quartz with chalcopyrite and molybdenite that occur across the breccia pipe. Hydrothermal alteration minerals are weakly developed silica, sericite, K-feldspar, chlorite, epidote, and tourmaline. Sericite, potassic, and silicic alterations occur in the center of alteration zone, and chlorite and epidote alteration occur along the periphery. Potassic alteration occurs mainly in the deeper part of deposit. The deposit is small with resources of 0.8 Mt. Cu grading 0.35% Cu.

2.3.5 Shuteen Cu-(Au) Mineralized Porphyry System

The Shuteen area is located about 450 km south-southeast of Ulaanbaatar in South Mongolia. The area contains the products of Paleozoic magmatic activity, as well as a large siliceous and advanced argillic-altered lithocap (e.g., Sillitoe 1995; Chang et al. 2011), making it a geologically attractive area in which to study magmatic-hydrothermal alteration and mineralization. Since the 1950s, joint Russian-Mongolian expeditions were undertaken to conduct reconnaissance exploration in the Shuteen region, and 1:200,000-scale geologic maps have been compiled for the

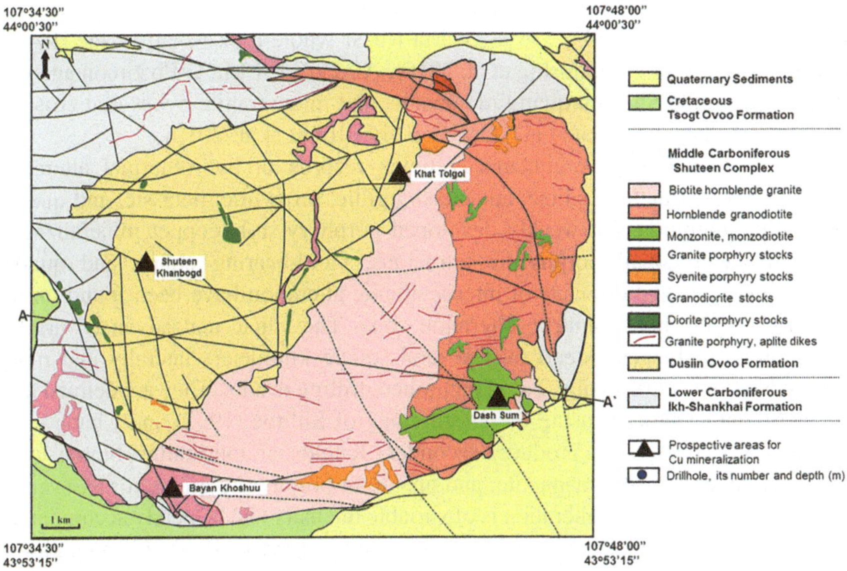

Fig. 2.10 Geologic map of the Shuteen area. The main copper prospects are denoted by triangles: Shuteen Khanbogd lithocap, Khar Tolgoi, Bayan Khoshuu, and Dash Sum (Batkhishig et al. 2014, with permission of SEG Publication)

area. In the 1980s, joint Czech-Mongolian projects studied the area in greater detail in relation to porphyry-style mineralization, identifying several prospective areas and compiling 1:25,000- and 1:10,000-scale geologic maps. Twenty-one drill holes were drilled as part of this work. Since 1997, geologic mapping, rock chip sampling, soil sampling, trenching, magnetic surveys, and local induced polarization (IP) surveys have been undertaken. This work helped to constrain the extent of hydrothermal alteration zones, especially the siliceous lithocap and zones of anomalous concentrations of Au, As, Se, Sb, and Mo. The exploration work has revealed high-temperature alteration zones, gold-bearing quartz veins, and IP and magnetic anomalies in the area. In 2002, trenches were excavated to better define the geology and structure in several key areas, and four diamond drill holes were drilled to a combined depth of ~2000 m. Prospecting and exploration results at Shuteen have identified a very large volume of hydrothermal alteration, but no economic mineralization has been found as yet, creating a debate as to whether Shuteen is barren, weakly mineralized, or well mineralized but underexplored (Batkhishig 2005; Batkhishig et al. 2005; Bignall et al. 2005).

The Shuteen area contains intermediate volcanic rocks of the Dusiin-Ovoo Formation and plutonic rocks of Shuteen pluton (also called Shuteen Complex) were emplaced over and into the Lower Carboniferous Ikh Shankhai Formation, respectively (Fig. 2.10). The Shuteen Complex is approximately 15 × 13 km in plain view (Fig. 2.10) and is characterized by comagmatic andesitic, dacitic, and rhyolitic volcanic rocks, diorite, and granodiorite (Batkhishig 2005; Batkhishig et al. 2010).

The Shuteen pluton has a well-defined isochron age of 321 ± 9 Ma, whereas the Shuteen andesites yielded a 336 ± 24 Ma Rb-Sr whole-rock isochron age (Iizumi and Batkhishig 2000; Batkhishig et al. 2010). A 325.5 ± 1 Ma U-Pb zircon age was obtained from the Shuteen quartz monzonite, and microgranite dykes that crosscut the intrusion have been dated at 325.4 ± 1 Ma (Blight et al. 2010).

The Shuteen Complex contains complex records of hydrothermal alteration assemblages that include silicic, advanced argillic, propylitic, potassic, and quartz-tourmaline alteration and weakly developed porphyry-style copper mineralization associated with small porphyritic intrusions. Gold-bearing quartz and quartz-tourmaline veins and domains of intense alunite alteration have been found on the periphery of the main Shuteen alteration zone. The silicic and advanced argillic alteration assemblages extend over several square kilometers laterally and many hundreds of meters vertically, based on limited drilling results. They are defined here as the Shuteen lithocap, using the terminology of Sillitoe (1995) and Chang et al. (2011). Potassic alteration produced secondary K-feldspar, tourmaline, biotite, chlorite, epidote, andalusite, magnetite, and apatite in porphyritic granodiorite, syenite, and diorite. The potassic alteration is of variable intensity and is locally accompanied by visible chalcocite mineralization and disseminated pyrite. Propylitic alteration is largely restricted to the outlying part of the Shuteen lithocap and is characterized by chlorite, epidote, calcite, sericite, pyrite, albitized feldspar, and magnetite. Propylitic alteration assemblages are most common in the plutonic rocks but have also been recognized locally in andesite and diorite porphyry, which retain their primary textures. Spatially restricted tourmaline-chlorite alteration zones are characterized by a sequence of alteration from quartz-tourmaline to tourmaline-sericite and tourmaline-chlorite. This style of metasomatic alteration is developed in all rock types (volcanic, plutonic, porphyritic, and sedimentary). Advanced argillic alteration developed mainly in the volcanic rocks that host the Shuteen lithocap. Short-wavelength infrared (SWIR) analyses using a PIMA and petrographic analysis have revealed several mineral associations, including andalusite-sericite-diaspore-pyrophyllite-topaz-alunite-pyrite, dickite-kaolinite-pyrite, and kaolinite-smectite-sericite-pyrite assemblages (Delgertsogt 2008). These advanced argillic assemblages have overprinted the propylitic and potassic alteration assemblages. The JICA, MMAJ (1995) reported a K-Ar age of 302 ± 9 Ma for argillized rocks from the Shuteen area.

Geologic, surface geochemical, and geophysical data have revealed intensely developed hydrothermal alteration zones and weak porphyry-style copper mineralization locally associated with porphyritic intrusions. Several prospects of copper mineralization are recognized within the complex, including Shuteen-Khanbogd, Khar Tolgoi, Bayan Khoshuu, and Dash Sum (Fig. 2.10). The Shuteen-Khanbogd prospect is the large silicic and advanced argillic alteration domain that defines the Shuteen lithocap. The lithocap is hosted by volcanic rocks and is 7.5 km long and 1.5–3 km wide (Fig. 2.10). The western contact of the Dushiin-Ovoo Formation and the Ikh Shankhai Formation is marked by quartz and quartz-tourmaline veins that contain chalcopyrite, galena, and pyrite mineralization. The sulfides are related to a diorite porphyry and quartz-sericite alteration. The rocks are altered to silicic, advanced argillic, and rare propylitic assemblages. Stream sediment geochemical anomalies of Au, Mo, Ag, As, Se, and Sb have been identified. Gold anomalies are low, ranging from 5 to 30 ppb (background is <5 ppb). Other geochemical anomalies (e.g., As, Mo, Se, Sb, and Au) overlap with each other. In 2002, four diamond drill holes were completed, but unfortunately, the drill holes did not intercept any significant gold and copper mineralization.

The Khar Tolgoi prospect covers approximately 6 km^2, and it contains the most intensely developed copper mineralization within the Shuteen Complex. Diorite porphyry, granodiorite porphyry, syenodiorite, and granosyenite have been affected by potassic and propylitic alteration. The mineralization is characterized by veinlets, disseminations, and breccias and includes chalcopyrite, pyrite, magnetite, carbonate, epidote, chlorite, actinolite, and tourmaline. Zones of oxidation contain chalcocite, covellite, secondary sulfates, and rare malachite. Khar Tolgoi has strong geochemical anomalies in Cu, Zn, and Mo, based on past drilling by a Czech geologic expedition team, which returned anomalous copper, including multiple intersections of 1000 to 2500 ppm Cu (rarely 6000 ppm) and (2 to 5 ppm Mo). The periphery of this zone contains quartz and quartz-tourmaline veins with minor chalcopyrite and rare galena.

Bayan Khoshuu is characterized by vein-type mineralization in a 0.15-km^2 area (Fig. 2.10). Partially weathered quartz tourmaline veins contain copper oxides and are hosted in granodiorite, diorite porphyry, and granosyenite porphyry. The host rocks have been affected by chloritic and potassic alteration. Ore minerals include pyrite, chalcopyrite, molybdenite, and galena, and secondary minerals include azurite, malachite, chalcocite, covellite, iron oxides, limonite, and jarosite. The mineralization is coincident with a well-defined IP resistivity high and has dimensions of approximately 300–500 m long and 140–260 m width.

Dash Sum is a 5 km^2 zone of intense faulting, with geochemical anomalies of Cu, Pb, Zn, and Co. Diorites, hybrid igneous rocks, and granodiorites have been silicified and host porphyry copper-style mineralization. The ore minerals include primary chalcopyrite and pyrite (locally hematite or magnetite) and hypogene bornite, chalcocite, and supergene malachite.

There are also areas of Au bearing quartz and quartz-tourmaline veins. Gold mineralization within the Shuteen Complex is associated with altered rocks and quartz veins. Gold, together with scheelite, wulfenite ($PbMoO_4$), and malachite, has been found in heavy mineral concentrates from Bayan Khoshuu (Fig. 2.10), and core samples have between 0.25 and 1.2 g/t Au. Gold, pyrite, chalcopyrite, and marcasite are found in quartz veins in the southwest part of the complex. Some surface samples taken from quartz veins contain 1 to 4 g/t Au. One sample contained 57 g/t Au (Batkhishig 2005; Batkhishig et al. 2005).

Currently, it is concluded that the Carboniferous Shuteen Complex hosts one or more fossil magmatic-hydrothermal systems. During the early stages of magmatic-hydrothermal fluid activity, weak porphyry copper-style mineralization associated with potassic and propylitic alteration formed in association with the emplacement of porphyry intrusions into the Shuteen Complex. Extensive exploration activity at Shuteen has not yet identified any economic resource, only large-scale geochemical anomalies and hydrothermal alteration zones have been identified. This may relate to limited diamond drilling. The porphyry copper-style mineralization remains to be discovered in the Shuteen area, and it is likely to be concealed beneath the Shuteen lithocap (Batkhishig 2005; Batkhishig et al. 2014).

2.3.6 Erdenetiin Ovoo Cu-Mo Porphyry Deposit

The Erdenetiin Ovoo deposit, at the Erdenet mine, is located in northern Mongolia, 240 km northwest of the capital, Ulaanbaatar. The deposit was discovered in 1964 and first explored by E. Kominek, G. Sanduijav, and I. Popoušek and later studied by V.S. Kalinin, L.I. Yakimov, I.D. Davydov, E.V. Mikhailov and A.E. Shabolovsky, Khasin et al. (1977), Sotnikov and Berzina (1989), Koval et al. (1985), Koval and Gerel (1986), Gavrilova et al. (1984), Gavrilova and Maksimyuk (1990), and others. The Erdenet mine is the largest Cu-Mo porphyry deposit in Mongolia, with estimated metal reserves calculated in 1991 of 7.6 Mt. Cu and 216,600 tonnes of Mo in approximately 1490 Mt. of ore @ 0.509% Cu and 0.015% Mo (EMC 2002). Estimated mineral resources in 2004 are 2370 Mt. @ 0.38% Cu and 0.013% Mo. Projected reserves of deeper ore are estimated to be 1.4 Mt. of contained Cu and 37,000 tonnes of Mo. Published mining reserves in 2002 were 766.7 Mt. @ 0.61% Cu containing 4.7 Mt. of copper. The Erdenetiin Ovoo deposit has been mined since 1978 at an annual rate of approximately 20 Mt. of ore from an open pit which currently covers an area of 2500 × 1150 m. The initial head grade in 1978 was 0.87–0.96% Cu. In 2002, 25.8 Mt. of ore at a head grade of 0.647% Cu was mined, to produce 131,705 tonnes of copper and 1590 tonnes of molybdenum in concentrate from a joint Mongolian-Russian ore dressing plant (EMC 2002). Metallurgical recovery in 2002 averaged 83.25% for Cu and 33.06% for Mo in concentrates. The composition of the copper concentrate is 27–35% Cu, 50–60 g/t Se, 50–70 g/t Ag, 8–9 g/t Te, and 0.3–0.5 g/t Au. The molybdenum concentrate from Erdenet contains 47–54% Mo, 450 g/t Re, 90 g/t Se, and 15 g/t Te. A small but increasing amount of tungsten is also produced. At the present day, it is a fairly large complex processing 26 million tonnes of ore per year and producing around 530.0 thousand tonnes of copper concentrate and around 4.5 thousand tonnes of molybdenum concentrates annually (EMC 2019).

Regional Geology. The Erdenet deposit occurs within the Selenge terrane of (Tomurtogoo et al. 1999; Badarch et al. 2002), which overlaps the cratonic Tarvagatay terrane and the Orkhon-Selenge trough of Mossakovsky and Tomurtogoo (1976). This terrane is dominated by Permian to lower Triassic volcano-sedimentary and intrusive rocks and upper Triassic to lower-mid Jurassic volcanics and sedimentary cover. The Orkhon-Selenge trough is part of the larger Mongol-Okhotsk fold belt which stretches over 3000 km from west to east, from the Khangai highlands in Central Asia to the Pacific Ocean (Zonenshain et al. 1990). Subduction and collision during the late Palaeozoic to early Mesozoic resulted in the uplift of the Khangai and Khentii highlands flanked by depressions, which included

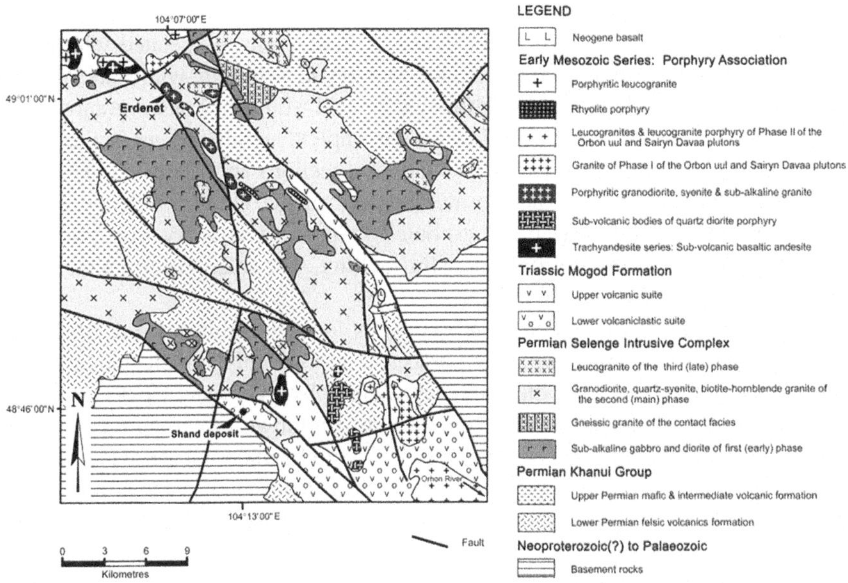

Fig. 2.11 Geological map of the Erdenet ore field (after Koval et al. 1985)

the Orkhon-Selenge trough and eastern Mongolian volcanic-plutonic belt, each with a different style of magmatism and mineralization. Porphyry copper mineral systems dominated in the north, within the Orkhon-Selenge trough, while polymetallic mineralization is characteristic of the Eastern Mongolian volcanic-plutonic belt in the south (Gerel 1990).

The Erdenet deposit is hosted by a late Permian to early Triassic intrusive complex (Fig. 2.11) which forms part of the extensional volcano-plutonic belt within the Orkhon Selenge trough.

The Orkhon-Selenge trough is the largest of the late Palaeozoic to early Mesozoic structures within the region, stretching for over 200 km in an east-northeast direction, with a width of 30–40 km, filled by Permian and early Mesozoic volcano-sedimentary rocks. The Permian volcanics belong to the Khanui Group, which is as much as 10–20 km thick in the core of the trough, and up to 5 km on its periphery. The Khanui Group is composed of four conformable formations. The lowest comprises mafic and intermediate volcanics and has a limited distribution. It is overlain by felsic volcanics of the succeeding formation, which are overlain in turn by volcanoclastics of the third. The fourth and uppermost of the formation is composed of mafic volcanics. According to Mossakovsky and Tomurtogoo (1976), the age of the Khanui Group varies from lower Permian in the west, to upper Permian in the east of the trough. The group is characterized by a sub-alkalic composition, with elevated K alkalinity and a bimodal nature (Kepezhinskas and Luchitsky 1973). It exhibits a lateral variation in character, with a volcanic suite in the western peripheral part the Orkhon-Selenge trough, grading to a volcanoplutonic association (Yashina and Matrenitsky 1978). Early Permian intrusions occur as a series of parallel dykes within felsic volcanics or as shallow stocks. Doleritic sills, associated with upper Permian basalts, form elongated belts in the central part of trough, while late Permian intrusive rocks, which include gabbro-syenite and younger granite-granosyenite associations known as the Selenge Intrusive Complex, are also found in the center and on the uplifted margins. The main faulting directions are northwest, northeast, and sub-latitudinal, creating the mosaic structure of the Orkhon Selenge trough, whose current geometry has been attributed to post-accretionary disruption and dislocation by extension and or strike-slip faulting. The main deformation age is mid-Jurassic (Mossakovsky 1975).

District Geology. The Erdenet ore district is located within a local 20–25 km diameter circular structure which embraces a number of subvolcanic intrusive centers, each of 2–2.5 km in diameter, represented by dykes and stocks of altered granodiorite. Four aligned intrusive centers form a northwest trending range of low hills, while a fifth is indicated by geophysical data to be under Quaternary sediment cover of the Erdenetiin Gol River. The five prospects from southeast to northwest are Oyut, Intermediate, Central, Northwest (the Erdenet mine), and Tsagaan Chuluut (Fig. 2.12).

The Erdenet ore field porphyries (hereinafter referred to as Porphyry Association) and the ore-bearing stockwork are intruded into biotite-hornblende granites, granodiorites and quartz syenites of the Selenge Intrusive Complex. The porphyries crosscut the third phase granites of the Selenge Complex, which in turn cuts the Permian sequence, including the uppermost mafic volcanic formation. The Porphyry Association intrusions (quartz-diorite to granodiorite porphyry and granite porphyry) postdate the Selenge Intrusive Complex intrusions and therefore cannot to be the last member of the Complex (Gerel and Munkhtsengel 2005).

The Porphyry Association includes shallow, subvolcanic and explosive stocks, dykes, necks and sheet-like bodies. They are variable in texture and composition, ranging from porphyritic quartz diorite and diorite, quartz syenite, granodiorite to granites and leucogranites, associated with explosive breccia, andesites and trachyandesites, dacites, rhyodacites and rhyolites, and trachytes and trachydacites. These rocks are predominantly younger than those previously described with K-Ar ages that vary from 235 to 185 Ma (Sotnikov et al. 1985). The porphyries are associated with volcanics of a trachyte-trachyandesite-alkaline basalt series occurring in volcanic, subvolcanic, and intrusive settings, forming sheets, volcanic cones, dykes, and sills. The evidence of a genetic relationship between the Porphyry Association intrusions and Mesozoic volcanics (Koval and Gerel 1986; Koval et al. 1985) can be summarized as follows: (a) the crosscutting relation between the Porphyry Association and the third and latest phase of the Selenge Intrusive Complex, showing that the former is the younger; (b) the disruption in the trend of magmatic evolution of the Selenge complex that would be required if the ore-related porphyries were included as members of the complex; (c) the difference in the structural trend of the porphyries and that of the Selenge complex; (d) the difference in the environment of emplacement, between the shallow setting with associated explosive facies that characterize the porphyries and the deeper hypabyssal

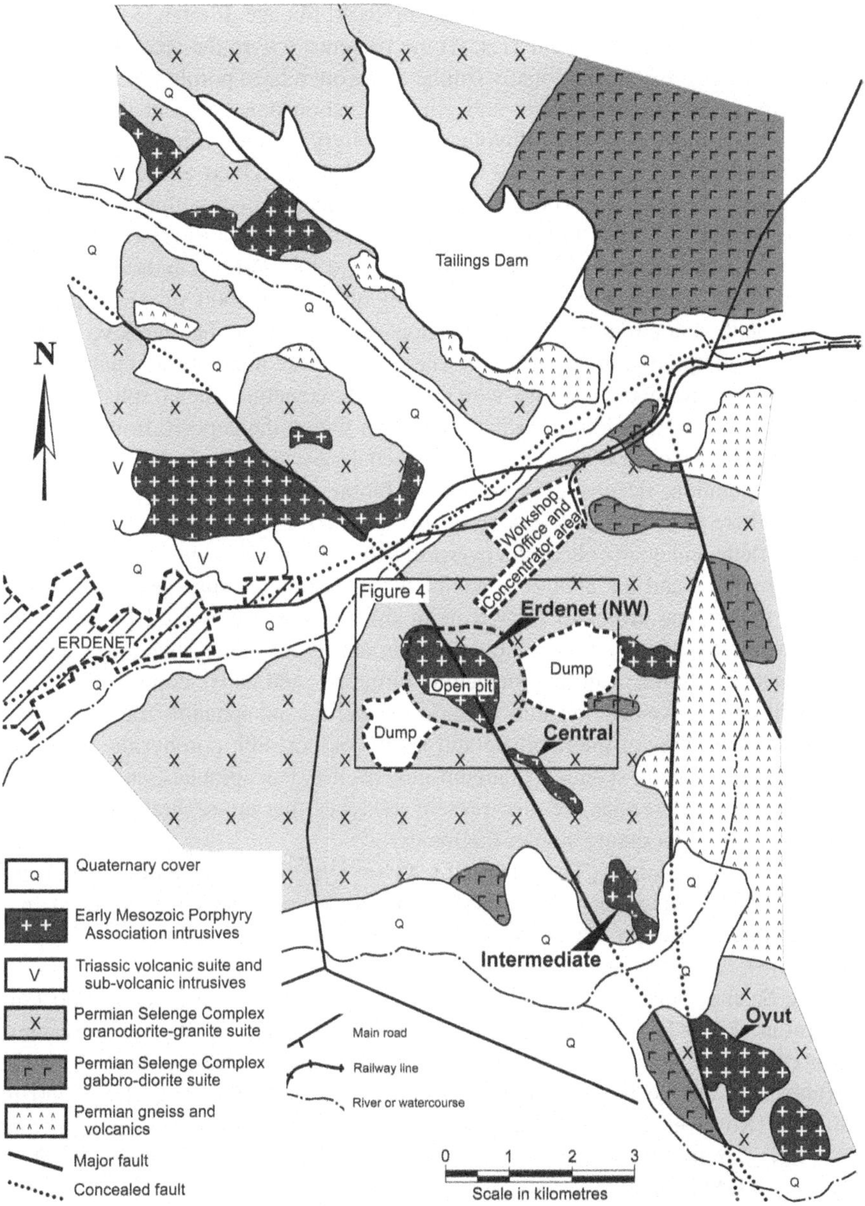

Fig. 2.12 Location plan and geological sketch map of the Erdenet district, northern Mongolia. Topographic and cultural features from satellite image, geology based on Watanabe and Stein (2000) and Fig. 2.11

conditions of the Selenge complex; (e) the up to 40 Ma age difference between the Selenge complex and the porphyries; (f) the presence within the area of Mesozoic volcanics which are compositionally similar to the ore related porphyries; and (g) the intense propylitic, quartz-sericite and silica alteration developed in the both the Mesozoic volcanics and the subvolcanic porphyry bodies. Similar suggestions were presented by Yakovlev (1977). The Erdenetiin Ovoo porphyry copper-molybdenum deposit is associated with, and concentrated within, subvertical, pipe-like intrusive bodies of late Triassic granodiorite porphyry of the Porphyry Association, with diameters of as much as 1.3–2.4 km. Drilling indicates that the mineralization persists over a vertical extent of at least 1000 m. Disseminated, fracture, and vein-controlled mineralization occurs within the intrusive bodies, associated with granodiorite porphyry. The principal ore minerals are chalcopyrite, chalcocite, bornite, and tennantite, with associated magnetite, galena, sphalerite, and quartz. There is a well-defined mineral zonation within the deposit, from pyrite on the periphery to a central high-grade core of chalcopyrite-bornite-chalcocite. Weak potassic alteration, occurring as secondary biotite, is followed by pink feldspar veining where fracturing and veining of the host rock is not intensive. Widespread quartz-sericite and quartz alteration overprints the vein assemblages and is related to major structures and the contact zones of the intrusions. A supergene enrichment blanket that ranges from 30 to 300 m in thickness is represented by a bornite covellite-chalcocite assemblage, overlain by an oxide suite including cuprite, delafossite, malachite, azurite, brochantite, turquoise, and native copper.

The Porphyry Association suite includes phaneritic and aphanitic rocks comprising a variable series of intermediate and felsic composition: porphyritic andesites, quartz diorites, quartz syenites, granodiorites, dacites, leucogranites, and rhyolites.

Hydrothermal breccias are also present as dykes that are several meters wide, containing irregular quartz veinlet fragments.

Intrusive Magmatism. The Erdenetiin Ovoo deposit is a cylindrical body of quartz- sulfide stockwork veins hosted by an ore-bearing stock which is centered on the multiphase pre-ore Erdenet Pluton of the Selenge Intrusive Complex (Fig. 2.13).

The Selenge Intrusive Complex, which occupies an area of around 2800 km^2 (Koval et al. 1982), includes multiple, mainly hypabyssal, and shallow intrusions, each covering a few km^2 to 1000 km^2. The individual intrusives have variable textures and compositions with K-Ar ages of intrusion that range from 250 to 210 Ma (Koval et al. 1982; Koval and Gerel 1986) and U-Pb ages ranging from 225.9 $\pm$ 3.6 to 252.3 $\pm$ 6.8 Ma (Berzina et al. 2012) and 249.7 $\pm$ 4.0 to 243.3 $\pm$ 4.6 Ma (Munkhtsengel 2007).

A series of mineralization related dykes and plugs, which are widespread within the Erdenet district and throughout the Orkhon-Selenge trough, was distinguished as the "Porphyry Association" by Koval et al. (1985), and Koval and Gerel (1986) or as the "Porphyry Complex" in the Erdenet ore field by Sotnikov and Berzina (1985). These intrusives are regarded by the same authors as being separate from and younger than the Selenge Intrusive Complex. Koval et al. (1982) regarded these intrusions to be of early Mesozoic age and included them within the Orkhon-Selenge

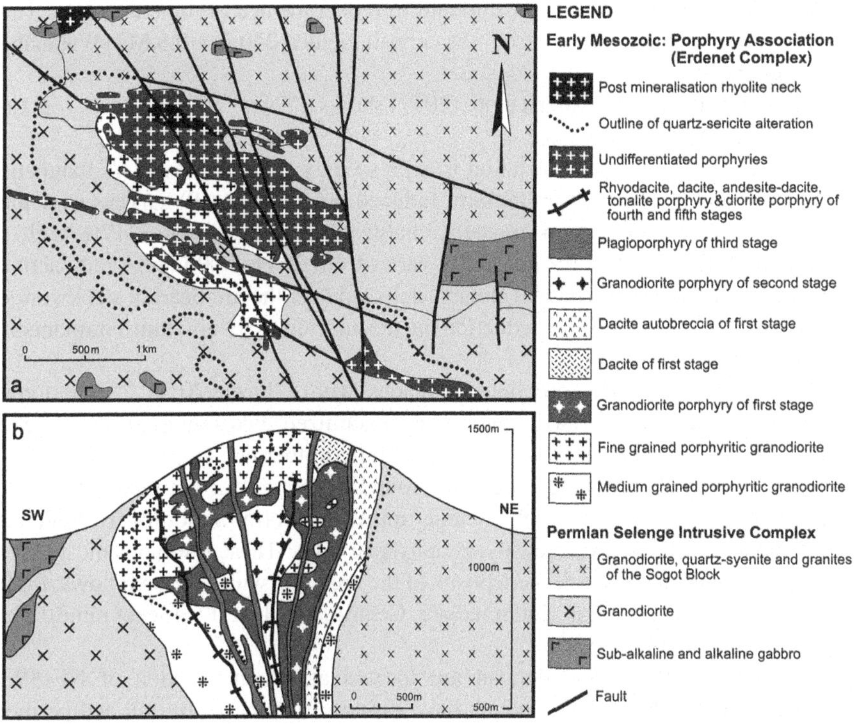

Fig. 2.13 The geology and schematic structure of the Erdenetiin Ovoo (Gerel and Munkhtsengel 2005, modified after Gavrilova and Maksimyuk 1990) (**a**) Surface map of the Erdenetiin Ovoo. (**b**) Schematic cross section

association, which comprises an "earliest" and two main granitoid phases, namely, (a) the "earliest" is represented by quartz-diorite, diorite and monzonite, (b) the first main granitoid phase comprises quartz-syenite, hornblende-biotite granite and granodiorite, and (c) the second phase which includes leucogranite and sub-alkaline granite (Gerel and Munkhtsengel 2005). Yashina and Matrenitsky (1978), however, include this "association" within their three separate subdivisions of the Selenge Intrusive Complex, which are (a) gabbro to syenite of middle Permian age; (b) syenite, granosyenite, and granite of middle Permian to early Triassic age; and (c) monzogabbro to diorite, syenite, granite, porphyry (Erdenet type) of middle Permian to early Triassic age.

Age of the Deposit. Sotnikov et al. (1995), using K-Ar age, dated the host Selenge complex at 290–260 Ma. They had previously K-Ar dated the Erdenet porphyry complex at 235–185 Ma (Sotnikov et al. 1985). Other K-Ar age dates for granites of the Erdenet Pluton yielded 226–245 and 226–221 Ma.

U-Pb age data for the Erdenet Porphyry Complex show 245.9 ± 3.3 to 235.6 ± 4.4 Ma (Munkhtsengel 2007). Lamb and Cox (1998) performed $^{40}Ar/^{39}Ar$ dating on the sericite from the quartz sericite alteration zone at Erdenet which yielded 207 Ma, which was thought to reflect the age of mineralization. The Re-Os age date on molybdenite from the deposit yielded 240.7 ± 0.8 Ma (Watanabe and Stein 2000).

The recent study of Kavalieris et al. (2017) shows a $^{40}Ar/^{39}Ar$ age of 239.7 ± 1.6 and 240 ± 2 Ma for muscovite.

Geophysics. Copper mineralization is expressed by a zone of lower nT magnetic field values and by a gravity minimum. Induced polarization (IP) surveys over the ore-bearing stock revealed an apparent chargeability anomaly of 3–17% with a maximum of 22.8%, while the resistivity varies from 200 to 800 Ωm. A radioactive anomaly in the potassium channel alone is detected over the ore-bearing stockworks, accompanied by a noticeable reduction in the thorium and uranium components (Jargalsaihan et al. 1996).

Deposit Structure. The ore field coincides with a local, 20 to 25 km long, structurally controlled corridor, containing mineralized dykes and stocks between 2 and 2.5 km in diameter, with associated hydrothermal alteration. Zones of propylitization and tourmalinization are found within rocks peripheral to these controlling structures. Four subvolcanic intrusives form a northwest trending ridge, while a fifth is predicted by geophysical data. These sub-volcanic bodies correspond to the main prospects/deposits of the ore field, which are as follows, from southeast to northwest, Oyut, Intermediate, Central, Northwest (Erdenet mine), and Tsagaan Chuluut (Fig. 2.12).

The Erdenet Pluton and deposit are located at the intersection of NE-SW, NW-SE, N-S, and E-W faults, which have produced strong fracturing and permeability at their intersection. Sotnikov et al. (1984) suggested these faults controlled the mineralization and hydrothermal alteration. Dextral movement on the N-S faults is interpreted to be syn-mineralization, while sinistral movement on the E-W and NE-SW faults is believed to be post-mineralization. The southwestern and northeastern parts of the Erdenet deposit have steep boundaries intruded by dykes, while NS and NE-SW trending faults roughly divide the Erdenet Pluton itself, into separate southeastern and northwestern zones, which contain intrusive breccias and mineralized stocks of the Porphyry.

Ore-Related Porphyries. The Erdenetiin Ovoo ore body coincides with "Porphyry Association" porphyry stocks and dykes and as a result of supergene processes can be divided into three vertical zones, comprising from top to bottom: a leached and oxidized zone, secondary sulfide enrichment, and primary ores. The "Porphyry Association" porphyries can be subdivided into:

First stage porphyries are the main syn-mineral phase and accompany the introduction of the main economic mineralization. Intrusives of this stage are dominated by hornblende-biotite granodiorite porphyry with lesser dacite. The intrusions of the first-stage porphyries were accompanied by explosive activity, producing eruptive breccias with granodiorite fragments cemented by granodiorite porphyries, and subsequently by dacites which cemented both brecciated granodiorites and granodiorite porphyries. Granodiorite porphyries comprise the main mass of the mineralized intrusive stocks, with associated apophyses, satellites, and dykes. The first stage porphyries form two connected isometric bodies.

Second stage porphyries are represented by granodiorite and granite porphyries. These porphyries crosscut both granodiorite porphyries and dacites of the first stage and quartz-sericite alteration.

Third stage porphyries are represented by biotite plagioclase, plagiogranite and granodiorite porphyries as well as leucocratic granite porphyries and leucogranitic plagioporphyries (Gavrilova and Maksimyuk 1990). These rocks occur within a sub-latitudinal zone up to 1.3–1.4 km wide and crosscut both first and second stage porphyries.

Gavrilova and Maksimyuk (1990) distinguished both a fourth and fifth stage porphyries. They observed that the third stage porphyries were crosscut by younger porphyries of the fourth stage, represented respectively by leucocratic porphyries and rhyodacites which occur as very rare dykes. The fifth stage porphyries are related diorite porphyries, amphibole-plagioclase andesites and granodiorite porphyries. Propylitic alteration is associated with this stage of intrusion. Post-ore stage intrusions include dykes and necks of latite, trachyandesite, rhyolite, and syenodiorite porphyries, with mainly sub-latitudinal orientations, dated at 185 Ma by K-Ar techniques (Sotnikov and Berzina 1989). Dykes of younger basalts, andesites, and basaltic trachyandesites have meridional trends.

Hydrothermal Alteration. Three principal alteration assemblages define a near symmetrical alteration zonation at the Erdenet deposit, centered on the granodiorite porphyry intrusion (of the Porphyry Association) and with decreasing intensity outward. The alteration zones, from the core to the periphery of the deposit are (a) sericitic (quartz-sericite) with local silicification, (b) intermediate argillic (chlorite-sericite), and (c) propylitic (chlorite and epidote-chlorite) and chloritic. The area covered by the total alteration halo (approximately 2.7 × 1.75 to 2.2 km) is more extensive than the porphyritic intrusion, which is around 2.4 × 1.4 km (Gavrilova and Maksimyuk 1990). Quartz-sericite alteration dominates in the core of the deposit, decreasing in intensity toward the periphery. It has affected between 25% and 45% of the host intrusions at Erdenet. In the central part of the ore body this alteration assemblage may comprise as much as 50–60%, and even up to 90% of the rock mass. It accompanies a stockwork of crosscutting and parallel fractures filled by thin ore bearing quartz veinlets, each fringed by selvages of quartz-sericite altered host rocks. The quartz-sericite vein fringes are from 0.3 to 6 cm wide, representing 5–10 times the thickness of the ore veins. Within the stockwork zone, the textures and structure of the original lithology are preserved. The quartz-sericite alteration grades outward into intermediate argillic (chlorite-sericite), to a propylitic (chlorite and epidote-chlorite) interval on the peripheries. At the distance of 1–1.5 km from the orebody, there is no evidence of alteration, except for some tourmalinization (Gavrilova and Maksimyuk 1990).

A second stage of alteration is characterized by brecciation, recrystallisation, and sericite-quartz, which overprints the previous stage of hydrothermally altered rocks. Silicification, with very rich chalcopyrite-bornite and molybdenite ores, is a characteristic and important component of this stage. Quartz-sericite and peripheral intermediate argillic alteration overprints an earlier phase of K silicate development, comprising quartz-K feldspar, biotite, and magnetite. This K silicate alteration is not well developed and preserved at Erdenet, having been overprinted and largely obliterated by up to five subsequent episodes of quartz-sericite and quartz-chlorite.

The width of the overall alteration zone tapers downward, e.g., at 1400 m above sea level the zone is 1.12 km wide, diminishing to 0.66 km at 920 m above sea level. The nature of the quartz-sericite alteration also varies with depth and can be subdivided into two zones. The deeper of these, between 900 and 1100 m elevation, represents a root zone, comprising separated, pillar-like, steep fracture-controlled masses, interpreted on the distribution of elemental concentrations, as representing a zone of vertical/steep angle channels where ore-bearing hydrothermal solutions were focused (Gavrilova et al. 1984). The upper zone, between approximately 1100 and 1600 m above sea level, represents the main bulk of the ore deposit and appears to be essentially flat lying and occurs within and surrounding all of the Porphyry Association stocks and satellite intrusions. It varies in thickness from 650 to 700 m in the SW, to 500 to 550 m in the core of the deposit, to 300 to 400 m to the NW. The overall structure of the hydrothermal system is reflected by the distribution of high silica content, which has been traced to a depth of 1600 m below the surface. Separate quartz-sericite alteration and silicification zones are also formed on the margins of the host porphyries, with a 650 m long and 50 to 150 m wide development along the northern contact of the main stock within host granodiorites, while on the southern contact only isolated lenses are recognized.

Mineralization. Khasin et al. (1977) detailed six stages of alteration and mineralization as follows: (a) pre-ore quartz-sericite, (b) quartz-pyrite, (c) quartz-pyrite-molybdenite, (d) quartz-pyrite-chalcopyrite, (e) quartz-pyrite-galena-sphalerite, and (f) post-ore gypsum-calcite with pyrite. Hypogene mineralization is characterized by bornite, chalcocite, and covellite, while oxidation produced Cu carbonates, oxides, phosphates and sulfate, native Cu, and ferrimolybdite (Fig. 2.14).

Gavrilova et al. (1984) distinguished one pre-ore and five distinct ore stages, namely:

Stage I—a quartz-pyrite association, represented by chalcopyrite-pyrite-quartz and molybdenite-quartz veinlets which are 1–2 cm, and rarely 3–5 cm in thickness.

Stage II—quartz-molybdenite veining.

Stage III—quartz-chalcopyrite-tennantite, characterized by sphalerite with associated chalcopyrite pyrite-quartz, molybdenite-quartz and pyrite-quartz.

Stage IV—quartz-galena-sphalerite.

Stage V—overprinting bornite-chalcocite-covellite (Table 2.2). Stages IV and V are only poorly mineralized. Coincident magmatic activity, hydrothermal alteration, and mineralization characterize all stages. Only the volume of intrusion differs for each stage, progressively decreasing from stage I to V.

The first two ore stages of Gavrilova et al. (1984) are dominated by vein stockworks, while the last three are localized by dykes and associated fracturing.

Sulfur isotope studies indicate two hydrothermal phases (Voinkov et al. 1977). After intensive fracturing, the first two ore stages formed the main ore body, dominated by zones of stockwork veining. Then, following a new fracturing event, related to the introduction of dykes, further mineralization was deposited

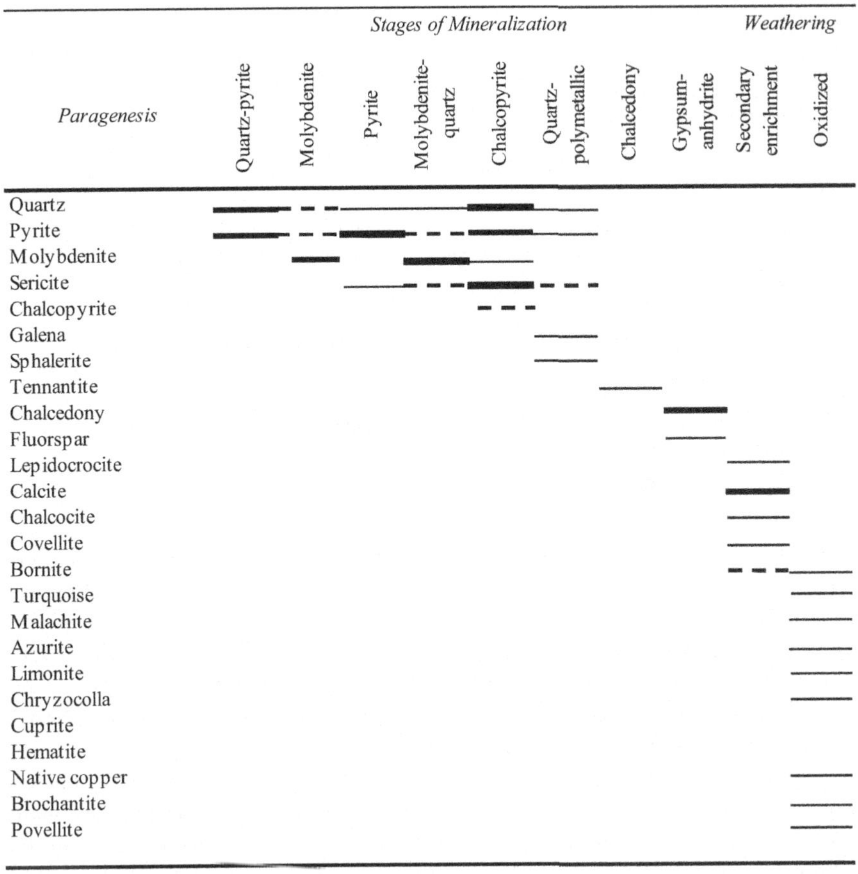

Fig. 2.14 Paragenetic sequence of ore and gangue minerals in the Erdenet porphyry Cu-Mo deposit (after Khasin et al. 1977)

characterized by fracturing and veining. Chalcopyrite accompanied both phases, while molybdenite was also introduced during the first phase. The increment of the total ore grade introduced in each stage decreased from 0.3 to 0.8% Cu in stage I, to 0.09–0.19% Cu in stage III. The bulk of the Cu mineralization is associated with stage I, Mo with stage II, and Zn with stage III, while the latter two stages are characterized by high Ag. However, mineralization is distributed irregularly, with the average Cu grade being 0.7–0.8%. Low-grade ores contain 0.2–0.3% Cu, while the richest are up to 1% Cu (Gavrilova et al. 1984).

Pyrite is the most common hypogene sulfide mineral. Gavrilova and Maksimyuk (1990) recognized five pyrite associations: (a) the earliest in quartz veinlets; (b) in molybdenite-quartz veinlets; (c) in association with chalcopyrite, tennantite, and sphalerite; (d) as meta-crystals of pyrite; and (e) in association with galena, sphalerite, and tennantite. Molybdenite occurs in association with pyrite and quartz and

Table 2.2 Paragenetic mineralogical associations (after Gavrilova et al. 1984)

Paragenetic mineral association	Form of ore minerals	Ore and gangue mineralogy	Characteristic trace elements
I. Quartz-pyrite	Impregnation, veinlets	Pyrite, quartz, sphalerite, chalcopyrite, molybdenite	Se, Ag, Co, Ni, Au
II. Quartz-molybdenite	Impregnation, veinlets	Molybdenite, quartz,	Mo, Re, Se
III. Quartz-chalcopyrite-tennantite Tennantite	Impregnation, veinlets, inclusions in minerals Inclusions	Chalcopyrite, tennantite, pyrite, sphalerite, enargite, galena Galena	Ag, Se, Co, Ni, Au
IV. Quartz-galena-shaplerite (polymetallic)	Veinlets, impregnation	Galena, sphalerite, tennantite, pyrite	Co, Ni, Mo, As, Ag, Bi, Cd
V. Superimposed bornite-chalcocite-covellite	Impregnation, veinlets, inclusions in other minerals	Bornite, chalcocite, covellite	Pb, Zn, Ag, As, Sb

rarely with chalcopyrite. Chalcopyrite is found in two paragenetic associations, namely, (a) quartz-chalcopyrite-tennantite and (b) with superimposed bornite-chalcocite-covellite. Tennantite also occurs in two associations: pyrite-chalcopyrite-tennantite sphalerite and tennantite-sphalerite-galena. Enargite is rare, mainly in association with pyrite and quartz. Galena occurs as impregnations that produce mono-mineral aggregates or veinlets after sphalerite and tennantite. Sphalerite is represented by marmatite and cleiophane.

Hypogene copper mineralization at the Erdenet deposit is dominated by chalcopyrite and bornite. Late stage covellite is also widespread, while bornite occasionally occurs as intergrowths with chalcocite. Besides Cu and Mo, the ore also contains the following elements: Se, Re, Au, Ag, W, Sn, Ga, Tl, Ge, Bi, Zn, Co, Ni, As, and Sb which occur as isomorphic traces in pyrite, chalcopyrite, molybdenite, tennantite, galena, and chalcocite (Gavrilova et al. 1984). The most common gangue mineral is quartz.

An upper oxide zone which is up to 70 m in thickness is characterized by an oxide-sulfide assemblage of cuprite, malachite, azurite, turquoise, brochantite, native copper, delafossite, and antlerite. Copper grades in the upper oxidized zone are very variable. No well-defined leached cap has been identified.

High-grade zones are associated with a 30–300 m thick supergene enrichment blanket, where developed, within structurally controlled sheeted veinlets, with supergene chalcocite, which is crystalline, amorphous, and collomorphic, replacing hypogene covellite and bornite. Supergene chalcocite is also developed on chalcopyrite within the stockwork zones.

Gavrilova and Maksimyuk (1990) provided additional information on the five stages of mineralization at Erdenet, which may be summarized as follows. During stage I, the main ore body formed by the impregnation of quartz chalcopyrite-pyrite and quartz-molybdenite veinlets within granodiorite porphyry stocks. Stage II is dominated by quartz-chalcopyrite-molybdenite and quartz-molybdenite veinlets with centimeters-size dimensions, quartz-pyrite, and pyrite veinlets which are 1–5 cm in thickness and quartz breccia and quartz veins. All of these veinlets are similar to the A- and B-type veins of Gustafson and Hunt (1975). Stage III represents a regressive event within the deposit, with a characteristic assemblage of sphalerite and galena, but only poor Cu mineralization (contributing between 0.09 and 0.19% Cu to the grade) and only occupies about 7–8% of the total orebody's volume as veins similar to the D-type of Gustafson and Hunt (1975). Stages IV and V are late mineralization phases recognized in drilling data. They consist of veinlets up to a few mm in thickness filled by quartz and sulfides. Alteration comprises sericite, chlorite, epidote, and carbonate, with relict biotite, amphibole, and plagioclase, while primary textures and preexisting magnetite are preserved. This latter assemblage represents the propylitic phase of alteration, which is crosscut by the D-type veinlets. Gavrilova and Maksimyuk (1990) concluded that the Erdenetiin Ovoo deposit had undergone three main alteration events, including (a) an early (pre-ore?) phase of magnetite rich K-silicate alteration (?), now largely obliterated by subsequent over-printing; (b) quartz-sericite (phyllic) and peripheral intermediate argillic sericite-chlorite alteration; and (c) propylitic alteration. Two retrograde stages are developed locally. The relative ages of porphyritic intrusives are determined by field observation of crosscutting relationships between veins and intrusions at intrusive contacts and by crosscutting relationships between veins of various types.

Kavalieris et al. (2017) described D veins that constitute more than 50% of the Cu grade (~0.5 wt % Cu) instead A and B veins of the early porphyry-style quartz veins locally infilled by pyrite-chalcopyrite, with subordinate bornite. After Kavalieris et al. (2017) the Erdenet porphyry system was overprinted by advanced argillic alteration, which outcrops 2 km northwest of the pit, and forms a lithocap that extends over 10×2.5 km. It is characterized by residual quartz, andalusite, Na-Ca and K-alunite, diaspore, pyrophyllite, zunyite, topaz, dickite, and kaolinite.

Sulfur Isotope Systematics. Tugarinov et al. (1974) studied the sulfur isotopic composition of Erdenet ore. The $\delta^{34}S$ value of the sulfides ranged from −3.3 to +2.7‰ with an average value of +0.4‰. Molybdenite is weakly enriched in $\delta^{34}S$ (+1 to +1.5‰) compared to copper sulfides and pyrite. The values show no differences in the vertical section. Such a homogenous sulfur composition is a result of homogenization of sulfur before mineralization. The value of $\delta^{34}S$ for the sulfates ranges from +9.1 to +12.5‰. The study suggested two stages of hydrothermal ore formation, as detailed above.

The $\delta^{34}S$ values of the primary sulfide minerals clustered near 0‰, indicating that the sulfur is mainly derived from a homogeneous magmatic source. By contrast, the $\delta^{34}S$ values of secondary Cu sulfide minerals ranged from −3.2‰ to −0.3‰, with an average of −1.6‰ (Kim et al. 2019). They also studied Cu isotopic signature of the Erdenetiin Ovoo deposit which gives 0.14‰ to 2.69‰ range of $\delta^{65}Cu$ values of Cu (I) sulfide minerals, suggesting that Cu predominantly originated from magmatic sources, whereas Cu (II) minerals such as chrysocolla, malachite and azurite presented much larger variations from −1.01‰ to 10.0‰ (Kim et al. 2019).

2.3.7 *Shand Deposit*

The Shand is located about 30 km south from Orkhon aimag. Geological setting, tectonics, and metallogenic features are similar to the Erdenet ore zone. The geology of the Shand area is characterized by late Permian volcanic, granitic rocks, dykes, and the Quaternary sediments. Ages of the granodiorite and the granodiorite porphyry indicate 235 and 239 Ma, respectively. Main geological structures are the NW-SE, the NS, and the NE-SW trending faults. The NW-SE trending fracture zones are estimated in the southern part of the area, from the outcrops of the syenite porphyry which are arranged in the NW-SE direction. The NW-SE trending lineaments are predominantly distributed in the southern part of the area, while the NS and the NE-SW trending lineaments are predominantly distributed in the northern part, based on satellite imagery (JICA, MMAJ 2003). The Shand deposit is located at the junction of the NS and the NW-SE trending lineaments, and it is a blind deposit covered by Quaternary deposits. Alteration mineral assemblage is sericite-chlorite similar to the alteration of the porphyry copper-molybdenum deposit in the Erdenet mine.

The greenish of the film-like malachite and azurite are observed in the potassium metasomatized medium-grained granodiorite, and the maximum ore grades carry 0.119% Cu, 0.036% Pb, 0.116% Zn, and 24 ppm Ag. The pyrite disseminations with the silicified zone occur in the western central part of the area. The oxidized zone is composed mainly of limonite. Quartz, K-feldspar, biotite, alunite, andalusite, and kaolin are present in the central part of the mineralization, and chloritization occur outward. Ore minerals such as spotty azurite and chalcopyrite, the disseminated pyrite, goethite, hematite, and limonite were observed. Ore grades returned 0.001–0.014% Cu, less than 0.001–0.003% Mo, 0.003–0.005% Pb, less than 0.001–0.002% Zn and 0.40–8.55% Fe (JICA, MMAJ 2003). Probable reserves of 500,000 tonnes Cu.

2.3.8 *Saran Uul Deposit*

The Saran Uul deposit (101 Mt. at 0.06% Cu, 0.5–4.9 g/t Au) is located 40 km south of the town center of Bayankhongor province, in Byankhongor zone in Central Mongolia that trending over 300 km NW-SE, and is 30–50 km wide. The Bayankhongor zone consists of a Paleozoic subduction system including Precambrian Zavkhan microcontinent, obducted ophiolites and accretionary and forearc sedimentary rocks intruded by Paleozoic granitoids. Many hydrothermal deposits such as South porphyry Cu-Au (240 ± 5 Ma), Khukh Bulgiin Khundii skarn (252 ± 5 Ma), Han Uul shear zone hosted Au (283 ± 6 Ma), and Taats Gol W-Au pegmatite (329 ± 7 Ma) are associated with Paleozoic granitoid intrusions. The ages of hydrothermal mineralization in the Bayankhongor zone vary from Early Carboniferous to earliest Triassic and related to Bayankhongor Au-Cu metallogenic belt (Jargalsaihan et al. 1996) with the abundance of Au-bearing quartz veins, placer Au, and Cu-Au porphyry and skarn deposits. A representative gold deposit in the area is gold-bearing quartz veins of mesothermal-epithermal the Tsagaan Tsakhir Uul deposit (Jargalan and Murao 1998).

The Cu-Au mineralization is associated with magnetite-series granitoids and includes porphyry, skarn, and vein types. The porphyry and skarn mineralization occurred during an increase in regional compressional stress in a non-extensional stress setting (JICA, MMAJ 1995).

The area of Saran Uul deposit is underlain by Early Silurian sedimentary rocks of sandstone, mud-stone, and red chert. A part of the sedimentary rocks is metamorphosed into muscovite schist. These rocks are intruded by the Late Paleozoic Saran Uul complex, which consists of tonalite and granite porphyry. Hornfelses occur in the sedimentary rocks near the contact of the granite porphyry (Takahashi et al. 1998). The mineralized granite porphyry in the prospect is approximately 1 km in diameter. The granite porphyry consists of large euhedral phenocrysts of plagioclase and K-feldspar with granular biotite, quartz, and amphibole. Post-mineralization latite dykes, several meters thick, intruded in the granite porphyry and sedimentary rocks. The granite porphyry has partly undergoing potassic alteration in which plagioclase is replaced by K-feldspar. Hydrothermal breccia, cemented with quartz and/or barite, is found in places along the margin of the granite porphyry. Sericite alteration is limited in the largest body of hydrothermal breccia in the northwest of the area. Quartz veins occur on the west of the largest body of granite porphyry. Malachite is disseminated along joints in the northeast of the intrusive body (Watanabe et al. 1999). Main ore minerals are chalcopyrite, bornite, molybdenite, gold, pyrite, sphalerite, galena, and fahlore and show 327 Ma (K-Ar muscovite). The average grade of Cu is 0.37% with prognostic reserves of copper (metal is 50,000 t).

2.3.9 Bayan Uul Occurrence

The Bayan Uul (>0.1% Cu, >0.002% Mo, >0.1 g/t Au) has been explored in the area of 0.6 by 2.3 km in Central Mongolia and is interpreted as a large, weakly eroded stratovolcano-intrusion type within the Early Mesozoic magmatic province of Mongol-Okhotsk belt. Porphyry Cu-Mo (±Au) mineralization is associated with andesitic volcanism of relatively high alkalinity surrounds internal zones of calc-alkaline magmatism of normal alkalinity. A typical magmatic environment in the ore regions includes an earlier granodiorite-granite association as well as trachybasaltic andesite-trachyandesite and porphyry associations providing a single magmatic series (Koval et al. 1988, 1989).

The Bayan Uul occurrence hosted intrusion is composed of quartz diorite and quartz monzodiorite within volcanic-plutonic Bayan-Uul ring structure. Mineralization is formed at the depth of 1.5–2 km and exceeds up to 500 m. The ensemble of hydrothermal-metasomatic rocks consists of diverse propylites, biotite-quartz-albite metasomatites, quartz-sericite and quartz-tourmaline metasomatites, quartz-tourmaline breccias, almost monomineralic quartz metasomatites, and argillic rocks. A zone of 300 by 900 m contains >0.3% Cu and 0.005% Mo. Abundant are quartz-tourmaline-chalcopyrite veins and quartz-tourmaline breccias The $^{39}Ar/^{40}Ar$ isotopic age is 220–223 Ma (Lamb and Cox 1998). The alteration zone is nearly oval in shape, is 3 km wide, and extends northeast for 5 km. The dominance of sericite, advanced argillic, and silica alteration at the surface suggests a relatively shallow porphyry Cu system. The deposit consists of stockwork veinlets, and veins of quartz, pyrite, chalcopyrite, and molybdenite that occur in or near porphyritic intrusions. High-level intrusive porphyry is contemporaneous with abundant dykes, faults, and breccia pipes (Nokleberg 2010).

2.3.10 Bayan-Airag (Golden Hills) Deposit

Regional Geology. The Golden Hills property is located in the western part of the Tuva-Mongol tectonic unit; the southwestern parts of the property overlap the Ozernaya unit (Sengor et al. 1993). These tectonic units have been further segmented by Badarch et al. (2002) in which the Golden Hills property occurs in the Zavkhan Cratonal block and the Lake island arc tectonostratigraphic terrane. The Golden Hills project area includes rocks of Neoarchean (PP), Stenian (MP), Ediacaran (NP), and Palaeozoic (dominantly Cambrian, some Devonian and Permian) age. These rocks are intruded by felsic intrusions, mostly of Mesoproterozoic age. Some Palaeozoic felsic intrusions are also present and consist of leucogranite and syenite. Minor mafic intrusions are thought to be of Neoproterozoic or Mesoproterozoic age. A key unit in the Golden Hills stratigraphy is the Mesoproterozoic Shuvuu Formation that hosts the copper-gold-silver massive sulfide and gold-silver oxide deposits. Within the Shuvuu Formation is a series of quartz-phyric felsic rocks which appear to be intimately related to the massive sulfide mineralizing event. Table 2.3 summarizes the relative ages or the various rock formations in the Golden Hills project area.

Local geology and mineralization. The Shuvuu Formation is a NW-SE trending belt of greenschist facies metavolcanics, marble, sandstones, and limestone. Exploration in the Golden Hills region has focused on the greenschist facies metavolcanic portion of the Ediacaran Shuvuu Formation (Altankhuyag 2005, Gantulga et al. 2018). The quartz-bearing schists of this formation are felsic metavolcanic and meta-epiclastic rocks (rhyolite in composition), with frequent pyroclastic textures. Based on the interpreted presence of pyroclastic rocks and the proximal occurrence of marble/limestone, the Shuvuu Formation is considered to have been deposited under shallow marine conditions. Within this formation numerous mafic sills and dykes occur. Most of these are post-deformation, but some appear to be coeval with the emplacement of the metavolcanics. The Shuvuu Formation contains the massive sulfide deposits and the potassium alteration zone. The rock stratigraphically above the main sulfide body (to the north) consists of variably schistose felsic volcanics cut by a large mafic sill emplaced roughly 100–120 m above the northern sulfide body (Fig. 2.15). In some areas, a syenite intrusion was emplaced to the north of this sill,

Table 2.3 Regional stratigraphy of the Bayan-Airag deposit, Golden Hills project

Age	Formation name	Lithology
Permian	Intrusions	Quartz syenite
Other Palaeozoic		Sediments
Ordovician	Intrusions	Granite and granodiorite
Cambrian		Sandstone, siltstone, conglomerate and carbonates
Neoproterozoic-Cambrian		Tuffaceous clastic and carbonates
Neoproterozoic		Dominated by mafic to intermediate volcanics and intrusions
Mesoproterozoic	Intrusions	Granite and granodiorite now metamorphosed to granitic gneisses
NP-Ediacaran	Tsagaan Olom F.	Contains dolomite, limestone, chert, and minor sandstone conglomerate
NP-Cryogenian	Zavkhan F.	Composed of sandstone, conglomerate, siltstone, basalt, andesite, dacite and rhyolite
NP-Ediacaran	Shuvuun F.	Described as consisting of greenschist facies-quartz bearing schist, marble, sandstone and limestones, intruded by mafic dykes and sills
MP-Stenian	Urgamal F., Tsagaannuur F.	Amphibolite, schists, quartzites and marble
Palaeoproterozoic	Buduun F.	Gneisses, amphibolite and biotite-amphibolite schist. Intruded by pyroxenite and hornblendite.

almost certainly post-deformation and fault related. The northern sulfide body is the main massive sulfide body at Golden Hills and is largely composed of pyrite (>90%). It is believed to have been deposited on the sea floor and, due to post mineral deformation, it currently strikes NW and dips about 65 degrees NE. The body is roughly 450 m long and about 40 to 80 m wide, with the narrow part in the NW. The depth extent is unknown. The upper 20–30 m of the sulfide body is relatively copper rich, with grades generally ranging from 1% to 4% Cu. The entire body is slightly enriched in gold, with assays ranging up to about 0.5 g/t. Zinc values range up to 8% occur but are irregular and often very low, with elevated zinc occurring variably at the top, bottom, or center of the high copper interval. The northern sulfide body does reach the surface as a gossan developed above this body cuts off abruptly at about 1790 m elevation. This may have been a result of faulting contemporaneous with the formation of the sulfides, or later faulting. The gossan between the top of sulfides and the 1790 m level is one part of the oxide gold resource. The rock beneath the northern sulfide body consists of a possible rhyolite dome which is relatively unaltered in what is now (post deformation) the deeper area of the deposit. Similar massive rhyolite units are observed in drill holes throughout the valley. Closer to the current surface, what appears to be an equivalent of the rhyolite unit is strongly altered, pyritized, and brecciated with pyrite cement. This may be a feeder zone for the northern sulfide body. Rhyolitic volcanics consisting largely of coarse lithic lapilli tuff units appear to be stratigraphically equivalent to the rhyolite dome in the area closest to the present surface, but the exact relationship is

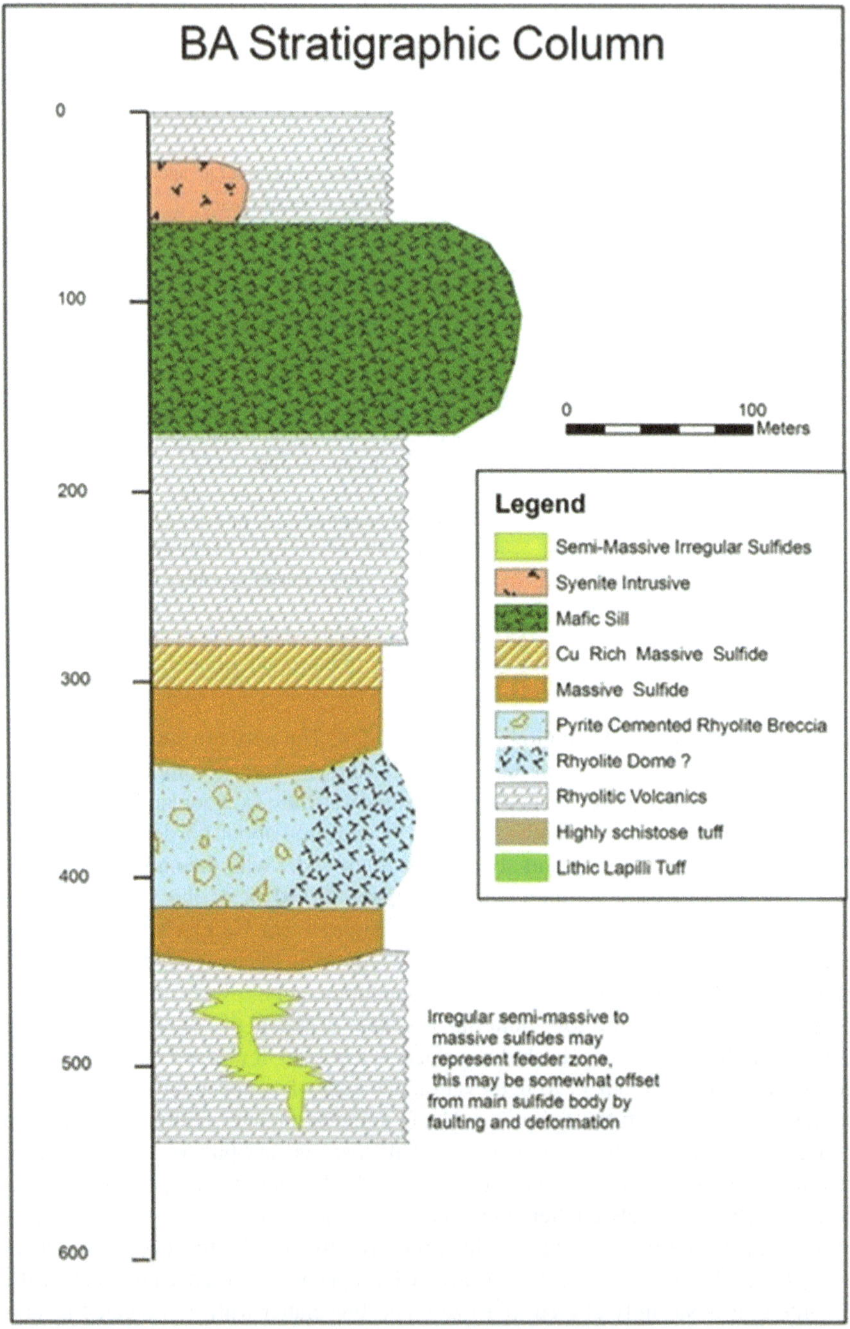

Fig. 2.15 Local stratigraphic column of the Bayan-Airag deposit, Golden Hills project

difficult to define. About 75 m beneath the northern sulfide body a second, smaller massive sulfide body (the southern sulfide body) is encountered. This is about 20–40 m thick and generally does not contain much copper, although it has low gold assays similar to those in the upper body. Beneath the southern sulfide body there is a series of felsic volcanic rocks, divided into an upper medium to coarse lithic lapilli tuff (about 100 m thick) and a lower finer grained crystal tuff (at least 100 m thick). In the western part of the deposit, these units are affected by variable sericitic and chloritic alteration with irregular bodies of pyrite containing variable gold and copper values. These are thought to have been formed by subsurface replacement of near seafloor strata, with alteration and mineralization spreading out from fractures that served as fluid channels into the stratigraphic units which were favorable hosts in terms of permeability and/or chemistry. The strongest pyrite is generally accompanied by intense chloritization in this area. It is possible that this area represents a feeder zone for the main sulfide bodies, which may have been slightly offset by later faulting. This mineralization continues to the present surface, and the oxidized equivalent constitutes the major oxide gold resource in the deposit. The oxidized lodes comprised of goethitic clays and gossan material with locally more hematitic+clay zones. Some of the volcanic rocks appear to be more schistose and in the case of the lithic lapilli tuffs the clasts seem to have been more flattened, probably corresponding to an area that absorbed the most strain during metamorphism.

To the east of the main sulfide bodies is an area of layered pyrite that is generally interbedded with strongly schistose altered crystal tuffs. Some gold and locally copper is associated with the pyrite, and elevated gold values are encountered in the oxidized equivalent contained in a relatively narrow stratigraphic interval, which does continue to the present surface.

This eastern area is one of the minor parts of the oxide gold resource. It is thought that this is a distal area of the deposit, probably equivalent to the sulfide bodies in the central area. The high schistosity may be due to the strongly altered units taking much of the strain during deformation.

In the main area of oxide gold mineralization, the gossan intervals are variably mineralized; "vuggy gossan" is commonly mineralized with 0.5–3 g/t Au but can also be only weakly mineralized. The clayey or earthy gossan is also significantly mineralized and contains most of the high grade intervals. It is thought that the general gold mineralization runs mostly under 2 g/t and is enriched in the higher grade streaks running through the deposit.

The base of the oxide zone is characterized by a leached zone of white sand/clay which can contain variably significant gold grades. This white material is believed to be the residuum from weathering of the pyrite beneath and is thought to comprise mainly quartz and possibly some barite – a normal product associated with exhalative centers associated with VHMS deposits. The transition from vuggy and earthy goethite to white sand and then into relative fresh massive pyrite is over about 4 m.

Patchy high-grade gold mineralization is found on both margins of the northern and southern sulfide bodies. The mineralization appears related to discontinuous zones of narrow (mm scale) quartz veinlets which have been identified to contain tellurium and are referred to as the quartz gold-telluride veins or QGT. The mineralization can occur right on the margins of the sulfide bodies, in some instances hosted by sulfides and in other instances the veins occur some distance into the hanging-wall host volcanics, up to 35 m away.

Relatively high silver values often accompany the gold. In the central part of the northern mineralization, it appears that the mineralization is relatively coherent and continues to surface as high grade gossan material, whereas there is no evidence of the accompanying sulfide body continuing to surface. This suggests a structural control for the mineralization.

In the western area high-grade gold-silver intervals are only found in rocks above the base of oxidation in both gossanous zones and clay altered volcanic rocks with a wide range of iron oxide mineral. The samples are usually comprised of a white clayey interval flanked by goethite and occasionally contain quartz veinlets. Some very high-grade mineralization is also found in gossan intervals, but this is probably less than a quarter of the high-grade intervals.

It is uncertain whether the high grade intervals in the oxide in the west and the veins around the sulfide bodies in the central area are related. Both are associated with high silver values, although nothing is known of the presence or absence of tellurium in the western area high-grade intervals.

Some high-grade gold-silver intercepts are found in the eastern area as well, generally associated with quartz veins, and not adjacent to pyrite bodies (unpublished communication by Bayan-Airag Exploration).

2.4 Concluding Remarks

The main economic copper deposits in Mongolia are porphyry copper that contain copper, molybdenum, and gold in a stockwork of small veinlets hosted in granodiorite, monzodiorite, monzonite, and basalts (Oyu Tolgoi) of the volcanic-plutonic belts formed in island arc or continental arc environment. Mineralization is represented by major chalcopyrite, bornite, molybdenite, followed by hydro-thermal alteration of K-feldapar, quartz-sericite and sericite-chlorite-epidote and late stage white mica, clay, and carbonate. Host rocks are mainly pyrite-rich. Age of mineralization is Late Devonian-Carboniferous in South Mongolia and Late Permian-Early Mesozoic in Northern Mongolia. Geochemistry of host rocks exhibit typical subduction-related geochemistry, e.g., enrichment in LILE and LREE, depletion in HFSE: Nb, Ta, and in many cases show adakitic signature with high Sr/Y and La/Yb ratio. The Sr-Nd-Pb-Hf isotopic data reflect predominantly depleted mantle source, and ore-bearing magmas generate in subcontinental lithosphere from juvenile material within the subduction-related setting.

Genesis of other types of copper deposits like massive sulfide is still under discussion and additional study is needed.

References

Altankhuyag D (2005) Baruun Mongolyn Bayan Airag tsul sulfide khuderjiltiin garal uusel (Genesis of the Bayan-Airag massive sulfide mineralization, Western Mongolia). Dissertation. Mongoian University of Sceince and Technology, Ulaanbaatar, 126 p.

Badarch G (2005) Tectonics of South Mongolia. In: Seltmann R, Gerel O, Kirwin DJ (eds) Geodynamics and Metallogeny of Mongolia with a special emphasis on copper and gold deposits. SEG-IAGOD Field Trip, 14–16 August 2005, 8th Biennial SGA Meeting, IAGOD Guidebook Series, CERCAMS/NHM, London, vol 11, pp 119–129

Badarch G, Cunningham WD, Windley BF (2002) A new terrane subdivision for Mongolia: implications for the Phanerozoic crustal growth of Central Asia. J Asian Earth Sci 21:87–110

Batkhishig B (2005) Magmatic-hydrothermal system of the Shuteen mineralized complex, South Gobi, Mongolia. Dissertation. Tohoku University, Japan, p 143

Batkhishig B, Bignall G, Tsuchiya N (2005) Hydrothermal quartz vein formation, revealed by coupled SEM-CL imaging and fluid inclusion microthermometry: Shuteen Complex, South Gobi, Mongolia. Res Geol 55:1–8

Batkhishig B, Ts N, Bignall G (2014) Magmatic-hydrothermal activity in the Shuteen area, South Mongolia. Econ Geol 109:1929–1942

Batkhishig B, Tsuchiya N, Bignall G (2010) Magmatism of the Shuteen complex and carboniferous subduction of the Gurvansaikhan terrane, South Mongolia. J Asian Earth Sci 37:399–411

Berzina AP, Lepekhina EN, Berzina AN, Gimon VO (2012) Zircons of igneous rocks at the Erdenetuin Obo Porphyry Cu–Mo deposit (Mongolia): U–Pb dating and petrological implications. Dokl Earth Sci 442(2):249–255

Bignall G, Batkhishig B, Tsuchiya N (2005) The Shuteen Cu–Au porphyry deposit. In: Seltmann R, Gerel O, Kirwin DJ (eds) Geodynamics and metallogeny of Mongolia with a special emphasis on copper and gold deposits. IAGOD guidebook Series, CERCAMS/HNM London, vol 11, pp 193–201

Blight JHS, Crowley QG, Petterson MG, Cunningham D (2010) Granites of the Southern Mongolia carboniferous arc: new geochronological and geochemical constraints. Lithos 116:35–52

Chang Z, Hedenquist JW, White NC, Cooke DR, Roach M, Deyell CL, Garcia JJ, Gemmell JB, McKnight S, Cuison AL (2011) Exploration tools for linked porphyry and epithermal deposits: example from the Mankayan intrusion-centered Cu-Au district, Luzon, Philippines. Econ Geol 106:1365–1398

Crane D, Kavalieris I (2012) Geologic overview of the Oyu Tolgoi porphyry Cu–Au–Mo deposits, Mongolia. Econ Geol Spec Publ 16:187–213

Dejidmaa G, Badarch G (2005) Summary of pre-accretionary and accretionary metallogenic belts of Mongolia. In: Seltmann R, Gerel O, Kirwin DJ (eds) Geodynamics and Metallogeny of Mongolia with a special emphasis on copperand gold deposits. SEG-IAGOD Field Trip, 14–-16 August 2005, 8th Biennial SGA Meeting, IAGOD Guidebook Series, CERCAMS/NHM, London, vol 11, pp 25–30

Dejidmaa G, Dorjgotov D, Gerel O, Gotovsuren A (2005) Preliminary description of mineral deposit models (types) for Mongolia. In: Seltmann R, Gerel O, Kirwin DJ (eds) Geodynamics and Metallogeny of Mongolia with a special emphasis on copperand gold deposits. SEG-IAGOD Field Trip, 14–16 August 2005, 8th Biennial SGA Meeting, IAGOD Guidebook Series, CERCAMS/NHM, London, pp 31–52

Delgertsogt B (2008) Umnud Mongoliin Shuteenii zes-porfiriin systemiin geologi, struktur (Geology and structural features of the Shuteen porphyry copper system in South Mongolia). Dissertation, Mongolian University of Science and Technology

Dolgopolova A, Seltmann R, Armstrong R, Belousova E, Pankhurst RJ, Kavalieris I (2013) Sr–Nd–Pb–Hf isotope systematic of the Hugo Dummett porphyry deposit (Oyu Tolgoi, Mongolia). Lithos 167:47–64

EMC (2002) Erdenet Mining Corporation, Annual Report 2002

EMC (2019) Erdenet Mining Corporation, Annual Report 2019

Fang W, Yang S, Liu Z, Wei X, Zhang B (2007) Geochemical characteristics and significance of major elements, trace element and REE in mineralized altered rocks of large-scale Tsagaan Suvarga cu–Mo porphyry deposit in Mongolia. J Rare Earths 25:759–769

Gantulga B, Tsogkhuu I, Baldorj B, Enkh-Amgalan S, Saikhanjargal S, Purevsuren J (2018) Results of operational exploration and resource estimation for the Bayan-Airag gold-copper ore deposit at Durvuljin soum of Zavkhan aimag in 2012-2017, Ulaanbaatar, Report No. 458452

Gavrilova SP, Maksimyuk IE (1990) Etapy formirovaniya Erdenetskogo molibden-medno-porfirovogo mestorojdeniya, Mongoliya (Stages of formation of the Erdenet molybdenum-copper porphyry deposit, Mongolia). Geologiya Rudnykh Mestorojdenii 6:3–17

Gavrilova SP, Maksimyuk IE, Orolmaa D (1984) Osobennosti magmatizma i sostava rud medno-molibdenovogo mestorojdeniya Erdenetiin-Ovoo (Features of magmatism and composition of ore of the copper-molybdenum deposit at Erdenetiin Ovoo), Endogenniiye rudniiye formatsii Mongolii Nauka, Moskva, 101–115

Gerel O (1990) Petrologiya, geokhimiya i rudonosnosti subshelochnogo mezosoiskogo magmatizma Mongolii (Petrology, geochemistry and mineralization of subalkaline Mesozoic magmatism in Mongolia), Dissertatsiya na soiskanie stepeni doktora geologo-mineralogicheskikh nauk, Irkutsk p 495

Gerel O, Batkhishig B, Munkhtsengel B, Chimedtseren A (2013) Porphyry type deposits associated with adakitic magma: case study of porphyry Cu-Au and Cu-Mo deposits in Mongolia. The 8th International forum on strategic technology, v. 1. MUST, 537–542

Gerel O, Munkhtsengel B (2005) Erdentiin Ovoo porphyry copper-molybdenum deposit in Northern Mongolia. In: Porter TM (ed) Super porphyry copper & gold deposits—a global perspective, vol 2. PGC Publishing, Adelaide, pp 525–543

Gustafson LB, Hunt JP (1975) The porphyry copper deposits at El Salvador, Chile. Econ Geol 70:857–912

Hou W, Nie F, Jiang S, Bai D, Liu Y, Yun F (2010) The geology and ore forming mechanism of the Tsagaan Suvarga large-size cu–Mo porphyry deposit in Mongolia. Sinica of China 31:307–320

Iizumi S, Batkhishig B (2000) Petrology of Carboniferous Shuteen pluton in the South Gobi fold belt, South Mongolia. In: Abstracts of the 107th Annual Meeting of Geological Society of Japan, Matsue City, Japan, 319

Jargalan S, Murao S (1998) Preliminary study on the characteristics of Tsagaan Tsakhir Uul gold deposit, Bayankhongor, southern Mongolia. Bull Geol Surv Japan 49(6):291–298

Jargalsaihan D, Kazmer M, Baras Z, Sanjaadorj D (1996) Guide to the geology and mineral resources of Mongolia, 111–221

JICA, MMAJ (1995) Japan International Cooperation Agency and Metal Mining Agency of Japan. Report on the mineral exploration in the Altan-Tal area, Mongolia: Phase-I: Tokyo, Japan

JICA, MMAJ (2003) Japan International Cooperation Agency Metal Mining Agency of Japan. Report on the mineral exploration in the western Erdenet area, Mongolia (Phase II), p 41

Kavalieris I, Khashgerel B-E, Morgan LE, Undrakhtamir A, Borohul A (2017) Characteristics and $^{40}Ar/^{39}Ar$ geochronology of the Erdenet Cu–Mo deposit, Mongolia. Econ Geol 112:1033–1053

Kepezhinskas VV, Luchitsky IV (1973) Permskie vulkanicheskie assotsiatsii Tsentrali'noi Mongolii (Permian volcanic association in Central Mongolia) In: Assotsiatsiya vulkanogennykh porod Mongoli'skoi Narodnoi Respubliki, ikh sostav i stratigraficheskoe polozhenie. Nauka, Moskva 71–93

Khashgerel B, Kavalieris I, Hayashi K (2008) Mineralogy, textures, and wholerock geochemistry of advanced argillic alteration: Hugo Dummett porphyry Cu-Au deposit, Oyu Tolgoi mineral district, Mongolia. Mineral Deposita 43:913–932

Khashgerel B, Rye RO, Hedenquist JW, Kavalieris I (2006) Geology and reconnaissance isotope study of the Oyu Tolgoi porphyry Cu-Au system, South Gobi, Mongolia. Econ Geol 101:503–522

Khashgerel B, Rye RO, Kavalieris I, Hayashi K (2009) The sericitic to advanced argillic transition: stable isotope and mineralogical characteristics from the Hugo Dummett porphyry Cu-Au deposit, Oyu Tolgoi district, Mongolia. Econ Geol 104:1087–1110

Khasin RA, Marinov NA, Khurtz C, Yakimov LI (1977) Medno-molibdenovoe mestorojdenie Erdenetyin-Obo v Severnoi Mongolii (The copper-molybdenum deposit at Erdenetiin Ovoo in northern Mongolia). Geologiya Rudnyh Mestorojdenii 6:3–15

Kim Y, Lee I, Oyungerel S, Jargal L, Ts T (2019) Cu and S isotopic signatures of the Erdenetiin Ovoo porphyry Cu-Mo deposit, northern Mongolia: implications for their origin and mineral exploration. Ore Geol Rev 104:656–669. https://doi.org/10.1016/j.oregeorev.2018.11.025

Kirwin DJ, Wilson CC, Turmagnai D, Wolfe R (2005) Exploration history, geology, and mineralization of the Kharmagtai gold–copper porphyry district, South Gobi Region, Mongolia. In: Seltmann R, Gerel O, Kirwin DJ (eds) Geodynamics and Metallogeny of Mongolia with a special emphasis on copper and gold deposits. SEG-IAGOD Field Trip, 14–16 August 2005, 8th Biennial SGAMeeting, IAGOD Guidebook Series, CERCAMS/NHM, London, vol 11, pp 175–191

Koval PV, Ariunbileg S, Libatorov YI, Maksimyuk UE (1988) The Bayan-Uul porphyry copper-molybdenum occurrence and relation to magmatism, central Mongolia. Geol Ore Deposits 3:24–35. https://doi.org/10.1016/0375-6742(89)90078-2

Koval PV, Gerel O (1986) Vulcanogennye assotsiatsii raionov medno-porfirovogo orudeneniya Mongolo-Okhotskoi vnutrikontinental'noi podvijnoi zony (Volcanic association of porphyry copper mineralization in the Mongol Okhotsk intraplate mobile zone) Geokhimiya vulkanitov razlichnykh geodinamicheskikh obstanovok. Nauka, Novosibirsk, 69–93

Koval PV, Gerel O, Smirnov VN, Tseden Ts (1985) Porfirovye intruzii Erdenetskogo rudnogo uzla: petrografiya i khimicheskii sostav (Porphyritic intrusion from the Erdenet ore deposit: petrography and chemistry) Voprosy geologii i metallogenii Mongolii. Tezisii doklady III nauchnoi konferentsii Kerulenskoi mejvuzovskoi geologicheskoi ekspeditsii, 59–111

Koval PV, Gerel O, Tseden Ts, Smirnov VN (1982) Assotsiatsii porfirovykh intruzii Erdenetskogo raiona (Association of porphyritic intrusions in the Erdenet area) Voprosy geologii i poleznykh iskopaemykh tsentralnoi I vostochnoi Mongolii. Tezisii doklady IV nauchnoi konferentsii Kerulenskoi mejvuzovskoi geologicheskoi ekspeditsii. Ulanbator, 15–17

Koval PV, Gotovsuren A, Ariunbileg S, Libatorov YI (1989) On prospecting for porphyry copper mineralization in intracontinental mobile zones (Mongol-Okhotsk belt, Mongolian People's republic). J Geochem Explor I32(1–3):369–380

Lamb MA, Badarch G (1997) Paleozoic sedimentary basins and volcanic-arc systems of southern Mongolia: new stratigraphic and sedimentological constraints. Int Geol Rev 39:542–576

Lamb MA, Cox D (1998) New ^{40}Ar/^{39}Ar age data and implications for porphyry copper deposits of Mongolia. Econ Geol 93:524–529

Lang JR, Titley SR (1998) Isotopic and geochemical characteristics of Laramide magmatic systems in Arizona and implications for the genesis of porphyry copper deposits. Econ Geol 93:138–170

Loucks RR (2014) Distinctive composition of Cu-ore-forming arc magmas. Aust J Earth Sci 61:5–16

Marinov NA, Khasin RA, Khurtz C (eds) (1977) Geologiya Mongolskoi Narodnoi Respubliki (Geology of the People's Republic of Mongolia), Vol. III mineral deposits. Nedra, Moscow

Mossakovsky AA (1975) Orogennie strukturii I vulkanizm paleozoid Evrazii i ikh mesto v protsesse formarivoniya kontinentalinoi zemnoi korii (Orogenic structures and volcanism of the Paleozoic in Eurasia and its position in the motion of continental crust). Nauka, Moskva, p 318

Mossakovsky AA, Tomurtogoo O (1976) Verkhnii paleozoi Mongolii (Upper Paleozoic of Mongolia). Trudii Sovmestnoi Sovetsko-Mongol'skoi Geologicheskaya Ekspeditsiya, Nauka, Moskva, 15: p 127

Munkhtsengel B (2007) Magmatic and Mineralization Processes of the erdenetiin Ovoo Porphyry Copper-Molybdenum deposit and Environmental Assessment, Northern Mongolia, Dissertation. Tohoku University, Japan, p 189

Nokleberg WJ (ed) (2010) Metallogenesis and Tectonics of Northeast Asia: U.S. Geol Surv Prof Paper 1765

Obolensky AA, Borisenko AS, Borovikov AA, Pavlova GG, Lebedev VI, Sherkhan O, Tsoodol B (1989) Metallogeniya rudnych rajonov Zapadnoi Mongoli: geologiya i razvedka territorii rudnych rajonov Mongol'skoi narodnoi Respubliky (Metallogeny of ore-districts in the western Mongolia: Geology and exploration of the territory of Mongolian Peoples Republic) Mezhdunarodnaya nauchnaya konferentsiya, posvyashchennaya 50-letiyu Geologicheskoy sluzhby Mongol'skoy Narodnoy Respubliki, Ulanbator, 88–89

Orolmaa D, Tungalag N (2015) Tsagaan Suvarga granitoid massif in the southern Mongolia: geology and geochemistry. In: Spiridonov IG, Kilipko VA (eds) Geochemical mapping, prospecting and geoecology. Proc Conf dedicated to AA Golovin. Instit Miner Geochem and Crystallochem of rare element, Moscow, pp 110–126

Oyunchimeg R (2008) Zes-alt porfiryn Khiyugo dammitt ordyn sulifidyn paragyenyezis ba altny khüderjilt (Sulfide Mineralogy and Gold Mineralization at Hugo Dummett Porphyry Cu-Au Deposit), Dissertation, Mongolian University of Science and Technology

Perelló J, Cox D, Garamjav D, Sanjdorj S, Diakov S, Schissel D, TO M, Oyun G (2001) Oyu Tolgoi, Mongolia: Siluro-Devonian porphyry Cu-Au-(Mo) and high-sulfidation cu mineralization with a cretaceous chalcocite blanket. Econ Geol 96:1407–1428

Peters B, Sylvester S (2014) Oyu Tolgoi Technical Report. Prepared for Turquoise Hill Resources Limited by OreWin Pty Ltd, Adelaide, p. 547

Porter TM (2016) The geology, structure and mineralization of the Oyu Tolgoi porphyry copper-gold-molybdenum deposits, Mongolia: a review. Geosci Front 7(3):375–407

Şengör MC, Natal'in BA (1996) Paleotectonics of Asia: fragment of a synthesis. Cambridge University Press, Cambridge

Sengor MC, Natal'in BA, Burtman VS (1993) Evolution of the Altaid tectonic collage and Palaeozoic crustal growth in Eurasia. Nature 364:299–307

Sillitoe RH (1995) Exploration of porphyry copper lithocaps: Australasian Institute of Mining and Metallurgy. In: Proceedings of the Pacific Rim Congress 95, Auckland, New Zealand, 19-- 22 November, 527–532

Sotnikov VI, Berzina AN, Zhamsran M, Garamzhav D, and Bold D (1985) Copper deposits of Mongolia: Novosibirsk, Nauka, Transactions of the Joint Soviet-Mongolian Scientific-Research Expedition, 43: p 223

Sotnikov VI, Berzina AP (1985) O meste rudnykh porfirov v skheme orogennogo magmatizma medno-molibdenovykh rudnykh uzlov Mongolii (on the place of ore porphyries in the scheme for orogenic magmatism of copper-molybdenum ore knots of Mongolia). Geol Geofiz:3–10

Sotnikov VI, Berzina AP (1989) Dlitelinoe diskretno-napravlennoe razvitie rudno-magmaticheskikh system na mestorojdeniyakh medno-molibdenovoi formatsii (prolonged discrete oriented development of ore-magmatic systems in copper-molybdenum formation). Geologiya i Geophizika 1:41–45

Sotnikov VI, Berzina AP, Bold D (1984) Zakonomernosti razmeshcheniya medno-molibdenovogo orudeneniya Mongolii (Regularity of distribution of copper-molybdenum mineralization of Mongolia) In: Endogennye rudnye formatsii Mongolii. Trudii Sovmestnoi Sovetsko-Mongol'skoi Geologicheskaya Ekspeditsiya, Nauka, Moskva, 38:89–101

Sotnikov VI, Panomorchuk VA, Berzina AP, Travin AV (1995) Geochronologicheskie reubeji magmatizma Cu-Mo-porfirovogo mestorojdenie Erdenetyin-obo, Mongoliya (Geochronological boundaries of magmatism in the Cu-Mo porphyry deposit at Erdenetiin Ovoo, Mongolia). Geologiya i Geophizika 36:78–89

Takahashi Y, Oyungerel S, Naito K, Delgertsogt B (1998) The granitoid series in Bayankhongor area, Central Mongolia. Bull Geol Surv Japan 49:25–32

Tomurtogoo O (ed) (1999) Geologic map of Mongolia: Institute of Geology and Mineral Resources Mongolian Academy of Sciences, and Mineral Resources Authority of Mongolia, Ulaanbaatar, scale 1:1 000 000

Tomurtogoo O, Badarch G, Orolmaa D, Byamba J (1999) Terranes and accretionary tectonics of Mongolia. Mong Geosci 14:5–10

Tugarinov AI, Voinkov DM, Grinenko LN, Pavlenko AS (1974) Izotopny sostav i istochniki sery molibdenovo-mednykh proyavlenii Mongolii (isotopic composition and sources of molybdenum-copper mineralization of Mongolia). Geokhimiya 2:171–178

Tungalag N (2014) Tsagaansuvargyn Cu-Mo-nii porfiryn ordyn granitoidyn gyeokhimi ba khüderjilt (Geochemistry of granitoid rocks and mineralization of the Tsagaan Suvarga Cu–Mo deposit), Dissertation, Mongolian University of Science and Technology

Tungalag N, Liu YH, Hsu JH, Huang LY, Yang HJ (2009) Mechanism and age constraints for Cu mineralization in the granitic rocks from Tsagaan Suvarga, South Mongolia, Abst Ann congress of Chinese Geophys Soc and Geol Soc of Taiwan, 91–92

Tungalag N, Sereenen J, Khashgerel B, Chuluunbaatar B, Kavalieris I (2018). Characteristics of the Late Devonian Tsagaan Suvarga Cu–Mo deposit, Southern Mongolia. Mineralium Deposita 1–12

Voinkov DM, Grinenko LN, Garamjav D, Jamsran M (1977) O stadiinosti rudoobrazovaniya na medno-porfirovom mestorojdenii Erdentyin-Obo v Mongolii: po izotopnym dannym sery (About ore stages of hydrothermal origin at the Erdenetiin Ovoo Cu porphyry deposit, Mongolia: by sulfur isotopes) Geologiya i razvedka, 1:77–80

Wainwright AJ (2008) Volcanostratigraphic Framework and Magmatic Evolution of the Oyu Tolgoi Porphyry Cu-Au District, South Mongolia Unpublished Ph.D thesis Universityof British Columbia, Vancouver, Canada p. 277

Wainwright AJ, Tosdal RM, Forster CN, Kirwin DJ, Lewis PD, Wooden JL (2011a) Devonian and carboniferous arcs of the Oyu Tolgoi porphyry Cu-Au district, South Gobi region, Mongolia Geological Society of America Bulletin, 123:306–328

Wainwright AJ, Tosdal RM, Wooden JL, Mazdab FK, Friedman RM (2011b) U-Pb (zircon) and geochemical constraints on the age, origin, and evolution of Paleozoic arc magmas in the Oyu Tolgoi porphyry Cu-Au district, southern Mongolia. Gondwana Res 19:764–787

Watanabe Y, Stein HJ (2000) Re–Os ages for the Erdenet and Tsagaan Suvarga porphyry Cu–Mo deposits, Mongolia, and tectonic implications. Econ Geol 95:1537–1542

Watanabe Y, Turmagnai D, Byambasuren D, Oyunchimeg G, Tsdenbaljir Y, Sato Y (1999) Geology and K-Ar ages of the South, Huh Bulgiin Hundii, Saran Uul, Taats Gol and Han Uul deposits in the Bayankhongor region, Mongolia. Resour Geol 9(3):120–130

Yakovlev VA (1977) Medi, svinets i zink (Copper, lead and zinc. In: Geology and Mineral Resources of Mongolian Peoples' Republic, Book III, Nedra, Moscow) 141–146

Yakubchuk AS (2004) Architecture and mineral deposit settings of the Altaid orogenic collage: a revised model. J Asian Earth Sci 23:761–779

Yakubchuk AS, Degtyarev K, Maslennikov V, Wurst A, Stekhin A, Lobanov K (2012) Tectonomagmatic settings, architecture, and metallogeny of the central Asian Copper Province. In: Hedenquist JW, Harris M, Camus F (eds) Geology and genesis of major copper deposits and districts of the world, A Tribute to Richard H Sillitoe, Society of Economic Geologists, vol 16, pp 403–432

Yakubchuk AS., Shatov VV, Kirwin D, Edwards A, Tomurtogoo O, Badarch G, Buryak VA (2005) Gold and base metal metallogeny of the central Asian orogenic supercollage. Econ Geol, 100th anniversary volume, 1035–1068

Yashina RM, Matrenitsky AT (1978) Petrokhimiya vulkanicheskikh i intruzivnykh porod Orkhon-Selenginskogo progiba: Mongoliya (Petrochemistry of volcanic and intrusive rocks of the Orkhon Selenge Trough: Mongolia) Izvestiya AN SSSR. Serii geologii 10:26–42

Zonenshain LP, Kuzmin MI, Natapov LM (1990) Mongolo-Okhotskii poyas (Mongol-Okhotsk belt) in: Tektonika litosfernykh plit territorii SSSR. Kniga 1, Nedra, Moskva, 282–319

Xanadumines.com/site/projects/kharmagtai/resources

Chapter 3
Lode Gold Deposits

Gunchin Dejidmaa and Uyanga Bold

3.1 Introduction

In the first half of the twentieth century (1900–1919), joint Mongolia-Russia exploration groups studied lode and placer gold mineralization in the Yeroo River and Boroo-Zuunmod districts of the Northern Khentii zone for the first time in Mongolia. As a result, placer gold deposits such as the Khuder, Tsagaannuur, Tolgoit, Ikh-Ulunt, Baga-Ulunt, Mogoi, Kharganat, Buural, Ikh Ajir, and Baga Ajir were found, and the richest ones were mined.

Mongolian government had carried out gold mining work between 1926 and 1956 and mined residual prospects from the previous mining activities in the Yeroo River district. Between 1948 and 1956, the richest parts of the Boroo-Zuunmod district were mined without any additional geologic and exploration work. The main period for gold exploration started in 1960 in Mongolia. The focus had been the Yeroo River and Boroo-Zuunmod districts and Bayankhongor zone. Throughout the history, not only Mongolian gold exploration geologists but Russian, East German, and Bulgarian geologists had done valuable amount of work. Furthermore, through regional geologic mapping work, many lode and placer gold occurrences were identified. Since the 1990s, Mongolian geologists took the lead by identifying many new placer gold deposits and have calculated the total gold resources, which led to active mining until today.

G. Dejidmaa
Geological Information Center, Mineral Resources and Petroleum Authority, Ulaanbaatar, Mongolia

U. Bold (✉)
Department of Research and Innovation Department, Mongolian University of Science and Technology, Ulaanbaatar, Mongolia
e-mail: uyangabold@must.edu.mn

O. Gerel et al. (eds.), *Mineral Resources of Mongolia*, Modern Approaches in Solid Earth Sciences 19, https://doi.org/10.1007/978-981-15-5943-3_3

Additionally, Japanese government had carried out gold exploration work in Mongolia between 1990 and 1996 through JICA projects, namely, the "Uudam Tal" and "Altan Tal," and focused primarily on lode gold mineralization. Through the "Uudam Tal" project (JICA 1992), gold mineralization potential of the Ulziit zone was extensively studied, Olon Ovoot gold deposit was fully explored, and fluid inclusion studies of the gold-bearing rocks were studied for the first time in Mongolia. Through the "Altan Tal" project, epithermal gold mineralization was studied for the first time in the Gurvantes-Nomgon metallogenic zone.

Between 1990 and 1996, location map of gold deposits and occurrences and metallogenic map in scale of 1:1,000,000 of Mongolia were published along with detailed descriptions, which have effectively contributed to and developed gold mineralization studies.

3.2 Gold Metallogeny

Seven gold-bearing metallogenic belts, the Mongol-Altai, Nuur, Northern Mongolia, Central Mongolia, Khangai-Khentii, Eastern Mongolia, and Southern Mongolia, have been recognized, with 25 zones and 70 districts (Fig. 3.1) currently identified in Mongolia.

Gold-bearing districts form the gold-bearing zones and are bounded by non-gold-bearing or poorly mineralized regions. Each district has mineralization of the same age. Depending on mineralization features such as associated mineral assemblages and types of gold deposits and occurrences, each district is defined as gold ore district, lode and placer gold district, and gold-copper ore district. Based on available geologic data, poorly studied regions with potential gold mineralization are defined.

3.3 Lode Gold Deposit Types in Mongolia

Based on previous studies, the following gold deposits, groups, types, and subtypes have been described in Mongolia. Many of them are based on gold occurrences only and hence require further work.

3.3.1 Gold Mineralization Hosted in Metamorphosed Conglomerate

This type of gold deposits was formed in Archean and lower Proterozoic conglomerates, for example, in the Witwatersand Basin in South Africa (e.g., Eriksson et al. 1981). Deposits of Devonian age are present and are primarily found in Russia.

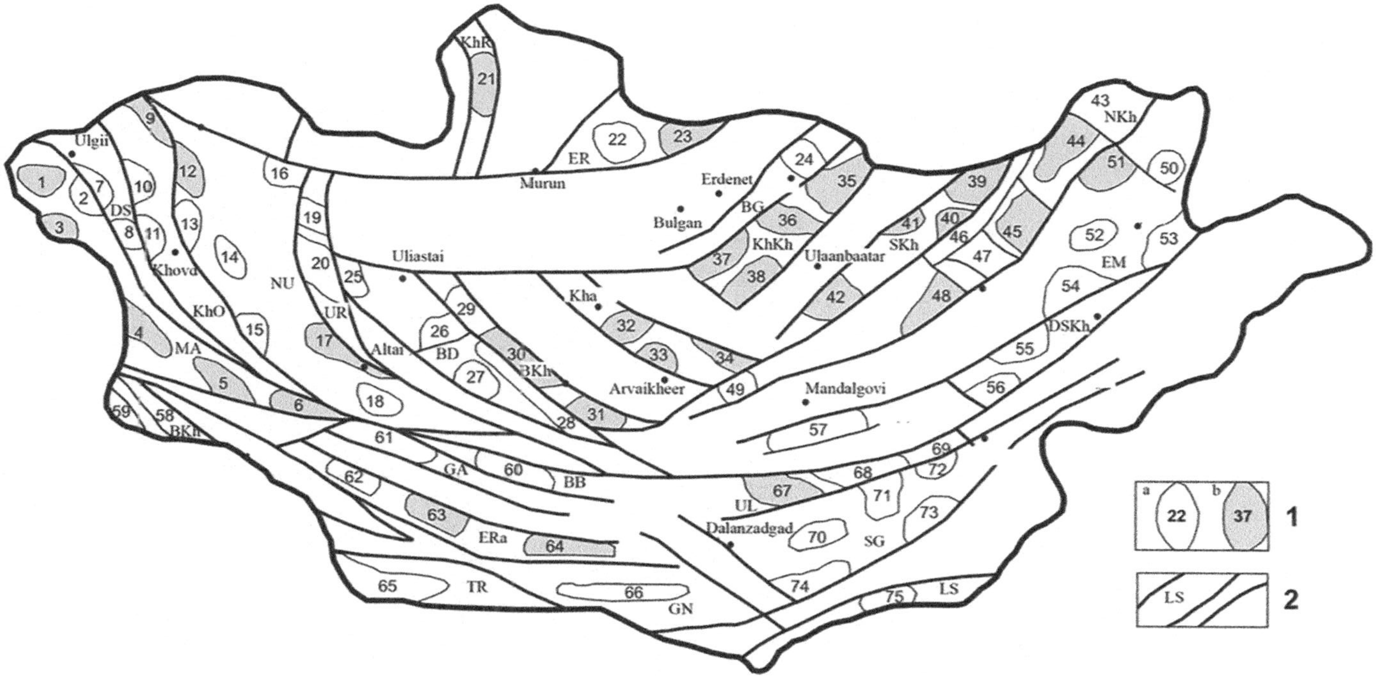

Fig. 3.1 Gold-bearing metallogenic zones and districts of Mongolia (Dejidmaa 1996). (1) Gold-bearing ore district: (a) lode gold mineralization district; (b) districts hosting both lode and placer gold mineralization. District numbers are the same as in Table 3.1. (2) Gold-bearing metallogenic zones (names are included in Table 3.1)

Table 3.1 Gold metallogenic belts, zone, and district

Metallogenic belt	Metallogenic zone	Numbers on Fig. 3.1	Gold district	Metal association
Mongol-Altai	Mongol-Altai (MA)	1	Dungerekh-Tsagaan river	W, Mo, Be, Sb, Au
		2	Sagsai Gol	W, Mo, Cu, Pb, Zn, Au
		3	Khurimt	Au
		4	Upper Bulgan	Au, Pb, Ag, W
		5	Uyench Bodonch	Au
		6	Shar Khooloi	Au
	Deluun-Sagsai (DS)	7	Ulgii	Au, Ag, Ba, Hg, W
		8	Tolbo Nuur	Ag, Sb, Co, Cu, Au
	Khovd (KhO)	9	Achit Nuur	Zn, Pb, Ag, Cu, Fe, Au
		10	Altantsugts	Au, Ag
		11	Khovd	Au
Nuur	Nuur (NU)	12	Burgastai Gol	Au, Ag, Cu
		13	Tsagaan Tolgoi	Au, Ag, Cu
		14	Baatar Khairkhan (occurrence)	Cu, Fe, Au
		15	Tsetserleg range (occurrence)	Cu, Zn, Au
		16	Khyargas ore	Cu, Au
		17	Sharga Gol	Cu, Zn, Au
		18	Gobi-Altai	Cu, Fe, Au
	Urgamal (UR)	19	Khuljgol-Bayankhairkhan	Au, Cu, Fe
		20	Urgamal-Zavkhanmandal	Au, Cu, Fe
Northern Mongolia	Khugiin gol (KhR)	21	Khugiin Gol	Au, Fe
	Egiingol (ER)	22	Teshig	Au, Ag, Cu, Fe
		23	Azarga Gol	Au, Ag, W, Mo
	Bayangol (BG)	24	Dulaankhaan	Cu, Mo, Pb, Zn, Au
Central Mongolia	Baidrag (BD)	25	Erdenekhairkhan	Pb, Zn, Fe, Cu, Au
		26	Unegt	Au, Ag, W, Mo, Fe
		27	Buutsagaan	Fe, Cu, Au
		28	Baidrag-Burd Gol	Au, Cu
Khangai-Khentii	Bayankhongor (BKh)	29	Galuut-Gurvanbulag	Sb, Au, Ag, W, Mo
		30	Duvunt	Au, Cu
		31	Tuin-Taats river	Au, W
	Khangai (Kha)	32	Battsengel	Au
		33	Uyanga-Taragt	Au
		34	Delgerkhaan	Cu, Au

(continued)

Table 3.1 (continued)

Metallogenic belt	Metallogenic zone	Numbers on Fig. 3.1	Gold district	Metal association
	Northern Khentii (KhKh)	35	Yeroo River	Au
		36	Boroo-Zuunmod	Au
		37	Zaamar	Au
		38	Ugtaaltsaidam-Argal range-Suvarga	Au
	Southern Khentii (SKh)	39	Balj Gol	Au
		40	Yol Uul-Bituu Khamar (occurrence)	Au, Ag, Fe
		41	Terelj	Au
		42	Baruun Urt	Au
Eastern Mongolia	Northern Kherlen (NKh)	43	Duchiin Gol	Au, Pb, Zn, Ag, Cu, As, Sb, Hg, CaF_2
		44	Turgen	Au, Ag, W, Mo
		45	Narsan Khundlun	Au, Ag, Fe, Sn, CaF_2
		46	Onon	Au, Ag, Hg, CaF_2, W
		47	Berkh	CaF_2, Au, Ag, Sn
		48	Kherlen	Au, Ag, Pb, Zn
		49	Erdenedalai	Au
	Eastern Mongolia (EM)	50	Tsav	Ag, Pb, Zn, Au, CaF_2
		51	Dornod	U, CaF_2, Ag, Pb, Zn, Cu, Au
		52	Bulgan (occurrence)	Au, Ag
		53	Choibalsan (occurrence)	Au, Ag
		54	Zaan Shiree-Munkhkhaan	Au, Ag
	Dundgobi-south Kherlen (DSKh)	55	Buyant	Au,
		56	Altanshiree-Delgerekh	Au, Fe, Cu
		57	Dundgobi (occurrence)	W, Mo, Au
Southern Mongolia	Baruun Khuurai (BKh)	58	Nukh Nuruu	Au, Ag, Pb, Zn
		59	Baitag	Au, Sb, As, Hg
	Bayangobi-Bayanlig (BB)	60	Bayangobi	Au, Cu, Zn, Pb, Fe
	Gobi-Altai (GA)	61	Takhilga Uul-Erdene	W, Mo, Cu, Au, Fe
	Edren range (ERa)	62	Eej Khairkhan-Senjit khyar	Au, Cu, Mo
		63	Edren Nuruu	Au
		64	Ongon Uul-Nemegt	Au

(continued)

Table 3.1 (continued)

Metallogenic belt	Metallogenic zone	Numbers on Fig. 3.1	Gold district	Metal association
	Tumurtei range (TR)	65	Taliin Meltes-Khatan Suudal	Au
	Gurvantes-Noyon (GN)	66	Gurvantes-Noyon	Au, Ag
	Ulziit (UL)	67	Mandal-Ovoo	Au
		68	Tsogt-Ovoo	Au
	South Gobi (SG)	69	Kharmagtai	Cu, Au
		70	Ikh Shankhai	Cu, Au
		71	Mandakh Shuteen	Cu, Au
		72	Saikhandulaan	Cu, Au
		73	Tsagaansuvarga	Cu, Mo, Au
		74	Bayan-Ovoo	Cu, Mo, Au
	Lugiingol-Sulinkheer (LS)	75	Nomgon	Sb, As, Au, Ag

Similarly in Mongolia, gold mineralization has been described in Carboniferous, Permian, Jurassic, and Cretaceous conglomerates in Khangai-Khentii metallogenic province (Table 3.1). However, not much geologic information is available at the moment. Gold mineralization has also been identified in Carboniferous and Cretaceous conglomerates in the Zaamar district, but their economic feasibility has not been concluded. Cretaceous conglomerates of the Yeroo River district are known for hosting gold mineralization and are assumed to have sourced placer gold deposits within the district. Near the Shar River coal deposit, in the northern part of the Shar River, conglomerates are found to have preserved gold and sourced placer deposits. To the west of the Delgerkhaan district, in Bayan-Undur soum of Uvurkhangai aimag, near the Kharzat River, Permian conglomerates have sourced small quaternary placer deposits. Moreover, it is assumed that Jurassic conglomerates may have sourced placer deposits in Uyanga and Taragt soums of Uvurkhangai aimag. In Mongolia, continental tectonism has dominated since the upper Paleozoic, which resulted in river, lake, and sea placer gold deposits which may have resulted in the development of lode gold mineralization. Gold-bearing conglomerates often interbed with gold-rich sedimentary layers, and it is also possible to form a large gold deposit with conglomerates with poor gold content if the extent of exposure is large.

3.3.2 Placer Gold Deposits

Placer gold deposits are common in Mongolia. Cenozoic age loose sediments are the prime targets. Moreover, ancient placer deposits of Carboniferous, Permian, Jurassic, Triassic, and Cretaceous rocks are also present, some of which are identified to have intermediately sourced from Jurassic and Cretaceous gold-bearing conglomerates (See the Chap. 4, "Placer Gold Deposits").

3.3.3 *Massive Sulfide Gold Deposits Formed in Association with Seafloor Volcanic Rocks*

In this type of deposits, gold becomes the primary enrichment, whereas copper, lead, and zinc become accessory. The Gozgor gold occurrence of the Burgastai River district of the Nuur metallogenic zone and the Erdenetolgoi occurrence of the Zavkhanmandal district are of this type (Table 3.1).

3.3.3.1 Gozgor Gold Occurrence

It is located 15.5 km to the northeast of Umnugovi soum of Uvs aimag. The occurrence was formed along the Tsagaanshiveet deep-seated fault in Neoproterozoic-lower Cambrian mafic and intermediate volcanic rocks of the Tsol-Uul Formation. Massive occurrences of pyrite, pyrrhotite, chalcopyrite, and malachite are documented in quartz veins. Malachite veinlet thickness ranges from 0.1 to 0.3 m and length from 0.4 to 1.5 m. Quartz veins are 0.5–1.0 m wide and 3.0–5.0 m thick. Sulfides are also preserved in quartz veins. During mineral separation, up to 78 gold grains were found in each sample, and corresponding gold content was between 0.06–20.9 g/t, 1.0–3.0% for copper, and 0.4% for barium.

3.3.3.2 Bayan-Airag Gold Occurrence

It is located at the Bayan-Airag mountain, 20 km north from Durvuljin soum center of Zavkhan aimag. The occurrence is in iron-mineralized quartzites associated with marbleized limestone and amphibolite, amphibole-chlorite, chlorite, quartz-sericite, and quartz-chlorite schist of the Neoproterozoic Shuvuu Formation. Native copper nodules occur in carbonate lenses, with chlorite alteration, and in amphibole-chlorite schist that are 5–10 m thick and 250 m long. The ore body is limonitic, and the gold content ranges from 3–6 to 60–100 g/t.

3.3.3.3 Erdenetolgoi Gold Occurrence

It is located in the same ore district as of the Bayan-Airag occurrence. The host rock is a shale that preserves a conformable argillaceous zone that is 1000 m long, 180–240 m thick, and with a dip of 55–60° to the northeast. This zone preserves two chalcedony-like beds that are 50 and 100 m thick, respectively, and that are deposited 25 m apart from each other. The mineral assemblage is characterized by quartz, albite, montmorillonite, kaolinite, sericite, chlorite, pyrite, hematite, and magnetite and contain gold nodules as large as 0.002–0.003 mm. The entire zone contains sulfides that are heavily oxidized. The oxidized zone contains 0.01–3.0 g/t of gold, and it is assumed that at depth, there may be mineralization with more gold (Altankhuyag 2005).

3.3.4 Epithermal Gold Deposits

Epithermal gold deposits formed in relation to continental volcanic activities are found in Eastern and Southern Mongolia (Table 3.1). Unfortunately, this group of deposits is poorly studied. Epithermal mineral systems are often rich in gold content, and hence gold exploration work has been recently implemented on potential regions in Mongolia. Depending on the geological setting and associated mineral assemblage, two subtypes are identified, namely, high sulfidation and low sulfidations.

3.3.4.1 Low Sulfidation Epithermal Gold Deposits

These deposits are formed in island arcs and continental rift zones. Mesozoic rifting has been an important tectonic event in Eastern Mongolia (Table 3.1), and it is assumed that it has a high potential for hosting epithermal gold deposits. This rifting event extends to southern Baikal Lake in Russia and Dalai Lake in China where epithermal gold deposits are known. For example, the Balei gold deposit in Russia and Erdenetolgoi silver-gold deposit in Inner Mongolia of China are associated with this Mesozoic rifting regime. In Russia, gold-bearing deposits are found in Jurassic and lower Cretaceous sedimentary rocks deposited in local grabens that follow deep-seated and regional faults. In Mongolia, the same age grabens have been formed especially in Onon and Ulz Rivers, where epithermal gold occurrences are identified. Identified ores are thought to have formed in low-temperature chalcedony-like quartz veins and quartz veinlets, stockworks, brecciated vein, and tube-like veinlets. Sulfides are present, but due to its grain size, it is difficult to be identified by naked eye. Due to this disseminated sulfide content, color of the quartz changes to dark gray. The Tsagaanchuluut Khudag 2, Baruun Khujirt River, and Zuun Khujirt River occurrences of the Turgen ore district of Northern Khentii metallogenic zone and Tsagaan and Bayanzurkh occurrences in the Berkh district are of this type (Table 3.1). In China, epithermal silver-gold mineralization found near Dalai River in Inner Mongolia in China may continue along the Main Mongolian Lineament (tectonic fault) in Mongolia. Moreover, in China, this type of silver-dominated gold deposits is found in Cretaceous volcanic rocks and Jurassic andesites, and the ore body is often silicified and forms metasomatized, sulfide-rich zones. The ore minerals are native copper, argentite, pyrargyrite, and electrum. Similar structure and mineralization are present in the Choibalsan group occurrences and in Munkhkhaan-

Zaanshiree ore districts. Some of these occurrences are silver-dominated, which provides an evidence to suggest that this type of gold deposits has high potential to be found in Mongolia.

Jivkheestei Occurrence

It is located in the eastern part of the Onon River in the Jivkheestei Range. The occurrence is in mineralized breccia that cuts light yellow-colored silicified shale. Its length reaches 150 m and width 1–20 m. Mineralized quartz veins are composed of agate-like, fine-grained yellow quartz grains and antimonite. The ore minerals are gold, antimonite, and pyrite. Gold content of 7 g/t has been measured.

Nirkhruu Occurrence

It is located 3 km northeast from the Jivkheestei occurrence. The occurrence is located in a 1000 m by 500 m region and consists of numbers of small quartz bodies. The ore minerals are gold, antimonite, pyrite, and marcasite with a gold content of up to 4 g/t.

3.3.4.2 High Sulfidation Epithermal Gold Deposits

The respective occurrences are identified in Umnugovi copper-gold and Gurvantes-Noyon metallogenic zones (Table 3.1), which are related to island arc, continental rift, and continental volcanism. It is suggested that these deposits are formed at the top of a porphyry mineralization system and found in regions were copper and copper-molybdenum porphyry mineralization are developed. Mineralization comprises sulfides, especially pyrite and enargite, and associated with hydrothermal alteration. For instance, cap-shaped hydrothermal alteration commonly occurs at the top as a silica cap followed downward by propylite, alunite, kaolin, and silica alteration. Gold mineralization is in the core of zoned hydrothermal alteration, whereas sulfides are concentrated in zones of silicic alteration and quartz breccia. The same structure is described in the Shuteen Khanbogd mountain of the Mandakh-Shuteen ore district in the Umnugovi metallogenic zone and in the Ikh Shankhai ore district. Heavy mineral separation carried out in valleys, where this structure is described, revealed the presence of gold grains, and the gold content is calculated to be 0.01–0.05 g/t in siliceous cap. Moreover, rims of the silica cap contain iron oxides (oxidized sulfides), and the siliceous bodies are vuggy in texture, which suggests gold mineralization potential. The metallogenic zone itself is associated with upper Paleozoic continental arc and overlap continental rift, and, hence, it has the potential to host gold deposits. The Oyutolgoi deposit is a good example occurring in this metallogenic zone (Kirwin et al. 2005; Jargaljav 2009).

3.3.5 Gold Deposits Hosted in Quartz-Carbonate Vein

This type of deposits is related to zones that are affected by regional metamorphism. Deposits described in Mongolia are hosted in two types of veins, quartz-carbonate, and sulfide-quartz. The host rocks are usually composed of greenschist facies sedimentary rocks with rare interbeds of volcanic rocks, and the ore bodies are often conformable to bedding. Intrusive rocks are present in some metallogenic districts, and the formation timing coincides with regional folding.

3.3.6 Gold Deposits Hosted in Quartz and Quartz-Carbonate Veins

This type of deposits is widely distributed. The ore bodies are primarily composed of quartz-carbonate and are sulfide poor. They are conformable to bedding of weakly metamorphosed terrigenous rocks and form saddle-like duplicated or step-like veins. Both these shapes are sometimes found in the same deposit. The veins are long and thick (see below). The ore minerals are pyrite, native gold, chalcopyrite, and rarely sphalerite, galena, arsenopyrite, and antimonite. Gold forms free grains in quartz associated with sulfide minerals. Gold content is usually greater than silver, and the Au:Ag ratio is 2:1. Silicification, sericitization, carbonatization, pyritization, and epidotization alterations are common. Where the veins are composed of quartz-carbonate, the host rock is often carbonatizated. This type of deposits is present in the Nukht range district of the Baruunkhuurai metallogenic zone and the Taliin meltes-Khatan suudal district of the Tumurtei metallogenic zone. In each district, several and closely located gold deposits and occurrences are found, which are shown by quartz veins in the Zaamar district (Table 3.1). In the Zaamar district alone, this type of gold mineralization sources the 2/3 of placer gold deposits known in Mongolia.

3.3.6.1 Bumbat Deposit

It is located in Zaamar soum of Tuv aimag, to the west of the central part of the Zaamar range. Two main veins form the ore bodies. The deposit is situated in the lower Paleozoic Zaamar Formation shale that hosts mafic metavolcanics and quartzite beds. Vein 118 strikes to the northeast, dips to the south by 62–75°, and is conformable to the host rocks. The vein length reaches 800 m and thickness 0.91–8.26 m. The depth reaches 300 m. The ore minerals are pyrite, gold, rarely chalcopyrite, and galena. Gold content ranges between 0.1 and 720 g/t and silver between 0.7 and 23.1 g/t. Vein 117 is located 300 m to the northwest. It is 300 m long, 1.36–1.63 m thick, and continues for 100 m in depth. The gold content ranges between 0.1–50.9 g/t and silver 0.5–8.9 g/t. The deposit is explored to 25 m of depth with a cutoff gold content value of 3 g/t. The average gold content is 10.82 g/t, and the total resources are estimated to be 346,635 tonnes.

3.3.6.2 Olon-Ovoot Deposit

It is located in Mandal-Ovoo soum of Umnugovi aimag, 13 km southeast from Mushgia Khudag. The host rocks are composed of Silurian sandstone, siltstone, and shale and Devonian gabbro and diorite bodies. Quartz veins follow a fault zone that exhibits pyritic, argillic, and silicic alteration. Gold-bearing quartz veins are 150–200 m thick and continue for 4 km along strike (Fig. 3.2).

The Olon-Ovoot gold mineralization is in an oxidized zone. The oxidized zone reaches depth of 70–80 m. Gold content ranges from 1 to 32.8 g/t with an average of 5 g/t. Gold grains range from 0.1 × 0.2 to 0.3 × 0.4 mm in size. In addition to gold, the ore minerals include malachite, galena, arsenopyrite, chalcopyrite, and rarely magnetite. Gangue minerals are fluorite, chlorite, calcite, and siderite.

3.3.7 Gold Mineralization in Sulfide-Quartz Veinlet

This is comparatively rare, but gold occurrences are described in the Bayangovi and Oortsog districts of the Bayangovi-Bayanlig metallogenic zone (Table 3.1). Gold is contained in sandstones hosted in shale and in step-like quartz veins. Gold content is variable and is directly related to number of quartz veinlets and pyrite.

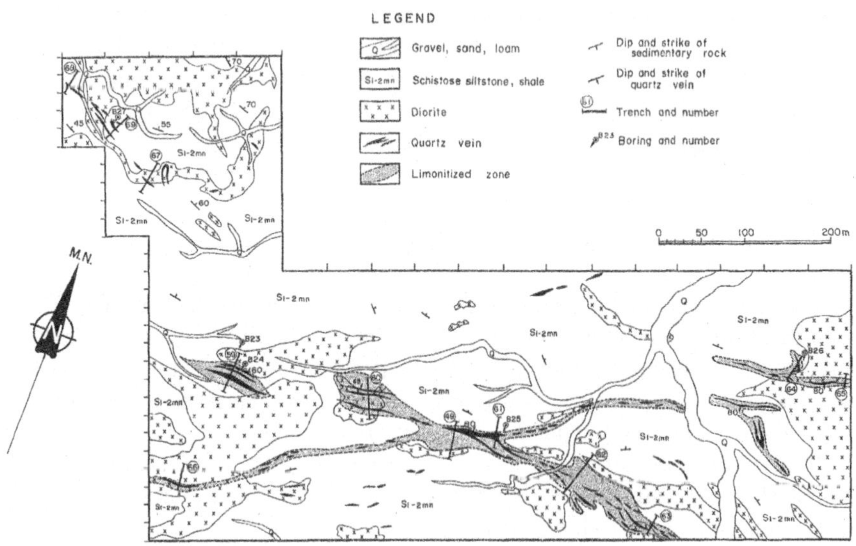

Fig. 3.2 Geological setting of Olon-Ovoot gold deposit (JICA 1992). (1) Unconsolidated sediments, gravel, sand, and mud; (2) sedimentary rocks, shale, siltstone, and mudstone; (3) intrusive complex, diorite; (4) gold-bearing quartz vein; (5) limonitic zone

3.3.8 *Intrusion-Related Hydrothermal-Metasomatic Gold Deposits*

This type of gold deposits related to granite intrusion are commonly distributed in Mongolia and are classified as quartz vein and quartz vein-metasomatic zone according to ore body shape and gold, gold-copper, gold-polymetallic, gold-silver-copper-arsenic-antimony, gold-tungsten-molybdenum, and gold-hematite-magnetite by elemental composition. Quartz vein ore is common, but the size becomes smaller compared to quartz veins associated with metasomatic alteration.

3.3.9 *Gold Deposits Hosted in Quartz Veins/Stockwork with Metasomatic Alteration*

This type of deposits has high industrial importance. Mineralization is within tectonic zones, and the host rock becomes mineralized and heavily altered enclosing gold-bearing quartz veins, veinlets, and stockworks. Ore body size ranges from several tens of meters to kilometers. Although the gold content is high (10–20 g/t) in quartz veins, the content decreases in altered host rock (0.1–5.0 g/t), and the overall gold content becomes low. Boroo gold deposit (Khishgee and Akasaka 2015), Tsagaan Gozgor, and Bayantsagaan occurrences of Boroo-Zuunmod ore district, Khyargas occurrence of Khyargas ore district, Tsagaanchuluut Khudag 1 occurrence of Turgen ore district, Kharguit, Dagai, and Urliin-Ovoo occurrences of Dornod ore district, Bor-Undur and Uvur Khooloi occurrences of Duchgol ore district, and Bayan-Ulz occurrence of Narsan Khundlun district belong to this type (Table 3.1). Based on mineral composition and primary and secondary metal assemblages, the main deposit type is divided into subtypes of Au (Ag), Au (Ag, Cu), Au (Ag, Pg, Zn), Au (Ag, As, Sb), and Au-Mo – W.

3.3.9.1 Boroo Gold Deposit

An industrially important part of the Boroo gold deposit is 2200 m long (from northeast to the southwest) with a maximum thickness reaching 600 m. Textures responsible for hosting the gold ore can be divided into two parts (Figs. 3.3–3.4). Ore body occupies a large area, consisting of hydrothermal-metasomatic alteration at large-scale and shallowly dipping faults (cataclastic zone).

Northeastern part of the deposit is 1200 m long and 600 m wide and hosted by the Ikh Tashir granite intrusion. The granite also enclosed xenoliths of host sedimentary rocks along its metamorphosed boundary. The region of interest is bound by an outcropping ore body to the east and the Kharaa Formation sedimentary rocks and the cross-cutting Ikh Tashir granite intrusion to the west. The southern boundary is defined by a fault called "Boroo no.2." Mineralized zone dies out to the north with

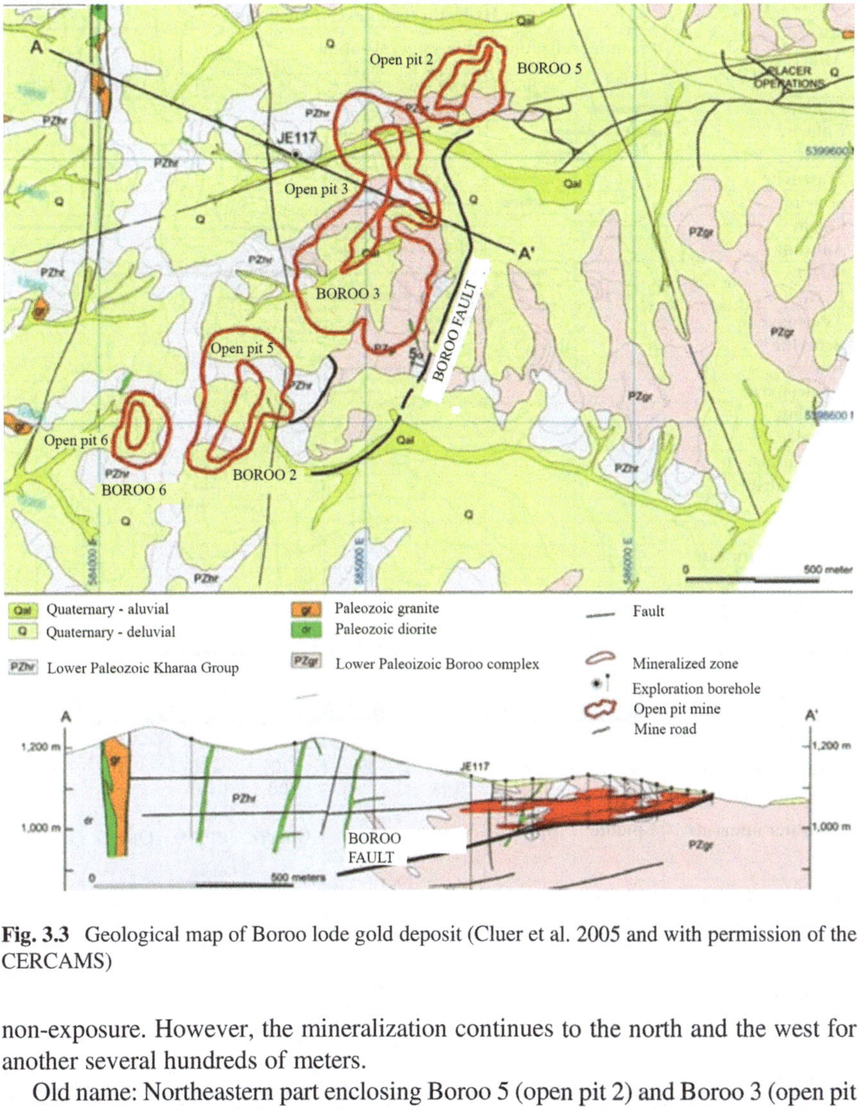

Fig. 3.3 Geological map of Boroo lode gold deposit (Cluer et al. 2005 and with permission of the CERCAMS)

non-exposure. However, the mineralization continues to the north and the west for another several hundreds of meters.

Old name: Northeastern part enclosing Boroo 5 (open pit 2) and Boroo 3 (open pit 3) and southwestern part enclosing Boroo 6 (open pit 6) and Boroo 2 (open pit 5).

Mineralized zone and mineralization are continuous and comparatively constant. No. 5 and no. 3 ores are the thickest and the richest in gold content. The ores are associated with fault systems developed from the northeast to the southwest. In the central region of the nos. 5, 3, and 2, ore body thickness reaches 20–40 m, and irregular listvenite lenses are preserved. Sulfidization is intensive in these lenses and are cut by various metasomatic quartz veins.

Metasomatic zones, possibly listvenite, are developed that are usually rich with gold content and heavily sulfidizated. Sedimentary rocks preserved next to the above ore-bounding structures are often not altered compared to granitic bodies. Large

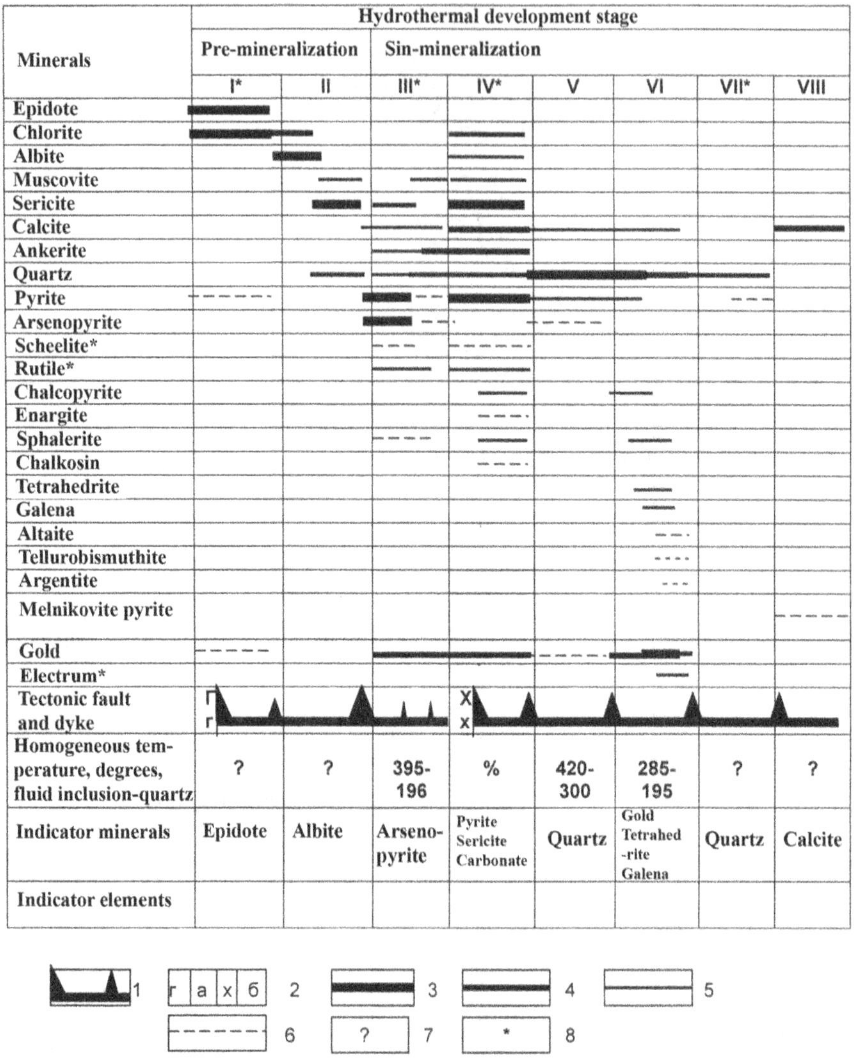

Fig. 3.4 Sequence of gold mineralization in the Boroo gold deposit (Dejidmaa 1985). (1) Discontinuous tectonic activation; (2) gabbroid veins ((a) diorite vein, (б) 3–6-dissemination of minerals); (3) river; (4) widely preserved; (5) auxiliary; (6) rare; (7) not studied; (8) mineral found at first; I–VIII: I, epidote-chlorite; II, albite-chlorite-quartz-sericite; III, gold-pyrite-arsenopyrite-quartz; IV, gold-listvenite; V, white quartz; VI, gold quartz; VII, chalcedony like quartz; VIII, carbonate

bodies of listvenite outcrop in Boroo nos. 5 and 2 due to exposure, up to 60 m depth, become oxidized. Metasomatic alteration is intensely developed in the western part of Boroo no. 3 with a great depth.

General trend of ore control textures in the entire deposit is from north-northeast to south-southwest.

Western part of the deposit occupies the metamorphosed boundary region of the Kharaa Formation that are preserved in over 1000-m-long and 500-m-wide region. Eastern and northeastern boundaries are defined by Ikh Tashir granite that cut the Kharaa Formation sedimentary rocks. The western boundary is defined by the Boroo no.6 zone. However, the northern boundary is not well-defined.

Due to variable metasomatic processes and active tectonics, the mineralization is quite irregular. The main difference from the northeast is that the enriched ore body does not form a single body, instead, forms highly enriched columns of the body separated by non-mineralized region. Western part of Boroo no.2 is in the northeastern part of the deposit. Boroo no.6 forms a circular structure at the surface and is affected by north to south and east to west trending vertical faults. The metasomatic zone thickness ranges from 0 to 30 m. Boroo no.6 quartz veins contain sulfides, and one of these veins had been explored until 1955.

3.3.9.2 Sujigtei Deposit

It is located 100 km north of Ulaanbaatar and at the eastern side of the Noyon mountain, 30 km from the Bornuur farm. Gold mining was conducted between 1920 and 1930. The ore body is associated with the Zuunmod rhyolite porphyry and dacite complex. Major trend of the associated faults is east and northwest, and the alterations are defined by silicification, pyritization, and listvenites. Dykes are composed of aplite and pegmatite. Host rocks are the Boroogol complex granite and granodiorite. The major structure of the deposit is the Sujigtei fault (almost vertical 80–85°, striking northeast) where the rocks are cataclastic, milonitizated, and hydrothermally altered forming a fault zone that is 60–180 m thick. Ore body is related to this fault system (Fig. 3.5).

There are total of 16 veins in the deposit, east to west and south to north trending. Veins are mostly vertical with a gentle dip to the southeast. The main vein is with two apophysis, and the sulfide veins are of economic importance. The Sujigtei deposit is a gold-sulfide-quartz deposit. Veins contain dust-like gold, pyrite, chalcopyrite, tetrahedrite, arsenopyrite, sphalerite, galena, burnonite, and altaite. Gangue minerals are quartz, sericite, calcite, and chlorite, and the secondary minerals are limonite, goethite-lepidocrocite, covellite, chalcocite, malachite, azurite, anglesite, cerussite, and romeite.

3.3.9.3 Khyargas Au (Ag, Cu) Occurrence

It is located 27 km east of the Khyargas soum of Uvs aimag, at the southern side of Khan-Khukhii Range, and to the east of the Icheet River. Main deposit content is gold, and it may contain copper as a secondary enrichment. The occurrence is hosted in basalt and serpentinizated ultramafic rocks that are 50–100 m wide with the length of maximum of 500 m that are part of Neoproterozoic to lower Cambrian ophiolite complex. Listvenitization is intensely developed in relation to gold mineralization.

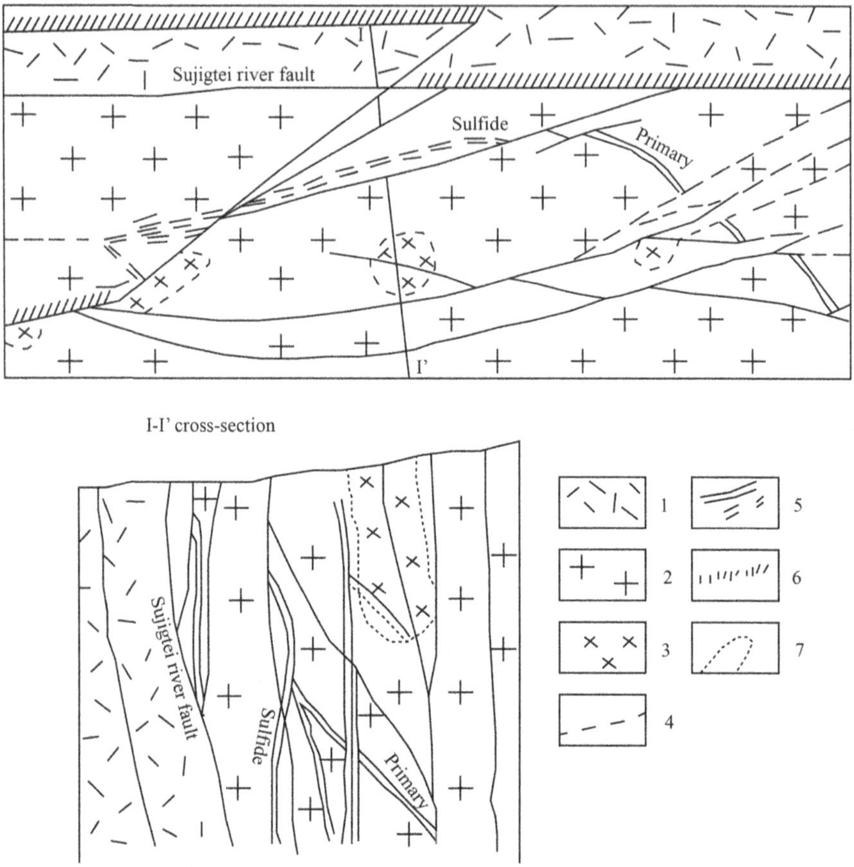

Fig. 3.5 Outline geologic map and cross-section of the Sujigtei deposit (Blagonravov and Shabalovskii 1977). (1) Devonian rhyolitic subvolcanic rocks; (2–3) middle-late Ordovician Boroogol complex (2-granite, 3-granodiorite); (4) fault; (5) gold-quartz and gold-sulfide-quartz veins; (6) silicified; (7) phase boundary

Listvenitizated body is fractured and is 400-m-wide and ~ 1000-m-long stockwork-like ore body. Gold content reaches 31 g/t in shallowly dipping fault zones, but the average gold content varies from 0.5 to 1.6 g/t in the entire stockwork. Pyrite, chalcopyrite, malachite, and iron oxides are the ore minerals. Copper and silver mineralization are not studied in detail, but in some spectral assays, 1.6% of copper, 0.3% nickel, and 20 g/t silver have been recorded.

3.3.9.4 Bayan-Uul Au (Ag, Pb, Zn) Occurrence

It is located in Dornog aimag, about 10 km east of gold-bearing Tsav polymetallic deposit. Main component of the deposit is gold, but it also contains silver, lead, and zinc as secondary metals. The ore body is within a metasomatic alteration zone. The main ore minerals are gold, pyrite, sphalerite, galenite, and chalcopyrite. The main ore mineral has gold content of 30–50 g/t in gold-pyrite, 10–30 g/t in chalcopyrite, and 0.4–2 g/t in galena. Moreover, 1–2 mm in size gold grains are found. Quartz-sericite-carbonate alteration zones especially the ones with pyrite contain gold content of 8–20 g/t. Sulfide-bearing 0.1–5 mm quartz veinlets contain 0.05 mm gold grains. Gold content in these veinlets reaches 100 g/t.

3.3.9.5 Bor-Undur Au (Ag, As, Sb) Occurrence

It is located 35 km north of Dashbalbar soum of Dornod aimag. The main ore metal is gold, but it also contains silver, lead, zinc, antimony, and arsenic. The ore body is located in the northeast trending fault in Triassic sedimentary rocks near Jurassic granitic body. 300-m-wide and 1000-m-long alteration zone is in this zone and contains up to 1-m-thick quartz-sulfide brecciated bodies, quartz-sulfide, and sulfide veins and veinlets. Quartz-sulfide veins, veinlets, and brecciated bodies contain pyrite and arsenopyrite. Gold is disseminated and dust-like. In quartz-sulfide vein, veinlets, and breccias, gold content reaches 2–15 g/t, silver 100–550 g/t, lead 1.5–6%, zinc 1.6%, bismuth 0.3%, antimony up to 2%, and arsenic 5.3%. 5–20-m-thick, quartz-sericite-carbonate alteration zones contain gold content of 0.1–0.5 g/t. Altogether, the gold content is 1.0 g/t in average and silver 5–15 g/t.

3.3.9.6 Bayan-Ulz (Au, W, Mo) Occurrence

It is located 12 km southwest from Bayan-Uul soum of Dornod aimag and on the southern side of Ulz River. The ore body was formed in relation to Jurassic granite intrusion that intruded through the upper Permian sedimentary rocks and forms a complex mineral system hosted in quartz stockwork and alteration zone. The ore has gold, tungsten, and molybdenum. Here 0.01–0.1-m-thick quartz veins are developed especially to the north-northeast and east-southeast of the granite intrusion forming a mineralized zone that is 300 m thick and 3000 m long. Ore minerals are pyrite, molybdenite, scheelite, bismuthite, pyrrhotite, rarely chalcopyrite, arsenopyrite, sphalerite, tetrahedrite, bismuth telluride, gold, tungsten, and magnetite. Gangue minerals are composed of quartz, feldspar, sericite, sometimes carbonate, chlorite, albite, and tourmaline. In core and hand samples, tungsten content is measured up to 1.0%, molybdenum 0.48%, gold 7.3 g/t, and silver 6.4 g/t. The host rock of the stockwork itself is silicified and sericitized and contains molybdenum.

3.3.9.7 Delberekh Bulag Au (Ag, Pb) Occurrence

It is located 44 km southwest from Bayan-Uul soum of Dornod aimag and on the southern side of Ulz River. Gold makes up the main ore component in this occurrence as well as silver and iron. The main characteristics of this occurrence are that the ore minerals contain magnetite and hematite. Silicification, kaolinitization, and limonitization are intense in this zone and hosts veins and veinlets containing hematite, quartz-hematite-magnetite, chlorite-hematite-quartz, and quartz-tourmaline-hematite-magnetite forming a stockwork. The mineralized zone occupies 30–300-m-wide and 250–1500-m-long region. It is intensively oxidized at the surface with hematite and limonite. The ore minerals are composed of hematite, limonite, magnetite, galena, cerussite, arsenopyrite, molybdenite, bismuthite, cassiterite, scheelite, pyrite, and gold. Gold content reaches 10 g/t in quartz veins and veinlets, but the overall content of the stockwork becomes 0.01–5.0 g/t.

3.3.9.8 Urliin-Ovoo Occurrence

It is located in the Eastern Mongolia ore district (Table 3.1) in Gurvanzagal soum of Dornod aimag. The main feature of this occurrence is the presence of gold telluride minerals and sulfides in addition to gold. Several other similar occurrences such as Dagai and Kharguit are located not far from this occurrence. Ore body forms a metasomatic alteration zone hosting quartz veins. Gold ore bodies are formed in close relationship with Jurassic micro-syenite, lamprophyre, and diabase dykes in lower Paleozoic granodiorite. Host rock is altered to clay minerals and quartz-sericite-carbonate. The ore minerals are pyrite, chalcopyrite, galena, sphalerite, bornite, tennantite, tetrahedrite, molybdenite, free gold, free silver, gessite, veissite, sylvanite, krennerite, rikkardite, and telluride with iron. Quartz veins are usually 100–200 m long and 0.01–0.25 m thick surrounded by up to 5-m-thick metasomatic alteration zone. Gold content increases in relation to sulfide content increase. Average gold content is 5–8 g/t and sometimes reaches 20–43 g/t. Gold occurs in three variations, as a free grain (1–3 mm in size), as dispersed in sulfide, and as telluride complex. Free gold contains 0.5% bismuth, 0.05% antimony, and 0.01% copper.

3.3.10 Gold Deposits Hosted in Quartz Veins

These deposits form small to large gold deposits composed of one to several quartz veins. Sulfide content is variable in the veins and mainly ranges from 2% to 15%. The gold content is comparatively high in quartz veins that are associated with intrusive rocks. Host rocks are usually hydrothermally altered. These types of deposits are usually mined underground. Gold enrichment is high in this type of deposits. The well-known examples are the Narantolgoi, Tsagaanchuluut, Boroo 7, Sujigtei, Urt, Ereen, and Baavgait deposits of the Boroo-Zuunmod ore district, Ereen deposit of Teshig ore node, and Tsagaan Tsakhir-Uul deposit of the Baidrag ore district (Table 3.1). Element associations are Au (Ag), Au (Ag, Cu), and Au (Ag, As, Sb).

3.3.10.1 Narantolgoi Au (Ag) Deposit

It is located in the Jargalan soum of Tuv province in the Boroo Narantolgoi ore node of Boroo-Zuunmod gold ore district. The main component is gold with minor silver. The deposit is composed of east-west trending, parallel, and vertically dipping quartz veins. The mineralized area is characterized by Mesozoic granodiorite that intrudes through lower Paleozoic Kharaa Formation sedimentary rocks. The main vein is 1800 m long, and its average thickness ranges from 1.05 to 0.4 m from south to north. The main ore body contains 9 g/t gold in average, which totals up to 725,101 tonnes of ore hosting 6.5 tonnes of gold. The less enriched body has a gold content of 8.2 g/t, 177,343 tonnes of ore, and 1.5 tonnes of gold. The ore minerals are arsenopyrite, pyrite, enargite, sphalerite, chalcopyrite, tennantite, galenite, petsite, and pure gold. Total content of sulfide is 5%. Host rock is often weakly altered to quartz-sericite-carbonate. Gold forms isolated grains that are 0.01–0.12 mm in size.

3.3.10.2 Ereen Au (Ag, Cu) Deposit

It is located in the Teshig soum of Bulgan aimag in the Teshig ore district. The main component of the deposit is gold, but it also contains silver and copper. Ore bodies generally trend north and consist of vertical veins of sulfide-quartz hosted in lower Paleozoic granodiorite body (Fig. 3.6).

Seventy-two veins have been identified that make up an area of 27 km^2 with a total length reaching 10 km. The main ore minerals are pure gold, pure silver, pyrite, chalcopyrite, sphalerite, galenite, and molybdenite. Gold content is variable but averages at 0.1–230 g/t. The silver content reaches 61.2 g/t and copper 1.94%. The inferred gold reserve reaches 8500 tonnes with 90 tonnes of gold, 250 tonnes of silver, and 70,000 tonnes of copper.

3.3.10.3 Tsagaan Tsakhir Deposit

It is located in the Baidrag ore node in Bumbugur soum of Bayankhongor aimag. Ore body is hosted by lower Paleozoic granite, granodiorite, upper Paleozoic diorite, and minor dykes (Fig. 3.7). The quartz veins trend to northwest and west and have preserved ancient mining activities. Gold-bearing veins are present in four areas. In the southwestern part of the deposit, veins 1, 2, 3, and 6; to the west of the Ulaantolgoi mountain, veins 7, 8, 52–54, and 43; in the northern part of Ulaantolgoi,

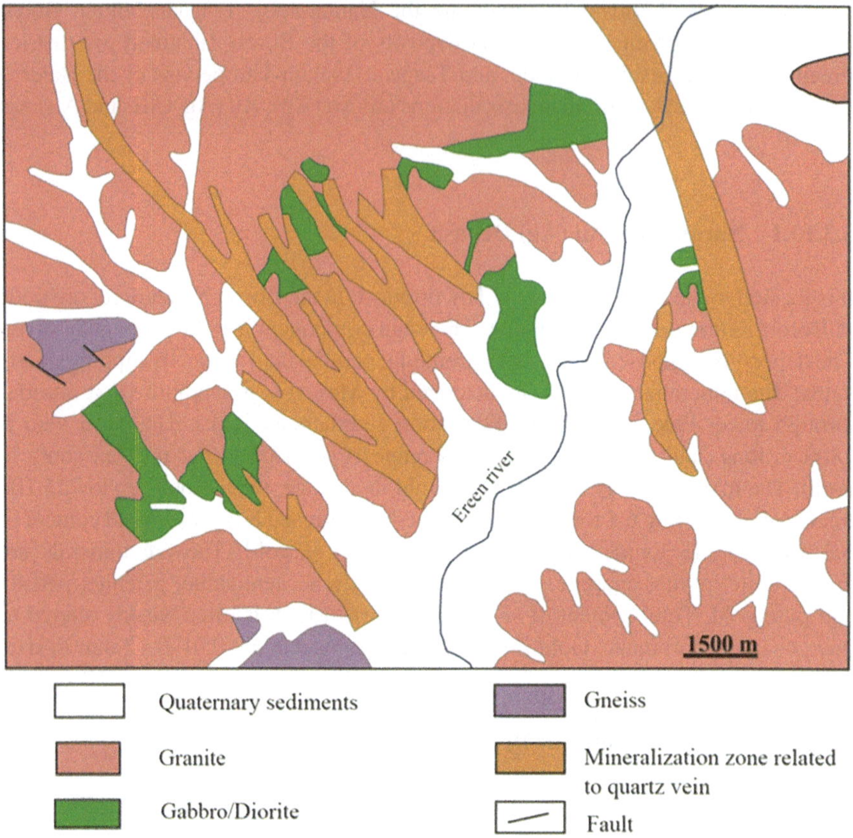

Fig. 3.6 Geology and mineralization of Ereen deposit

veins 9, 10, 14–23, 47–51, and 55; and near the Guchin River, veins 4, 24–29, and 40–42 are located. The ore minerals are pyrite, arsenopyrite, sphalerite, chalcopyrite, tetrahedrite, galena, burnonite, gold, altaite, gessite, and tellurobismuth. Veins 1–3, 6, 9, 10, 14, and 15 are 800–1000 m long, 0.19–0.95-m-thick veins 1, 2, and 6 contain 22 g/t, 15 g/t, and 7 g/t gold, respectively, and 2800-m-long and 0.14–0.53-m-thick vein 10 contain 11 g/t of gold. Veins 3, 9, 14, 15, and 16 contain 1.5–2.5 g/t of gold. The gold content is less than 1 g/t in other veins. Although there is no direct measurement available, silver may be enriched in this deposit. In the veins 1–3, 6, 10, and 16, the reserve is estimated to be 240,233 tonnes with 7.6 tonnes of gold.

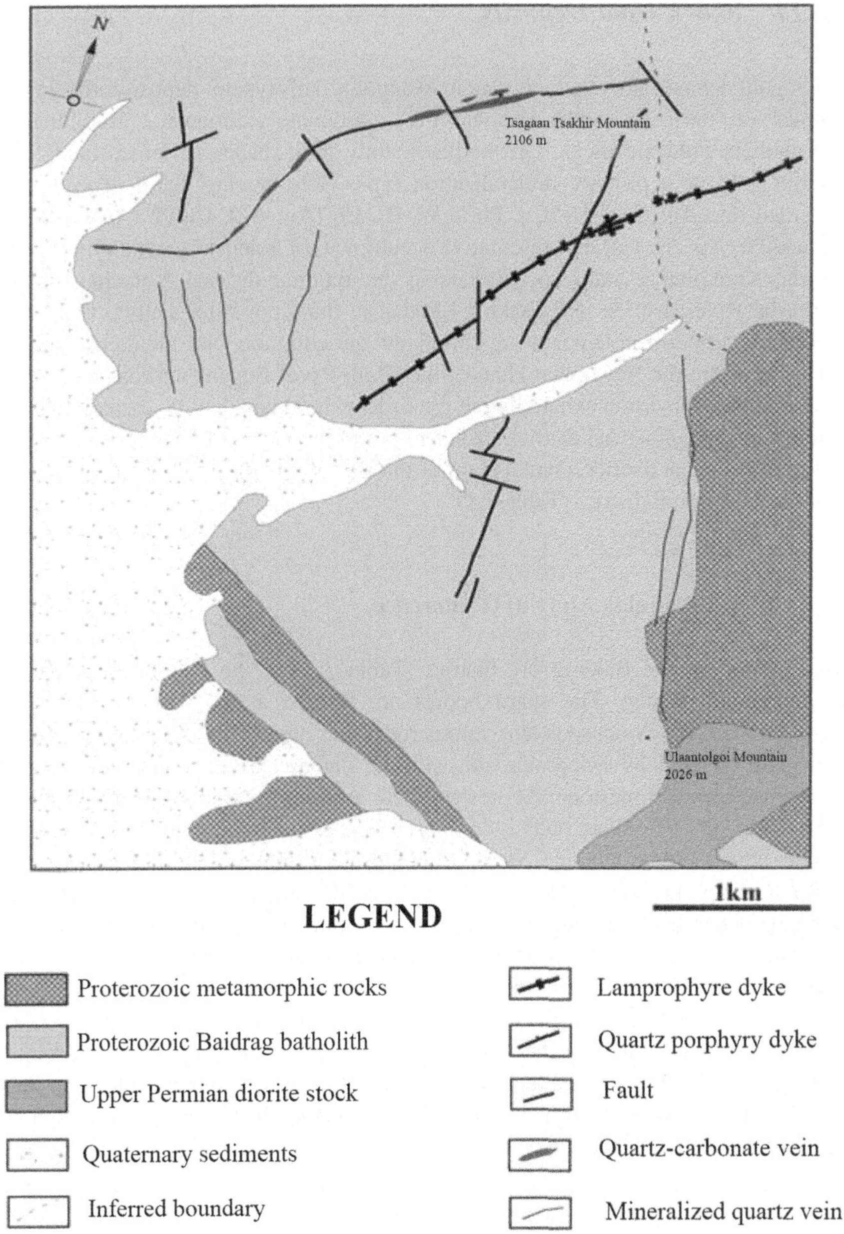

Fig. 3.7 Geologic overview map of the Tsagaan Tsakhir-Uul gold deposit (Jargalan 2002)

3.3.11 *Skarn Gold Deposits*

Skarn gold deposits are relatively rare in Mongolia. This type of deposit forms in the contact of intrusion with carbonaceous, carbonate-sedimentary, and mafic-intermediate volcanic rocks. Ore bodies mainly take shapes of lenses, and they contain gold as a primary mineralization type. Gold-bearing skarn deposits are classified into Au (Cu), Au (Cu, Pb, ±W, B), Cu (Au, Ag), Cu-Pb (Au), and Pb, Zn, and Pb (Au, Ag, Cu). Representative occurrences of gold and secondarily copper are the Khukhbulag valley occurrence in the Baidrag district, Yolochka in the Burgastai River district, and Tsakhir Khudag in the Gobi-Altai district. Gold and secondary copper-iron deposits are comparatively common, and the representative occurrences are the Bayankhairkhan of the Khulj River-Bayankhairkhan district of the Urgamal zone, Erdenekhairkhan of the Erdenekhairkhan district, eastern Tsakhir Khudag of the Gobi-Altai district, Erdenetolgoi of the Duchgol district, Buutsagaan of the Buutsagaan district, Oyut Tolgoi of the Bayangol zone, and Teshig 1, 2, and 3 of the Teshig ore district (Table 3.1).

3.3.11.1 Khukhbulag Au (Cu) Occurrence

It is located in the Baidrag ore district (Table 3.1) of the Bumbugur soum of Bayankhongor aimag. The skarn bodies are formed at the boundary between Neoproterozoic carbonaceous terrigenous rocks and upper Paleozoic diorite-granite body and covered by Neoproterozoic sill-like diabase. Skarn bodies are formed along marmorized limestone. At present, there are four ore bodies identified below the diabase sills. The skarn body is composed of grossular, clinopyroxene, vesuvian, calcite, albite, and hornblende. Skarn is composed of light gray to dark gray quartz, feldspar, scapolite, and epidote. Gold and sulfides are part of the ore. The ore minerals occur as grains and in veins. The ore minerals are chalcopyrite, bornite, cubanite, pyrrhotite, gold, and rarely arsenopyrite, bismuthite, pure bismuth, pyrite, sphalerite, enargite, cassiterite, millerite, and linneite. Gold grains are fine-grained, and 94% of them are less than 0.1 mm in size with the biggest grain reaching 0.5 mm. Oxidized zone near the surface is rich with malachite, covelline, chalcocite, and rarely azurite, goethite, lepidocrocite, and bismuth. Copper content reaches 1.0%, bismuth 0.01%, tin 0.03%, zinc 0.008%, and silver 20 g/t. The gold content ranges from 1 to 20 g/t.

3.3.11.2 Buutsagaan Au (Pb, Cu) Occurrence

It is located 10 km southwest of Buutsagaan soum of Bayankhongor aimag. It was formed at the boundary between Neoproterozoic marmorized limestone, schist, and volcanic rocks cut by upper Paleozoic granite. The skarn bodies are 300 to 600 m long and 0.3–4.5 m thick in an area that is 250 m wide and 750 m long and exhibit a distinct zonal arrangement. These skarn zones are defined as granite-pyroxene-plagioclase, pyroxene-spinel, and pyroxene-spinel-forsterite-bearing skarn. Gold content in sulfide-bearing parts reach 150 g/t, cadmium 300 g/t, copper 2%, zinc 1.5%, and silver 50 g/t.

3.3.11.3 Teshig 1 Au (Pb, Cu) Occurrence

It is located 15 km northwest from Teshig soum of Bulgan aimag at the boundary between the Neoproterozoic-lower Cambrian unit composed of mafic-intermediate volcanics, sandstone, limestone, and carbonaceous-sandstone and lower Mesozoic granite intrusive. Copper-gold-bearing magnetite-garnet-epidote skarn strikes northwest, 1.5 km long, and 25–80 m wide. The ore minerals are malachite, azurite, chalcopyrite, pyrite, bornite, and covellite with the gold being hosted by skarn, magnetite, and limonite-magnetite body. Gold grain size ranges from 0.001 to 0.7 mm and is 0.05–0.2 mm in average. The mineralized zone is divided into three parts by vertical faults. 700-m-long Eastern block contains 0.01–0.9 g/t gold and 0.01–1.0% of copper. Here, 700-m-long and 25-m-thick ore body containing 0.35% of copper is identified. In the magnetite ore, gold content ranges from 0.5 g/t to 4.15 g/t, and in the magnetite-limonite and limonite ores, gold content ranges from 0.95 to 15.35 g/t averaging at 4.1 g/t. The central block is 300 m long and composed of two to six magnetite ore bodies that are 0.3–4.0 m thick with a gold content of 0.3% in average. In these gold-bearing bodies, the gold content is 0.1–1 g/t and copper 0.3%.

3.3.12 Porphyry Gold Deposits

Porphyry gold deposits are common in Mongolia, and the copper-molybdenum Erdenet and Tsagaansuvarga mines are the main representatives. Only some of the subtypes of this type of deposits preserve gold. Gold content is related to the characteristics of the porphyry magmatism, and the general trend is defined by a direct relationship between gold and copper. For example, in the Kharmagtai, Khunguit, and Saikhandulaan districts, gold mineralization follows copper-dominated ores and in some cases become almost equivalent to copper content. On the other hand, in the Erdenet mine, although the mine is copper-molybdenum type, it is gold poor. Whereas in the Tsagaansuvarga and Avdartolgoi deposits, gold contents are up to 0.2 g/t. The first phase of the porphyry intrusive of the Erdenet minc is granodiorite porphyry, and it is enriched in sodium. However, the Tsagaansuvarga intrusive is more alkaline and enriched by potassium, and the Avdartolgoi intrusive is also rich in potassium even though the main phase is granodioritic. Interestingly the Umnugobi copper-gold metallogenic zone porphyry is enriched by sodium, but the first phase is dioritic in composition. Gold-bearing porphyry deposit types are further divided into Au-Cu (Ag), Cu (Au, Ag), and Cu-

Mo (Au, Ag). Kharmagtai, Ovoot Khyar, and Ukhaa Khudag occurrences of the Umnugobi metallogenic zone, Khadat occurrence of the northern Khentii zone, Saran-Uul occurrence of the Baidrag zone, and Biger occurrence of the Nuur zone (Table 3.1) are classified as Au-Cu subtype. In Cu-Mo (Au, Ag) deposits, the representative ratios of Cu, Mo, and Au are equal, and Ag becomes secondary. The main deposits are the Tsagaansuvarga and Oyutolgoi group deposits, and Oyut and Khatavch occurrences of South gobi zone, Avdartolgoi deposit, and Lutaa occurrences of the Duchgol ore district of the Northern Khentii zone are of this type. In gold-bearing copper porphyry, copper is the main component with gold, silver, and molybdenum as secondary components. The Bayan-Uul occurrence of the Delgerkhaan ore district is the main representative. In the boundary region of the porphyry deposits, in which gold is associated with quartz vein and quartz stockwork, a zone of hydrothermal alteration can be formed. The latter mechanism is responsible for forming the Avdartolgoi deposit of the Duchgol district, Kharmagtai, Ukhaa Khudag, Ovoot Khyar, and Khatsar occurrences of the Kharmagtai district, Oyut and Khatavch occurrences of the Bayan-Ovoo district, and the Bayan-Uul porphyry occurrence of the Delgerkhaan ore district (see the Chapter "Copper Deposits" in this book).

3.3.12.1 Kharmagtai Au-Cu (Ag) Occurrence

It is located 60 km southwest of Manlai soum of Umnugovi aimag. The occurrence is hosted in lower Devonian Uguumur Formation clayey – and siliceous rocks, and upper Paleozoic Dush-Ovoo Formation andesite, which are cut by upper Paleozoic diorite porphyry, quartz-diorite subvolcanic bodies, and granodiorite porphyry. Two breccia pipes are identified in the granodiorite porphyry and are heavily mineralized. The breccia clasts are 1–5 cm in size and are composed of andesite, diorite porphyry, and quartz cemented by heavily silicified andesite. The breccia itself is mineralized, and the host granodiorite porphyry contains sulfide-magnetite-tourmaline-quartz veinlets. The host rocks exhibit epidote, chloritic, carbonate, and occasionally argillic and tourmaline alteration. Quartz-tourmaline-hematite veins and brecciated bodies are well represented. Sulfide-magnetite-tourmaline-quartz veinlets surround the breccia pipes and in the northwestern and southern parts have 400 by 300 m and 550 by 220 m stockworks. These stockworks have 0.015–0.3% copper. The southern stockwork is comparatively rich, and the copper content reaches 0.33%. Ore minerals are chalcopyrite, pyrite, chalcocite, bornite, galena, sphalerite, cuprite, malachite, azurite, and gold. The gold content reaches 0.1–1.3 g/t with an average of 0.2 g/t, and silver reaches 0.1–14 g/t. Gold mineralization potential is still poorly known at the time of writing.

3.3.12.2 Avdar Tolgoi Cu-Mo (Au, Ag, W) Deposit

It is located in the Dashbalbar soum of Dornod aimag at the crossing of the Ulz and Duchgol districts. The deposit is hosted in Jurassic granodiorite porphyry and form well-defined breccia pipes. Breccia clasts reach up to 20 cm and are composed of sedimentary rocks, granodiorite porphyry, and clasts of granodiorite and syenite porphyries derived from nearby dykes. The breccia cement is composed of fine grains of adularia, albite, quartz, carbonates, sericite, pyrite, chalcopyrite, molybdenite, and scheelite. Copper-molybdenum and scheelite mineralization are present in both the breccia pipes and host rocks. In the breccia, molybdenum content reaches 0.03–0.661%, copper 0.1–1.0%, and tungsten 0.01–0.015%. The main ore minerals are the molybdenite, chalcopyrite, pyrite, scheelite, bismuthite, pure arsenic, vulangerite, galena, and sphalerite. The gold content reaches 1 g/t and silver 10–40 g/t.

3.3.13 Copper and Iron Massive and Disseminated Sulfide Deposits

Representative deposits are mainly found in the Nuur metallogenic province (Nuur and Urgamal metallogenic zones) and rarely in Mongol-Altai (Mongol-Altai metallogenic zone) and South Mongolia (Bayangobi-Bayanlig, Gobi-Altai, Edren Range) metallogenic provinces (Table 3.1). The ore bodies form conformable bodies as lenses within seafloor-related volcanic basalt, andesite, and their tuff. Based on the major component, it is divided into copper, copper-zinc, and lead-zinc massive and disseminated sulfide deposits. The main representatives are the Borts-Uul deposit located in the south of the Khan-Khukhii range of the Nuur metallogenic province and the Bayantsagaan occurrence of the Bayangobi-Bayanlig zone.

3.3.13.1 Borts-Uul Deposit

It is located to the 25 km southwest from the Tsagaan Khairkhan soum of Uvs aimag. Although the deposit is not large, both massive and disseminated sulfides are developed. The deposit is hosted in lower Cambrian basalt-andesite-rhyolites. The gold mineralization is developed in three areas. In the north, the mineralization follows fractures within andesite, basalt, lava-breccia, tuff, and tuffite. The host rock is chloritizated, epidotizated, and carbonitizated. There are total of three ore bodies that are conformable to the host rock. The ore body thicknesses reach 17 m with an average length of 1.4 km. The ore minerals are chalcopyrite, bornite, and pyrite and locally hematite and magnetite. Copper content varies dramatically with a highest value of 4%. Silver content is 60 g/t, and gold is 0.4 g/t. 1.7 km to the south from the northern part defines the center of the deposit and is occupied in andesite,

andesite-basalt porphyry. There are four mineralized zones identified with a length of 0.5 km. Thickness ranges from 2.0 to 15.0 m. Here the copper content averages at 1.3% and silver 5.0 g/t. The ore is composed of chalcopyrite, bornite, chalcocite, covellite, and pyrite. The host rock is chloritized and epidotizated. The third part of the deposit is dominated by pyrite, located 1.5 km east from the northern region, and composed of dacite porphyry and associated tuffaceous rocks. Here, a 100-m-wide and 200–250-m long zone is heavily limonitizated and rich with pyrite. The copper content reaches 0.1%, gold 0.4 g/t, and silver 5.9 g/t. Resource is estimated in the northern part only to a depth of 100 m. Ore body is estimated to be 28,200 tonnes with an average gold content of 1.0–1.5%.

3.3.13.2 Bayantsagaan Occurrence

It is located 30 km northeast from the Bayan-Undur soum of Bayankhongor aimag in the Bayangovi-Bayanlig metallogenic zone. The occurrence is hosted by lower Devonian metamorphosed intermediate and mafic volcanic rocks. Two ore bodies are identified; each continues for 1 km in length. The lower body contains 1.5–2% copper and 0.03–1.0 g/t gold. The upper ore body is hosted in quartzite and hematite quartzite bed and contains 0.05–0.3% copper, 0.1–1.0% zinc, and up to 1.0 g/t gold and 0.5% lead. 1 km to the south from this occurrence hosts 200-m wide and 1-km-long zone that hosts chalcopyrite nodules, malachite, and azurite. Here the copper content reaches 1% and above.

3.4 Concluding Remarks

Within the seven gold-bearing metallogenic belts, the Mongol-Altai, Nuur, Northern Mongolia, Central Mongolia, Khangai-Khentii, Eastern Mongolia, and Southern Mongolia, 25 zones, and 70 districts, gold deposits and occurrences hosted in metamorphosed conglomerate, Cenozoic unconsolidated sediments as placer gold deposits, volcanic rocks, quartz-carbonate veins, quartz veins/stockwork, intrusive rocks-skarn/porphyry, and other sedimentary rocks. Gold deposits associated with metamorphosed conglomerates are commonly distributed in Khangai-Khentii province in Carboniferous, Permian, Jurassic, and Cretaceous strata. Volcanic rocks in the Burgastai River and Zavkhanmandal districts often host massive sulfide, epithermal, and copper-iron massive and disseminated sulfide deposits. Various types of quartz veins/stockwork that are affected by regional metamorphism become targets for gold exploration work. Moreover, hydrothermal-metasomatic, skarn, and porphyry-type gold mineralization favor intrusive rocks and are distributed commonly within Mongolia. However, gold mineralization in other sedimentary rocks such as sandstone and shale as well as detrital-enriched deposits formed in association with volcanic rocks need to be studied in detail.

References

Altankhuyag D (2005) Bayan-Airagiin tsul sulphidiin huderjiltiin garal uusel (Origin of Bayan-Airag massive sulfide mineralization). Mongolian University of Science and Technology, Dissertation, Ulaanbaatar

Blagonravov BA, Shabalovskii AE (1977) Zoloto (Gold). In: Marinov NA, Khasin RA, Khurts Ch (eds) Geology of the Mongolian Peoples' Republic, Mineral Resources, vol III. Nauka, Moscow, pp 217–263

Cluer J, Kotlyar B, Gantsetseg O, Togtokh D, Wood G, Ullrich T (2005) Geology of the Boroo gold deposit, northern Mongolia. SEG-IAGOD Guidbook Series 11:105–117

Dejidmaa G (1985) Geochemicheskye osobennosti zolotorudnogo polya Boroo v Mongolii (Geochemical features of the Boroo gold field in Mongolia). Dissertation, Novosibirsk

Dejidmaa G (1996) Gold metallogeny of Mongolia. Mong Geosci 1:6–29

Eriksson KA, Turner BR, Vos RG (1981) Evidence of tidal processes from the lower part of the Witwatersrand Supergroup, South Africa. Sediment Geol 29(4):309–325. https://doi.org/10.1016/0037-0738(81)90078-6

Jargalan S (2002) Petrogenesis of plutonic rocks of Tsagaan Tsahir Uul area, Mongolia: Implications to tectonic evolution. Dissertation, Tohoku University, Japan

Jargaljav G (2009) Ores and metasomatites of the gold-copper deposit Central Oyu (South Mongolia). Dissertation, Irkutsk

JICA (1992) Report on the mineral exploration in the Uudam Tal area, Mongolian People's Republic (phase 1). Tokyo, Japanese International Cooperation Agency, 124

Khishgee C, Akasaka M (2015) Mineralogy of the Boroo Gold deposit in the North Khentei Gold Belt, Central Northern Mongolia. Resour Geol 65(4):311–327. https://doi.org/10.1111/rge.12073

Kirwin D, Forster C, Kavalieris I, Crane D, Orssich C, Panther C, Garamjav D, Munkhbat T, Niislelkhuu G (2005) The Oyu Tolgoi copper-gold porphyry deposits, south Gobi, Mongolia. Geodynamics and metallogeny of Mongolia with a special emphasis on copper and gold deposits SEG-IAGOD field trip, 14–16

Chapter 4
Placer Gold Deposits

Tankhain Semeihan and Uyanga Bold

4.1 Geological Setting of Placer Gold Deposits in Mongolia

The placer deposits refer to concentrations of minerals that are denser (>3.0 g/cm^3) and more resistant to chemical and physical destruction than sediments normally transported and deposited. These dense and resistant minerals are called "heavy minerals" which, apart from gold (the subject matter of this chapter), also include ilmenite, rutile, monazite, zircon, diamond, sapphire, ruby, cassiterite, magnetite, chromite, garnets, etc. (Taylor and Eggleton 2001).

In Mongolia, placer gold deposits are comparatively common due to an abundance of primary gold mineral systems. In terms of primary gold mineralization, Precambrian to late Mesozoic geotectonic events play an important role. Occurrences of Mongolian placer gold deposits correlate well with regions and ore junctions with high gold mineralization potential (Marinov et al. 1977). In order to explore placer gold deposits, detailed geological studies on formation types of primary gold mineralization, three-dimensional characteristics of the primary ore, size of gold nuggets, their ease of potential to be released, geomorphological features of each given region, and topographic features are crucial. Main primary ore types that supply gold are gold-quartz, gold-quartz sulfide, gold-skarn, and gold-silver. Moreover, Carboniferous, upper Permian, upper Jurassic, and lower to upper Cretaceous conglomeratic successions act as an intermediate source.

Placer gold deposits are divided into four main types: (1) eluvial, (2) deluvial, (3) proluvial, and (4) alluvial, with the most common being the proluvial and alluvial placers. Upper Cretaceous sedimentary successions are of economic importance, whereas Miocene-early Pliocene and Quaternary successions have the potential to

T. Semeihan · U. Bold (✉)
Geologic Information Center, Mineral Resources and Petroleum Authority, Ulaanbaatar, Mongolia
e-mail: uyangabold@must.edu.mn

O. Gerel et al. (eds.), *Mineral Resources of Mongolia*, Modern Approaches in Solid Earth Sciences 19, https://doi.org/10.1007/978-981-15-5943-3_4

become economically viable. The majority of the placer gold deposits of Mongolia corresponds to late Cenozoic tectonic events and is often located in mountainous regions of the Mongol-Altai, Gobi-Altai, Khangai-Khentii, Khuvsgul, and Tenger mountains. Based on regional mineral exploration work carried out in the entire territory of Mongolia, regions with primary gold mineralization are divided into Mongol-Altai, Kharkhiraa, Khuvsgul, Zed, Bayankhongor, North Khentii, Khangai, Uvur Altai, southern Khentii, northern Kherlen, and Umnugobi zones (Dejidmaa 1996), which overlap well with the gold placers of Mongolia.

4.2 Origin and Shape Characteristics of Placer Gold Deposits

The majority of the economically important gold placers are located in northwestern Khentii and southern Khangai ranges. The placer gold deposits are classified into eight categories, based on their origin and geomorphology (Semeihan and Altantsetseg 1994).

4.3 Main Types of Placer Gold Mineralization

4.3.1 Eluvial-Weathering Zone Type

Eluvial and weathering zone placer gold deposits are common in districts with known primary gold mineralization. Although this type of deposits is rare in Mongolia, due to their formation near their source, these gold placers often have the greatest industrial potential. Our study suggests that physical weathering is more important for the modern eluvial placers to form whenever chemical weathering becomes crucial in Miocene-Pliocene placers.

The main product of weathering zone environment is usually very clayey and is commonly found in remnant surfaces at watershed regions near Jargalant, Baga-Ulunt, Tolgoit, and Bugantai Rivers in the Yeroo River basin. The strata are red in color and 5 m thick in general. Gold mineralization has been measured at 20–50 mg/m^3 in the lower part of these red beds with gold nuggets between 0.1 and 0.4 mm in size and is often silicified. These red beds are Neogene (Devyatkin et al. 1987) in age and cover tens of km^2 of area in the gold mineralized districts. The best example is the red beds in the Ikh Dashir placer in Boroo district where the gold content reaches 4.8 g/m^3.

Present-day and Quaternary eluvial gold placer mineralization that is related to weathering process is identified as granular, fragmented, and clayey strata that are present in modern watershed regions. Thickness of the related strata reaches 10-m maximum. Oxidation (chemical weathering) is common in primary gold deposits in wet and mountainous regions, as exemplified in the Shaazgait, Naran Tolgoi, and Boroo gold deposits. In these deposits, oxidation reached 20 m of depth from the surface, and the mineral assemblage is characterized by malachite, azurite, and limonite related to primary sulfide composition of the ore body. Moreover, enlargement of pyrite hosted accessory gold was documented, which is related to recrystallization in oxidized environment and becomes round and droplet shaped.

4.3.2 Deluvial and Deluvial-Alluvial Placers

As mountainous regions occupy most of the territory of Mongolia, mountain slopes become the major accumulation surfaces for placer gold deposits. Mongol-Altai, Khangai, Khentii, and Khuvsgul ranges are high mountains and are often wet throughout most of the year, which correspond to active development of river system where heavy fractions of weathered material find their way to travel down slope. However, in Gobi-Altai and southern part of Mongolia where the weather is generally dry and hot, deluvial deposits become more common, and wind-born sediment transport dominates the placer gold accumulation process.

Deluvial and deluvial-alluvial placer deposits with high Au content are located in northwest Khentii, Zaamar, Boroo, and Zuunmod districts and are all close to the source deposits. For example, near the Boroo primary gold deposit, in the mouth of Ikh Dashir valley, gold nuggets as large as 0.52 mm have been recovered where Au content of the host sediment reached 0.1–1 g/m^3. Here, gold was found along with cinnabar, scheelite, barite, cirtolite, xenotime, orangite, monazite, ilmenite, rutile, and anatase, reflecting the mineral composition of the primary ore. The second most representative placer gold deposit is Dutluur-Am, located near the western bank of the Tuul River about 600 m long, in a 40–50 m wide gorge. The gold-bearing sedimentary rock is a Jurassic conglomeratic unit with an average gold content of 0.1–0.5 g/m^3 (Fig. 4.1).

4.3.3 Proluvial Placers

Proluvial placer gold deposits are formed in the mountain foothills, where there is no constant water flow, margins of wide valleys, and alluvial fans. Steepness varies along the margins and in some cases, smaller river valleys develop in the center. Sediments are not sorted and are composed of clay, sand, gravel, irregular rock fragments, lenses, and beds of tens of meter thick sedimentary successions. Proluvial placer gold deposits are not widely recognized in Mongolia with a few occurrences in small river banks and alluvial cones; the corresponding placers deposits are small in scale.

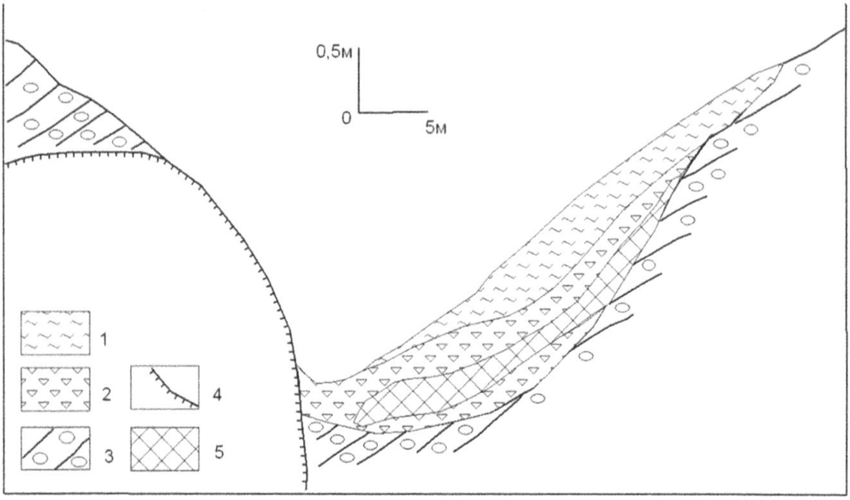

Fig. 4.1 Cross section of Dutluur-Am deluvial placer gold deposit. 1 clay, 2 silts and clay, 3 upper Jurassic conglomerate, 4 lower Paleozoic shale, 5 gold-bearing unit of economic significance

Currently identified proluvial gold placers are the Nariin Khundii, Nergui, and Ikh Dashir deposits in the Zuunmod gold district, Ar Khar Chuluut (Fig. 4.2), the Bukht placers in the Bayankhongor gold district, and Zaisan Salaa in the Burgastai gold district.

Ar Khar Chuluut valleys is as wide as 20–100 m and as long as 2.5 km with a Y-shaped topography characterized by a slope that steepens by 100 m in 1 km. Gold-bearing sediments were accumulated due to temporary flooding intervals. Modern and Quaternary sediments are the main hosts for the gold nuggets that are 1–3 mm in size, poorly rounded, vuggy, and light yellow in color. The heavy minerals are galena, cinnabar, scheelite, cerussite, chromite, ilmenite, rutile, pyrite, magnetite, and martite. The Ar Khar Chuluut deposit was first found in 1930 in a joint Russian-Mongolian exploration work, which led to mining activities that continued from 1963 until 1964.

4.3.4 Alluvial Placers

Alluvial gold placers are the most common and are generally rich in Au content. A typical deposit is represented by the Tuul alluvial placer gold, which is located in the Tuul River valley (Fig. 4.3), to the west of the Zaamar range in the Khentii zone and formed along a deep-seated Bayangol fault. The fault itself continues east for many kilometers hosting primary gold mineralization.

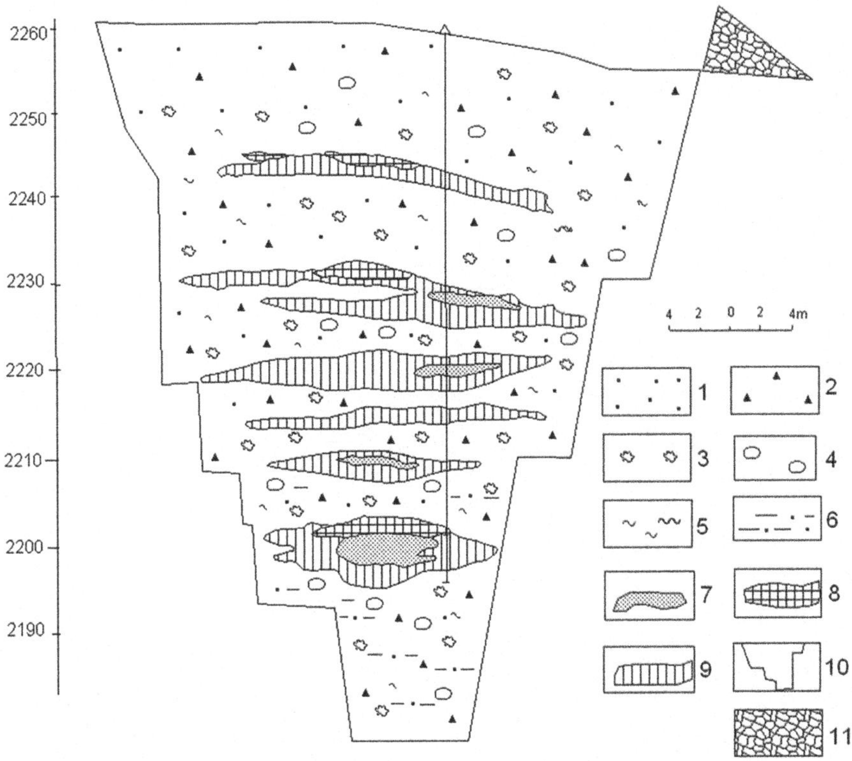

Fig. 4.2 Cross section of the Ar Khar Chuluut placer gold deposit. Sedimentary lithology: 1 sand, 2 grains, 3 angular rock fragments, 4 gravel, 5 clay, 6 sandstone. Au content: 7, 2 g/m^3 and up; 8, 1.0–2.0 g/m^3; 9, 1.0 g/m^3 and lower; 10, open pit wall; 11, ore dump

Gold-bearing sediments are of alluvial in origin and composed mainly of variously colored gravels with 20–35 cm large boulders and can be as thick as 13 m. The basement is a lower Cretaceous well-cemented conglomerate. The most economic gold accumulation is found in the mid-depth of the entire sediment package that are 9.9 km in length, 665 m in width, and 7.3 m in depth. Average Au content is 0.5 g/m^3 but not consistent throughout the deposit.

Gold nuggets of the Tuul placer are small to medium in size, from 0.62 to 1.7 mm. During exploration work, 14 gold nuggets (20 × 15 mm) that are more than 0.5 g each were acquired, and during mining, 700 g and 1.6 kg gold nuggets were found. Gold nuggets are often platy, thread-like, occasionally round or dendritic in shape. The nuggets are not fully rounded and have vugs that are filled with silica. Some of the nuggets are coated with iron oxides and the colors vary from yellow to green yellow.

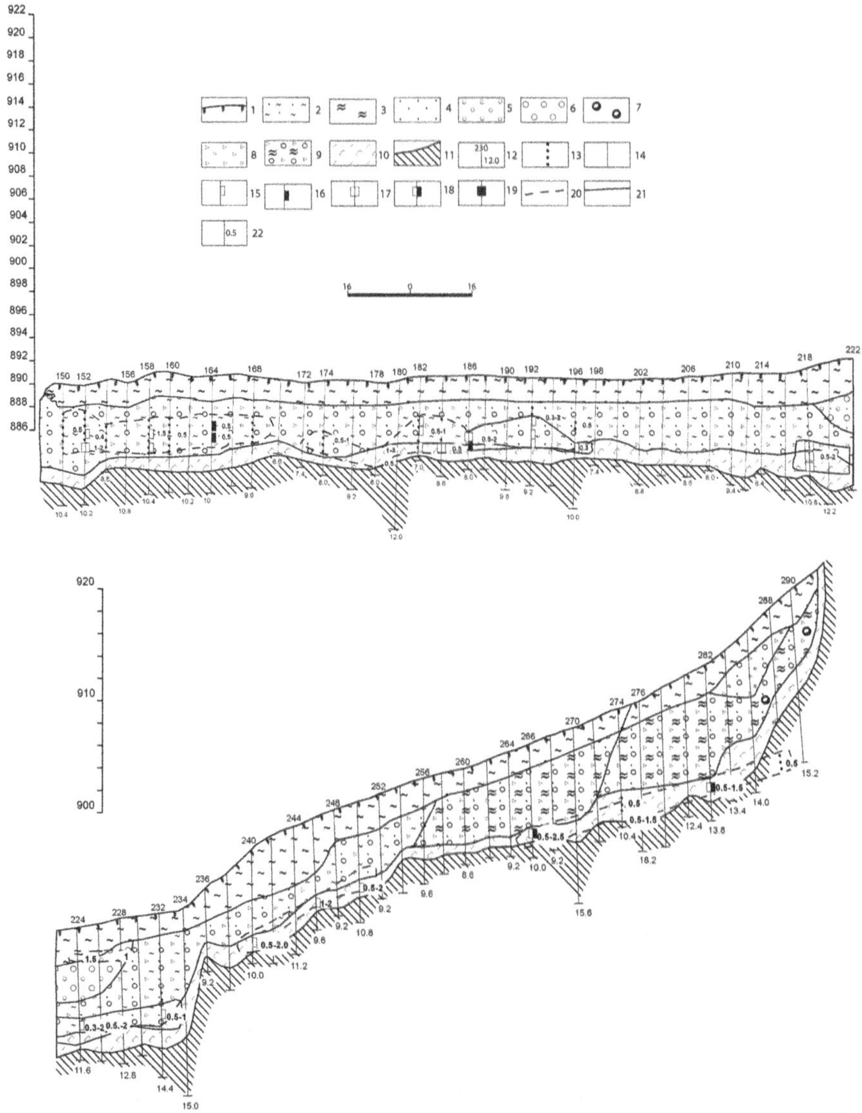

Fig. 4.3 Cross section of the Tuul River valley. Sediment composition: 1 soil, 2 gray clay and sand, 3 crimson clay, 4 variously sized sand, 5 sand-pebble. 6 cobble, 7 boulder, 8 fragmented boulder, 9 crimson clayey sand-pebble, 10 angular cobble-eluvial, 11 various shale, 12 borehole number, depth (m), Au content (g/m^3), 13 < 0.1, 14 0.1–0.9, 15 0.2–0.499, 16 0.5–0.999, 17 1.0–4.99, 18 5–10, 19 >10. Gold-bearing sediment boundary: 20 > 0.2 g/m^3, 21 < 0.2 g/m^3, 22 gold nodule size (mm) written on the western hand side of borehole lines

4.4 Other Types of Placer Gold Deposits

4.4.1 Terrace Placers

Terrace placer gold deposits are mostly located in Yeroo, Tuul, Baidrag, Onon, and Burgastai River valleys. The Yeroo terrace gold placer was first mined in 1900, and the leftover dumps are found in Ulaantolgoi and Khunguchi River tributaries. The terrace is 20 m thick, and the sediments are characterized by gray-colored sand-gravel with interbeds of boulder-bearing sediments near the top. Size fractions decrease down section and become sandy and clayey. Here the gold content varies from 0.3 to 2.9 g/m^3, and the richest horizon is immediately above the basement in fine-grained sediments. Heavy fractions are composed of iron oxides and pyrite. Gold-bearing sediment thickness varies from 0.5 to 1.0 m.

The richest terrace deposit is also found in the Tuul River valley (Fig. 4.4) in the western side of the Zaamar range. Here the basement and overlying terrace sediments are discussed in five terrace treads (*) that are studied in Khailaast and Toson River branches in the Tuul River.

Terrace Tread I. It is 7.4 km long and 510 m wide, and the total thickness of the sediments is 12 m. Basement consists of metamorphosed shale and the overlying sediments that are composed of well-rounded cobbles and pebbles with rare boulders (3–5%). Gold-bearing sediments are at the bottom comprising lens-shaped bodies that continue for 6.8 km in extent. Gold content varies from 0.3 to 16.6 g/m^3 with an average of 2.6 g/m^3. Soil thickness is 13.5 m, and the gold-bearing sediment is 1.4 m.

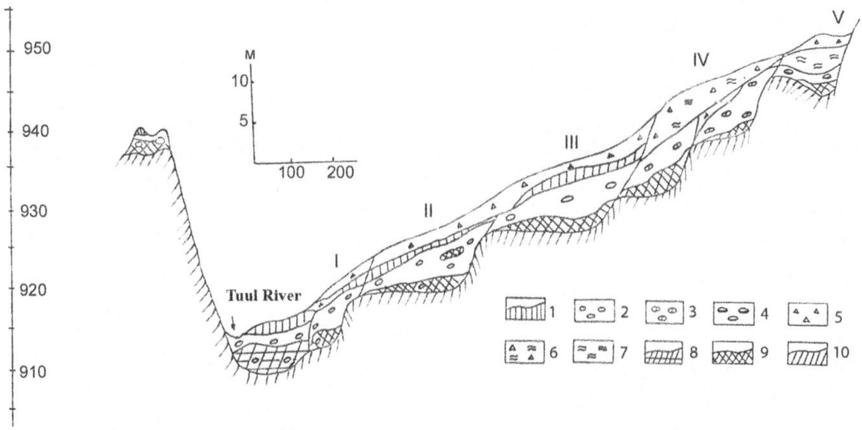

Fig. 4.4 Five treads of the Tuul River terrace. 1 Alluvial sediments. 2 Gray-colored channel alluvial sediments. 3 Brown-colored channel alluvial sediments. 4 Yellow-colored channel alluvial sediments. 5 Deluvial sediments. 6 Clayey sediments related to landslide. 7 Crimson clay. 8 Gold-bearing sediments of Holm sediments. 9 Gold-bearing sediments of the terrace. 10 Metamorphosed basement shale. т-Holm. I–V terrace tread numbers. A terrace consists of a flat or gently sloping geomorphic surface, called a tread

Terrace Tread II. It continues for 7 km, and seven separate bodies are identified. Gold content varies both vertically and horizontally from 0.1 to 39 g/m^3. Soil thickness ranges from 4.7 to 22.8 m with 0.8–1.8 m thick gold-bearing sediments that comprise gravels and cobbles.

Terrace Tread III. Two large separate bodies were identified along the valley and are 600 m and 400 m long, respectively. Gold-bearing sediments consist of gravels and pebbles with rare boulders. The gravels are cemented by brown clay and silt. Gold content ranges from 0.7 to 37.3 g/m^3 in gold-bearing sediments that are as thick as 1.1–1.8 m.

Terrace Tread IV. Lithology is the same as that of Tread III. Gold-bearing strata are 80–160 m wide and continue for 170–630 m along strike. Gold content ranges from 0.4 to 4.5 g/m^3 with an average of 3.3 g/m^3. The total sediment thickness varies from 4.4 to 22.4 m, and the gold-bearing sediment reaches 1.6 m in thickness.

Terrace Tread V. It is 180 m long and 20 m wide, and the gold content is 1.37 g/m^3 in average. Total sediments are 15 m thick, and the gold-bearing sediment is 1.4 m thick. This step is present on both sides of the Tuul River valley, exposure poor on the west and uplifted on the left where the basement is exposed in some places. Alluvial sediments are yellow in color and composed primarily of clay and small gravels.

4.4.2 Glacier and Water-Glacier Type

Glacier-related placer gold deposits have not been found in Mongolia yet; however, Khangai and Khentii mountain ranges have experienced a long-lasting glaciation. Reconnaissance geologic work has been done in these regions and several placer gold mineralization points in glacial moraines of the Khangai Range have been found.

Glacier, water-glacier-type placer gold deposits are defined in Russia relating to a mechanism where glacially derived water transports gold mineralization downstream, which are often overlain by glacial deposits related to deglaciation. Hence, there is a potential to find placer gold deposits in the regions in Mongolia where glaciation took place in its geologic history.

4.4.3 Ancient Placers

Historic placer gold deposits are found in ancient gold mines. Gold deposits mined in the twentieth century such as Tolgoit, Sangiin, Ikh-Ulunt, Gozon Shar, Niilkh deposits in the Khoit Khentii zone, Ikh, Baga Adjir, Bukhlei River valley, Mogoi River, and Kharganat deposits near Ikh Alt region can be included in this type. The gold-bearing sediments are the remnant sediments from mining activities and are often washed with low Au content. The highest Au content of 0.2 g/m^3 was documented in the waste dumps of the Buural and Tolgoit deposits and was mined again.

This type of deposit is common in Bayankhongor ore district where the locals had historically mined gold. Many ancient mining-related diggings are found in Baidrag valley, Ulziit River terrace, Khukh bulag valley, Bumbat, Mukhar Ereg, and Uvur Chuluut regions. The Altan-Uul gorge of the Nemegt Range also preserved evidence of ancient gold mining activities as in the Sarga River valley of Khuvsgul. Since 1990, mining of placer gold deposit has been active and has resulted in opportunities to rewash remnants of Khuder, Yeroo, and Zaamar district gold deposits. As modern technology advances and becomes capable of washing placer gold nuggets that are <0.1 mm, many of our remnants from gold mining may become a target for future exploration of gold placers. Many studies suggest that ancient gold deposits may preserve 30–35% of the total gold resources of the parent mine.

4.4.4 Gold-Bearing Conglomerate

Many conglomerate sequences become a primary source for placer gold mineralization in many countries in the world including South America, South Africa, and Canada. A typical example are the Archean conglomerates of the Witwatersrand gold deposit in South Africa, which has produced 50,000 tonnes of gold in the last 130 years (Frimmel et al. 2005). In Mongolia, although many conglomerate successions are present, not much work has been done to assess their gold mineralization potential.

Lower Carboniferous gold-bearing conglomerate: These deposits are present in the river systems of the Zaamar ore district. Conglomerate cement is composed of chlorite-sericite, calcite, and sericite-silica and is often metamorphosed and weathered. There are three gold occurrences identified with gold content of 0.2 g/t. Gold nuggets are 0.1–0.2 mm in size.

Upper Carboniferous-Lower Permian gold-bearing conglomerate: Eastern side of Toson River and Sharkhub valley of Zaamar district preserve a gold-bearing conglomerate. Conglomerate is enriched with Fe showing epidote alteration. Only 1–2 gold nuggets were recovered from exploration work.

Lower-middle Jurassic gold-bearing conglomerate: Toson River and eastern bank of Tuul River preserve Lower-Middle Jurassic conglomerates. The gravels are well-rounded and are rarely composed of boulders (Devyatkin et al. 1990). Gold was recovered at the base of the conglomerate bed, and up to 20 gold nuggets that are 0.2–0.3 mm in size were found. The surface of the gold nuggets is irregular and platy. Au content was determined in this quartz cobble of the conglomerate at 8.6–13.6 g/t.

Upper Jurassic-Lower Cretaceous gold-bearing conglomerate: This conglomerate is abundant in the western bank of the Tuul River. Every bed of the sedimentary strata yielded 30–50 gold nuggets, and the Au content is up to 0.2 g/t.

Lower Cretaceous gold-bearing conglomerate: It is present in the Yeroo and Shariin River valleys. Using borehole sampling, 1.7 g/m^3 gold content was obtained in the Sangiin River deposit. These conglomerates act as intermediate source for the

Bugant River group deposits (Tolgoit, Sangiin, and Namagtai). Lower Cretaceous conglomerates are widely distributed in the Shariin River system and are the source for Tsamkhag, Ulunt, Khuiten, and Mogoi River deposits. In the Sariin River system, the gold-bearing conglomerates are mostly composed of boulders and angular cobbles with cements that are sandy and silty.

Upper Cretaceous gold-bearing conglomerate: Upper Cretaceous gold-bearing conglomerates were first identified in the Altan-Uul district in Umnugobi province by Devyatkin et al. (1987). Locals have historically mined these conglomerates. Example occurrences are the Tamgat and Zost gorges.

4.5 Distribution of Placer Deposits in Geomorphological Zones in Mongolia

Current geomorphological zones of Mongolia are defined after Mesozoic features that are enriched by Cenozoic tectonic reactivation (Devyatkin 1970). The first comprehensive geomorphological study of Mongolia was done by Soviet scientists in 1972. However, N.A. Florensov, and S.S. Korzhuyev, published the first geomorphological map of Mongolia at the scale of 1:1,500,000 as part of a book.

In the geomorphological map of Mongolia (Fig. 4.5), specific conditions for the placer gold deposits to form and the respective genetic types are shown. Moreover, regions with potential placer gold mineralization in relation to both endogenic and exogenic processes of neotectonic events are shown. In the map, four gold mineralization zones are recognized; they are (1) Khangai Khentii, (2) Altai, (3) Gobi, Southern Mongolia, and (4) Eastern Tenger mountain.

Most economically significant placer gold deposits are found in mountainous regions of Altai, Khangai, and Khentii ranges of Mongolia, which are mostly related to alluvial accumulation (Fig. 4.6). Surrounding these mountain ranges, proluvial-alluvial placer gold deposits are formed, which change to more proluvial type in dry and hot regions. There is a potential to find glacial-type placer deposits in Quaternary sediments of Mongol-Altai, Khangai, and Khentii ranges where recent glaciation took place. Eluvial and eluvial placer gold deposits are mostly formed in a landscape characterized by low hills.

4.5.1 *Mongol-Altai Gold Mineralization Zone*

Mongol-Altai gold mineralization zone occupies the northwestern part of the Mongol-Altai range and continues 450 km to the southeast with 30–60 km width. Many gold mineralization sites have been identified especially in the Upper Bulgan and Khurimt-Sagsai districts.

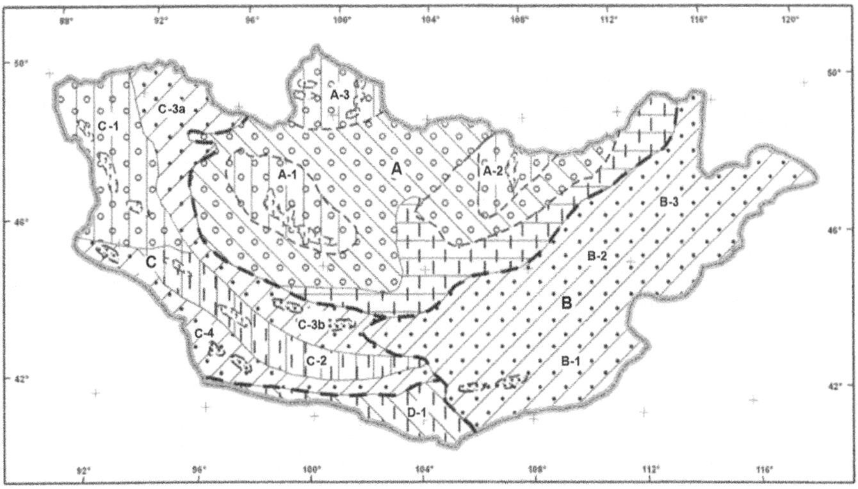

Fig. 4.5 Geomorphological map of Mongolia. *Geomorphological subdivision*. A- Khangai Khentii zone. Subzones: A-1, Khangai central uplift; A-2, Khentii central uplift; A-3, Khuvsgul rift related uplift. B- Gobi, southern Mongolia zone. Subzones: B-1, Gobi; B-2, Middle Khalkh low hills; B-3, Southern Mongolia plateau. C- Altai zone. Subzones: C-1, Mongol-Altai central uplift; C-2, Gobi-Altai linear-central uplift; C-3, Lake zone (a, Ikh Nuur; б, Lake depression); C-4, Inner Gobi basin of Altai. D- Eastern Tenger-Uul zone. D-1, Gobi Tenger-Uul uplift

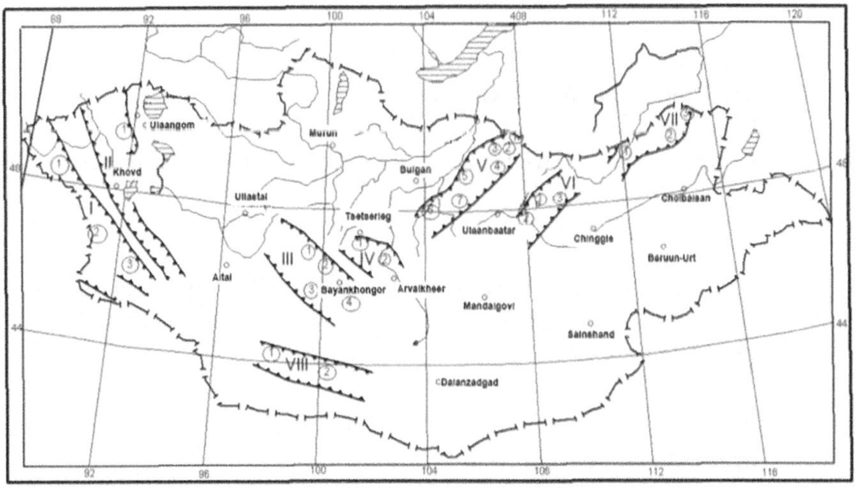

Fig. 4.6 Location map of placer gold deposits and occurrences of Mongolia. Primary and placer gold mineralization zones: I-Mongol-Altai (1 Khurimt Sagsai, 2 Upper Bulgan, 3 Bodonch – Uyench); II-Kharkhiraa (1 Burgastai group); III-Bayankhongor (1 Baidrag, 2 Jargalant, 3 Burd River, 4 Tuin River); IV-Ulziit-Teel-Ult (1 Kharguit, 2 Ult); V-North Khentii (1 Khuder River, 2 Bugant, 3 Bukhlei, 4 Yeroo River, 5 Tseel, 6 Zaamar, 7 Boroo-Zuun mod); VI-South Khentii (1 Terelj, 2 Baruun-Urt, 3 Ulz River); VII-VIII-Nemegt-Sevrei (1 Edren Range, 2 Altan-Uul)

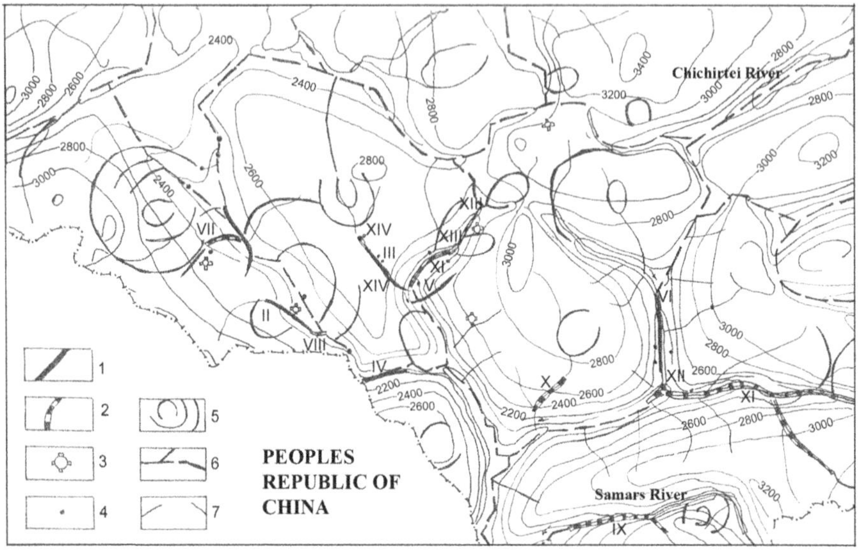

Fig. 4.7 Location map of the Khurimt group deposits /Scale 1: 200,000/. 1 Placer gold deposits: I-Nergui, II-Khuiten River, III-Pochkin River branch, IV-Songint River, V-Altan River branch, VI-Yult River. 2 Placer gold occurrence with economical significance VII-Nergui, Baleika, VIII-Lower valley of Khuiten River, IX-Samarsan gorge, X-Toitogt River, XI-Khurimt River, XII-Lower and Upper parts of Yolt River, XIII-Lower and upper parts of Yamaat River, XIV-Lower part of Pochkin channer, XV-Upper part of Songint River. 3 Primary gold occurrences. 4 Gold nodules recovered during exploration. 5 Central and circular structure. 6 Faults and fractures bounding morpho-blocks. 7 Tectonomorpho-isoline

4.5.1.1 Khurimt Group Placer Deposits

Khurimt group placer deposits (Fig. 4.7) occur in the Khar-Erchis River. The Yamaat, Sangiin River, Pochkin, Khuiten, and Altan Salaa placers are widely known due to active mining carried out by the Chinese explorers since 1943. Length of a typical deposit ranges from 3.5 to 5.0 km with a total sediment thickness reaching 15 m and are often found along the first tread of a terrace. Recovered gold nodules reach 0.3 mm in size, and the heavy mineral assemblage is scheelite, cassiterite, bismuth, and barite.

4.5.1.2 Upper Bulgan Group Gold Placers

These deposits are found in the upper reaches of the Bulgan River (Fig. 4.8). Detailed geological exploration work was only carried out in the Khuurai Salaa deposit, and it is currently being mined.

Khuurai Salaa gold placer deposit: It is located in the Khuurai Salaa gorge of the eastern tributary of the Ikh Jargalan River 28 km west of Bulgan soum of Bayan-

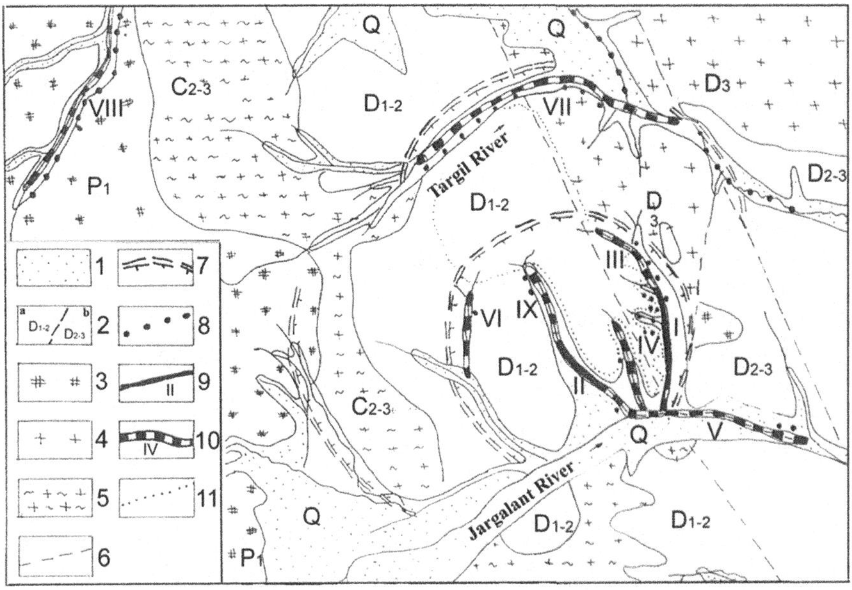

Fig. 4.8 Location map of the Upper Bulgan placer deposits /Scale 1:100000/. 1 Quaternary sediments (Q). 2 Lower-middle Devonian (D_{1-2}) felsic and intermediate volcanic rocks (a), Middle-Upper Devonian (D_{2-3}) Tsagaansalaa Formation conglomerate, sandstone, and carbonates (b). 3 Lower Permian (P_1) muscovite-biotite granite. 4 Biotite-porphyritic granite. 5 Middle-Upper Carboniferous (C_{2-3}) biotite-bearing gneissic granite. 6 Fault. 7 Circular fault related to central uplift tectonics. 8 Regions were gold nuggets recovered during exploration. 9 Placer gold deposits, I-Khuurai salaa, II-Indert. 10 Placer gold occurrence, III-Upper part of Khuurai salaa valley, IV-Nergui, V-Ikh Jargalant, VI-Togoit, VII-Targil, VIII-Near the country border, IX-Upper part of the Indert valley

Ulgii province. The valley is 5 km long and 100–400 m wide and was formed along Middle-Upper Devonian fault system. Gold-bearing sediments are found 0.5 to 4 m depth from the surface, contain rare boulders, and are primarily composed of sandstone. As the clay content increases, the overall color of the sediments becomes lighter and yellowish. Placer gold was sourced from primary gold deposit hosted in quartz those that were present prior to faulting. Au content ranges from 2.1 to 5.6 g/m^3 with an average gold-bearing sediment thickness of 1.16 m. Gold nuggets are 2 mm in average, occasionally reaching 3.5 mm, platy, angular, and poorly rounded.

Placer gold deposits and occurrences in Bodonch and Uyench Rivers: The Bodonch and Uyench River valleys are deeply incised to a depth of 1200–2000 m and are located in the Mongol-Altai mountain range. The gold-bearing sediments are of Cenozoic in age and related to Quaternary glacial activities. Terrace treads are well-developed in both river valleys. Tamt gorge placer deposit is identified in the Bodonch river valley and Ikh Khar Chuluut and Nariin gorge deposits in the Uyench River valley.

Table 4.1 Characteristics of placer gold deposits of Kharkhiraa gold district

№	Placer gold deposit name	Length (km)	Width (m)	Thickness (m)		Average gold content (g/m^3)	Au content (kg)
				Cover	Gold-bearing sediment		
1	Burgastai river	2.5	170	2.5–5.0	1.6–2.5	0.1–2.7	1060
2	Zaisan Salaa (Meren)	4.6	50–100	2.0–4.4	4.5	2.7	424
3	Taliin Salaa	1.2	10–20	0.6–1.0	1.8–2.8	0.204–0.456	1430
4	Khorkhoit zaag	1.8	140	5.5–9.0	1.2–3.2	0.265	20.7
5	Eljgen river		100		0.5–1.5	0.5–8.0	150
6	Ulaan nachin	0.6–1.0	20–40	1.3	0.5	0.5–1.0	10.0
7	Khukh Del am	4.0	40–60	3.0	1.0	0.1–0.8	
8	Baadai am	1.0	70	0.6–1.4	0.6–0.8		
9	Burgas am	1.2	40	0.8–1.6	0.2–2.8	0.06	
10	Ikh Ulunt	0.825	56	0.2–1.0	0.8–1.8	0.6	21.6

Ikh Khar Chuluut gold placer is 40–100 m long and 100–200 m wide in the Uyench River valley. Gold-bearing sediment has a thickness reaching 2 m with a gold content of 0.5–1 g/m^3. Gold nuggets are well-rounded, isometrical, and platy and up to 5 mm in size. The gold was sourced from primary gold deposits that existed prior to the faulting of the Turgen River deep-seated fault system. Moreover, gold-bearing pegmatites have been documented in the same region, which may have supplied the heavy minerals such as rutile and ilmenite including gold.

In general, only recently, detailed exploration work has been carried out in the Bodonch and Uyench valleys, and the placer gold deposits are considered as new discoveries.

Placer gold deposits in the Kharkhiraa zone: The Kharkhiraa zone (Table 4.1) occupies a large region including the Turgen Range and the eastern part of the Kharkhiraa mountains. These mountains are moderately elevated between 2500 and 3000 m. Erosion depths range from 500 to 1200 m with gentle slopes. River systems are well-developed such as the Burgastai, Tsagaan, and Orlogo River valleys.

In the Burgastai River, skarn and gold-barite-carbonate-quartz primary gold deposits and occurrences are present. Ore bodies are hosted in skarn, quartz vein, quartz-carbonate vein, and silicified zones. Typical Au content is between 0.1–7.7 g/t and up to 34 g/t.

4.5.2 *Bayankhongor Gold Mineralization Zone*

The Bayankhongor gold mineralization zone occupies a large region encompassing the southern side of the Khangai range and is 600 km long and 80–100 km wide. Northwest trending faults are common, and Proterozoic-Paleozoic carbonaceous and volcanic, upper Ordovician, Silurian-Devonian, Carboniferous, and Permian

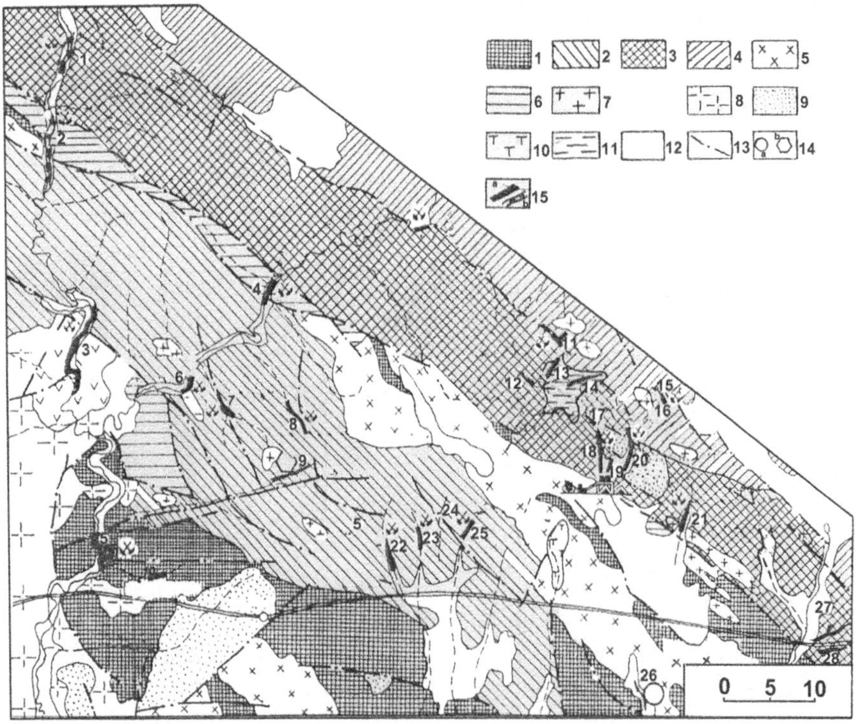

Fig. 4.9 Geologic setting of the Bayankhongor zone and location map of the gold deposits. 1 Archean-Lower Proterozoic gneiss, amphibolite, crystalline schist, and marble. 2 Neoproterozoic metamorphosed sandy shale. 3 Neoproterozoic-lower Cambrian carbonaceous terrigenous-volcanic rock assemblage that host Cambrian ophiolite mélange. 4 Lower Paleozoic greenstone belt. 5 Lower Paleozoic granite, granodiorite, diorite, and gabbro. 6 Lower-middle Paleozoic carbonaceous-terrigenous-volcanic rocks. 7 Carboniferous granitoids. 8 Lower Paleozoic granite, granodiorite, gabbrodiorite. 9 Lower Cretaceous sandstone, shale with claystone and coal beds. 10 Miocene alkaline basalt. 11 Pliocene boulders, sandy terrigenous sedimentary rocks. 12 Quaternary alluvial and proluvial sediments. 13 Fault. 14 Primary gold deposit (a, hydrothermal; b skarn). 15 Placer gold deposit (a), occurrence (b). Baidrag Group: Rashaan Bulag (1), Artsag (2), Urt (3), Ulziit (4), Baidrag (5), Ulziit River (6), Dalt valley (7), Deep valley (8), Khukh Bulag valley (9), Builast Khudag (10); Jargalant Group: Dalt (11), Altan-Am (12), Bayan-Am(13), Khooloi Khudag (14), Bukht (15), Ar-Chuluut (18), Sair Khudag (17), Mukhar-Ereg (18), Zuun Mukhar-Ereg (19), Uvur-Chuluut (20), Jargalant (21), Sair valley (22), Bumbat (23), Ar Gashuun (24), Togoot-Am (25); Burdgol Group: Guchin (26), (27)

terrigenous sedimentary rocks are widely distributed (Fig. 4.9). Magmatism was active in the Cambrian, and small bodies of ultramafic rocks, gabbro, gabbro-diorite, and felsic granitoids are present. Active fault-induced metamorphism was prevalent and responsible for formation of small quartz veins and veinlets throughout the region, which have supplied the placer gold accumulation. Placer gold deposits (Table 4.2) are identified in Baidrag, Burd, and Tuin River valleys of Jargalant Range, associated with Proterozoic and Paleozoic shales strata, of alluvial, proluvial,

Table 4.2 Placer gold deposit characteristics of the Bayankhongor zone

No.	Placer gold deposit name	Length (km)	Width (m)	Thickness (m) Cover	Thickness (m) Gold-bearing sediments	Average gold content (g/m^3)	Estimated Resource (kg)
Jargalant range group							
1	Mukhar-Ereg	4.9	80	0.81	1.45	0.760	1550
2	Zuun Mukhar-Ereg	3.7	80		1.37	0.930	
3	Uvur-Chuluut	3.4	80	0.5	1.84	0.870	458.5
4	Sair khudag-upper		160	3.19	1.14	1.100	492.6
	Lower		160	5.8	1	4.540	
5	Khooloi Khudag	4.75	75	0.6	1.2	1.250	536.3
6	Altan-am	4.45		2.8	0.7	1.550	
7	Dalt	6.85	31.5	2.2	1.6	1.400	277.2
8	Jargalant	5.0	60		1.3	1.640	197
9	Ar-Chuluut	2.5	50–100	5	2	3–40	1047
10	Bukht	2–3	200–400	6–8	0.15–0.30	5–10	
11	Bumbat	19.0	20–86	1.56	1.08	0.37	323
12	Togoot	7.0	40–50	2.6	0.62	0.530	39.5
13	Ar-Gashuun	7.5		1.48	1	0.699	116
Baidrag group							
1	Baidrag river	50.0		0.5–9.0		0.3–4.0	
2	Baidrag river tributaries: Khukh Bulag valley	12.0	100	3.5	1	0.627	
3	Sair valley	30	100–200	2.1	0.6	1.080	
4	Dalt valley	2.0	200–300			0.5–1.0	
5	Gun valley	20.0	80	2.3	0.7–1.0	0.780	96.2
Burdgol group							
1	Burd river	30	500–2000	4–5		0.250	
2	Guchin river	3.5	110	1.07	1.23	0.360	143.2
3	Altan-us gorge	7.7	72	1.07	0.67	0.560	209.5
4	Tashaa gorge	5.8	46	0.72	0.4	0.484	42.9

and deluvial-proluvial in origin, and found in channels, terraces, alluvial cones, and dry gorges. Devyatkin et al. (1990) have suggested that the placer deposits were formed in Oligocene, Pliocene, and Quaternary. Mining has started early and was active between 1939 and 1953.

4.5.3 North Khentii Gold Placers

North Khentii gold placer zone is located in the western part of the Khentii uplift that is bound by deep-seated faults of the Bayan and Yeroo River valleys. Placer gold deposits found in this zone are divided into subzones based on metallogenic features and origin of the deposits, which are the Bukhlein of Yeroo River, Tseel, Boroo-Zuunmod, Zaamar, and Tsaidam districts. Yeroo River and Zaamar districts preserved many placer gold deposits, whereas primary gold deposits such as Boroo, Sujigt, and Naran Tolgoi characterize Boroo-Zuunmod district, and a very few primary and placer gold deposits are identified in the Tseel and Tsaidam districts.

4.5.3.1 Yeroo River District

This district hosts many primary gold occurrences of gold-quartz, gold-sulfide-quartz, and gold-magnetite-quartz. Related to and sourced from these rocks, since 1901, placer gold deposits were actively studied. Until 1920, a foreign company named "Mongolor" had mined Khuder, Tsagaan-Uzuur, Tolgotoi, Ikh Baga Ajir, Khoziil shur, Niilekh, Mogoi, Ikh Kharganat, Khuiten, and Khushuu placer deposits. Detailed geological exploration work resumed from 1960 until 1994, and as a result, Tolgoit placer gold deposit was discovered in 1974 and Ikh Alt deposit in 1984. In the Yeroo River district, scale of 1:100,000 geologic map was finished recently, and many placer gold deposits are newly found (Table 4.3).

4.5.3.2 Bukhlei District

It consists of the Bukhlei River (Table 4.4) basin located to the northwest of the Yeroo River district, Shar River, and Khuiten River valleys. The district follows the Bayangol Fault and continues for 40–50 km with a width of 45–50 km with an area of 0.9 thousand km^2. Resources are estimated on Ulunt, Tsamkhag, Ikh Kharganat, Baga Noyon, and Khuiten gold placer (Table 4.4). Several other placer gold occurrences such as of Bukhlei, Ikh Noyon, and Mogoi valleys occur in the district. Ikh Kharganat, Mogoi, and Ikh Noyon valley deposits were mined by "Mongolor LLC." Between 1996 and 2007, the Ulunt, Tsamkhag, Ikh Kharganat, and Khuiten deposits were mined out.

4.5.3.3 Zaamar District

It is located in the western part of the North Khentii gold mineralization zone, to the west of the Zaamar Range, and downstream of the Tuul River basin. The district is 130 km long and 90–100 km wide, and based on the distribution of placer gold deposits, the district is divided into Zaamar, Tsaidam, and Tseel group deposits.

Table 4.3 Characteristics of the placer gold deposits of the Yeroo river district

No.	Name	Length (km)	Width (m)	Thickness (m) Cover	Thickness (m) Gold-bearing sediments	Average content (g/m^3)	Estimated Resource (kg)
1	Tolgoit	8.0	300		17.7	0.31	6883
2	Sangiin	2.4	60	12.0	2.9	1.72	543
3	Nariin-I	2.2	65		18.8	0.27	584
4	Nariin-II	0.5	20	2.9	1.8	2.24	41
5	Bolot	0.7	35	10.0	2.1	1.76	181
6	Ikh-Ulunt	3.9	40	4.0	1.1	2.26	234
7	Bugantai	4.3	35	3.9	1.1	1.65	254
8	Baga-Ulunt	2.5	55	8.4	1.2	4.46	486
9	Niilekh	0.7	30	4.2	1.4	1.09	34
10	Yalbag	4.6	155	3.6	1.3	1.28	1417
11	Tsagaanchuluut	2.7	55	3.2	1.0	1.43	136
12	Khushuut	0.4	80	2.2	1.6	1.17	30
13	Nariin-III	2.1	30	6.2	1.4	2.40	148
14	Yeroo-north (Ikh-alt)	3.7	285		9.0	0.25	2332
15	Yeroo-south (Ikh-alt)	3.8	80		6.6	0.32	657
16	Ikh-Adjir	4.0	175		8.8	0.31	1884
17	Buural	3.7	70		5.4	0.38	473
18	Altan	0.8	30	3.1	1.1	1.12	31
19	Baga-Adjir	0.8	30	4.0	1.2	2.00	54
20	Yeroo-Kherein	5.8	65		11.6	0.19	702
21	Khuder	6.6	65	4.0	1.1	1.16	1066
Total		64.2					18,170

Table 4.4 Characteristics of placer gold deposits of the Bukhlei district

No.	Name	Length (km)	Width (m)	Thickness (m) Cover	Thickness (m) Gold-bearing sediments	Average content (g/m^3)	Estimated Resource (kg)
1	Ulunt	4.5	100	12.5	4.3	1.17	1740
2	Tsamkhag	4.0	75	12.6	2.7	1.25	819
3	Kharganat	4.8	70	3.2	1.5	0.95	360
4	Khuiten	12.4	45	3.5	1.2	1.02	718
5	Baga-Noyon	3.7	50	3.0	1.3	0.85	159
6	Shar river	3.0	150	12.0	2.9	2.44	3214
7	Burkhant	2.8	70	3.9	1.9	2.32	994
8	Ikh-Uvuljuu	1.6	80	5.9	2.5	0.90	505
9	Khavchuu	0.9	70	9.0	2.8	0.99	374
Total		37.7					8883

Table 4.5 Characteristics of placer gold deposits of the Zaamar district

No.	Name	Total valley length (km)	Placer gold deposit length (km)	Resource (ton)	Metal content (t/km)
1	Toson	10	1.2	0.2	0.17
2	Tsagaanbulag	10	1.7	0.3	0.18
3	Urt-Sair	4	2.4	0.2	0.08
4	Bayangol	13	2.5	0.7	0.28
5	Khailaast	14	8.1	6.0	0.74
6	Baga-Khailaast	7	3.2	1.3	0.41
7	Tsagaanchuluut	6	2,0	2.4	1.2
8	Ar-Naimgan	13	6.8	2.4	0.35
Total		77	27.9	13.5	0.5

Major river systems are the Tuul River and its western tributaries such as the Ar Naimgan, Khailaast, Bayangol, and the Toson River.

The largest primary gold deposit of the Zaamar district (Table 4.5) is the Bumbat deposit. In general, primary gold deposits are characterized by gold-quartz and gold-sulfide-quartz, which occur as quartz veins, and gold-bearing metasomatites in this district. These primary deposits feed the gold placers. Moreover, gold mineralization is documented in Lower Carboniferous and Upper Permian conglomerates preserved in the Zaamar Range and Lower-Middle Jurassic, Upper Jurassic-Lower Cretaceous conglomerates of the Tuul River valley. Gold nuggets size of these conglomerate-hosted deposits ranges from 0.1 to 1 mm and serve as an intermediate source for the placer deposits. Gold mineralization was identified during 1:50,000 and 1:200,000 scale geologic exploration work, and based on the results, Semeihan et al. (2002) began gold-oriented exploration work since 1980. Valley, terrace, and channel-type placer gold deposits are common in the district, and one of the largest is the Tuul valley deposit (Table 4.6), which includes Tuul, Bayangol, Tokhoirol, and Toson gorges (Fig. 4.10).

4.5.3.4 Tseel District

The Tseel district is bound by the Orkhon River in the northwest, Kharaa valley in the northeast, and Kharnuur and Zuukhii valleys in the southeast. Landform is low-moderate mountainous (1100–1250 m a.s.l.) with gentle slopes and 100–250 m incised valleys, which are suitable for the accumulation of gold in alluvial placers. Tsagaan Gozgor primary gold occurrence is one of the largest in the district, and because the gold nuggets are large in size, proluvial type placer gold deposits have formed. Gold ore bodies are generally small in size and often characterized as gold-quartz and gold-sulfide-quartz types. Au content ranges from 0.005–1 to 0.1–1.5 g/t. (Table 4.7).

Table 4.6 Characteristics of placer gold deposits of Tuul river

Name and corresponding number on the map	Length (km)	Width (m)	Cover soil depth (m)	Average content (g/m^3)	Resource (ton)	Length (кm)	Width (m)	Cover thickness (m)	Gold-bearing sediment (m)	Average content (g/m^3)	Resource (ton)
Toson (4)	11.0	200–300	8.1	0.21		4.0	120	7.2	1.6	1.35	
Bayangol (3)	10.0	60–800	7.4	0.5	5.7	8.0	40–680	11.2	1.3	2.5	12.5
Tokhoirol (2)	14.0	120	12.0	0.25							
Tuul (1)					13.9		40–400	10.0	2.0	1.2	0.5
1st part	4.4	870	6.9	0.526	13.9						
2nd part	7.6	360	8.9	0.369	9						

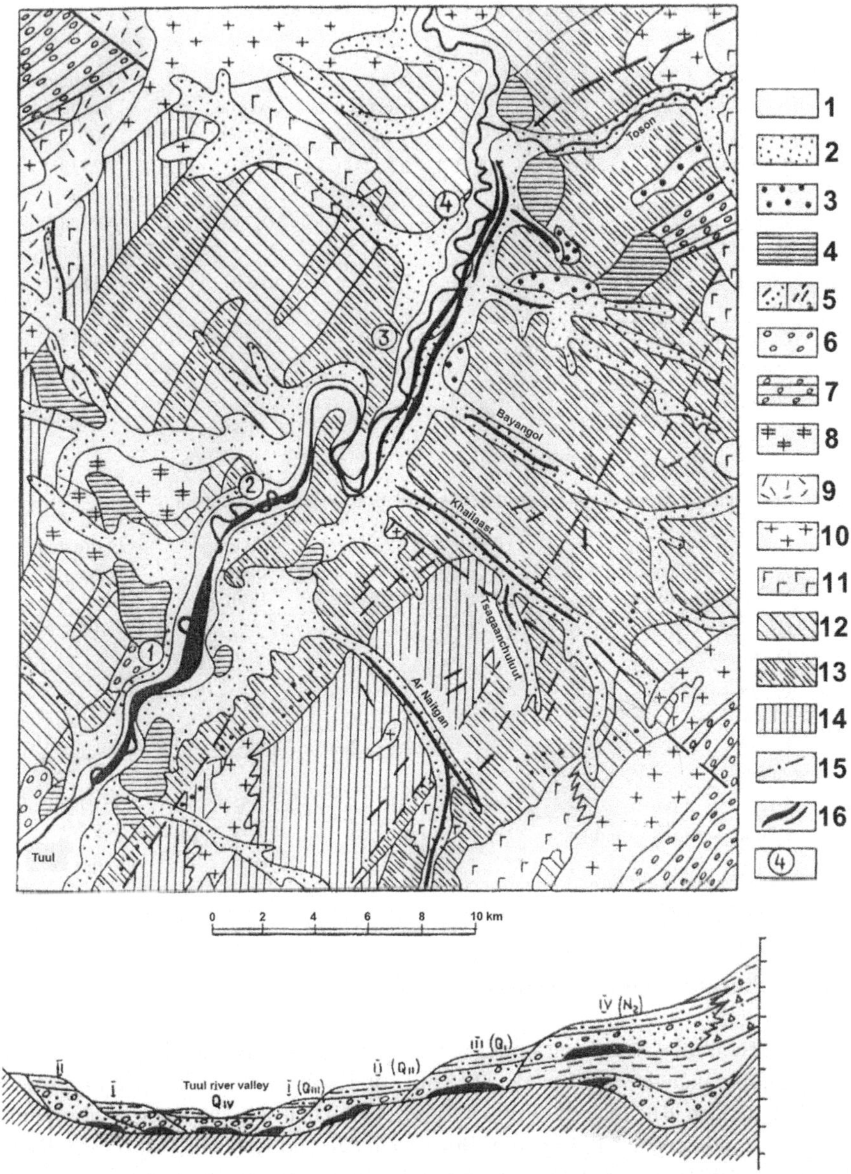

Fig. 4.10 Location map of placer gold deposits of the Zaamar district. 1–2 Quaternary sediments. 1 Alluvial. 2 Tuul River terrace alluvial sediments and their cover. 3 Pliocene red-colored clay with rare gravels. 4 Lower Cretaceous, loosely cemented, cobble conglomerate, siltstone. 5 Upper Triassic-Lower Jurassic Del complex – Diorite, porphyrite, diabase (1) quartz vein (2). 6 Middle-Upper Triassic conglomerate, sandstone. 7 Lower Carboniferous conglomerate, sandstone, and siltstone. 8–9 Devonian: 8 Granite, 9 Rhyolite, dacite. 10–11. Upper Ordovician-Lower Silurian: 10 granite, granodiorite, 11 gabbrodiorite. 12–14 Upper Cambrian-Lower Ordovician Kharaa complex: 12 upper carbonaceous

Table 4.7 Characteristics of placer gold deposits of the Tseel district

No.	Name	Length (km)	Width (m)	Thickness (m)		Average gold content (g/m^3)	Estimated Resource (kg)
				Cover	Gold-bearing sediments		
1	Tsagaan Gozgor	210	60	0.5–2.0	1.5–2.0		425.00
2	Bosgo-am	2.3	54	4.90	0.98	0.350	43.80
3	Chono-am	2.0	77	1.26	1.98	0.270	41.47
4	Eastern Chono-am	0.88	40	1.33	1.15	1.112	40.00

Table 4.8 Characteristics of placer gold deposits of Boroo-Zuunmod district

No.	Name	Length (km)	Width (m)	Thickness (m)		Average content (g/m^3)	Estimated Resource (kg)
				Cover	Gold-bearing sediments		
1	Ikh-Dashir	1.7	100	4.0–10.0	1.0–3.0	2.5	2614.6
2	Nergui valley	0.8	40	3.2	0.4–0.8	0.5–1.0	
3	Nariin valley	3.0	40		0.6–3.2	0.2	
4	Gatsuurt	4.6	140	3.8–9.2	2.1	1.5–2.4	3510.4
5	Biluut	2.0	110	2.5	0.1–2.4	0.15–2.1	153
6	Bulag-am	2.0	140			0.3	250.0

4.5.3.5 Boroo-Zuunmod District

Historically, mining had been important especially in the twentieth century. Landform features include medium elevated hills of 1200–1800 m a.s.l., with ravine of 300–600 m deep. The district (Table 4.8) is a region that was subjected to uplift, and the placer gold deposits are found in Boroo, Zagdal, and Kharaa River basins. River valleys are often U shaped, and the total sediment thickness reaches 90 m.

Fig. 4.10 (continued) sandstone, 13 Middle – interbedded sandstone ad phyllite, 14 Lower – interbedded sandstone and phyllite. Quartz-actinolite-sericite-zoisite-biotite schists dominate in the western part of the Tuul River. 15 Fault. 16 Placer gold. 17 Mining steps of Tuul valley deposit: 1 Tuul, 2 Tokhoirol, 3 Bayangol, 4 Toson

Ikh-Dashir, Nergui, Nariin, and Gatsuurt deposits are discovered in the district, and it is certain that the Ikh Dashir placer gold deposit was sourced from the Boroo primary gold deposit.

4.5.4 South Khentii Gold Mineralization Zone

South Khentii zone is located in the eastern and southeastern parts of the Khentii Range and is tectonically related to recent increases of Neotectonic events (Selivanov 1971, 1972). Gold mineralization is associated with small bodies of Mesozoic gabbro-diorite, fault-related dykes, quartz veins, and fractures. Similar to the North Khentii zone, primary gold-quartz and gold-sulfide-quartz deposits are also common. Gold-silver occurrences are found in the Ulz, Onon, and Turgen River basins. This primary gold mineralization was the source of placer gold accumulation in the region. The placers are divided into Baruun-Urt, Terelj, and Ulz-Onon groups.

Terelj group deposits are located in Kherlen River basin. Remnants of "Mongolor LLC" mining activities of eluvial gold placers in the Terelj River, a western tributary of Kherlen River, are present. The mined valley length is 3 km with a width of 80–100 m. Soil thickness is 1–2 m, and gold-bearing sediment thickness reaches 5 m. Average of 150 kg of gold is estimated to have been recovered during this mining work. Since 1990, detailed geologic work has been done in the same region, and as a result, many placer gold deposits were found (Table 4.9).

Baruun-Urt group: This group of deposits is located in the southwestern part of the South Khentii gold mineralization zone. Landform is characterized as low hills (400–1500 m) that are slowly uplifted due to Neotectonic events. Quaternary alluvial and modern deluvial-proluvial sediments are distributed in these regions, and the gold deposits and occurrences are related to gold-quartz and gold-sulfide-quartz primary mineralization. The discovered placer gold deposits, Baruun-Urt, western Baruun-Urt, Khushuu-Uul, and Salkhit, are low in Au content and are in the last stage of mining (Table 4.10).

Ulz group: This group of deposits is located in the eastern tributary of the Ulz River and Turgen River basin and in the south of Ereendavaa Mountain pass. The general landform is characterized by a moderately elevated region (1000–1400 m, a.

Table 4.9 Characteristics of Terelj group placer gold deposits

No.	Name	Length (km)	Width (m)	Thickness (m) Cover	Thickness (m) Gold-bearing sediments	Average content (g/m^3)	Estimated resource (kg)
1	Altan sudal	10.0	40–60	2.0	1.2	0.7	600.0
2	Jamba valley	5.0	60	2.0	0.8	0.5	120.0
3	Zambal valley	12.0	120	4.0	1.0	0.6	500.0

Table 4.10 Characteristics of Baruun-Urt Group placer gold deposits

No.	Name	Length (km)	Width (m)	Thickness (m)		Average content (g/m^3)	Estimated Resource (kg)
				Cover	Gold-bearing sediments		
1	Baruun-Urt	7.0	100	0.8	1.8	0.6	566.5
3	Khushuu-Uul	2.0	50	1.7	0.7		
4	Buural Khangai						294.01

Table 4.11 Characteristics of Ulz group placer gold deposits

No.	Name	Length (km)	Width (m)	Thickness (m)		Average content (g/m^3)	Estimated Resource (kg) B + C
				Cover	Gold-bearing sediments		
1	Tsagaanchuluut khudag	6.0	250	1.6	2.0	0.5	946.4
2	Tsagaanchuluut khooloi	19.0	140	9.0	1.1	1.1	1297.6
3	Khuurai						
4	Rashaant	6.0	20–40	3–12	2.0	1.4	249.9
5	Dagshaa		80		0.50	0.3	
6	Uvuljuu	1.5	40–120	0.5–10	1.00	0.6	285.0

s.l.) with moderate incisions of 200–300 m. Primary gold deposits are gold-quartz-sulfide and gold-silver which acted as feeders for the placer gold deposits of Tsagaanchuluut Khooloi and Tsagaanchuluut Khudag (Table 4.11).

4.5.5 Edren and Altan-Uul (Nemegt Range) Gold Mineralization Zone

Since the twentieth century, gold mining has been known in the central part of the Edren Range and in the Altan mountain that are located in the western continuation of the Nemegt Range. Detailed geologic exploration work was carried out by Devyatkin et al. (1987). The Altan Mountain placer gold deposit was related to Neogene-Quaternary conglomerate. However, a more recent view considers resedimentation of upper Cretaceous gold-bearing conglomerates during upper Quaternary (Devyatkin et al. 1987).

Altan-Uul group: These deposits occur in moderate elevations along mountain slopes. As Upper Cretaceous and Paleogene gold-bearing conglomerates were transported downward, placer were established in gorges such as Alag-Shand,

Toromkhon, Zost, and Tamgat. Gold-bearing sediments are proluvial in origin and found at <1.5 m depth with a thickness of 0.6–1.0 m. Au content is estimated as 0.07–0.850 g/m^3, and the gold nuggets are often platy and well-rounded with a diameter reaching as large as 4.5 mm. No further gold-oriented geologic exploration work has been done in the Nemegt Range.

Edren group: Gold-bearing sediments are composed of Lower-Middle Devonian volcanic rocks, silicified terrigenous rocks, and cross-cutting felsic sills, resulted in the formation of numerous quartz-veins in the region. Proluvial-alluvial type placer gold deposits are present in Khoshuut, Khur-Uul, and Khuurai Khundii gorges in Neogene to lower Quaternary gravel-cobble dominated alluvial fan that covers 40,000 km2 of area. Alluvial fan thickness is 3.5 m with 2 m thick gold-bearing sediments. Au content is 2 g/m^3. Gold exploration work has not been done in the Edren Range.

4.6 Concluding Remarks

In relation to primary gold mineralization formed through Precambrian to late Mesozoic geotectonic events, placer gold deposits are abundantly formed in Mongolia. The main genetic types of these deposits are classified into eluvial, deluvial, proluvial, and alluvial placer, and the richest accumulation is often related to Cretaceous sedimentary successions as exemplified in Mongol-Altai, Gobi-Altai, Khangai Khentii, Khuvsgul, and Tenger mountain belts. Eluvial gold placer deposits are formed near their primary source and commonly found in Yeroo River basin. Deluvial deposits occupy mountainous regions, and high Au content has been revealed in Khentii, Zaamar, Boroo, and Zuunmod districts. Placer gold deposits found in mountain foothills are classified into Proluvial placers and define the genesis of gold mineralization in the Zuunmod, Bayankhongor, and Burgastai gold districts. The richest Au content is often associated with alluvial placer gold deposits, which are found in Khentii zone. Placer gold deposits can also be formed in river terraces, glacial sediments, and conglomerates.

References

Dejidmaa G (1996) Gold metallogeny of Mongolia. Mong Geosci 1:6–29

Devyatkin YV (1970) Kainoziskaya geologiya Vostochnoi Mongolii (Cenozoic geology of Western Mongolia). In: Mesozoic and Cenozoic geology of Western Mongolia Nauka, pp 44–102

Devyatkin YV, Purevdorj T, Semeihan T, Shabalovskii NY, Shuvalov VF (1990) Mezozoyskiye allyuvial'nyye otlozheniya Mongolii (Mesozoic alluvial deposits of Mongolia). In: Mongolian gold, Ulaanbaatar, pp 18–20

Devyatkin YV, Semeihan T, Shuvalov VF (1987) Stratigraficheskiye urovni i geneticheskiye tipy rossypey v mezozoye i kaynozoye Mongolii (Stratigraphic levels and genetic types of placers in the Mesozoic and Cenozoic of Mongolia). In: The VIII meeting on placer geology, Kiev

Frimmel H, Groves D, Kirk J, Ruiz J, Chesley J, Minter W (2005) The formation and preservation of the Witwatersrand goldfields, the world's largest gold province. Econ Geol 100:769–797

Marinov NA, Khurts Ch, Kleiner YM (1977) Problemy formirovaniya i zakonomernosti rasprostraneniya rossypey (Problems of formation and patterns of distribution of placers). In: Geology of MPR, Mineral Resources, vol III. Nedra, pp 574–586

Selivanov YI (1971) Neotektonicheskaya karta Mongolii v masshtabe 1: 5,000,000 (Neotectonic map of Mongolia in a scale of 1: 5,000,000). In: Second Conference on the problem of geology and mineral resources of foreign countries, Moscow, pp 54–55

Selivanov YI (1972)) Neotektonika i geomorfologiya Mongolii (Neotectonics and geomorphology of Mongolia). Nedra, Moscow

Semeihan T, Altantsetseg D (1994) Atributy geomorfologicheskoy karty Mongolii v masshtabe 1:1,000,000 (Attributes of geomorphological map of Mongolia in a scale of 1:1,000,000). Ulaanbaatar

Semeihan T, Jargalan S, Fujimaki H (2002) Geomorphologic features and its relation with placer gold distribution of Mongolia. Research results of Northeastern Asian research center, series 6.

Taylor G, Eggleton RA (2001) Regolith geology and geomorphology. John Wiley & Sons Ltd., New York

Chapter 5
Rare Metals: Tin, Tungsten, Molybdenum, Lithium, Tantalum, and Niobium Deposits

Ochir Gerel

5.1 Introduction

Rare metal mineralization in Mongolia is represented by deposits and occurrences of rare metals (Sn, W) and rare elements (Ta, Nb, Zr, Be, Li, Cs, Sr, and REE) which can be divided into two groups: magmatic and hydrothermal, according to their composition and genesis (Khasin 1977; Kovalenko and Yarmolyuk 1995; Gerel 1995, 1998; Koval 1998). The magmatic group includes deposits of magmatic genesis, and ore mineralization is associated with igneous rocks such as rare metal granites (e.g., Ta-bearing lithium-fluorine granites and Nb- Zr- REE-bearing peralkaline granites), pegmatites, ongonites, Cs glasses, Be-bearing tuffs, Zr-Nb-REE-alkali rhyolites, rare metal albite-bearing nepheline syenites, REE-bearing carbonatites, and magnetite-apatite alkaline volcano-plutonic rocks. The hydrothermal group includes hydrothermal deposits of Sn, W, and Mo with significant enrichment of rare elements (Ta, Li, Be) and fluorine. Rare metal mineralization of both groups is closely related to igneous rocks of a relatively narrow compositional range including highly differentiated leucogranites, peralkaline granites, lithium-fluorine granites, leucogranites, nepheline syenites, and coeval volcanic rocks.

Geological Information Center lists about 20 tin-tungsten greisen, stockwork and quartz vein-type deposits and more than 160 occurrences and placers distributed in the Mongol Altai, North Mongolia, Central Khangai, Khentii, East Mongolia, and Gobi-Khyangan metallogenic belts (Fig. 5.1, Table 5.1). Tungsten, tin, and molybdenum very often form complex deposits, where they are present in various combinations of themselves and other elements.

O. Gerel (✉)
Geoscience Center, School of Geology and Mining, Mongolian University of Science and Technology, Ulaanbaatar, Mongolia
e-mail: gerel@must.edu.mn

O. Gerel et al. (eds.), *Mineral Resources of Mongolia*, Modern Approaches in Solid Earth Sciences 19, https://doi.org/10.1007/978-981-15-5943-3_5

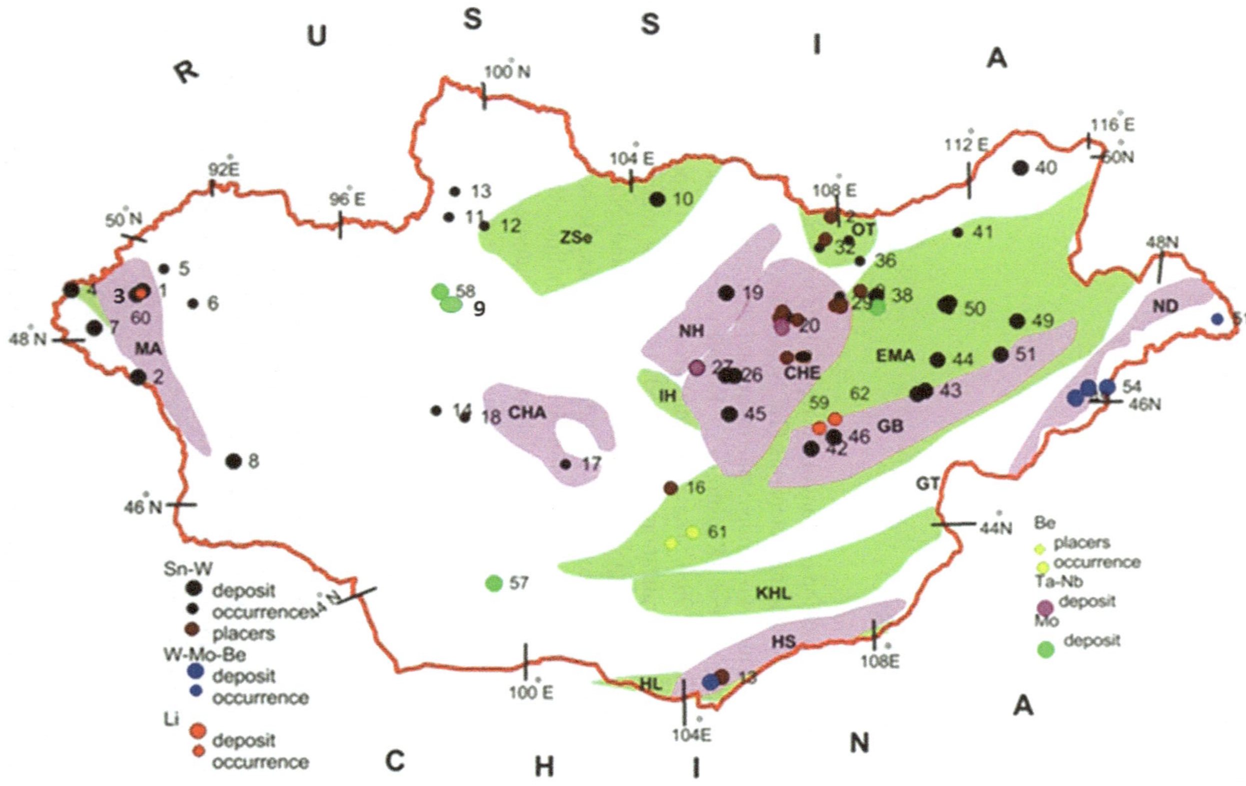

Fig. 5.1 Distribution of rare metal deposits. Metallogenic belts: *MA* Mongol Altai, *ZSe* Zed-Selenge, *OT* Onon-Tura, *CHA* Central Khangai, *CHE* Central Khentii, *EMA* Eastern Mongolian, *GB* Gobi Ugtaal-Baruun Urt, *ND* Nukhet Davaa, *IH* Ikh Khairkhan, *KHL* Mushgai Khudag-Ulgiikhiid, *HL* Khanbogd-Lugiin gol, *HS* South Gobi-Khyangan. Number of deposits. Note: occurrences are not shown by number in the figure (see Table 5.1)

Table 5.1 Rare metal mineral deposits and occurrences

Deposit name	Deposit type (model)	Host rocks	Alteration	Isotope age	Mineralogy	Resources and reserves
Ulaan Uul (Kyzyl tau) (1)	W-Mo-Be greisen/stockwork and quartz vein	Devonian felsic and intermediate volcanic rocks cut by Jurassic granite pluton	Quartz-muscovite	Granite age 180–200 Ma (K-Ar), 170–180 and 196 Ma (Rb-Sr)	Wolframite, beryl, molybdenite and Y-fluorite, sulfides	285,500 t @ 1.1% and 3138 t WO_3 (remaining) Inferred resources 5870 t, 500 kg Y
Sagsai (2)	W-Mo-Be greisen/stockwork and quartz vein	Permian Sagsai granite pluton intruded Devonian volcanic rocks	Chlorite, sericite, mica	Granite age Permian	Molybdenite, wolframite, chalcopyrite, bismuthite, scheelite	0.15% Mo
Tsunkheg (3)	W-Mo-greisen/stockwork and quartz vein	Upper Ordovician-lower Silurian silty schists and gabbro-diabase	Carbonate and sulfide	Devonian	Wolframite, scheelite, chalcopyrite, siderite, chalcocite, arsenopyrite, pyrrhotite, sulfosalts bismuthite	80,000 t @ 3.33% WO_3; 2682 t WO_3, 405,000 t @ 2.29% WO_3; 9267 t WO_3
Nuuriin Gol (4)	W-Mo-Be greisen/stockwork and quartz vein	Middle-upper Ordovician siltstones and sandstones	Quartz-muscovite	Late Devonian	Wolframite, scheelite molybdenite, sulfide minerals, beryl, fluorite	30,000 t @ 1.82% WO_3, 600 t WO_3
Achit Nuur (5)	W-Mo-Be greisen/stockwork and quartz vein	Upper Ordovician tuffaceous siltstone and sandstone	Quartz-muscovite	Devonian?	Wolframite, beryl molybdenite, chalcopyrite, fluorite	58,000 t @ 0.05–3.29% WO_3, @ 0.01–2.7% BeO
Umnu Khutul (6)	W-Mo-Be greisen/stockwork and quartz vein	Granite intruded Altai group sandstone, siltstone and shale	Quartz-sericite		Wolframite, molybdenite fluorite	3000 t @ 0.2–1% WO_3, 600 t @ 0.1% Mo

(continued)

Table 5.1 (continued)

Deposit name	Deposit type (model)	Host rocks	Alteration	Isotope age	Mineralogy	Resources and reserves
Khovd Gol (7)	Epithermal W-Sb	Upper Cambrian-lower Devonian Mongol Altai group (siltstone and sandstone)	Propylitic limonite	Devonian?	Ferberite, antimonite, carbonate minerals	48,070 t @ 10% WO_3; 4736 t WO_3 (remaining); 16, 850 t @ 10% WO_3, 1107 t WO_3(remaining)
Uudav occurrence (8) Bodonch ore field	W-Mo-Be greisen/stockwork and quartz vein	Mesozoic granite intruded the Mongol Altai group	Greisenization	Granite age Mesozoic	Wolframite, beryl, fluorite, gold	240 t @ 0.02–5.6%, averaging 3.2% WO_3, 0.05–1.0 ppm Au
Zoost Uul (9) occurrence	Porphyry Mo	Granite porphyry	Quartz-sericite alteration	Granite age Triassic	Molybdenite, chalcopyrite, sphalerite, galena, fluorite	0.01–0.2% Mo, 0.01–0.3% Cu
Bulagtai (10)	Mo-W-Sn greisen, stockwork, and quartz vein	Lower-mid Cambrian shale, metavolcanics	Quartz-sericite, silica, feldspathic	Granite age Jurassic	Wolframite, molybdenite, pyrite, chalcopyrite, galena, arsenopyrite, fluorite	Mo 0.04%, WO_3 0.004%, Mo 819,000 t, W-819 t
Tsagaan Uul occurrence (11)	Sn-W greisen, stockwork, and quartz vein and W skarn	Contact between Devonian granite with limestone dolomite-bearing sandstone and siltstone sequence	Quartz-sericite skarnization of limestone	Granite age Devonian	Wolframite, scheelite	0.03% WO_3 in greisen, and 0.03 WO_3 in skarn Bi 0.1% in skarn
Tsagaan Tolgoi occurrence (12)	W-Mo greisen	Devonian granite	Greisenization	Granite age Devonian	Molybdenite, wolframite	
Altan boom occurrence (13)	Ta-Nb	Leucogranite	Quartz-albite alteration	Granite age Devonian	Tantalite, cassiterite	0.05% Ta_2O_3. 0.03% Nb_2O_3, 0.03% Sn, 0.002–0.28% Mo

Beliin Bulag occurrence (14)	W-Mo-Be greisen, stockwork, and quartz vein	Permian granite pluton cut Proterozoic metamorphic rocks	Greisenization	Granite age Permian	Wolframite, chalcopyrite, bismuthite, molybdenite, fluorite	8000 t @ 0.02% WO3 @ 0.007% Bi
Ulziit occurrence (15)	W-Mo-Be greisen, stockwork, and quartz vein	Jurassic granite stock intruded Devonian sandstone and shale	Muscovite, quartz	Granite age Jurassic	Wolframite, cassiterite	4400 t @ 0.1–0.2% WO_3
Dulaan Ovoo occurrence (16)	Be vein	Jurassic granite stock with Li-F granite intruded Devonian sedimentary-volcanic rocks	Quartz, muscovite greisenization	Jurassic	Beryl, phenakite, fluorite	@ 3.6 Be
Kharaat Yagaan Tolgoi occurrence (17)	W-Au quartz vein	Jurassic granite intruded Proterozoic metamorphic rocks	Silica	Early Jurassic	Wolframite, gold	2200 t @ 0.001–0.1% WO_3, Au 0.1%
Salkhit Khudag occurrence (18)	W-Mo-Be greisen/stockwork and quartz vein	Granite pluton	Albitization, limonitization, pyritization	Jurassic	Wolframite, beryl Molybdenite, galena, bismuthite, chalcopyrite, fluorite	0.06–0.2% Mo, 3% Pb, 100 g/t Ag, 0.2% Li, 200 g/t Cd
Tsagaan Davaa (19)	Sn-W greisen/stockwork and quartz vein	Granite pluton intruded Cambrian-Ordovician sedimentary rocks of Kharaa group	Greisenization silica	Granite age 190.49 ± 4.7–224 Ma (K-Ar) 199.36 ± 0.73 (LA-ICP-MS zircon)	Wolframite, cassiterite molybdenite, galena, chalcopyrite, beryl, fluorite	90,000 t WO_3. @ 0.44% to 1.2% WO_3, 0.05% Sn, 3000 t WO_3 mined out
Janchivlan occurrence (20)	Sn-W greisen, stockwork and quartz vein	Porphyritic coarse-grained biotite granite of Janchivlan multi-stage pluton	Potassic gresienization	Granite age 217 Ma	Wolframite, cassiterite, molybdenite, topaz, tourmaline, galena	0.01–0.8% WO_3, 1300 ppm Nb, 2400 ppm Pb

(continued)

Table 5.1 (continued)

Deposit name	Deposit type (model)	Host rocks	Alteration	Isotope age	Mineralogy	Resources and reserves
Tsogt occurrence (21)	Sn-W greisen, stockwork and quartz vein	Porphyritic coarse-grained biotite granite of Janchivlan multi-stage pluton	Greisenization	Granite age 217 Ma	Cassiterite, wolframite, topaz, fluorite	0.03% Sn, 0.05% As
Ulaan Burd occurrence (22)	Sn-W greisen, stockwork and quartz vein	Medium-grained granite of Janchivlan pluton	Greisenization	Granite age 217 Ma	Cassiterite, wolframite fluorite, tourmaline	0.01–0.5% Sn
Elstei occurrence (23)	Sn-sulfide	Medium-grained granite of Janchivlan pluton	Greisenization	Granite age 226 Ma	Cassiterite, fluorite, arsenopyrite, sphalerite, chalcopyrite, topaz	0.01–0.32% Sn, 0.01–0.4% Cu, 0.01–1.0% Ag
Urt Gozgor occurrence (24)	Sn-Ta-Nb	Microcline-albite and lepidolite-albite granite of Janchivlan pluton	Albitization	Granite age 195 Ma	Cassiterite, columbite, topaz, fluorite	Ta 60 ppm, Li 600 ppm, Rb 800 ppm Sn 50 ppm
Buural Khangai occurrence (25)	Ta-granite	Amazonite-albite granite of Janchivlan pluton	Albitization	Granite age 195 Ma	Cassiterite, topaz, fluorite columbite, tantalite	2–5 ppm Sn,
Ikh Khairhan (26)	W-Mo-Be greisen, stockwork, and quartz vein	Contact of granite pluton and Triassic andesite	Silica, sericite, pyrite, and greisen	105.9 Ma, (K-Ar muscovite) 126 + 5 Ma (W veins) Host granite 161–133 Ma (K-Ar)	Hubnerite, scheelite, molybdenite, sphalerite, chalcopyrite, Cassiterite, fluorite	150,000 t @ 1.3% WO_3 65,200 t @ 0.8% WO_3 840 t WO_3 (mined) 550 t WO_3 (remaining)
Ongon Khairkhan (27)	W-Mo-Be greisen, stockwork, and quartz vein	Contact zone of granite pluton and Devonian-	Silica, sericite, pyrite, and greisen	K-Ar age of granite 183–131 Ma	Hubnerite, scheelite molybdenite,	7.4Mt @ 0.15% WO_3 5.3 Mt. @ 0.05–0.10% WO_3

		carboniferous sandstone, siltstone and shale			sphalerite, chalcopyrite, cassiterite, fluorite	11,136 t WO_3 3.993 t WO_3
Khar Chuluut (28)	W-Mo-Be greisen, stockwork, and quartz vein	Granite pluton intruded Triassic andesite	Silica, sericite, pyrite, and greisen	Late Mesozoic granite	Hubnerite, scheelite, molybdenite, sphalerite, chalcopyrite, cassiterite, fluorite	100,000 t @ 0.66% WO_3 697 t WO_3
Modot occurrence (29)	Sn-W greisen and quartz vein	Modot granite pluton cut Proterozoic schists	Greisenization	Late Triassic-early Jurassic granite	Cassiterite, wolframite, topaz, fluorite, chalcopyrite	
Bayanmod Occurrence (30)	Sn-W greisen and quartzvein	Modot granite pluton cut Proterozoic schists	Greisenization	Late Triassic-early Jurassic garnite	Cassiterite, wolframite, topaz, fluorite, chalcopyrite	Placer 8, 516 t Sn, 1, 297 t WO_3 (mined out)
Beis occurrence (31)	W skarn	Proterozoic schists, amphibolite, gneiss, marble	Skarnization	Host rocks Proterozoic	Scheelite, garnet, vesuvianite, actinolite	0.01–1.26% WO_3 Y 4 g/t
Upper Iluur occurrence (32)	W-Mo greisen and quartz vein	D-C rocks and Mesozoic granite pluton	Greisenization	Late Triassic-early Jurassic garnite	Molybdenite, wolframite, pyrite	36,000–38,000 t @ 1.57% Mo
Bayan Undur (34)	Sn-W greisen and quartz vein	Mesozoic granite pluton	Greisenization	Late Triassic-early Jurassic garnite	Cassiterite, wolframite	Placer 960 g/m^3 Sn, 50 g/m^3 WO_3
Saikhan (Bor Khujir) (35)	W-Mo-Be greisen, stockwork, and quartz vein	Granite pluton	Greisenization	Late Triassic-early Jurassic garnite	Beryl, wolframite, cassiterite	@ 0.58% REE. 12,432 t WO_3, 5940 t Be, 13,355 t Sn, 1215 t Ag
Baruun Khujkhan (33)	Sn-W greisen, stockwork, and quartz veins	Granite pluton	Greisenization	Late Triassic-early Jurassic granite	Cassiterite, wolframite, sulfide minerals, topaz, fluorite	Placer – 4.3 Mt. m^3 sediment of 1725 t Sn, 375 t WO_3
Ikh Zelend (36)	W-Mo-Be greisen stockwork, and quartz vein,	Contact of granite intrusion and D-C rocks	Argillic, silica, and sericite	K-Ar age of granite 131 Ma	Wolframite, scheelite, fluorite, chalcopyrite, beryl, arsenopyrite	438.9 t @ 0.23% Sn; 5.8 t WO_3

(continued)

Table 5.1 (continued)

Deposit name	Deposit type (model)	Host rocks	Alteration	Isotope age	Mineralogy	Resources and reserves
Zuun Tarts Gol (37)	Sn-W greisen/ quartz vein	Leucogranite, tourmaline granite, granite porphyry	Gresienization	Granite age late Triassic-early Jurassic	Wolframite, cassiterite, scheelite, beryl, fluorite	Placer – 156–147 g/ m^3 Sn, 13–56 g/ t WO_3
Undurtsagaan (38)	W-Mo-Be greisen stockwork, and quartz vein	Lower Devonian sedimentary rocks metamorphosed to greenschist facies and granite exposed in the depth of 600 m	Chloritization, limonitization, epidotization	Granite age 192–179 Ma	Wolframite, molybdenite, beryl, topaz, chalcopyrite	175, 990 t WO_3 t Mo, 43,350 t BeO, 13,460 t Bi. 134 Mt. @ 0.127% WO_3, 0.036% Mo, 0.03% BeO, 0.01% Bi
Tsagaanchuluut (39)	W-Mo-be greisen stockwork, and quartz vein	Carboniferous granodiorite, granite, and Devonian sedimentary rocks	Greisenization	Granite age carboniferous	Molybdenite, wolframite	34,300 t @ 0.005–0.19% Mo, 54, 000 t @ 0.08% W
Chuluunkhoroot (40)	W-Mo-Be greisen/stockwork and quartz vein	Devonian and Triassic sandstone and shale	Greisenization	Granite age late Jurassic	Wolframite, molybdenite, beryl	500 t WO_3 300 t WO_3 (remaining) 229 t WO_3 (mined)
Oortsog (42)	Sn-polymetallic skarn	Upper Proterozoic rocks intruded by early Mesozoic granite pluton	Epidote, actinolite, limonite	Granite age Triassic	Cassiterite, sphalerite, chalcopyrite, bismuthinite, galena, fluorite	33, 000 t Sn, 13, 137 t Zn, 5292 t Cu, 1000 t Pb
Narsan Khundlun (41)	Sn-sulfide-silicate vein/ stockwork	Devonian volcanics	Silica, chlorite, sericite, tourmaline	Granite age Jurassic	Cassiterite, fluorite, tourmaline, chalcopyrite, topaz, bornite	30,000 t @ 0.4–0.5% Sn. 25,000 t @ 0.3–10% Sn.
Buyant (43)	W-Mo-Be greisen/stockwork and quartz vein	Jurassic granite	Greisenization	Granite age Jurassic	Wolframite, scheelite, fluorite	

Burentsogt (44)	W-Mo-Be greisen/stockwork and quartz vein	Granite pluton intruded Triassic sedimentary rocks	Greisenization	Granite age 140–168 Ma	Wolframite, beryl, fluorite, scheelite, chalcopyrite	0.8–1.0 Mt. @ 0.36–1.63% WO_3; 3000 t WO_3 (mined out) 465 t (remaining)
Baga Gazar (45)	Sn greisen/ stockwork and quartz vein	Granite pluton	Greisenization	Granite age 235–192 Ma greisen 200 Ma	Cassiterite, wolframite, topaz, fluorite, and sulfide minerals	2685.6 t Sn @ 0.18–0.5%, 0.02% WO_3
Ikh Nartiin Khiid (46)	Sn greisen/ stockwork and quartz vein	Granite pluton	Greisenization	Granite age 134 ± 1,6 Ma	Cassiterite, scheelite, fluorite, chalcopyrite	600 t @ 0.4–0.6% Sn
Tumentsogt (47)	W-Mo-Be skarn	Neoproterozoic rocks and Permian volcanics	Greisenization	K-Ar age of granite 140–139 Ma	Wolframite molybdenite, fluorite, beryl	60 Mln.t @ 0.22–0.5% with 143 t WO_3 mined out
Baruun (Western) (48)	W-Mo-Be greisen/stockwork and quartz vein	Contact of granite pluton and upper Proterozoic sedimentary formation	Greisenization	Late Triassic-early Jurassic	Wolframite, galena, arsenopyrite, bismuthite beryl, fluorite	880, 000 t @ 0.28% WO_3, 2000–3000 t WO_3
Ariin Nuur (49)	Mo porphyry	Permian and Jurassic granite	Feldspatization, albitization, greisenization	Permian and Jurassic	Molybdenite, sphalerite, chalcopyrite, ilmenite, chalcosite, galena, cassiterite, wolframite, and oxidized zone minerals	19,850 t @ 0.06–0.1%, Mo, @ 0.3% Cu 21,850 t and 41,850 t
Sharkhad (50)	W-Mo-Be greisen/stockwork and quartz vein	External contact of the Sharkhad alaskite granite pluton	Quartz-muscovite-fluorite	Granite age Jurassic	Wolframite, beryl, molybdenite, fluorite bismuthite, scheelite	
Salaa (51)	W-Mo-Be greisen/stockwork and quartz vein	Granite pluton	Greisen alteration	Late Triassic-early Jurassic	Wolframite, molybdenite, scheelite, chalcopyrite pyrite	230 t Mo, @ 1.4–2.79% WO_3

(continued)

Table 5.1 (continued)

Deposit name	Deposit type (model)	Host rocks	Alteration	Isotope age	Mineralogy	Resources and reserves
Yeguzer (49)	W-Mo-be greisen, stockwork, and quartz vein	Contact zone between granite porphyry stock and Ordovician sediments	Quartz-muscovite and chlorite-carbonate alteration	Granite age late Triassic-early Jurassic	Wolframite, beryl, molybdenite, fluorite topaz, aquamarine, heliodor	21.58 Mt. @0.19–0.23% WO_3, 0.05–0.08% Mo, 0.08–0.12% BeO 45,530 t WO_3, 12,030 t Mo, 41,000 t BeO
Tuviin (central) (50)	Sn-W greisen	Li-F granite stock	Greiseniztion	Granite age late Triassic-early Jurassic	Cassiterite, wolframite beryl, topaz	3.8 Mt. @ 0.13% WO_3, 0.14%BeO, 0.04% Sn, 0.015% Mo
Numrug occurrence (51)	W-Mo-Be greisen/stockwork and quartz vein	Granite stock	Greisenization	Granite age late Triassic-early Jurassic	Wolframite, molybdenite, cassiterite, beryl, fluorite	0.08–0.09% WO_3
Ar Bayan, Uvur Bayan, and Bayan Uul occurrences (53)	W-Mo-Be greisen/stockwork and quartz vein	Granite stock	Greisenization	Granite age late Triassic-early Jurassic	Wolframite, molybdenite, beryl	18, 900 t WO_3, 8.3Mt @ 0.19% WO3, 1.2 Mt. @0.16 WO_3
Batgua occurrence (52)	W-Mo-Be greisen/stockwork and quartz vein	Granite stock	Greisenization	Granite age late Triassic-early Jurassic	Wolframite, cassiterite, molybdenite, galena, chalcopyrite, fluorite, topaz	150 t WO_3
Nukht / Uvurbayan (53)	W-Mo-Be greisen/stockwork and quartz vein	Granite stock	Greisenization	Granite age late Triassic-early Jurassic	Molybdenite, wolframite galena, cassiterite pyrite	1,788,500 t @ 0.22% Mo (4, 030 t Mo)
Umnut (54)	W-Mo quartz vein	Granite stock	Quartz-sericite alteration	Granite age late Paleozoic	Wolframite, molybdenite	3840 t @0. 0.1–2.1% WO_3 200 t @0.01% Mo

Kharmorit (55)	Sn-W greisen/ stockwork and quartz vein	Triassic granite pluton intruded Permian sandstone-shale sequence	Chloritization, argillic, greisenization	Late Triassic-early Jurassic	Cassiterite, wolframite (ferberite), chalcopyrite, sphalerite, beryl, fluorite	780 t of Sn and 65 t of WO_3
Tsagaan Chuluut (56)	Porphyry Mo	Carboniferous granodiorite, granite, cut by granite porphyry dikes	Potassic, chlorite, silica	Granite porphyry dike age Jurassic	Molybdenite, wolframite	34.3 Mlt @ 0.12 (average) Mo and 54,000 WO_3
Zuun Mod (57)	Mo porphyry	Syenite 320–315 Ma, monzonite, granodiorite	Phyllic, potassic, greisen, argillic prophyllitic, silicic, tourmaline	Microgranite 302 Ma; sulfide age 298 Ma	Molybdenite, chalcopyrite	218 Mt. t @ 0.057% (average) Mo and 0.069% Cu
Zost Uul (58)	Mo porphyry	Granite porphyry	Phyllic, potassic silicic	Granite age late Permian-early Triassic	Molybdenite chalcopyrite sphalerite, fluorite	102.0 Kt @ 0.01–0.2, Mo, Cu @ 0.01–0.3%
Munktiin Tsagaan Durvuljin (59)	Li-pegmatite	Contact between Neoproterozoic and early Paleozoic granite pluton	Silica, sericite	Early Paleozoic?	Spodumene, lepidolite	2,276,989 t @ 0.65 Li_2O, 4286.1 t @ 0.15% Rb, 0.622 t @ 0.03% Cs
Buraat Uul (60)	W-Mo-Be greisen/stockwork and quartz vein	Granite stock intruded Devonian andesite	Quartz, sericite, greisenization	Late Paleozoic?	Beryl, molybdenite, wolframite, columbite, tantalite, fluorite	301 t @ 0.22 1.38% (average 0.5%) BeO
Durvun Dert Teg Uul (61)	Be tuff	Be tuff with fragment of ongoryholite		Late Mesozoic	Quartz, feldspar, fluorite B-bearing bertrandite	0.05% Be
Khukh Del Uul occurrence (62)	Li-pegmatite	Precambrian schist and granite pluton	Greisenization	Devonian?	Lepidolite, spodumene, petalite, beryl, cassiterite, fluorite, Li-mica	0.006% Ta_2O_5 76 t Ta_2O_5

(continued)

Table 5.1 (continued)

Deposit name	Deposit type (model)	Host rocks	Alteration	Isotope age	Mineralogy	Resources and reserves
Khukh Del (63)	Sedimentary Li	Conglomerate, gravelstone, sand-stone, siltstone, clay	Silica, sericite limonite, argillic, carbonate,	Host rocks mid-upper Jurassic	Li, K, Cs, Rb, and light REE	992.2 tons of Li metal in 644.3 mil, t ore @ 0.153%, Li Ta 25 g/t 34.4 t Ta
Urt Gozgor, Buural Khangai (64)	Li-F granite Nb(Ta)-Zr-REE alkali granite	Leucogranite and alaskite	Albitization, potassic, greisenization	Granite age 195 Ma granite	Lepidolite, cassiterite, tantalite, columbite	Li-F granites 289–878 ppm, greisen 1315 ppm, albitite 648 ppm Li
Ongon Khairkhan (65)	Li-ongonite	External contact of Ongon Khairkhan pluton	Albitization	120 ma	Albite, microcline, lepidolite, quartz topaz	965–3900 g/t Li 0.005% Ta_2O_5

Economically viable tin deposits occur as a small-sized eluvial and alluvial placer (<10,000 tonnes of Sn) in Modot district (Modot, Khentii province), Janchivlan, with most of the cassiterites mined out in the 1950s. A total of about 20 tungsten small deposits (50–100 tonnes of WO_3) and 98 occurrences have been listed by the Geological Information Center.

Beryl is a common mineral in pegmatites and in W-Mo-Be greisen/stockworks and quartz vein deposits and occurrences. There are about 21 Be occurrences mainly in pegmatites with Be content of 0.2–0.4% Be. Total estimated resources are 36.2 tonnes of Be. Beryllium is also a by-product in the W-Mo greisen/stockworks and quartz vein deposits and occurrences, such as Buraat Uul (0.22–1.38% Be, 301 tonnes BeO), Yeguzer (0.086% BeO, 40.000 tonnes), Undurtsagaan (0.03% BeO, 43.913 tonnes BeO), and others. Beryl is also known from Be-bearing tuffs in Dorvon Dert with content of about 0.05% BeO.

Lithium is known from pegmatites, Li-F granites, ongonites (average Li content of 2.780 ppm), and a recently discovered sedimentary lithium in the Khukh Del deposit in the middle Gobi. The Khukh Del Ulaan Li-pegmatite contains 76 tonnes of Ta_2O_5 grading 0.006% Ta_2O_5, Urt Gozgor placer – 34.4 tonnes Ta_2O_5 with content of 25 g/t. The tantalum and niobium deposits associated with alkali granites contain in Khalzan Buregtei 90.2–0.3% Nb and 0.015–0.025% Ta, in Shar Tolgoi (22.500 tonnes of Ta_2O_5 grading 0.02% Ta_2O_5). Other deposits such as Ulaan Tolgoi, Tsakhir in Lake zone, and Ta-Nb bearing alkali granites could be an additional source for these metals (see Chap. 6, Rare Earth Mineral Deposits).

5.2 Rare Metal Metallogeny

Rare metal deposits and occurrences are known in Mongol Altai, Central Khentii, North Mongolia (Zed-Selenge), Central Khangai, East Mongolian (South Agiin), and South Gobi-Khyangan metallogenic belts (Khasin 1977; Dejidmaa et al. 2002; Rodionov et al. 2004 and extensively reported in Nokleberg 2010).

5.2.1 Mongol Altai Metallogenic Belt

Mongol Altai (MA) metallogenic belt of W-Mo-Be greisen, stockworks, and quartz vein deposits (Western Mongolia). This late Triassic-early Jurassic metallogenic belt includes small plutons of leucogranite that intrude the Altai and Khovd turbidite terranes. The belt is hosted in granitoids that intruded along the major northwest-striking Khovd regional fault zone (Dandar 2012) and is formed during Mesozoic intraplate rifting related to magmatism along extension zones (Nokleberg 2010). In this region various REE deposits are related to middle Devonian collisional, Carboniferous post-collisional, and Permian and early Jurassic late-stage and post-orogenic granitoids (Dandar et al. 1999). The belt extends from northeast to southeast. Three major ore fields (Nuuriin Gol, Ulaan Uul, and Bodonch) and Achit Nuur ore district are distinguished with medium- to small-sized W deposits and a large number of occurrences. The major deposits are the Ulaan Uul and Tsunkheg W-Mo-Be deposits; the W-Mo occurrences in the Ulaan Uul ore field; and a number of occurrences in Achit Nuur district and Bodonch (W-Sn) ore field.

5.2.2 Central Khentii Metallogenic Belt

Central Khentii (CHE) metallogenic belt of Sn-W and W-Mo-Be greisen, stockworks, and quartz vein, REE-Li-pegmatite, and Ta-Li-ongonite includes four ore districts: South Khentii, Ikh Khairhan, Modot, and Upper Onon with Tsagaan Davaa, Ikh Khairhan, Ongon Khairkhan W deposits; Janchivlan Sn, W, Ta, and Li occurrences; Li-ongonite; Modot Sn-W vein/stockwork and Sn placer deposits; Zuun Tarts Sn-W occurrences and placer deposits; Undurtsagaan W-Mo; and Tsagaan Chuluut Mo deposit. This metallogenic belt was formed during final closure of the Mongol-Okhotsk Ocean (Zonenshain et al. 1990; Kovalenko and Yarmolyuk 1995; Koval 1998) followed by a high volume of collisional granitoids. The deposits are related to small plutons emplaced during a continental post-collisional event. The margins of this metallogenic belt are northeast-trending faults that may also be favorable for epithermal Au deposits and intrusion-related sedimentary-hosted deposits (Gerel et al. 1999). The Sn-W deposits and occurrences are hosted in a late Triassic and early Jurassic granodiorite and granite belt that forms the Khentii megadome (600 km long and 200–220 km wide) trends northeast. This dome is the Mongolian part of the Khentii-Daurian megadome that has been uplifted from the early Mesozoic through the recent. The Khentii megadome contains Devonian and Carboniferous turbidite intruded by Paleozoic and Mesozoic granitoids. The host granites have a K-Ar isotopic age of 190.49 $\pm$ 4.7 Ma and an Rb-Sr isotopic age of 225 to 188 Ma and consist of three types: (1) coarse-grained porphyritic biotite granite and rare amphibole-biotite granite, (2) medium-grained two mica granite, and (3) K-feldspar biotite granite (alaskite) and Li-F granite including microcline-albite, amazonite-albite, and lepidolite-albite granites. Many granites are peraluminous with typical higher alkalinity (Gerel 1995; Gerel et al. 1999). The first granite type contains many unique miarolitic pegmatites (Gorkhi, Zuunbayan, and Janchivlan) with piezoelectric quartz mined out at the 1960s. The second granite type contains W-Sn veins (Modot, Bayan Mod, and Khujikhan) and rare scheelite skarn. The third granite type contains Ta-bearing granite (Urt Gozgor, Buural Khangai, Borkhujir), W-Sn vein and Be greisen deposits (Tsagaan Davaa). Numerous Sn placers, including the very large Sn placer deposit in Modot district, occur nearby. Greisen bears biotite (zwitter) and contains topaz-quartz, tourmaline-quartz, and muscovite-quartz zones. In the Ikh Khairkhan ore district, W-Sn greisen/stockwork and quartz vein deposits and Li-ongonite are associated with late Jurassic-early Cretaceous granites (Nokleberg 2010).

5.2.3 North Mongolian Metallogenic Belt

North Mongolian metallogenic belt is a part of large Zed-Selenge (ZSe) metallogenic belt (Nokleberg 2010). In northern Mongolia, the district is related to granitoids intruding the Ediacaran-early Paleozoic Zed island-arc terrane. The metallogenic belt is hosted in the Transbaikal sedimentary-volcanic-plutonic belt. The belt is formed during subalkaline and alkaline granitoid magmatism associated with transform continental-margin faulting (Mongol-Okhotsk and related faults) and magmatism during late-stage closure of the Mongol-Okhotsk Ocean. The granitoids consist of small stocks of granite porphyry and leucogranite, dikes of aplite, fine-grained granite, syenite, syenite porphyry, and granite porphyry. The district contains W-Mo-Be greisen, stockworks, and quartz vein and W-Mo and W-Mo-Be skarn. The major deposit is the Bulagtai vein and stockworks W-Mo deposit, Altan Boom Ta-Nb occurrence, Tsagaan Uul and Tsagaan Tolgoi, and other Mo and W occurrences. These mineral systems are closely related to late Jurassic leucogranite stocks of isotopic age of 180–170 and 145–140 Ma (Nokleberg 2010).

5.2.4 Central Khangai Metallogenic Belt

Central Khangai (CHA) metallogenic belt of Sn-W, W-Mo-Be greisen/stockworks, and quartz vein deposits includes about 30 occurrences of Sn, W, Mo, Be, and Li associated with leucogranites mainly of Triassic-early Jurassic age cut granites of Permian Khangai Batholith. The belt is formed during the closure of Mongol-Okhotsk Ocean and post-collisional events in the late Triassic-early Jurassic time.

5.2.5 East Mongolian or South Agiin Belt

East Mongolian or South Agiin (EMA) belt is a part of large East Mongolian-Priargunskiy metallogenic belt (Nokleberg 2010). The belt is formed during middle Jurassic-early Cretaceous extensional tectonism. The belt includes a few cassiterite-wolframite-quartz and wolframite-cassiterite-beryl-quartz veins that are related to small plutons composed of biotite, two mica, and muscovite-fluorite leucogranite with a K-Ar isotopic age of 146 Ma intensely altered to greisen (Koval 1998). Major W-Mo deposits (Undurtsagaan, Chuluunkhoroot, Narsan Khundlun), (Tsagaanchuluut Mo stockwork and greisen deposit), deposits, and occurrences in South Kherlen (Tumentsogt, Burentsogt W-Mo deposits) and in Middle Gobi (Sn-polymetallic skarn Oortsog Ovoo, Sn greisen Baga Gazar deposit, Ikh Nartiin Khiid, Buyant W vein, Khukh Del Uul pegmatite) are all associated with late Jurassic leucogranite and granite porphyry stocks. This type of deposit occurs mainly in small outcrops that are overlain by a late Jurassic through early Cretaceous basalt and rhyolite bimodal sequence. Most deposits were exhausted before 1970, but the Undurtsagaan W-Mo and Tsagaanchuluut Mo stockworks and greisen deposits are new discoveries.

5.2.6 South Gobi-Khyangan Metallogenic Belt

South Gobi-Khyangan (SH) of Sn-W greisen, stockworks, and quartz vein metallogenic belt (Khasin 1977; Dandar 2012) or Kharmorit-Khanbogd-Lugiingol (HL) belt of Sn-W greisen, stockworks, and quartz vein, REE (±Ta, Nb, Fe) carbonatite, peralkaline granitoid-related Nb-Zr-REE, and REE-Li-pegmatite deposits and Khyangan metallogenic belt of stockworks, quartz vein; cassiterite-sulfide-silicate vein and stockworks; W-Mo-Be greisen, stockworks, quartz vein deposits (Nokleberg 2010), or Gobian rare metal-rare earth belt (Kovalenko and Koval 1984) includes large number of rare metal deposits and occurrences. The major Sn-W greisen deposits in this metallogenic belt are Sn occurrences in Kharmorit ore field associated with the Khalzan Uul leucogranite Complex (Rb-Sr isotopic age of 194 ± 9.06 Ma) and Sn placers. The Kharmorit-Khanbogd-Lugiingol belt is formed during the emplacement of collisional granitoids during final closure of the Mongol-Okhotsk Ocean and the formation of Mongol-Transbaikal arc. The Khyangan metallogenic belt is related to Khyangan continental-margin arc (Nokleberg 2010). The isotopic ages of granites hosting or associated with the deposits range from 100 to 75 Ma. The largest is the Yeguzer W-Mo-Be greisen/stockwork and vein deposit.

5.3 Description of Selected Sn-W Deposits

The Sn-W deposits are classified as Sn-W greisen, stockworks, and quartz vein; Sn pegmatite; W-Mo-Be greisen, stockworks, and quartz vein; and cassiterite-sulfide-silicate vein and stockworks and Sn (W) placers (Nokleberg 2010; Dejidmaa et al. 2005).

5.3.1 Sn-W Greisen, Stockworks, and Quartz Veins

The Sn-W greisen, stockworks, and quartz vein deposits consist of individual veins and multiple vein systems, very common, but many are of small size. The deposits occur in or near granitic intrusions emplaced at relatively shallow levels and structurally controlled by deep faults and fissures. The tectonic environment is interpreted as a continental collision belt. The principal ore minerals are cassiterite, wolframite, and scheelite. Associated minerals are molybdenite, chalcopyrite, sphalerite, pyrite, arsenopyrite, muscovite, fluorite, beryl, topaz, feldspar, quartz, and tourmaline. Hydrothermal alteration assemblages consist mainly of quartz-white mica (greisen) with fluorite, topaz, tourmaline, Li-mica, muscovite, biotite, quartz, and other minerals such as albite and microcline. Cassiterite, wolframite, chalcopyrite, sphalerite, pyrite, and other sulfide minerals are disseminated in greisen-altered rocks.

5.3.1.1 Janchivlan Ore District

Janchivlan ore district of the Central Khentii metallogenic belt hosts W, Sn, W, Ta-Nb, Be, and Li greisen and quartz vein mineralization associated mainly with alaskite and Li-F granites and a number of Sn placers. The Janchivlan pluton is a multiphase intrusion of about 1000 km^2 and is composed of stage I porphyritic coarse-grained granite with a number of miarolitic pegmatites, the stage II two mica medium-grained granites with W-Sn greisen and placers (Elstei, Elst, Zuun Elst, Bayandavaa), W-Sn quartz veins (Janchivlan, Bayandavaa, Urt Gozgor), and placers, alaskite, and Li-F granites: microcline-albite, amazonite-albite, and lepidolite-albite with Sn-Ta-Li mineralization (Urt Gozgor, Buural Khangai). Sn-W greisen and quartz vein occurrences in Janchivlan ore district are mainly of non-commercial value but are the source for placer deposits.

5.3.1.2 Modot Sn-W Deposit

Modot Sn-W greisen, stockworks, and quartz vein deposit (Khasin 1977; Jargalsaihan et al. 1996) consists of Sn-W quartz veins related to Mesozoic granite pluton with a K-Ar isotopic age of 199–175 Ma. The pluton intrudes the Vendian-Early Cambrian metamorphic rock, Paleozoic granitoids, and Permian molasses. The deposit occurs along the pluton margins, in the pluton, or in adjacent hornfelses. The veins dip gently and strike northwest to north. Some veins dip steeply. The ore minerals are cassiterite, wolframite, arsenopyrite, pyrite, galena, sphalerite, and chalcopyrite. Greisen alteration occurs. The deposit is small and has produced 300 tonnes of WO_3.

5.3.1.3 Tsagaan Davaa Deposit

Tsagaan Davaa deposit (Fig. 5.2) is situated about 100 km NW of Ulaanbaatar. This deposit (Ivanova 1976; Khasin 1977; Jargalsaihan et al. 1996) consists of quartz-wolframite veins and zones that occur in a multistage late Triassic-early Jurassic Tukhum (Tsagaan Davaa) granite pluton intruded into slightly metamorphosed Cambrian to Ordovician sedimentary rocks of Kharaa Group. The host granite has a K-Ar isotopic age of 190.49 $\pm$ 4.7–224 Ma (Smirnov et al. 1977) and LA-ICP-MS zircon isotopic age of 199.36 $\pm$ 0.73 Ma (unpublished data, 2018). Host granite

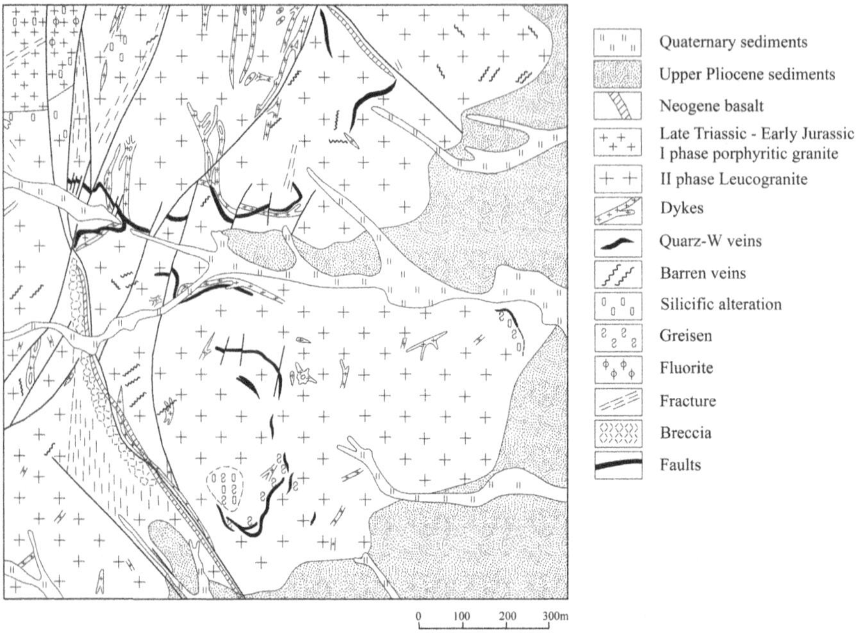

Fig. 5.2 Schematic geological map of the Tsagaan Davaa tungsten deposit (from Khasin 1977)

rocks consist of three types: (1) coarse-grained porphyritic biotite granite and rare amphibole-biotite granite; (2) medium-grained two mica granite; and (3) K-feldspar biotite granite (alaskite) and Li-F-enriched granite including microcline-albite and amazonite-albite granites. Granites are slightly peraluminous, calc-alkaline series with high-K and obviously enriched in Rb, Th, K, and Pb and depleted in Nb, Ti, La, and Ce with relative enrichment of Zr and Hf. Ore veins mainly occur in the central upper part of the pluton that consists of fine- to medium-grained biotite and leucocratic granite. The veins are 2 km long and 200–500 m wide and occur at different hypsometric levels. The veins form subhorizontal bodies dip gently south, southeast, and southwest, parallel with pluton roof. Ore minerals are wolframite, cassiterite, molybdenite, and beryl, with rare chalcopyrite and pyrite. Gangue minerals are garnet, fluorite, and biotite. Greisen and silica alterations are common. Assemblage of biotite and fluorite is characteristic in the deposit. Besides wolframite and cassiterite, the veins also contain minor amounts of scheelite and chalcopyrite and abundant supergene azurite and malachite along fractures.

The deposit is of medium size and has a resource of 3139 tonnes of WO_3 in ore grading 1.52% WO_3 and 738 tonnes of WO_3 in ore grading 0.72% WO_3. The deposit went into production in 1978, and about 3000 tonnes of wolframite concentrate were produced. According to the latest ore reserve calculation, another 1000 tonnes of wolframite will be available for mining. According to BGR analysis, one strip sample of the tailings contains 0.44% WO_3 and 0.05% cassiterite. The tailings could contain some 400 tonnes of WO_3 and 45 tonnes of cassiterite (Elsner et al. 2011).

5.3.1.4 Ikh Khairkhan Ore District

Ikh Khairkhan ore district located 160–170 km southwest of Ulaanbaatar includes Ikh Khairkhan, Ongon Khairkhan, and Khar Chuluut tungsten deposits and a number of occurrences associated with granites of Late Mesozoic age. Plutons occur in the Kharkhiraa uplift that divided the Upper Paleozoic structures of Khentii and Khangai. The geology of this area is dominated by mid-Paleozoic sandstone, siltstone, and quartzite, covered by volcano-sedimentary sequence of Ikh Khairkhan basin and Cenozoic sediments. The Silurian to Devonian volcanic and sedimentary rocks of Kharkhorin Group are intruded by late Mesozoic plutons of Ikh Khairkhan, Ongon Khairkhan, Baga Khairkhan, Maikhan, and Bayan Baraat with hornfels halos and Cretaceous ongonite dikes. Plutons have isometric shapes up to 500 km^2 in size or form asymmetric cupolas. Based on a gravity study, most of the plutons are an apical part of a large plate-shaped intrusion (Turutanov and Zorin 1978). The dominant main type is gneissic fine- to medium-grained biotite granite, sometimes porphyritic (Ongon Khairkhan pluton); a second type (making up no more than 12% of the intrusion area) is represented by small pegmatoid granite and fine- to medium-grained leucogranite, Li-F granite, ongonite, and granite porphyry.

Ikh Khairkhan deposit (Fig. 5.3) is associated with the Baga Khairkhan pluton of K-Ar age of 154–121 Ma (Koval 1998) and occurs in the outer contact in a northeast-trending of 2–3 km fracture zone, within replaced to hornfels andesitic volcanics.

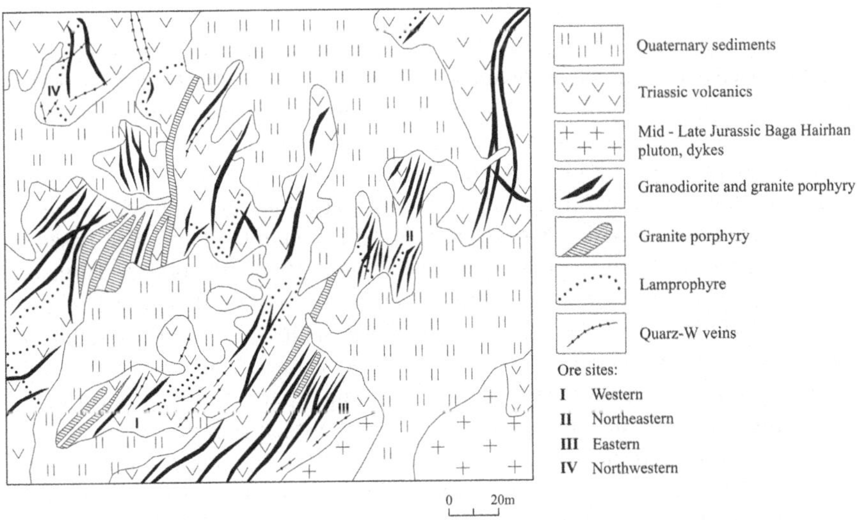

Fig. 5.3 Schematic geological map of Ikh Khairkhan tungsten deposit (Source Ivanova 1976)

Numerous dikes of diorite porphyry, granite porphyry, lamprophyre, and the diabase of Mesozoic age are widely distributed in the deposit area. The deposit is divided into western, northeastern, eastern, and northwestern prospects, of which the most economically important is the western prospect. Fourteen ore bodies were discovered in the host Triassic andesitic volcanic rocks, but only six quartz-wolframite veins, from 160 to 900 m long, were economic. The major vein is 900 m long, dipping 70–80° to the NE over 550 m, and is from 0.3 to 8.0 m thick. Veins generally are 60–550 m long and 0.2–5.0 m thick and are up to 300 m deep. Ore bodies have complex morphology, irregular thickness, and mineralization (Ivanova 1976). Ore styles are massive, less often banded. Ore minerals consist of hubnerite, scheelite with minor sphalerite, molybdenite, chalcopyrite, cassiterite, and fluorite.

Four stages are distinguished: hubnerite, scheelite, sulfide, and quartz-fluorite-calcite (Ivanova 1976). The first hubnerite stage is represented by main quartz-hubnerite and subordinate fluorite-muscovite-feldspar-hubnerite associations. Hubnerite crystals are up to 10 cm large and have 73–90% $MnWO_3$. Fluorite is very common, and beryl is rare. The scheelite stage is limited. In the sulfide stage, two associations sphalerite-chalcopyrite and sulphobismuthite-chalcopyrite have been distinguished. In the latest stage, quartz-fluorite and fluorite-calcite associations were formed. The homogenization temperature of zonal fluorite varies from 235 to 190 °C in core. A supergene zone is composed of Fe, Mn, W, Bi, Zn, and Cu oxides and hydroxides developed on hubnerite, pyrite, and other minerals. The deposit was mimed underground in 1962–1972. Calculated reserves are 480 tonnes of WO_3 in ore grading 1.32% WO_3 based on cutoff grade of 0.4% WO_3 and a minimum vein's thickness of 0.8 m.

5.3.1.5 Ongon Khairkhan Deposit

Ongon Khairkhan deposit (Fig. 5.4; the name ongonite is derived from Ongon) is a medium-sized deposit associated with up to 200 km^2 in size of the Ongon Khairkhan pluton of K-Ar age of 183–131 Ma, intruded into mid-Paleozoic metamorphosed sedimentary rocks. The ore deposit has a complex structure with pre-ore dislocations, represented by fractured, brecciated, and silicified zones.

The deposit consists of five ore bodies with stockwork and vein-type mineralization located in host rocks, 3–6 km northeast from granite pluton, that based on gravimetric data is not exposed and could be in the depth of 500–600 m (Turutanov and Zorin 1978). Vein type is located in the southern part of the deposit, where main quartz-wolframite veins are concentrated within a thick zone of fractured and silicified sandstones and siltstones. Veins are bead-shaped, from 30 to 100 m long, divided by zones of greisenized sandstones with thin (up to 3 cm) quartz-wolframite veinlets. All veins are steeply dipping (70–85°). The stockwork bodies occur in the central part of the deposit and are 60–500 m long and 0.2–5 m thick. Similar to the Ikh Khairkhan deposit, four mineralization stages are distinguished. The homogenization temperature in the earliest fluorite of hubnerite stage is 295–235 °C, and the latest is 170–103 °C (Ivanova 1976). Kovalenko et al. (1971) comparing

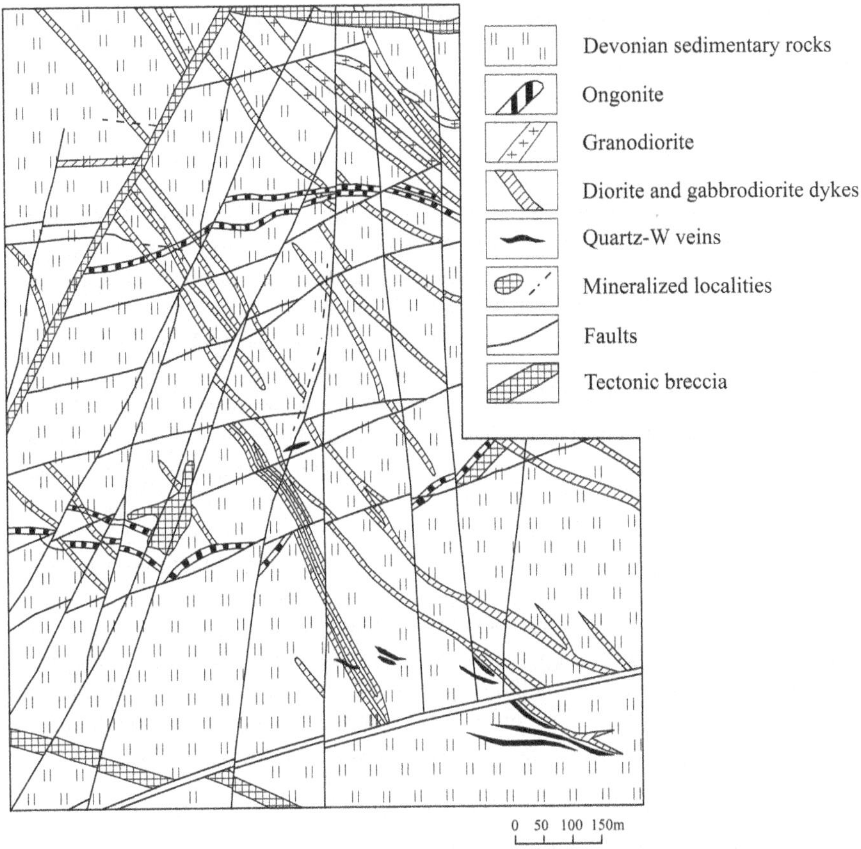

Fig. 5.4 Schematic geological map of the Ongon Khairkhan W deposit (Source Dandar 2012)

mineralization with ongonite dikes suggested two different stages of mineralization: quartz-wolframite vein type is associated with the Ongon Khairkhan granite pluton, while stockworks are associated with a not exposed Li-F granite intrusion. Estimated resources 7.4 Mt. graded 0.15%WO_3; 5.3 Mt. graded 0.05–0.10% WO_3 with tonnage of 11.136 tonnes of WO_3, the test mining in the late 1980s and early 1990s is 3.993 tonnes of WO_3.

5.3.1.6 Khar Chuluut Deposit

Khar Chuluut deposit lies 15 km south of the Ikh Khairkhan deposit and is associated with the Bayanbaraat granite pluton. The mineralization occurs within medium-grained to slightly porphyritic granites and in the northwestern part within host hornfelses after sedimentary rocks and Triassic volcanics. Stockworks and greisen zones are developed along a shear zone. A total of 20 ore bodies were discovered, including six larger bodies. Resources are calculated 104, 900 tonnes of ore grading 0.66%WO_3, about 700 tonnes of WO_3.

5.3.1.7 Kharmorit Deposit

Kharmorit deposit (Fig. 5.5) is situated in South Gobi province within South Gobi tin district. The deposit (Ivanova 1976; Dandar 2012) consists of greisens and veins in the apical part of a Li-F granite porphyry stock and in adjacent host rocks. The granite has an Rb-Sr isochron age of 194 ± 9.06 Ma.

The greisen zones extend from 100 to 500 m long and as much as 3 m wide. The deposit has two types: (1) cassiterite-wolframite-quartz vein and cassiterite-wolframite-zinnwaldite-quartz greisen and (2) cassiterite-sulfide with Sn, Cu, Pb, and Zn. The ore minerals are cassiterite, pyrite, arsenopyrite, galena, sphalerite, and chalcopyrite and rare scheelite and wolframite. Gangue minerals are quartz, muscovite, zinnwaldite, beryl, tourmaline, sericite, and chlorite. Very common minerals are topaz and fluorite. A well-developed supergene zone contains relics of sulfides and secondary minerals. Tin content is very irregular and sometimes very high. Thereby, the deposit exhibits a complex mineralization and includes Sn-sulfide, Zn-Pb and Be, Sn-W greisen, and Sn-W vein stages. The various stages are zoned and developed in the altered cupola of the stock with wolframite and cassiterite, in hornfels with cassiterite and sulfides, and with cassiterite in host sandstone and shale. Associated Sn placer deposits are also common. The deposit is small and has resources of 780 tonnes of Sn and 65 tonnes of WO_3 (Nokleberg 2010).

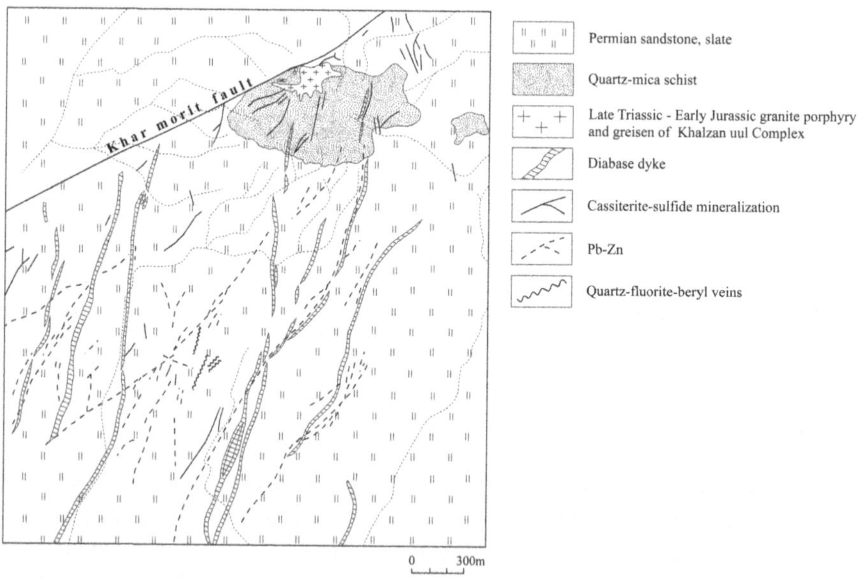

Fig. 5.5 Distribution of ore zones in the Kharmorit W-Sn deposit (from Khasin 1977)

5.3.2 W-Mo-Be Greisen, Stockworks, and Quartz Vein Deposits

These mineral systems occur mainly in small outcrops that are intruded by granitoids and/or overlain by a late Jurassic through early Cretaceous basalt and rhyolite bimodal sequences. Major W-Mo vein and stockworks are associated with late Jurassic leucogranite and granite porphyry stocks: Chuluunkhoroot, Undurtsagaan, Tumentsogt, Burentsogt deposits in northeastern Mongolia, Ikh Nartiin Khiid deposits in Middle Gobi, and Bulagtai in northern Mongolia. Most deposits were exhausted before 1970, but the Undurtsagaan W-Mo and Tsagaanchuluut Mo stockworks and greisen are new discoveries. The W-Mo vein deposits consist of wolframite, molybdenite, and beryl that occur in various veins and breccia zones. The veins generally also contain various combinations of muscovite, fluorite, carbonate, scheelite, pyrite, arsenopyrite, chalcopyrite, sphalerite, and bismuthinite. The main examples are Undurtsagaan, Yeguzer, Burentsogt, and Chuluunkhoroot. Some W-Mo deposits (Yeguzer, Khujkhan, Umnu Khutul) have aplitic cupolas, sheeted quartz veins, and hydrothermal breccia pipes with unidirectional solidification texture (UST) (Kirwin 2005).

5.3.2.1 Bulagtai Deposit

Bulagtai deposit is located in North Mongolia near the border with Russia within Zed-Selenge metallogenic belt. This deposit is associated with Jurassic (180–170 Ma) and early Cretaceous granite plutons (145–140 Ma) which intruded Cambrian shale and metavolcanic rocks. Mineralization styles are stockworks, greisen, and veins. Stockworks comprise quartz-wolframite, quartz-molybdenite-wolframite, quartz-wolframite-sulfide (pyrite, chalcopyrite, galena, and arsenopyrite), and fluorite-bearing veinlets. Quartz-wolframite veins extend to the Russian territory and were mined out. Veins are 0.1–0.2 m thick with 1–2 cm wolframite crystals that form nest-like accumulations. Tungsten content is 0.004–0.1% and Mo 0.04%. There are also five occurrences of Sn and W greisen and Mo skarn. The Bulagtai deposit's resources were calculated as 4000 tonnes of W and 1400 tonnes of Mo grading 0.03% WO_3 and 0.01% Mo, respectively.

5.3.2.2 Undurtsagaan Deposit

Undurtsagaan deposit (Fig. 5.6) is the largest in Mongolia and a world-class tungsten deposit with a relatively low grade but high tonnage, containing beryllium, molybdenum, and tin. It occurs in the Khentii province 80 km NW of the Undurkhaan province center. The deposit was discovered in 1977 by an International geological expedition and prospecting and exploration work carried out in 1978–1982. The deposit is related to upper Onon ore district of the Central Khentii Sn-W

Fig. 5.6 Geological map of the Undurtsagaan W-Mo-Be deposit (from Dandar 2012)

metallogenic belt. The Undurtsagaan W-Mo deposit consists of stockworks and greisen hosted in the lower Devonian greenschist facies metamorphosed shale with a gabbro sheet of 2 km long and 500 m wide. Mineralization is hosted in the gabbro, which is replaced by chlorite, epidote, and limonite alteration.

At the depth of 440 m in drill core mineralized granite porphyry and leucogranite of 192–179 Ma, K-Ar age has been discovered. Muscovite K-Ar age from vein yields 175 $\pm$ 6 Ma. A large number of quartz-wolframite, quartz-wolframite-molybdenite, quartz-beryl-topaz, quartz-muscovite-beryl, quartz-feldspar, and quartz veinlets form a stockwork system. The stockwork is 1500 m long and 1000 m wide and elongated to NW. Veinlets are from 0.1–3 cm up to 20 cm thick. There are also scheelite- and sulfide-bearing quartz-carbonate veinlets. Veinlets with wolframite and beryl are observed in the core of stockwork while veinlets with sulfides in the margins. Alteration assemblages after gabbro are composed of quartz-biotite-sericite, pyrite-chlorite-actinolite, and fluorite-wolframite-biotite. Tonnage are 175, 990 tonnes of WO_3, 26,175 tonnes of Mo, 43,350 tonnes of BeO, and 13,460 tonnes of Bi. Calculated resources are 134 Mt. grading 0.127% WO_3, 0.036% Mo, 0.03% BeO, and 0.01% Bi.

In the ore field of Undurtsagaan deposit, the Mungun Undur Ag-Pb-Zn deposit occurs in association with dikes of porphyritic syenite, diabase, and basaltic andesite.

5.3.2.3 Salaa Deposit

Salaa deposit lies 16 km east of the Baruun Urt town in Eastern Mongolia. The deposit is associated with late Triassic (K-Ar age of 205 Ma) leucocratic granite pluton, forming of 12 quartz-wolframite veins (65% $MnWO_3$). Host granites are altered to quartz-muscovite greisen. Quartz-wolframite veins are from 100 to

400–500 m long and 0.5–1.3 m thick, sometimes with bulges up to 2.5–3.3 m. Some veins of economic interest were traced to depths of 120 m and 40–60 m. Wolframite in the veins is irregularly distributed locally forming crystals up to 20 cm in size. Besides wolframite, there are rare pyrite, chalcopyrite, and molybdenite. Resources are estimated of 230 tonnes grading 1.4–2.79% WO_3.

5.3.2.4 Chuluunkhoroot Ore Field

Chuluunkhoroot ore field is situated in Dornod province, 20 km north of Dashbalbar, 250 km north of Choibalsan town. The ore field includes one with the same name W-Mo-Be deposit and a number of occurrences.

Chuluunkhoroot deposit is associated with small late Jurassic Chuluunkhoroot intrusion of fine-grained granites intruded into Devonian and Triassic sandstone, siltstone, and slate and concentrated in the apical greisenized part of granite intrusion. Many pegmatites and quartz-beryl-wolframite, quartz-fluorite, and quartz veins are exposed in two hills: Vein and Ore, divided by Chuluunkhoroot valley. Quartz-wolframite veins are from 20 to 100 m long and from 0.1 to 0.3 m wide. The main vein is 150–160 m long, 0.2–0.4 m thick, and up to 60 m deep. Wolframite aggregates reach 20–30 kg even up to 500–100 kg. Mineralization is irregularly distributed. The average grade is 1.36% WO_3, 0.06% Sn, 0.14% Bi, and 0.05% Mo. The total calculated reserves are 500 tonnes of WO_3 with remaining reserves of 300 tonnes of WO_3. A total of 229 tonnes of WO_3 have been mined.

5.3.2.5 Yeguzer Ore Field

Yeguzer ore field is located in southeastern part of Mongolia, 270 km from Choibalsan town, and includes the Yeguzer medium-sized deposit together with other small deposits and occurrences. The ore field is dominated by intrusions of Paleozoic and Mesozoic age. Three stages (molybdenite, wolframite, and sulfide) of mineralization were distinguished (Ivanova 1976). The molybdenite stage includes three mineral associations: quartz-muscovite-molybdenite, quartz-muscovite-beryl-molybdenite, and quartz-biotite-beryl-fluorite. These minerals occur as veins and veinlets of massive structure in contact with host rocks. The wolframite stage includes quartz-muscovite-wolframite and quartz-topaz-wolframite associations. In some deposits hubnerite stage is observed to be characterized by quartz-rhodochrosite and quartz-arsenopyrite-hubnerite associations in the brecciated and greisenized granite porphyries but rare in the hornfelses. The sulfide stage is developed in veinlets and brecciated zones of greisenized granite porphyries and greisens and is characterized by quartz-polymetallic associations (quartz, galena, and sphalerite) that overprint the hubnerite mineral association. With galena and sphalerite, chalcopyrite and fahlore are also present. The latest are carbonates with siderite, ankerite, and fluorite-calcite associations.

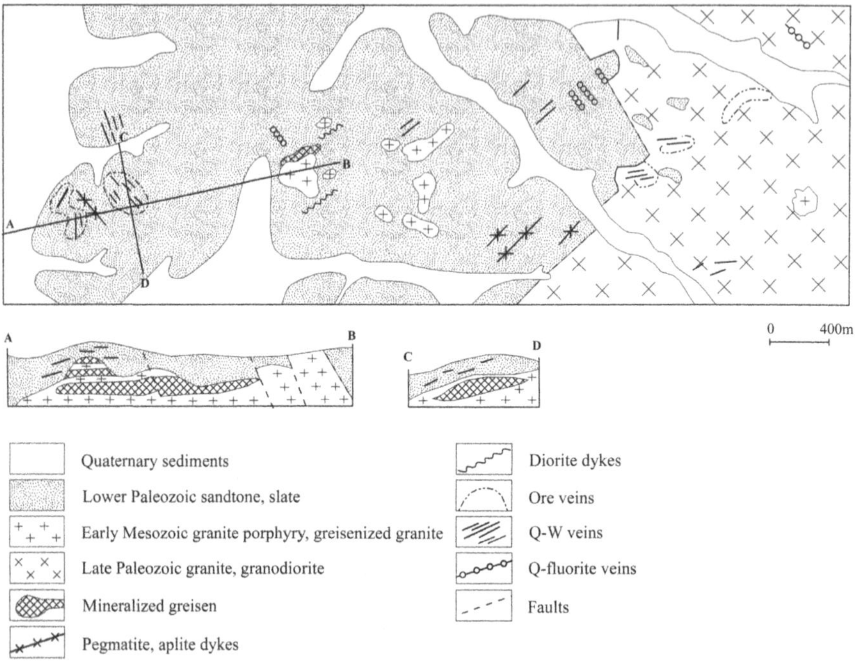

Fig. 5.7 Geological sketch map of the Yeguzer W-Mo-Be deposit (Batjargal et al. 2017)

Yeguzer deposit (Fig. 5.7) was mined in 1963–1966. Mineralization is represented by Mo-bearing greisens and greisenized granite porphyries in the apical part of the late Triassic-early Jurassic granite pluton composed of fine- to medium-grained leucogranite and alaskite, intruded into lower Paleozoic terrigenous sequence. Leucogranites are depleted in Ba, Sr, Eu, and Nb. Wolframite-bearing greisens are located in the apical part of a crescent-shaped granite stock with surface dimensions of approximately 2 km^2 (Jargalsaihan et al. 1996). The UST quartz layers were described by Batjargal and Lkhamsuren (1979). The host rocks are aplitic porphyries and pervasively sericitized, and the USTs are semi parallel bends of quartz, mm to cm in width. The quartz-wolframite veins are localized in host sedimentary rocks. There are about 120 quartz-wolframite veins. Gruner et al. (1984) identified a greisen zone 3000 m long, 1500 m wide, and 150–180 m deep with four greisen orebodies with an average grade of 0.19–0.23% WO_3 and 0.05–0.08% Mo, 0.08–0.12% BeO, and 0.13–0.21% Bi. The greisen zone is located in the contact between granite porphyry and sedimentary rocks metamorphosed to hornfelses. The thickness of the greisen varies from 5–6 to 84 m, with average of 21 m.

Ore veins are 100–500 m long, 0.5–14.3 m thick, and 50–250 m deep with very high content of WO_3 ranging between 2% and 8% and contain wolframite, molybdenite, beryl, fluorite, and topaz with a minor amount of aquamarine and heliodor. Wolframite is partly oxidized and very fine-grained. Oxygen isotope $\delta^{18}O$ data in quartz-molybdenite veins is 10.8‰, in quartz-wolframite vein 10.9‰ (Batjargal et al. 2017), showing very close origin for both ore minerals. Homogenization temperature in quartz veins varies between 296 and 384 °C. Estimated resources reach 21.58 Mt. of grade 0.19–0.23 WO_3, 0.05–0.08% Mo, and 0.08–0.12% Be O and with 45,530 tonnes of net WO_3 and 12,030 tonnes of net Mo and 41,000 tonnes considerable reserves of BeO and 912,000 tonnes of Bi (Gruner et al. 1984).

5.3.2.6 The Tov (Center) Deposit

The Tov (Center) deposit is located close to the Yeguzer deposit and comprises 38 greisen ore bodies of 20–480 m length, 1–20 m thick, and about 100–250 m in depth. Another 40 lenticular bodies of 60–120 m length, but only of 1–2 m in thickness, were discovered during trenching in historic times. The main ore minerals are disseminated wolframite, scheelite, beryl, cassiterite, and molybdenite accompanied by minor sulfides. The total calculated reserves are 5340 tonnes of BeO with a grade of 0.2%, 4990 tonnes of WO_3 with a grade of 0.06%, 1360 tonnes of Sn with a grade of 0.08%, and 530 tonnes of Mo with a grade of 0.015%.

5.3.2.7 Uvurbayan Deposit

Nukht/Uvurbayan deposit is located 36 km south from Yeguzer. The vein-type deposit was explored in 2006–2007. The ore partly contains massive molybdenite and pyrite as well as wolframite, cassiterite, and galena. The reserves are 1,788,500 tonnes of ore at 0.22% Mo (4030 tonnes of Mo) and 2,512,500 tonnes of ore at 0.21% Mo (5260 tonnes of Mo, C1). An analysis of the flotation test concentrate gave grades of 24.0% Pb, 21.5% Mo, 8.3% Zn, and 1.6% Bi and minor amounts of Cu, Sb, and Sn (Elsner et al. 2011). Wolframite was extracted separately. The wolframite from ore samples shows increased contents of niobium.

Ar Bayan deposit and the Bayan Uul occurrence are located 40 km NE of Yuguzer deposit. Mineralization is confined to greisen and stockwork zones within late Triassic-early Jurassic granites. The stockwork mineralization grades 0.18% WO_3 with minor amounts of Be, Sn, Pb, Li, and Ag. Elevated Mo grades of 0.03% may also be of interest. The orebodies are 4–40 m thick and up to 250 m long but not well defined.

The total WO_3 resources are 18,900 tonnes. Of these, Bayan Uul contains 2195 tonnes of WO_3 and 2919 tonnes of Mo.

5.3.2.8 Batguai Deposit

The Batguai wolframite-molybdenite deposit lies about 8 km NE of Yeguzer. The main minerals are wolframite, molybdenite, scheelite, beryl, fluorite, bismuthinite, and arsenopyrite. Two veins of 20–60 m length may contain reserves of 150 tonnes of WO_3.

5.3.2.9 Umnut Deposit

The Umnut tungsten deposit lies about 70 km SE of Erdenetsagaan, close to the Chinese border. Tungsten mineralization is related to shear and greisen zones within Carboniferous granite and occurs in 15 quartz-wolframite veins (minor molybdenite) with a length of up to 60 m, an estimated depth of up to 60 m, and a WO_3 content of 0.14–0.54%. The estimated reserves are only 18.4 tonnes of WO_3. Other information suggests reserves of 164 tonnes of WO_3.

5.3.2.10 Umnu Khutul Deposit

Umnu Khutul deposit is located 20 km south of the Umnugobi soum, Uvs aimag in Western Mongolia. Mineralization occurs in granite stock hosted in Altai Group turbidite composed of siltstone, sandstone, and shale. Quartz-wolframite-fluorite-molybdenite mineralization forms stockwork 500 m long and 80 m wide with a UST zone of 200 m and up to 15 m in thickness. The textures vary considerably from parallel quartz bands to a highly contoured layer including atoll structures (Kirwin 2005). The thickness of individual quartz bands ranges from less than one of mm to tens of cm. WO_3 content in veins is 0.2–1%, Mo-0.1%, Sn 0.6%, and Li-0.2% and resources, 3000 tonnes of WO_3 and 600 tonnes of Mo.

5.3.2.11 Tumentsogt Ore District

Tumentsogt ore district includes the Tumentsogt, Western and Sharkhad deposits, and several occurrences associated with Mesozoic granite plutons. Tungsten mineralization is accompanied by fluorite mineralization.

Tumentsogt deposit (Fig. 5.8) (Khasin 1977; Kovalenko et al. 1988) consists of wolframite and molybdenite related to a Mesozoic granitic pluton of coarse-grained porphyritic granites and fine-to medium-grained leucocratic granite. The deposit is hosted in granite porphyry and fine-grained granite that are the latest stage of pluton. Greisen is formed during alteration of coarse-grained porphyritic granite and rarely fine-grained granite. Greisen bodies are irregular and extend to a depth of 150 meters. Quartz-muscovite and muscovite assemblages form the core of greisen bodies with muscovite-fluorite, molybdenite-muscovite, and rarely beryl-muscovite. Fluid inclusion temperatures reach 340–260 °C (Ivanova 1976). Pyrite and scheelite are also observed. Muscovite contains high amounts of Li, Rb, and Cs. Pegmatite, quartz veins, and albitite are common. The deposit is small and has resources of 1043 tonnes of WO_3 in the main deposit. The West Tumentsogt deposit has resources of 2302 tonnes of WO_3.

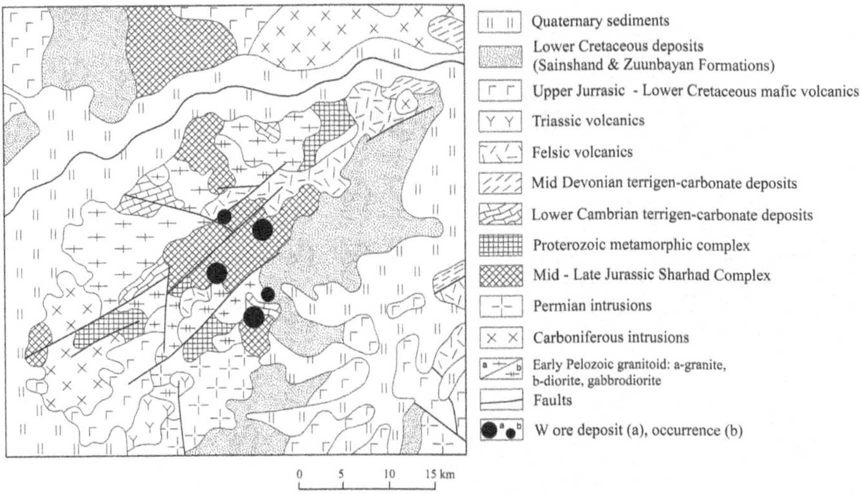

Fig. 5.8 Geological sketch map of the Tumentsogt deposit (from Khasin 1977)

5.3.2.12 Baruun Deposit

Baruun (Western) deposit occurs 8 km SW from the Tumentsogt deposit at the SW contact of a granite pluton within the Upper Proterozoic sedimentary rocks. The mineralization occurs in quartz veins and stockworks within coarse- to medium-grained leucocratic granite. The deposit is controlled by a NE-trending fault zone of 1.7 km long and 0.7 km wide. The deposit is divided into western and eastern zones being dominated by quartz-muscovite greisen and linear stockworks, respectively. The greisen zones vary from 20 to 300 m with 80 m in width. Numerous quartz-wolframite veins vary from 10 cm to 2.3 m in width and extend from 10 to 800 m. Main minerals are represented by wolframite, arsenopyrite, bismuthite, beryl, fluorite, and galena. Fluid inclusions in wolframite (hubnerite) are homogenized in temperature of 260–292 °C and in fluorite from greisen in temperature of 345–348 °C (Ivanova 1976).

5.3.2.13 Ulaan Uul Ore Field

Ulaan Uul ore field of the Mongol Altai metallogenic belt is located in the Delyun Yusted depression bordered by Tolbo Nuur and Khovd deep faults and filled by Ordovician conglomerate siltstone, and limestone and Devonian volcanics, intruded by Carboniferous and Permian gabbro, diorite, and granodiorite plutons. This ore field includes the Ulaan Uul (Kyzyltau) and Tsunkheg deposits, Buraat, and other small occurrences.

Ulaan Uul deposit. This deposit (Fig. 5.9) is located 50 km NE of Ulgii town in Bayan-Ulgii province, Western Mongolia, and consists of about

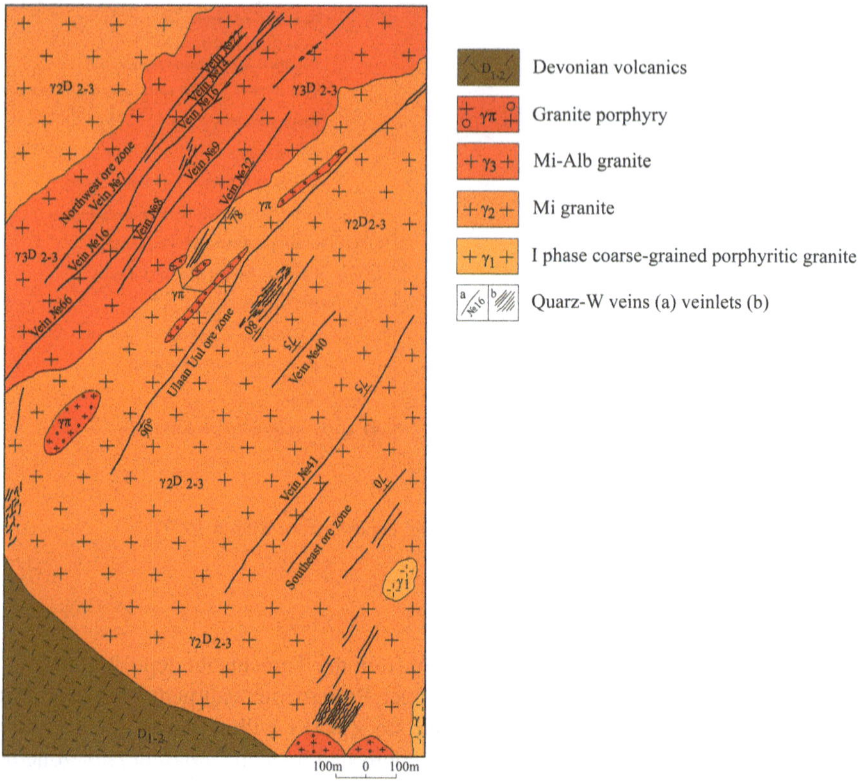

Fig. 5.9 Geological sketch map of the Ulaan Uul deposit (Dandar 2012)

40 northeast-trending quartz-wolframite veins located in the western part of the Jurassic Ulaan Uul leucocratic granite pluton. The veins are as much as 1000 m long and from 0.1 to 1.5 m wide and contain beryl, molybdenite, Y-bearing fluorite, and sulfides. The Ulaan Uul pluton is approximately 2.5 km in diameter elongated northwest along the major Khovd fault. Host rocks are Devonian andesite, dacite, and rhyolite volcanics. K-Ar isotope ages range from 200 to 180 Ma, and Rb-Sr isochron ages are 180 to 170 and 196 $\pm$ 20 Ma. The granite pluton consists of porphyritic coarse-grained biotite granite, medium-grained microcline granite, and microcline-albite leucogranite. The deposit is small and has reserves of 2280 tonnes of WO_3 and resources of 5870 tonnes of WO_3, 8.4 tonnes of Nb, and 500 kg of Y. According to a recent feasibility study, the remaining reserves are 285,000 tonnes of ore containing 3.138 tonnes of WO_3 with a grade of 1.1% WO_3 (Elsner et al. 2011).

5.3.2.14 Tsunkheg Deposit

This deposit (Jargalsaihan et al. 1996) is situated 15 km W of the Ulaan Uul deposit and consists of complex vein and W stockwork that are hosted in a northeast-trending zone of Ordovician-Silurian sandstone, siltstones, tuffs, and tuffaceous siltstone. Host rocks are altered to sulfides and carbonates, contact metamorphosed, and intruded by minor bodies of gabbro and diabase. Three steeply dipping quartz-wolframite veins extend for 200 to 300 m along strike, more than 100 m down dipping, and range from 0.3 to 0.45 m in thickness. The northeast-trending zone extends for 950 m and is as much as 270 m wide and extends to a depth of 300 m. The ore mineral assemblages are scheelite-quartz-feldspar, molybdenite, wolframite-quartz-pyrite, pyrrhotite-scheelite, chalcopyrite, and sporadic quartz-carbonate. The deposit is large with resources of 80,000 tonnes grading of 3.33% WO_3 (B) and 405,000 tonnes grading 2.29% WO_3 with 2682 to 9267 tonnes of WO_3 and 100 g/t Ag, 0.5–1.0% Cu, 1% Sb, 0.5–1.0% Zn, and 2% As.

5.3.2.15 Buraat Uul Occurrence

Buraat Uul occurrence is located within a granite stock cut Devonian andesitic volcanic rocks. Quartz-wolframite veins from 5 to 30 cm thick contain wolframite, fluorite, beryl, and molybdenite. Estimated prognostic resources 2500 tonnes of 0.5 grade Mo, 300 tonnes of grade 0.07% Be, and 2500 tonnes of grade 1.5–19.3% WO_3.

5.3.2.16 Nuuriin Gol Deposit

Nuuriin Gol deposit is situated 250 km northwest of Bayan-Ulgii province and associated with Devonian granite porphyry intruded the upper Cambrian to lower Ordovician sandstone, siltstone, and shale. Mineralization occurs in quartz veins along fracture zone. Veins are from 30 to 270 m long and from 0.1 to 0.6 wide. Two mineralized zones were identified. The largest zone is 400 m long and 20 m thick. Within this zone about 100 quartz-wolframite veins and stockworks were found. Ore minerals are represented by wolframite (hubnerite), scheelite, and accessories: molybdenite, beryl, and fluorite. Calculated resources estimate 32,000 tonnes of ore grading 1.82% WO_3 and a total amount of 600 tonnes WO_3. Additional placers in the area contain wolframite 30 g/m^3 and scheelite 10 g/m^3.

5.3.2.17 Achit Nuur Ore District

Achit Nuur ore district includes of about 20 W-Mo-Be greisen, Sn greisen, Sn-W vein, and skarn occurrences associated with leucogranites and granites of late Paleozoic and Mesozoic age.

Achit Nuur occurrence is located within Upper Ordovician tuffaceous shale and sandstones hosting 200 m long and 20 m wide quartz-wolframite-beryl longitudinally oriented vein zone. The veins are from 10 to 30 m long and from 0.1 to 0.5 m wide. Main minerals are wolframite, beryl, also molybdenite, chalcopyrite, scheelite, and fluorite. Estimated resources are of about 58, 000 tonnes grading of 0.05–3.29% WO_3 and 0.01–2.7% BeO.

5.3.2.18 Bodonch Ore Field

Bodonch ore field contains about seven W and W-Sn occurrences associated with small bodies of Mesozoic granite.

Uudav occurrence is located 58 km SW of Must Uul soum in Western Mongolia. The deposit is associated with leucocratic granite pluton intruded into metamorphosed Upper Cambrian Altai Group that composed of metasandstones and sericite-chlorite slates. Granites are medium-grained greisenized forming quartz-muscovite greisens in the northern part of the pluton. Mineralization occurs in 13 quartz-wolframite veins some gold, 15–300 m long and 0.1–0.4 m up to 3 m thick. Veins have NW strike and are steeply dipping (75–80°). Wolframite (hubnerite-90% $MnWO_3$) is found as 3 × 5 cm elongated crystals and contains Nb 2000 ppm and Bi 700 ppm. Beryl and fluorite are common. Estimated prognostic resources are 240 tonnes of 0.02–5.6%, averaging 3.2% grade of WO_3, and 0.5–1 ppm gold. Northwest of the occurrence, a wolframite placer was found.

5.3.2.19 Undur Tsakhir Deposit

The Undur Tsakhir deposit is located in Taishir soum of Gobi-Altai province, approximately 1050 km from Ulaanbaatar city, 45 km from Gobi-Altai province. According to the exploration survey, the total proven and probable reserve amount is 18,757.09 tonnes of ore with average content of 0.083% Mo. This deposit also has Re with average content of 0.04 g/t Re and WO_3.

5.3.2.20 Ikh Nartiin Khiid

Ikh Nartiin Khiid deposit is located in Dundgovi province within a 122–131 Ma pluton composed of coarse-grained porphyritic granite of first stage with beryl-bearing pegmatite, quartz-wolframite greisen, and veins associated with second stage medium-grained granite. About 53 veins are accounted and 17 of them are wolframite-bearing. Veins are from 25 to 500 long and from 0.4 to 6.4 m wide, strikes NE and dips SE with angle of 30–90°. Wolframite-hubnerite forms 5 × 10 cm large crystals. Besides wolframite there are scheelite, fluorite, chalcopyrite, and pyrite. Host granites are silicified and argillized. In the eastern part of pluton, quartz-muscovite greisen with wolframite was also discovered. The 600 tonnes grading of 0.4–0.6% WO_3 resources were calculated.

5.3.2.21 Burentsogt Deposit

Burentsogt deposit is located in Khentii province 70 km SE of province center town Undurkhaan. The deposit is divided into western with main resources and eastern parts. Two quartz-wolframite veins associated with a 140–168 Ma greisenized granite pluton were found in the eastern part. Granite pluton is 3.7 × 2 km in size and is composed of porphyritic amazonite and amazonite-albite granite with pegmatite bodies in the core and equigranular biotite granite along the margins of pluton. The pluton and ore deposit hosted rocks are composed of Triassic conglomerate, sandstone, and siltstone cut by series dikes of aplite, granite porphyry, amazonite granite, diorite porphyry, and lamprophyre with a longitudinal orientation. Mineralization is controlled by NE- and E-W-oriented faults. A total of 30 veins with NE, NW, and NS direction are from 3 to 250 m long and 0.2–03 m, sometimes up to 1.5–2 m wide, and traced to 300 m in depth. In the eastern part of the deposit, one vein within amazonite granite stock is 400 m long, extending to country rocks. Greisen alteration is common. Wolframite crystals reach up to 15–20 cm, muscovite to 1–7 cm, and beryl to 10 × 1 cm. Fluorite is present in two generations. Besides wolframite rare scheelite, chalcopyrite, pyrrhotite, and cassiterite are present. Fluid inclusions in quartz, beryl, and fluorite contain CO_2 and are homogenized in temperature of 244–322 °C (Ivanova 1976). The deposit was mined in 1970–1972 in underground.

5.3.2.22 Buyant Deposit

Buyant Deposit is located in Khentii province, 70 km south of Burentsogt deposit hosted in Jurassic granite stock intruded the older Devonian granites. Two ore zones, Buyant I and Buyant II, are recognized. Buyant I is represented by quartz-wolframite veins of NE and NW direction within medium-grained mid-late Jurassic granite pluton. Veins are from 50 to 650 m long, from 0.1 to 2.0 m wide, and steeply dipping from 40 to 80°. The Buyant II is located 6 km SE from Buyant I within Paleozoic granites forming NE-trending quartz-wolframite veins from 10 to 20 m long and from 0.1 to 0.3 m wide. Besides wolframite, fluorite and scheelite are present. Veins were formed at temperatures of 206–236 °C (Ivanova 1976). The ore zones are associated with greisen alteration.

5.3.3 Sn Vein/Stockwork Deposits

This type is associated with Mesozoic granites and occurs in Eastern and Central Mongolia within East Mongolian metallogenic belt.

5.3.3.1 Narsan Khundlun Sn Deposit

Narsan Khundlun Sn deposit (Fig. 5.10) is situated in Dornod aimag, 23 km SW of Bayan Uul soum. The deposit is related to the Ereendavaa terrane in Eastern Mongolia and was discovered in 1981–1985 during 1:50,000 scale mapping in the Ulz rare metal belt with other Sn, Mo, Ag, and Au occurrences (Dejidmaa et al. 2002). The rocks in the area consist of lower-mid Devonian felsic volcanics, tuff, Permian sandstone, conglomerate of Ulz Formation, Devonian Tsagaan Undur, and

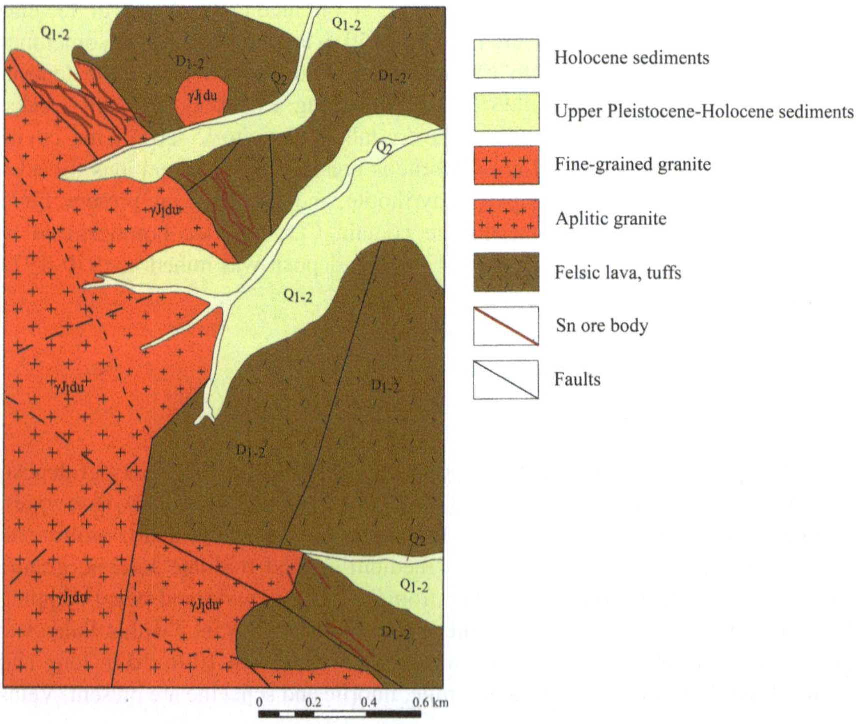

Fig. 5.10 Geological sketch map of the Narsan Khundlun tin deposit (Bayarsaikhan et al. 2017)

early Jurassic Dun Uul granitoid complexes. Mineralization is controlled by a NW-oriented fracture system and associated with early Jurassic Dun Uul granite pluton of 4x2 km in size elongated to the northwest and composed of pinkish porphyritic and medium-grained leucogranite hosted in Devonian volcanic rocks. The granites are silica (SiO_2 76.5–76.8%) and alumina-saturated A-type, depleted in Ba, Sr, Eu, and Nb. The REE distribution is close to symmetric slightly enriched in LREE and flat HREE (Bayarsaikhan et al. 2017). The mineralization occurs in quartz-tourmaline and quartz-chlorite metasomatic bodies. A total of about 20 mineralization zones are distinguished within host Devonian volcanics and Jurassic granite. Five prospects north, west, east, south, and "heart" are recognized. Ore bodies are composed of tourmaline-fluorite-cassiterite and chlorite-cassiterite-sulfide associations and contain quartz, chlorite, tourmaline, muscovite, fluorite, cassiterite, pyrite, chalcopyrite, and rare topaz. Ore bodies are concentrated in the inner and outer contact of pluton, filling northwest-oriented faults falling of 65–75° to northeast. Alteration is represented by quartz, chlorite, sericite, and tourmaline. Latter are chalcopyrite and pyrite and the latest is bornite. There are four ore bodies 300–400 m long and 1.25–5.1 m thick. The tin grades are up to 3.42% with average 0.85%. Reserves were estimated 5635 tonnes of Sn at 0.61%. The calculated resources are 30,000 tonnes with grade of 0.4–0.5% (Jargalsaihan et al. 1996).

5.3.3.2 Baga Gazar Sn Deposit

Baga Gazar Sn Deposit consists of complex Sn mineralization related to the Baga Gazar granite pluton of early Mesozoic age (192–235 Ma, K-Ar age), hosted by South Khentii late Paleozoic-early Mesozoic depression, filled by terrigenous rocks interbedded with volcanics. The Sn mineralization is defined by: (1) quartz-topaz and biotite-greisen occurred in the contact of the pluton with host rocks and in central part of pluton and (2) thick and short quartz-mica veins with topaz, cassiterite, and wolframite confined to the fracture zones. The quartz-topaz and biotite-greisens are dark in color and have a zonal structure: biotite-bearing alaskite, microclinized granite, biotite-greisen quartz-topaz, and quartz zones. Late hydrothermal mineralization forms thick quartz veinlets containing wolframite and sulfides (pyrite, chalcopyrite, arsenopyrite, bismuthinite, and sphalerite). Reserve estimation is 889,344 tonnes, with 2685 tonnes at a grade of 0.18–0.5% Sn.

5.3.4 Placer Sn (+W) Deposits

Cassiterite is very resistant, and placer tin deposits occur in many areas with primary Sn-W greisen/stockworks and quartz vein deposits. The Modot district is the best studied and contains the Bayanmod and Khujkhan tin placer deposits and a number of Modot and Bayanmod quartz Sn-W vein and greisen occurrences that are the main source for placers but have a very low grade and not economically mineable.

Commercial cassiterite deposits and occurrences occur as small-sized eluvial and alluvial placers (<10,000 tonnes of Sn). The major mining area was Modot district, but most cassiterite was mined out in the 1950s. Today cassiterite is mined on a small scale in the Modot valley. In Khujkhan placer, cassiterite is mined together with wolframite on a small scale. Other areas like Janchivlan were intensively mined in the past. About 20 placers with grades of 100–600 g/m^3 SnO_2 of total 6000 tonnes of tin are known. Placers extend over 5–10 km with width of a few tens meters up to 500 m and a thickness of 0.5–15 m. The placers contain cassiterite, accessory wolframite, monazite, topaz, fluorite, and zircon.

5.3.4.1 Modot Ore District

Modot ore district (Fig. 5.11) includes a number of commercial Sn placers (Bayanmod and Khujkhan) and small Sn-W greisen and quartz vein (Bayanmod, Modot) associated with late Triassic-early Jurassic Modot granite pluton. Many placers were mined in the last century, but a reassessment of old dumps may be worthwhile to estimate the economic potential for extracting ores from the dumps (Elsner et al. 2011).

In this area about 13 cassiterite placer deposits are known, and the largest are Bayanmod and Khujkhan. The placers are located within the late Triassic to early Jurassic granite intrusion. First exploration work was done by Mongolian-Russian Dornod Metal Expedition in 1947–1950. The sources for these placers are the W-Sn greisen/stockwork and quartz vein deposits and occurrences in this area. Between 1949 and 1955, concentrates with 8516 tonnes of Sn and 1297 tonnes of WO_3 were produced from the main about 10 km long Modot valley. In the period of 1973–1990, about 856 tonnes of Sn were produced.

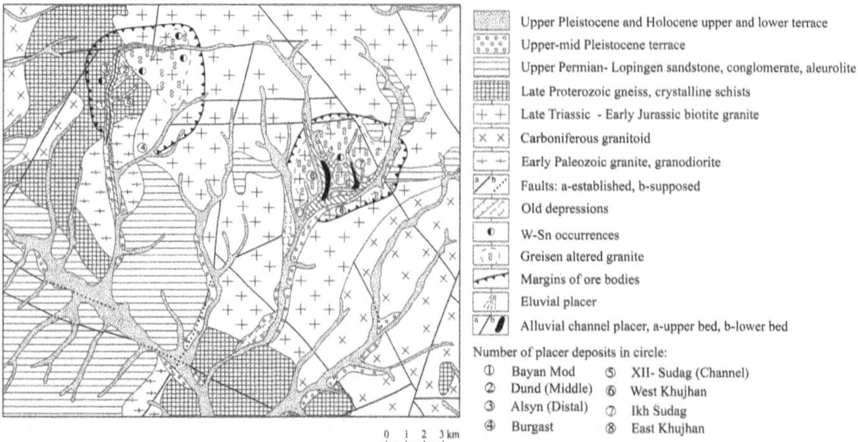

Fig. 5.11 Distribution of tin-tungsten placers in the Modot ore district (Khasin 1977)

5.3.4.2 Bayanmod Sn-W Placer Deposit

Bayanmod Sn-W placer deposit is alluvial, the thickness of the fluvial sediments varies between 5 and 50 m. Exploration work was carried out in 1947–1950 by Mongolian-Russian Dornod expedition, in 1985–1988 by Mongolian-Czechoslovakian Intergeo expedition. The 8516 tonnes of Sn and 1297 tonnes of WO_3 were produced in 1973–1977, 108 tonnes of cassiterite concentrate containing 30% Sn produced in 1973–1977, and in 1979–1990 about 712 tonnes of Sn. The deposit has two ore-bearing layers: upper and lower. The upper cassiterite layer is hosted in quaternary gray sandy loam and loam in the depths of 1–5 m. The layer is 5850 m long, 16–375 m wide, and 0.5–8 m thick. The cassiterite content reaches 2565 g/m^3 with average content of 473 g/m^3 and tungsten content 916 g/m^3 with average of 39 g/m^3. The lower layer is hosted in Pliocene red loam matrix within a medium to coarse pebble-bearing sequence. The layer is 2800 m long, 19–171 m wide, and 0.5–8 m thick, with average thickness of 3.2 m. The cassiterite content varies from 194 to 10,353 g/m^3 with average of 1436 g/m^3, and tungsten average content is 14 g/m^3. Heavy concentrate contains cassiterite, wolframite, scheelite, quartz, feldspar, fluorite, tourmaline, topaz, garnet, andalusite, mica, sillimanite, rare hematite, ilmenite, magnetite, chalcopyrite, pyrrhotite, and accessories: sphene, rutile, corundum, zircon, and epidote. The cassiterite grains are up to 8 mm and wolframite 0.1–0.5 cm in size.

5.3.4.3 Khujkhan Sn-W Placer Deposit

Khujkhan Sn-W placer is divided into Eastern and Western. Eastern Khujkhan deposit extends to the Ikh Sudag placer. The cassiterite-bearing layer is 2680 m long, 178 m wide, and 1.5 m thick. The cassiterite content is 435 g/m^3 and wolframite 14 g/m^3 with estimated reserves of 232 tonnes of Sn and 95 tonnes of WO_3. Previously calculated reserves were about 4.3 million m^3 sediment containing 1724 tonnes of cassiterite and 375 tonnes of WO_3. The Western Khujkhan is situated near the greisen/stockwork Sn-W deposit. A valley is asymmetric and contains numerous small placers. A lower layer composed of red Neogene sediments has thickness of 1.1 m and contains 362 g/m^3 cassiterite and 16 g/m^3 wolframite, with 75 tonnes of Sn and 3 tonnes of WO_3. An upper layer is gray and contains 246 g/m^3 cassiterite and 20 g/m^3 wolframite, with 36 tonnes of Sn and 3 tonnes of WO_3. A new reserve estimated a total amount of 2000 tonnes of Sn + WO_3. Three layers were mined each about 1.5 m thick. The upper layer has a low grade, and the other two layers are high grade.

The origin of all these deposits, namely, Bayanmod, Khujkhan, and Ikh Sudag is interpreted as formed during three stages from Pliocene to Holocene. In the Pliocene, brick-red layers with pebbles and rock debris with cassiterite-bearing lower layers from 5 to 10 m, rarely from 20 to 30 m thick, were formed (Bayanmod, Western Khujkhan, Ikh Sudag). In Pleistocene, gray alluvial deposits and upper layer of placers (Middle Ikh Sudag) were formed, followed by formation of eluvial and deluvial (Bayanmod middle) and deluvial-proluvial deposits in Holocene (Dandar 2012).

5.3.4.4 Janchivlan Ore Field

Janchivlan ore field contains about 20 placers (including Elstiin, Urt Gozgor, and Avdrant) located in the Central Khentii metallogenic belt. Cassiterite- and wolframite-bearing quartz veins, greisens, and pegmatites are the source for colluvial and alluvial placers, mainly with cassiterite and accessory wolframite, topaz, monazite, fluorite, and zircon. Single placers extend up to 5–10 km and are from few tens of meters to 500 m wide and 0.5–15 m thick. The cover is 1.3 m to 2.2 m thick. The cassiterite-bearing layer is from 1.3 m to 2.0 m thick with cassiterite grade between 403 and 589 g/m^3. The total reserves are 3.88 million m^3 of ore containing 904 tonnes of Sn. About 20 placers were exploited in the past that carried between 100 and 600 g/m^3 cassiterite and contained about 6000 tonnes of tin.

5.3.4.5 Elset and Eastern Elset Placer

The Elset and Eastern Elset placer is 4.7 km long and 60–400 m wide with two cassiterite-bearing layers. Upper layer contains sand, pebbles, sandy loam, and rarely bulges; the bottom layer contains sand, rare pebbles, and Pliocene red deposits with rare bulges. Both layers vary in thickness and width. The main heavy mineral in the placer is cassiterite associated with wolframite, hematite, bismuthinite, and arsenopyrite. The cassiterite content is 200 to 600 g/m^3. The total reserves are 4288 tonnes of cassiterite. Based on placer sand reserves of 13 million m^3, the placers contain 3000–3500 tonnes of Sn.

5.3.4.6 Urt Gozgor Placer

Urt Gozgor placer is 2.0–2.5 km long and 60–300 m wide. The heavy mineral bearing layer is 0.5–2.5 m thick. The cassiterite grade average is 286 g/m^3, sometimes 1000 g/m^3 with reserves of 80–286 tonnes of Sn (220 tonnes of Sn). Heavy mineral concentrate contains cassiterite, wolframite, columbite-tantalite, scheelite, ilmenite, zircon, and monazite. Grade of Ta is 25 g/m^3 and estimated reserves are 34.4 tonnes of Ta.

5.3.4.7 Avdrant Group Placer Deposit

Avdrant group placer consists of six placers. The placers are 1.8–4.5 km long, about 10–450 m wide, and 1–2 m thick. The cassiterite average grade is 300–550 g/m^3, with estimated reserves total of 1.8 Mt. of Sn with the largest placer containing 200 tonnes of Sn.

5.3.4.8 Upper Onon Group Placers

Upper Onon group placers located in North Khentii near the Russian board include a number of placer deposits and occurrences. Deed Khomor placer contains 1073 g/m^3 of 1099.5 tonnes of Sn and 1129 g/t of 1156.6 tonnes of WO_3 and 54 tonnes of scheelite at grade of 54 g/m^3. The Zuun Tarz Gol placer with two layers up to 2 m thick contains 156–417 g/m^3 Sn and 13–56 g/m^3 wolframite.

5.3.4.9 Kharmorit Group Placer Deposits

Kharmorit group placer deposits in the South Gobi near the Chinese border include a number of placers. The main placer is 4 km long, 0.1–5.0 m thick, and 24–110 m wide. Reserves are 714 tonnes of Sn at 485–3082 g/m^3 Sn and about 20 tonnes of WO_3 at 210 g/m^3 WO_3.

There are other Sn and W placers in Mongolian Altai and other areas with W-Sn vein and greisen/stockwork occurrences, but these are not reported here.

5.3.5 W-Mo Skarns

Tungsten-molybdenum mineralization consists of various types of scheelite skarn that occurs along the margin of granite plutons that intrude metamorphosed Precambrian carbonate-aluminosilicate rocks. W skarn deposits in Mongolia are not of economic value at the present time and are associated with Mesozoic granitoids in northeastern and southeastern Mongolia. The main deposits are Bayan-Ovoo, Mandal, and Tumentsogt deposits and other occurrences.

5.3.5.1 Mandal Deposit

Mandal deposit is located in Tsagaan Uul soum, Khuvsgul province. The deposit is associated with Permian granite pluton. Host rocks are related to Darkhad formation composed of black marble limestone and light limestone and silicified and metamorphosed to hornfels black shales of Neoproterozoic-early Cambrian age cut by small outcrops (270 × 70–100 m) of Permian (K-Ar age of muscovite 275 ± 5.7 Ma) porphyry stock and dikes of Mandal Complex (Darambazar et al. 2017). In the apical parts of the stock silicified granite, 50–70% quartz and UST are observed. Granites are depleted in Ba, Nb, Sr, Eu, and heavy REE. Quartz-sericite altered granite porphyry body and dikes cut by molybdenite and scheelite- bearing veinlets.

Molybdenite Re-Os age is 281.6 Ma. On the outer contact skarn has formed and is bordered by hornfels zone or alternation of skarn and hornfels. There are two ore bodies: (1) scheelite-molybdosheelite-bearing silicified carbonate, Fe skarn, and hornfels zone and (2) quartz-molybdenite stockwork. Measured, proven, and probable reserves of mineable resource in the Mandal deposit are 4873.66 tonnes of Mo equivalent, 3003.68 tonnes of Mo, and 2792.83 tonnes of W in the 19.9 Mt. of total ore with average content of 151.14 g/t Mo and 140.53 g/t W. According to preliminary survey findings of Baruun Mandal, the reserve amount of Baruun Mandal is 57,000 tonnes of W and 3500 tonnes of Mo.

5.3.5.2 Tumentsogt Skarn Occurrence

The main tungsten mineralization is W-Mo-Be greisen/stockwork and quartz vein type as described above, and skarn mineralization was observed at the contact with Devonian limestone with alteration consisting of silica and epidote forming epidosite. Skarns are pyroxene-garnet, pyroxene-garnet-vesuvianite, and pyroxene type.

5.3.6 *Sn-Polymetallic Skarn Deposit*

5.3.6.1 Oortsog Ovoo Deposit

This deposit (Podlessky et al. 1988; Jargalsaihan et al. 1996) consists of steeply dipping sheets like bodies of skarn along the contact between Late Permian-Triassic granite pluton and marble with beds of calc-silicate schist. The granitic rocks occur as shallow bodies with miarolitic structures, represented by alkali K-feldspar granites of A-type (Gerel 1990). The skarn sheet bodies range from 200 to 1500 m long, 5 to 80 m wide, and comprise as much as 25 lenticular ore bodies composed of garnet, pyroxene, magnetite, cassiterite, and base metal ore minerals. Three stages are recognized: (1) early stage of pyroxene-garnet and magnetite; (2) cassiterite, stannite, loellingite, Zn sulfide, Pb sulfide, Cu sulfide, and Fe sulfide; and (3) less common fahlore, enargite, bismuthinite, and scheelite. Also occurring are supergene cerussite, smithsonite, anglesite, greenockite, martite, montmorillonite, kaolinite, and gypsum. Four ore belts are distinguished: (1) central magnetite, pyroxene-garnet skarn within carbonate rocks, NE-trending, 240 m long and 80 m wide, with 5526 tonnes of Sn and (2) eastern zone 250 m long and 160 m wide in the contact between granite pluton and carbonate rocks. Grades are 0.02–1.28% Sn, 0.001–0.06% W, 0.02–1.28% Zn, and 0.01–0.9% Cu. Calculated reserves are 39,200 tonnes. The calculated resources are 33, 000 tonnes of Sn, 13,137 tonnes of Zn, 5292 tonnes of Cu, and about 1000 tonnes of Pb.

5.3.7 W Skarns

This type of mineralization consists of various types of scheelite skarn situated along the margin of Mesozoic plutons that intrude metamorphosed Precambrian carbonate-alumosilicate rocks and Paleozoic carbonate-terrigenous and volcanic-terrigenous rocks. W skarns generally form layered bodies that may extend for as much as 1.2–2 km along intrusive contact and range from 3.5 to 7.0 m wide (Podlessky et al. 1988). The skarn exhibits a very complicated structure with rhythmically repeated zones of pyrite, arsenopyrite, and Sn-W and Cu-sulfides. The varieties of skarn include clinopyroxene-plagioclase, banded garnet-clynopyroxene, and garnet-vesuvianite types. Scheelite with small amounts of molybdenite, sphalerite, and pyrite is a common ore mineral and may occur in both endoskarn and exoskarn. Gangue minerals are quartz, plagioclase (andesine), epidote, and actinolite. Scheelite may occur as disseminations in quartz-feldspar veins in exoskarns. The tungsten skarn may contain late-stage quartz and hematite alteration and superimposed sulfides (chalcopyrite and sphalerite). Calcareous skarns form stratabound bodies up to 400 m long and up to 50 m wide.

W and Sn skarn with scheelite and cassiterite are interpreted as forming during by bi-metasomatic processes along the contact between calc-alkaline felsic plutons and carbonate-terrigenous and volcanic rocks. Examples are Bayan-Ovoo, Beis, Erdene Tolgoi, and Bayandun occurrences.

5.3.7.1 Beis Skarn Deposit

Beis skarn deposit occurs in the contact between the Modot granite pluton of 220 ± 2.8 Ma (U-Pb, LA) with Proterozoic metamorphosed to amphibolite facies rocks. Skarn bodies are composed of thin rhythmic alternations of pyroxene-plagioclase and pyroxene-garnet often with vesuvianite exoskarns. In contact with pluton, granitic injections into metamorphic schists were observed. The Sn-W and Mo quartz vein and greisen mineralization are also associated with the Modot pluton. Quartz veins in apical part of pluton and in the inner contact of skarn bodies are followed by quartz-muscovite quartz-albite-muscovite greisen and greisenization of granites. Skarn bodies are thin but very long up to 1.5–2.0 km and occur in the steep contact of intrusion with host rocks. Silica alterations are common and sometimes totally substitute pyroxene-plagioclase rocks, and only relics of plagioclase and muscovite are preserved. Besides silica alteration and feldspatization, epidotization with scheelite-sulfide mineralization is common.

5.3.8 The W-Sb (Hg) Mineral Systems

These mineral systems consist of veins, stockworks, and breccias with quartz (occasionally with chalcedony and opal) and carbonate (dolomite, siderite, ankerite, magnesite, and calcite). The ore minerals are wolframite, scheelite, minor pyrite, marcasite, arsenopyrite, chalcopyrite, pyrrhotite, sphalerite, galena, cinnabar, and gold. The deposits occur in regions of low-grade metamorphism along the margins of continental terranes and are controlled by shear zones, nappe systems, or large faults. The wall rocks are usually argillized and sericitized. An example is the Khovd Gol deposit and Mushguu occurrence in Mongolian Altai (Kempe et al. 1994; Getmanskaya and Chernov 1996; Ivanova 1976).

5.3.8.1 Khovd Gol Deposit

Khovd Gol deposit (Fig. 5.12) is situated 100 km west-southwest of the center of Bayan-Ulgii province, the Ulgii town. Deposit was discovered in 1986. A total of 23 ore-bearing zones and 20 ferberite-quartz veins shallow dipping (5–20°) and NE-trending have been discovered. The largest ore bodies have thickness of 450 m and are 1–2 km long. Other veins are 50–420 m long and 0.1–0.5 m thick and were traced to the depth up to 280 m. Ore veins are rich in ferberite. Quartz is acicular in shape and forms druse. Other ore minerals are limonite, scheelite, calcite, rare pyrite, antimonite, cinnabar, and gold. This deposit is epithermal W-Sb type (Kempe et al. 1994; Getmanskaya and Chernov 1996). The up-to-date, proven W reserves estimate 4176.5 tonnes in approximately 70,000 tonnes of ore. The average W grade is 6% WO_3. The probable reserves are 1246 tonnes of WO_3 in about 19,000 tonnes in three blocks with an average content of 6.1%; thus the total reserve is 5422.5 tonnes of WO_3 in the 885,000 tonnes of ore.

5.3.9 Sn-Silicate-Sulfide Vein Deposits

Tin silicate-sulfide vein deposits type is represented by cassiterite-tourmaline mineralization in veins and veinlets of outer contact zones of multiple granite and granodiorite intrusions (Kovalenko et al. 1988). The composition of these granitoids ranges from granodiorite and granite of calc-alkaline series to subalkaline granite. The minerals are cassiterite, quartz, tourmaline, chlorite, carbonates, chalcopyrite, arsenopyrite, sphalerite, and molybdenite. The main examples are at Khukh Uul and Zaan Shiree occurrences.

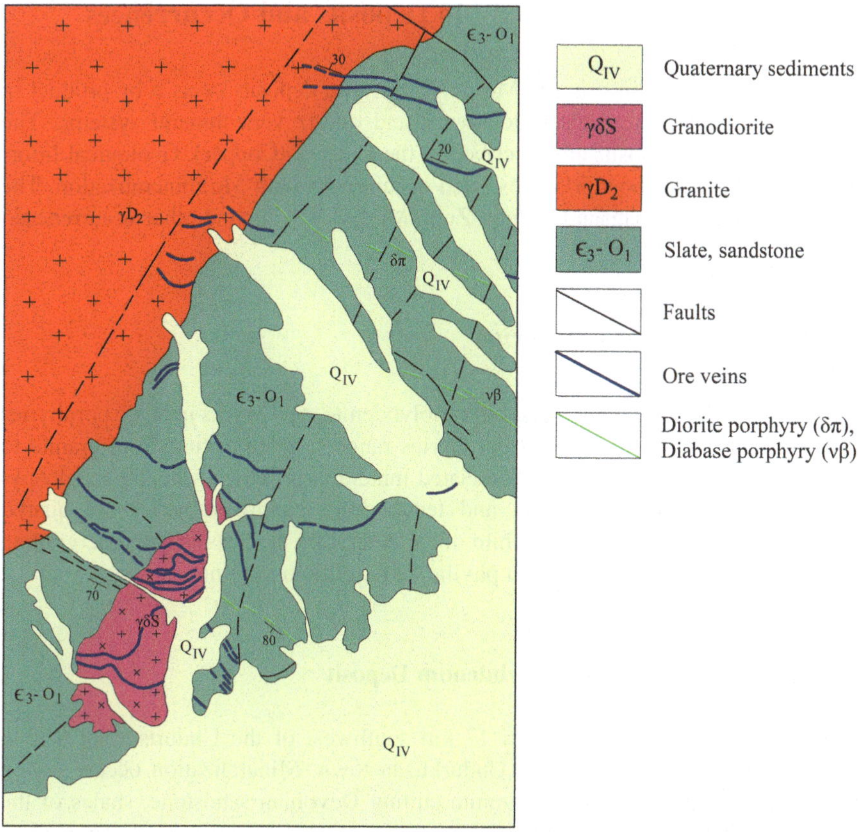

Fig. 5.12 Geological sketch map of Khovd Gol W-Sb (Hg) vein and stockwork deposit (from Dandar 2012)

5.3.10 Sn-Sulfide Vein Deposit

This type consists of elongated, brecciated zones along the exocontacts adjacent to subalkaline and Li-F hypabyssal granitic plutons (Khasin 1977). The deposit type is interpreted as having been generated in a post-orogenic setting. The main minerals are cassiterite, galena, sphalerite, muscovite, fluorite, tourmaline, chlorite, carbonate, pyrite, arsenopyrite, and chalcopyrite. Alteration minerals are tourmaline, chlorite, muscovite, and quartz. The main example is the Kharmorit deposit, which is associated with muscovite-microcline pegmatite and Sn-W greisen. The Sn-sulfide mineralization may be part of the Sn-silicate-sulfide deposit type (Kovalenko et al. 1988) (see in Sect. 5.3.1, Kharmorit deposit).

5.4 Description of Selected Mo Deposits and Occurrences

Molybdenum occurs in primary Mo porphyry-type deposits or as a by-product in Cu-Mo and W-Mo-Be greisen/stockwork and quartz vein mineral systems. The porphyry Cu-Mo deposits are described in the Chap. 2 (Copper). Geological Information Center lists about 50 deposit and occurrences with Mo mineralization. The main examples are Tsagaan Chuluut, Zuun Mod, Zost Uul, and other occurrences.

5.4.1 Porphyry Mo Deposits

Porphyry Mo deposits consist of quartz-molybdenite stockworks in felsic porphyries and adjacent country rocks. The porphyries range in composition from granite to granodiorite and monzogranite. Associated minerals are pyrite, scheelite, chalcopyrite, rare cassiterite, wolframite, and tetrahedrite; gangue minerals are quartz, K-feldspar, biotite, calcite, and white mica. Alteration consists of potassic grading outward to propylitic, locally with phyllic and argillic overprints.

5.4.1.1 Tsagaan Chuluut Molybdenum Deposit

Tsagaan Chuluut Mo deposit lies 17 km southwest of the Undurtsagaan W-Mo deposit and 50 km northwest of Undurkhaan town. Mineralization occurs within Carboniferous granodiorite and granite cutting Devonian sandstone, shales of the Kherlen trough in the zone of a NW-trending fault, accompanied by numerous granite porphyry dikes. Molybdenum mineralization is controlled by Jurassic granite porphyry dikes and provably has genetic relation with them. Stockworks' Mo mineralization measure 700 × 800 m is springer and stringer-impregnated and traced to a depth of 300 m. Mineralization is accompanied by potassic, chlorite, and silica alteration developed in granodiorites. Molybdenum grade varies from 0.005% to 0.19% averaging 0.12% with cutoff grade 0.08%. The probable reserves are estimated to be 34.3 Mt. of Mo and 54,000 WO_3.

5.4.1.2 Zuun Mod Molybdenum Deposit

Zuun Mod Mo deposit is located within 200 km of China's border in Bayankhongor province, 900 km southwest of Ulaanbaatar. The Zuun Mod deposit was generated in a continental arc environment and comprises lower Devonian-early Carboniferous volcanics, cut by Carboniferous-Permian syenites of 320–315 Ma, monzonites, and granodiorites. Molybdenum mineralization is associated with 302 Ma microgranite and alkali granite with porphyry-style veining and disseminated sulfides (298 Ma). Alteration is quartz-sericite (phyllic), potassic, greisen, argillic, and propylitic and occasionally silica and tourmaline. Molybdenum mineralization forms disseminations or stockworks. In the oxidized zone, the Mo grade is 0.5–2%, in quartz-sericite-pyrite alteration zone within granodiorite sulfide content increased to 3–5%. Sulfide contents increase with depth. In a three-km long portion of the porphyry system, three mineralized zones identified were known as the Racetrack North, Racetrack South, and the Stockwork. Initial drilling of these zones intersected thick intervals of significant mineralization proximal to surface and open at depth. The Zuun Mod molybdenum deposit has measured and indicated resources of 218 Mt. with an average grade of 0.057% Mo and 0.069% Cu, at a 0.04% Mo cutoff grade. This equates to 124,057.5 tonnes of contained Mo metal and 150,002.9 tonnes of contained Cu metal. In addition, Zuun Mod has an inferred resource of 168 Mt. averaging 0.052% Mo and 0.065% Cu (erdene.com/en/projects/zuun-mod).

5.4.1.3 Zost Uul Porphyry Molybdenum Deposit

The Zost Uul porphyry Mo deposit in Ider terrane, northern Mongolia, comprised of Proterozoic metamorphic rocks and Late Permian volcanics. Late Permian-early Triassic intrusions are represented by fine-grained granodiorites and granites. Molybdenum mineralization is hosted in granite porphyry stockwork (1.8 × 2.0 km). Stockworks have three ore-bearing zones, characterized by alteration assemblages: quartz-sericite-pyrite with rare quartz-molybdenite veins and thin quartz and quartz-pyrite veinlets with disseminated molybdenite, rarely chalcopyrite. Molybdenite occurs as thick flakes along selvages of veinlets or as disseminations in quartz. Molybdenite-bearing quartz veinlets are cut by pyrite veinlets and cut by quartz-carbonate veinlets. Also present are skarns containing garnet and epidote with pyrite-chalcopyrite mineralization. Alteration is mainly sericite (phyllic), potassic, and silicic. Mineralized zones contain quartz-molybdenite and quartz-pyrite stringers with chalcopyrite, sphalerite, sericite, feldspar, and fluorite. Mineralized zone is 100–200 m wide and extends for a distance up to 1800 m, Mo content 0.01–0.2% and Cu 0.01–0.3%. Inferred resources are 102,000 tonnes of Mo.

5.4.1.4 Ariin Nuur Molybdenum Deposit

Ariin Nuur Mo deposit is located 120 km south of the Choibalsan town and 65 km north of Baruun Urt. The deposit's area is dominated by Permian leucocratic granites; granosyenites; granodiorites; dikes of aplite, pegmatite, granite, and granodiorite porphyries; lamprophyres; and post-ore dikes of andesite (Ivanova 1976; Jargalsaihan et al. 1996). Potassic, albite, and greisen alteration of granites are common. Economic mineralization is of stockwork type. Four stockwork zones are recognized, and the largest one is central stockwork, 1000 m long and 400–600 m wide, elongated in a meridional direction. The core part of the stockwork consists of greisenized granites with quartz-potassic altered rocks developed at its periphery. The principal latitudinal trending fault divides the stockworks into two blocks. The southern block, containing the bulk of mineralization, is uplifted, while the northern block is down faulted. The richer mineralization is expected to occur at depths exceeding 300 m. Other stockwork zones are 300–600 m long and 50–100 m wide. The oxidized zone extends to a depth of 10–20 m, followed by zone of secondary enrichment to a depth of 30 m. The major ore minerals are pyrite, molybdenite, and chalcopyrite less common sphalerite, galena, magnetite, fahlore, chalcocite, wolframite, and cassiterite. The oxidation (supergene) zone is dominated by limonite, jarosite, covellite, malachite, azurite, and molybdenum ochres. The gangue minerals are quartz, feldspars, muscovite, little fluorite, epidote, topaz, and garnet. Probable reserves are estimated to be 41,850 tonnes of Mo grading 0.06–0.1% Mo, averaging 0.08%. Copper is subordinate with a grade of 0.02–0.154%.

5.5 Description of Selected Be Occurrences

Beryl is a common mineral in pegmatites W-Mo vein deposit and occurrences and W-Mo greisen/stockworks and vein systems. Geological Information Center lists 21 beryl occurrences in Mongolia. Beryllium occurrences were discovered in rare metal pegmatite dikes in Baruunkhuurai in the southern part of Mongolian Altai. In the Alag Tagt, pegmatite-hosted schists contain 0.3% BeO against the 0.03% in pegmatites. The Bayan Uul Be-bearing occurrence contains 20–30% Be with total resources of 1.3 tonnes. Another Be deposit is the Munkhtiin Tsagaan Durvuljin in Dundgovi province. The biggest W-Mo-Be vein/stockwork occurrence is Buraat Uul in Bayan-Ulgii province (BeO grade varies from 0.22% to 1.38% with 0.5% on average, total resources 301 tonnes), Yeguzer (0.086% BeO, 40,000 tonnes of BeO), and Undur Tsagaan (0.03% BeO of 43,913 t of BeO). Other Be deposits and occurrences are Khoit Khuld (0.01% BeO, 33.3 tonnes of BeO) in Bayan-Ulgii aimag and Khurheree in Selenge aimag (0.01% BeO, 27.7 tonnes of BeO). Grades of Be in pegmatites vary between 0.001% and 13.3% with resources between 0 and 12.0 tonnes of BeO. Be-bearing tuffs were discovered in Durvun Dert and Teeg Uul in Umnugobi (Southgobi) province with a Be average content of 0.05%. Beryllium tuffs consist of bedded and graded-bedded tuff containing fragments of ongorhyolite, rhyolite, quartz, feldspar, fluorite, and Be-bearing bertrandite (Kovalenko and Koval 1984).

5.5.1 Buraat Uul Occurrence

Buraat Uul occurrence is the W-Mo-Be greisen/stockwork and quartz vein type described above. The occurrence is hosted in granite stock intruded Devonian andesitic rocks. Granites are greisenized forming a typical quartz-muscovite greisen. Beryl is a minor component in 10 quartz-beryl veins with a length of 50–200 m on average 100 m, a 0 thickness of 0.3–1.0 m on average 0.85 m, and 50 m extension to depth. The veins dip 70–85° NW. The main minerals, besides beryl, are molybdenite, wolframite, fluorite, columbite, and tantalite. The grades of BeO vary from 0.22% to 1.38% with 0.5% on average. The total resources are 301 tonnes of BeO.

5.6 Description of Selected Li Deposits and Occurrences

Lithium mainly occurs together with other rare metals, such as Nb, Ta, Be, Cs, etc. Lithium occurrences in Mongolia are hard rock-related types: rare metal lithium-bearing-pegmatite (about 30 occurrences), ongonite, and lithium-fluorine granite, and newly discovered sedimentary Li deposit of Khukh Del.

5.6.1 Lithium Pegmatites

Li-pegmatites in Mongolia comprise late Paleozoic and Mesozoic REE pegmatites that are mainly associated with calc-alkaline and Li-F leucocratic granites. Three subtypes of rare metal pegmatites are defined (Kovalenko and Koval 1984): Li-mica, muscovite (muscovite-albite), and muscovite-microcline. The Li-mica subtype contains Ta-Nb minerals, cassiterite, Li-mica, quartz, albite, microcline, apatite, tourmaline, topaz, beryl, and other minerals. These can be found at Khukh Del Uul and Unjuul. The muscovite-albite subtype consists of columbite, tantalite, quartz, albite, microcline, and muscovite. The main example is the Berkh. The muscovite-microcline subtype includes cassiterite, wolframite, quartz, microcline, and muscovite. This subtype occurs at Tumentsogt, Bayandelger, Bayan-Ovoo, and Khalzan Uul. Many of the known rare metal pegmatite deposits occur in Eastern Mongolia. The Li-mica subtype occurs in Li-F granite plutons of elevated alkalinity, and the muscovite subtype occurs in calc-alkaline granite series (Kovalenko and Koval 1984). Muscovite subtype contains columbite-tantalite and beryl with albite, microcline, quartz, and muscovite. Examples of the Li-mica subtype are at Janchivlan, Borokhujir, Baga Gazar, Avdrant, Baruuntsogt, Yugzer, and others (Table 5.1). Examples of the muscovite (muscovite-albite) subtype are at Yeguzer, Tumentsogt, Baruuntsogt, Yargait, and others. Ta-Be and W-Mo mineralization are also associated with the Ta-granite deposit type. Albite alteration is associated with these granites. Li-mica subtype contains Ta 40–150 g/t; Li content is up to 10–20%. Li reserves are not yet known. The Khukh Del Uul, Unjuul, and Berkh rare metal pegmatite deposits occur as dike-like or lenticular bodies that range in size from few meters to hundreds of meters long and from 1 to 10 m wide. Associated Li-Sn-Be pegmatites (Kovalenko and Yarmolyuk 1995) consist of Li-mica, Ta, and Sn-W minerals that are hosted in intrusions that postdate the peak of batholiths emplacement. The associated granites are mainly calc-alkaline and Li-F leucogranites and their volcanic and subvolcanic analogues.

5.6.1.1 Khukh Del Uul Li-Pegmatites

Pegmatites of Khukh Del Uul are situated 250 km southeast from Ulaanbaatar (in Middle Gobi). In the area of 6 km^2, there are about 25 pegmatite bodies from 50 to 300 m long and from 1 to 10 m wide, trending to northeast or northwest. Host rocks are represented by Precambrian schists: chlorite, chlorite-sericite, sillimanite, tourmaline, and rarely Devonian porphyritic granite. Syenite porphyry, quartz porphyry, and felsic rocks cut Devonian granites. Five types of pegmatite are distinguished: (1) schorl-microcline-quartz; (2) Li-muscovite-topaz-microcline-quartz; (3) lepidolite-albite-quartz; (4) Li-muscovite-lepidolite-albite-quartz; (5) rubellite-clevelandite-quartz. Rubellite-clevelandite- quartz pegmatite is the most common characterized by lack of well-developed zoning. Medium-grained texture with coarse-grained zone of quartz-topaz-lepidolite composition. An early stage of lepidolite bearing pegmatite consists of: microcline, quartz, topaz, clevelandite, lepidolite, rare Li-muscovite, rubellite, beryl and cassiterite. The core of pegmatite bodies is composed of quartz-topaz overprinted by metasomatic albitite, and greisen-like rocks. Greisens are composed of quartz and lepidolite or Li-muscovite, and form veins or schlieren from cm to tens of cm. These greisen-like bodies are cut by albite-like and albitite veinlets. Schorl bearing pegmatites are totally overprinted by albite alteration. The latest veinlets are composed of jilberdite, quartz, and fluorite.

5.6.1.2 Munkhtiin Tsagaan Durvuljin Lithium Deposit

The Munkhtiin Tsagaan Durvuljin Li deposit is located in Shiveegovi, Bayanjargalan soum, Govisumber and Dundgovi aimag. The area is dominated by Neoproterozoic Oortsog Formation slightly metamorphosed volcano-sedimentary sequence, cut by early Paleozoic plagiogranite, granite-gneiss and microgranite, diorite and aplite dikes. The Oortsog Formation is cut by E-W trending faults and folded, and the pegmatite bodies occur at the contact of early Paleozoic granites and marble and slate of the Oortsog Formation. Li-bearing pegmatite bodies contain lepidolite and spodumene.

5.6.2 Lithium-Fluorine (Li-F) Granites

Lithium-fluorine (Li-F) granites are highly evolved post-orogenic granites described in the literature as rare metal granites with elevated content of Li, F, and other lithophile elements. Li-F granites include microcline-albite, amazonite-albite and lepidolite-albite types (Kovalenko et al. 1971). Li-F granites are widespread in Eastern Mongolia within large provinces of post-orogenic subalkaline granites, which form multistage plutons with three phases of granites (Janchivlan) or simple small plutons (Avdrant) mainly composed of one stage granites. Their subvolcanic and volcanic are ongonite (from Ongon Khairkhan in Central Mongolia) and volcanic-ongorhyolites (Teg Uul) in South Mongolia.

Li-bearing rocks are lepidolite-albite granites. They have a porphyritic texture with main minerals represented by albite, quartz, K-feldspar, lepidolite, and topaz. Porphyritic texture is shown by round quartz grains of 3–5 mm in size, rarely feldspar set in fine-grained sugar-like groundmass, composed mainly by albite. Post-magmatic rocks associated with rare metal granites are greisen with dark biotite, quartz-lepidolite greisen, and albitite. Li-containing mineral is lepidolite. Li content in different Li-F granites is Janchivlan lepidolite-albite granite, and greisen is 878–1315 ppm, in Baga Gazar 130–630 ppm, in Avdrant 338–396 ppm, and in Baruun Tsogt 129–150 ppm.

5.6.2.1 Urt Gozgor and Buural Khangai Li-F Granites

Urt Gozgor and Buural Khangai Li-F granites occur in the southern part of the multiphase Janchivlan pluton situated 90 km southeast from Ulaanbaatar. This Mesozoic (187–220 Ma) pluton occurs in the southern part of the Khentii uplift and hosted in Paleozoic metamorphic-sedimentary rocks, occupying an area of about 1000 km^2. It consists of three phases: (1) porphyritic course-grained biotite granite with miarolitic pegmatites; (2) medium-grained biotite-muscovite granite with Sn-bearing greisen; and (3) biotite-alaskites and leucogranites. Lithium-fluorine granites are represented by microcline-albite, abundant amazonite-albite, and lepidolite-albite. The latter could be a source for Li. Lithium content in amazonite-albite and microcline-albite granites varies from 289 to 428 ppm, and in lepidolite-albite granite 878 ppm, in quartz-lepidolite greisen 1315 ppm, and in albitite 348 ppm (Kovalenko 1977).

5.6.3 Li-Ongonite Type

Ongonite was discovered in Central Mongolia and named after Ongon Khairkhan W deposit near the external contact of late Mesozoic Ongon Khairkhan granite pluton (Kovalenko et al. 1971). Ongonite is white or light blue rock, massive with con-choidal fractures, mainly porphyritic with variable number of phenocrysts or from distinct porphyritic to aphyric without phenocrysts. Groundmass in porphyritic and aphyric ongonites is aphanitic. Main minerals of ongonite are albite, potassium feldspar, quartz, lepidolite, and topaz, and accessory minerals are fluorite, garnet, zircon, monazite, ilmenite, columbite-tantalite, cassiterite, and pyrite. Similar subvolcanic ongonite and ongonite-like dikes are known from other areas in Khentii Batholith and surrounding rifts. The Li_2O content in the largest “Amazonite dike” in Ongon Khairkhan (see Fig. 5.4) is 0.47–0.35%, and in the dike of Baga Gazar, granite pluton is 0.03–0.05% Li. Ongonite dike in Ongon Khairkhan is emplaced at 120 Ma in a high structural level in the crust. The dike is strongly peraluminous, characterized by high alkalis, enrichment in Rb, Cs, Ga, and Ta, depletion in Mg, Ca, Zr, Ba, Sr and Eu, and has anomalous K-Rb, Rb-Sr, Zr-Hf, and Nb-Ta ratio (Dostal et al. 2015). Ongonite dikes in the "Stockwork" prospect from Ongon Khairkhan W deposit are cut by quartz-mica veinlets with wolframite. These veinlets differ from quartz-wolframite veins in the Ongon Khairkhan wolframite deposit, and researchers recognized two ore stages. Ongonites are cryptocrystalline, with massive or fluidal structure. Li content in ongonite groundmass varies from 965 to 3900 ppm; microcline also contains Li from 55 to 270 ppm and albite from 315 to 985 ppm.

The ongonite-hosted Ta-Li deposit type is subdivided into two facies: (1) volcanic and (2) plutonic (subvolcanic) both of which are volcanic analogues of Ta-bearing granites and pegmatites (Kovalenko et al. 1971; Kovalenko and Kovalenko 1986). The main ore elements are Ta with minor Rb, Nb, Be, Li, and Sn. Plutonic ongonites (Ongon Khairkhan, Baga Gazar, Janchivlan, Khoit Tamir) are rich in Ta (up to 130 ppm, with average content of 88 ppm) and Li (average content of 2780 ppm) and Rb (average content of 2380 ppm). Ongonite is porphyritic with albite, quartz, K-feldspar, topaz, and Li-fengite phenocrysts set in fine-grained matrix, composed mainly of albite. Volcanic ongonites (Teg Uul) are poorer in abovementioned elements (Ta 37 ppm, Nb 170 ppm, Rb 1040 ppm, and Be 90 ppm) than plutonic ongonites but form large bodies: volcanic cones, stratified bodies, and sheets. The volcanic ongonites are vitreous. The main occurrences of volcanic ongonites are at Teg Uul, Dorit Uul, and Durvent Uul in southern Mongolia. The Nb content in this deposit type ranges from 0.05% to 0.8%, Zr from 0.5% to 5.0%, and REE from 0.3% to 4.5%. High concentrations of Li, Be, Sn, and Zn are also characteristic. As an example, the Be occurrence of Teg Uul extends over 1 km^2 and is composed of tuff with a thickness of up to 10–20 m. Associated rocks are late Mesozoic rhyolites and ongorhyolites that form volcanic necks.

5.6.4 *Sedimentary-Hosted Lithium Deposit*

5.6.4.1 The Khukh Del Lithium Deposit

The Khukh Del lithium deposit (Fig. 5.13) is situated in the Dundgovi province approximately 550 km southwest from Ulaanbaatar and 98 km east of the Mandalgovi city. The Khukh Del sedimentary lithium deposit is located in the Tugrug valley filled by middle-upper Jurassic coarse conglomerate and sandstone, fine- to medium-grained sandstone, calcite-bearing siltstone, and clay. In the ore field, silica, limonite, argillic, carbonate, sericite, and pyrite are common alteration minerals, especially in sandstone (Chuluun and Odontuya 2017). Mineralization concentrated in siltstone reaches 500–900 ppm and up to 0.1% Li. The identified ore body deposit is 6 km long, 5 km wide, and 1.11 m thick. In addition to Li, the deposit is enriched in K, Cs, Rb, and light REE. The deposit reserves are 443.6 thousand tonnes of lithium in 283.9 Mt. of ore with an average content of 0.156%. The measured reserve amount is 37.7 thousand tonnes of metal lithium, and the

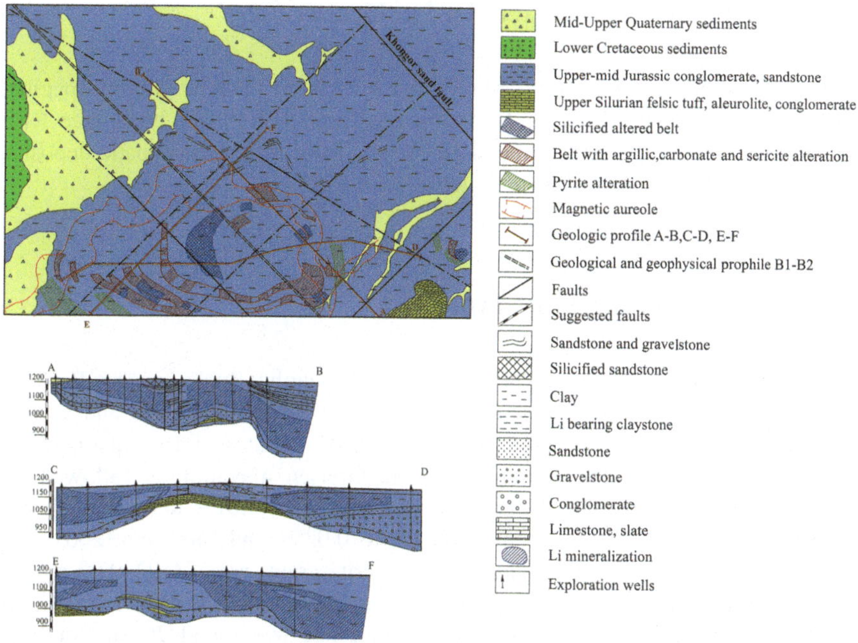

Fig. 5.13 Geological map of Khukh Del sedimentary lithium deposit (after Chuluun and Odontuya 2017)

proven reserve amount is 151.3 thousand tonnes. In addition, the measured and proven reserve has 173.3 thousand tonnes of Sr with average content of 0.145% and 4970.5 thousand tonnes of K with average content of 3.7%. Identified inferred mineral is 992.2 thousand tonnes of lithium metal in 644.3 Mt. of ore with 0.153% average content of lithium.

5.6.5 *Lithium-Bearing Lake Brines*

Other possible sources for lithium-bearing lake brines could be of high potential for future exploration work (Vakhromeev 2009). Analyses of lake waters in Eastern Mongolia show elevated Li concentration of 6.91–15.3 μmol/L in groundwater; from wells Li concentration is 1.8–6.93 μmol/L (Linhoff et al. 2010). In Southern Mongolia, Cretaceous oil basins could be other areas for research and exploration. Sampling from salt marsh and spring water in oil basins shows high concentration of Li, Br, and other elements. Deep well data in Zuunbayan oil deposit also confirm such high concentration of these elements. Studies of lake water from Western Mongolia also show higher concentration of Li (Ariunbileg et al. 2009)

5.7 Description of Selected Ta-Nb Occurrences

Tantalum and Nb mineralization are reported from many alkaline granites, pegmatites, and metasomatized rocks. The main types are REE-Nb-Zr types (see REE deposits Chap. 6, Rare Earth Mineral Deposits), Ta-bearing metasomatic deposits, Ta-bearing Li-F granites, and pegmatites.

5.7.1 Ta-Bearing Deposit Type

Altanboom Ta occurrence is situated in northern Mongolia. Rock types are Riphean schists and late Paleozoic leucocratic muscovite granite. Two stocks and several dike-like, albitized, greisenized, and silicified granites form quartz-albite metasomatitic rocks with Ta mineralization. Greisen alteration is followed by molybdenum mineralization. Ta_2O_3 content varies from 0.01% to 0.05%, Nb_2O_3 from 0.0002% to 0.03%, Sn from 0.0001% to 0.03%, and Mo from 0.002% to 0.289%. Ta-bearing mineral is tantalite. Cassiterite grains are of 0.02–0.03 mm in size. Molybdenum minerals are represented by molybdenite, powellite, and wulfenite. Similar to the Altanboom rare metal occurrence was found 20 km south of Altanboom.

5.7.2 Ta-Bearing Li-F Granite Deposit Type

The Ta-bearing Li-F granite deposit type occurs in the apical and outer contact zones of Li-F or subalkaline leucogranite plutons or in small stocks and dike-like bodies (Kovalenko et al. 1971). The main ore minerals are columbite and tantalite with microlite, Pb-pyrochlore, and cassiterite. The major gangue minerals are albite, microcline, quartz, Li-mica, topaz, pyrochlore, and beryl in granite with muscovite. The Ta-bearing granite deposit type is divided into two subtypes: with Li-mica and with muscovite.

5.7.3 Ta-Nb-REE Alkaline Metasomatite Deposit Type

Janchivlan Ta-Nb-REE alkaline metasomatite deposit is hosted in albite-lepidolite and amazonite-albite granite that occur along the southwest part of Mesozoic Janchivlan granite pluton and are controlled by the northwest-trending Ulaandavaa fault (Kovalenko et al. 1971). In association with granite, quartz-lepidolite greisen, albite metasomatite, quartz-muscovite greisen, and quartz veins are developed. Granites are composed of albite, quartz, lepidolite, amazonite, microcline, and topaz and accessory minerals as fluorite, columbite, monazite, Pb-pyrochlore, zircon, and cassiterite. Tantalum grade from surface to depth of 100 m is 60 g/t Ta (Ta/Nb = 1.2), with 600 g/t Li, 800 g/t Rb, and 50 g/t Sn. The average grade is 0.001–0.011% Ta.

5.7.4 Avdrant Peralkaline Granitoid-Related Nb-Zr-REE Deposit Type

This deposit (Kovalenko et al. 1971) is hosted in albite-amazonite granite that occurs in the upper part of Mesozoic granite plutons with a K-Ar isotopic age of 222–172 Ma and in dikes in adjacent host rock. The albite-amazonite granite is medium-grained and composed of amazonite, albite, quartz, and zinnwaldite. The amazonite-albite granite contains 330–1400 ppm Li, 6–75 ppm Ta, and 76–350 ppm Nb. The average grade is 0.007% Ta and 0.008% Nb. Li-ongonites contain about 0.005% Ta_2O_3. In placers the grade increases to 0.016% Ta_2O_3. The Urt Gozgor Sn placer is 2.0–2.5 km long and about 60–300 m wide. Established reserves are of 80–286 g/m^3 Sn (220 tonnes of Sn), Ta grades average 25 g/m^3 indicating 34.4 tonnes of Ta.

5.8 Concluding Remarks

Rare metal mineralization in greisens, stockworks, and quartz vein is common in Mongolia and associated mainly with highly evolved Mesozoic granites. Tungsten deposits in Mongolia are generally of small size; the largest world-class deposit is Undurtsagaan and medium-sized Yeguzer and Ongon Khairkhan. Rare metal deposits are classified as Sn-W and W-Mo-Be greisen/stockwork and quartz vein types, W and W-Mo skarn, Sn-polymetallic, and Sn-W-sulfide. The deposits have complex mineralization, with many types of rare metal deposit, and occurrences could be found in ore districts. In all areas with Sn-W and W-Mo-Be greisen/stockwork and quartz vein deposits, Sn-W placers are developed forming small- and medium-sized deposits that were mined in the past. Other rare metals such as Li, Be, and Nb occur in pegmatite or are Be by-products of W-Mo-Be type. Sedimentary Li deposit type is a new discovery and could be an important source of Li, like many lakes in Western and Eastern Mongolia. Other noncommercial, but interesting types are Li-ongonite, Li-F granites, Be-tuffs, and brine.

References

Ariunbileg S, Isupov VP, Bladimirov AG, Krivonogov SK, Shatskaya (2009) Mikrokomponentny sostav mineralizovannych ozer Vostochnoi Mongolii (microcomponent content of mineralized lakes in Eastern Mongolia). Mongolian Geosci 35:115–116

Batjargal S, Lkhamsuren J (1979) Specifics of mineralization stages at Yeguzer deposit. Mongolian Tech Inst 6:87–103. (In Mongolian)

Batjargal S, Monkhbat B, Gandigara B (2017) Geology and ore mineralization stages, ore associations of Yeguzer rare metal deposit. In: Geology, mineralogy and metallogeny of Mongolia's ore deposits. MUST Printing House, Ulaanbaatar, pp 7–18. (In Mongolian)

Bayarsaikhan N, Bat-Ulzii D, Naidan N (2017) Geology and ore body study of the Narsan Khundlun tin deposit. In: Geology, mineralogy and metallogeny of Mongolia's ore deposits. MUST Printing House, Ulaanbaatar, pp 49–61

Chuluun O, Odontuya O (2017) Geology, structure and ore mineralization of the Khukh Del sedimentary lithium deposit. In: Geology, mineralogy and metallogeny of Mongolia's ore deposits. MUST Printing House, Ulaanbaatar, p 62

Dandar S (2012) Volfram, tsagaan tugalganii ord, ilreluud (Tungsten, tin deposits, and occurrences). In: Lkhamsuren J (ed) Metallic mineral deposits, V 6, 2nd ed, Soyombo Printing, Ulaanbaatar. pp 120–192

Dandar S, Enhbaatar Sh, Dejidmaa G, Monhbat Yo, Navchgerel Ch, Khurelbaater L, Amar O, Enhjargal M (1999) Results of 1: 200,000 scale metallogenic investigation carried out in Mongol Altai: Geologic Information Center, Ulaanbaatar, Open-File Report 5306 (in Mongolian)

Darambazar D, Gangush B, Niislelkhuu G (2017) Geology and mineralization of the Mandal Mo-W deposit. In Geology, mineralogy and metallogeny of Mongolia's ore deposits. Ulaanbaatar. 49–62 (In Mongolian)

Dejidmaa G, Bujinlkham B, Ganbaatar T, Oyuntuya N, Enkhtuya B, Eviikhuu A, Monkh-Erdene N (2002) In: Lkhamsuren J, Dejidmaa G, Gerel O, Batjargal S, Bold-Erdene B, Batbold D, Begzsuren B, Bat-Erdene B (eds) Distribution map of mineral deposits and occurrences in Mongolia. Mineral Resources Authority of Mongolia, Geologic Information Center, Ulaanbaatar, Scale 1:1000000

Dejidmaa G, Dorjgotov D, Gerel O, Gotovsuren A (2005) Preliminary description of mineral deposit models (types) for Mongolia. In: Seltmann R, Gerel O, Kirwin DJ (eds) Geodynamics and metallogeny of Mongolia with a special emphasis on copperand gold deposits. SEG-IAGOD Field Trip, 14–16 August 2005, 8th Biennial SGA Meeting, IAGOD Guidebook Series, CERCAMS/NHM, London, pp 31–52

Dostal J, Kontak DJ, Gerel O, Shellnutt GJ, Fayek M (2015) Cretaceous ongonites (topaz-bearing albite-rich microleucogranites) from Ongon Khairkhan, Central Mongolia: products of extreme magmatic fractionation and pervasive metasomatic fluid: rock interaction. Lithos. 09/2015 236–237:173–189

Elsner H, Buchholz P, Schmitz M, Tamiraa A (2011) Industrial minerals and selected rare metals in Mongolia. An Investors' Guide. p 322

Gerel O (1990) Petrologiya, geochimiya i rudonosnost' subschtelochnogo Mezozoiskogo magmatisma Mongolii (Petrology, geochemistry and mineralization of subalkaline Mesozoic magmatism in Mongolia): U.S.S.R. Academy of Sciences, Irkutsk, Doctor of Science Dissertation, p 395

Gerel O (1995) Mineral resources of the Western part of the Mongol-Okhotsk foldbelt. Resour Geol Spec Issue 18:151–157

Gerel O (1998) Phanerozoic felsic magmatism and related mineralization in Mongolia. Bull Geol Surv Jpn 49(6):239–248

Gerel O, Kanizawa S, Ishikawa K (1999) Petrological characteristics of granites from the Avdrant and Janchivlan plutons. In: Problems of Geodynamics and Metallogeny of Mongolia. 13:30–34

Getmanskaya TI, Chernov VS (1996) Geologo-geneticheskie modeli wolframovych greisenovych mestorozhdeniy (Geology-genetic models of tungsten greisen deposits) In: Geology of ore deposits. 2:137–145

Gruner H, Sansoni G, Tolke A (1984) Final report. Tungsten-Molybdenum Ore Yeguzer-MPR. Report of the Forschungsinstitut fur Aufbereitungstechnik Frieberg (unpublished)

Ivanova GF (1976) Mineralogiya i geochimiya vol'framovych mestorozhdenii Mongolii (mineralogy and geochemistry of the tungsten deposits in Mongolia) M. Nauka, p 257

Jargalsaihan D, Kazmer M, Baras Z, Sanjaadorj D (1996) Guide to the geology and mineral resources of Mongolia. Geological exploration, Consulting and Services (GCS) Co Ltd. Ulaanbaatar

Kempe U, Dandar S, Wolf D (1994) Y-Anreicherungen mit Y/Sc- und Y/Ho-Fraktionierungen in den Mineralisationen von Kyzyltau (Mongolischer Altai) abs.: Berichte Deuts. Mineral. Gesel, Beihef. Eur J Mineral 6:136

Khasin RA (1977) Olovo, vol'fram i molybden (Tin, tungsten and molybdenum) In Geology of Mongolian People's Republic, Nedra, Moscow, 3:270–436

Kirwin DJ (2005) Undirectional solidification textures associated with intrusion-related Mongolian mineral deposits. In: Seltmann R, Ochir G, Kirwin D (eds) Geodynamics and Metallogeny of Mongolia: with a special emphasis on copper and gold deposits. International Assocciation on the Genesis of Ore deposits (IAGOD), London, pp 63–85

Koval PV (1998) Regional'ny geochemicheskiy analyz granitoidiv (Regional geochemical analysis of granitoids), Novosibirsk. p 490

Kovalenko VI (1977) Petrologiya i geochimiya redkometal'nych granitoidov. (Petrology and geochemistry of rare metal granitoids). Nauka. Siberian Branch. Novosibirsk, p 205

Kovalenko VI, Koval PV (1984) Endogennye redkozemel'nye I redkometal'nye rudnye formatsii Mongolii (Endogenic rare earth element and rare metal ore formation in Mongolia. In: Endogenic ore formations of Mongolia. 38:50–75

Kovalenko VI, Koval PV, Yakimov VM, Sherhan O, Yarmolyuk VV (1988) Rdekometall'naya metalogeniya Mongol'skoi Narodnoi Respubliky na osnove potentsial'noi rudonosnosti magmaticheskych associatsii (rare metal metallogeny of Mongolian People's Republic on basis of potential ore-bearing magmatic rocks). In: Ore-bearing magmatic associations. Nauka, Moscow, pp 114–159

Kovalenko VI, Kovalenko NI (1986) Ongonity (topazonsnye kvartsevye keratophyry) -subvulkanicheskie analogi redkometall'nych lnmni-ftoristych granitov (Ongonites (topaz-bearing quartz keratophyre) subvolcanic analogous of rare metal lithium-fluorine granites. M. Nauka. p 127

Kovalenko VI, Kuzmin MI, Zonenshain LP (1971) Redkometallnye granitoidy Mongolii (rare-metal granitoids of Mongolia: petrology, trace elements and genesis), vol 5. Nauka, Moscow, p 239

Kovalenko VI, Yarmolyuk VV (1995) Endogenous rare metal ore formations and rare metal metallogeny of Mongolia. Econ Geol 90(3):520–529. https://doi.org/10.2113/gsecongeo.90.3.520

Linhoff BS, Bennett PC, Tamir P, Gerel O (2010) Geochemical evolution of uraniferous soda lakes in eastern Mongolia. Environ Earth Sci. https://doi.org/10.1007/s12665-010-0512-8

Nokleberg WJ (ed.) (2010) Metallogenesis and Tectonics of Northeast Asia: U.S. Geological Survey. Professional Paper 1765, p 624

Podlessky KV, Vlasova DK, Kudrya PF (1988) Skarny i svyazannoe s nimi orudenenie Mongolii (Skarns and connected ores of Mongolia): Joint Soviet Mongolian Scientific-Research Geological Expedition, Transactions, Moscow, 45, p 149

Rodionov SM, Obolenskiy AA, Dejidmaa G, Gerel O, Hwang DH, Miller RJ, Nokleberg WJ, Ogasawara M, Smelov AP, Yan H, Seminskiy ZV (2004) Descriptions of metallogenic belts, methodology, and definitions for Northeast Asia mineral deposit location and metallogenic belt maps: U.S.Geological Survey Open-File Report 2004–1252 explanatory text, p 442

Smirnov VN, Koval PV, Tsypulov YP, Kovalenko VI, Antipin VS (1977) K-Ar vozrast granitoidnych associatsii Khenteya (K-Ar) age of granitoid associations in Khentei (Mongolia) Dokl. Russ Acad Sci 232(1):192–195

Turutanov EK, Zorin YA (1978) Glubinnye structury granitnych plutonov Mongolii I Zabaikaliya (deep structure of granite plutons in Mongolia and Transbaikalia). Nauka, Novosibirsk, p 182

Vakhromeev AG (2009) Forming regularity and conception of adoption of commercial brines (on example of Siberian platform, southern part). Abstract of Doctor of Sci. thesis. Irkutsk. p. 36 (in Russian)

Zonenshain LP, Kuzmin MI, Natapov LM (1990) Geology of the USSR: a platetectonic synthesis. In: Page BT (ed) Geodynamics series, geodynamics series, vol 21. American Geophysical Union, Washington, DC, p 242

Chapter 6
Rare Earth Mineral Deposits

Ochir Gerel, Yondon Majigsuren, and Baatar Munkhtsengel

6.1 Introduction

Since the 1970–1980s, several prospective areas for REE mineralization were recognized and prospecting activity had started. In Mongolia, the following types of REE mineral systems are found: (1) magmatic (metasomatic) rare earths deposits, light REE-P–Nb–F deposits and occurrences related to carbonatites in South Gobi (Mushgai Khudag, Khotgor, and Lugiin Gol deposits); (2) light and heavy REE-Nb-Ta-Zr-, Be-, and Y-bearing alkaline intrusive rocks (syenite, granite) in Western Mongolia (Khalzan Buregtei, Ulaan Del deposits and Tsakhir, Ulaan Tolgoi, Shar Tolgoi occurrences) and pegmatite-related REE in South Gobi, Khanbogd. REE-albite and REE albite-nepheline syenite deposit types form small occurrences. Another possible deposit type is REE-(Ti–P–Nb) residual mineralization in alkaline igneous and carbonatite complexes that up to now have not been studied. According to Geological Information Center, there are four major economic deposits of rare earths in Mongolia (Table 6.1) together with a number of occurrences (Table 6.2) interpreted as forming mainly in continental post accretion or rift tectonic environments. The economic REE mineral deposits and occurrences are mainly of Mesozoic age and formed within collisional intracontinental belts. Some also occur along a middle and late Paleozoic active continental margin.

O. Gerel (✉) · Y. Majigsuren · B. Munkhtsengel
Geoscience Center, School of Geology and Mining, Mongolian University of Science and Technology, Ulaanbaatar, Mongolia
e-mail: gerel@must.edu.mn; majigsuren@must.edu.mn; tsengel@must.edu.mn

O. Gerel et al. (eds.), *Mineral Resources of Mongolia*, Modern Approaches in Solid Earth Sciences 19, https://doi.org/10.1007/978-981-15-5943-3_6

Table 6.1 Major REE mineral deposits in Mongolia

Name	Age (Ma)	Mineralization	Reserves and resources
Mushgai Khudag Bayan Khoshuu	140 135	LREE Sr Ba Bastnaesite, apatite, parisite, synchysite	34 Mt., grade 1.36% REO
Khotgor	135	LREE Ap Bastnaesite, apatite, synchysite? parisite?	40 Mt., grade 1.22% REO
Lugiin Gol Ulgii	222–244	LREE	0.5 Mt., grade 2.67% REO
Khalzan Buregtei Tsakhir, Ulaan Del, Shar Tolgoi, Ulaan Tolgoi	393	REE, Zr, Ta, Nb	49 Mt., grade 0.6% REO

Table 6.2 REE occurrences in Mongolia

Site names and localities	Commodities	Comments
Tsagaan Chuluut	Monazite, 95–750 g/t	RE_2O_3 757.0 t
Khoit Khuldiin Gol	RE_2O_3, up to 1.25%	
Kholboo Khudag	RE_2O_3, 0.08–50%; Ta, 0.08%	
Khokh Tolgoi	Y, 0.1–0.2%; Nb_2O_5, 0.2%; Ta_2O_5, 0.05–0.2%; Nb_2O_5, 0.01–0.3%	Y, 100,000 t; Nb, 100,000 t; Ta_2O_5, 400,000 t
Khuldiin Gol	RE_2O_3, 0.08–1.28%; Ta, 0.03–0.1%; Nb, 0.01–0.3%	
Ar Gol	Y, 0.1%; La, 0.007%; Th, 0.02–0.1%	
Yarhis Gol	Nb, 0.01%; La, 0.3%; Ce, 0.5%; Y, 0.05%	
Urgun Khajuu	Zr, 0.007–0.1%; Sr, 0.03–0.7%	
Kharaa Gol	Th, 0.03–4.7%; Y, 0.2–0.3%	
Ulgii Khiid	La, 260 g/t; Ce, 300 g/t; Rb, 0.3%	
Navchin Gol	Monazite 0.5–120 g/m^3	
Shar Tolgoi	Y, 0.159–2.65%; Zr, 0.2–0.3%; Nb_2O_5, 0.052%	Zr, 1400.8 thous. t Nb, 199,000 t
Undur	Y, 0.04%; Zr, 0.4%; Ce, 0.1%	Zr, 600,000 t; RE_2O_3, 300,000 t
Aduun Gol	La, 0.1%; Nb, 0.006%; Sr, 0.02%; Y, 0.02%; Ga, 0.008%; Ce, 0.1–0.2%	
Ust Gol	Y, 0.01%; Ga, 0.1%; RE_2O_3, 0.01; Ce, 0.3%; La, 0.2%	
Mogoi Chuluut	RE_2O_3, 0.5%; Pb, 0.4%; Zn, 0.5%	RE_2O_3, 8 t; Pb, 1,6 t; Zn, 4,7 t
Zuun Doono Uul	Ce, 0.01%; Nb, 0.002%; Y, 0.003%; Zr, 0.016%	Ce, 36 t; Nb, 11,160 t; Y, 17,880 t; Zr, 97,140 t

6.2 REE Mineral Systems

6.2.1 REE in Carbonatites

This type includes REE-P-Sr-Ba-fluorite-Pb deposits and occurrences which are associated with alkaline volcanic-plutonic complexes of Late Mesozoic age (Samoilov and Kovalenko 1983). These alkaline complexes occur in the intraplate environment and are controlled by main faults and rift zones. The host rocks are represented by potassic alkaline rocks with carbonatite such as melanephelinite, melaleucitite, phonolite, trachyte, latite, trachybasalt, and syenite. Main examples of this type are the Mushgai Khudag and Lugiin Gol REE deposits in southern Mongolia. The deposit subtypes are (1) carbonatite and eruptive trachyte breccia with a carbonatite matrix containing 0.1–0.8% REE or enriched in light REE (LREE) and containing up to 18% Sr; (2) magnetite-apatite rocks with REE content of 1.0–14.5%; and (3) bastnaesite carbonatite with a REE content of 1–18% (Kovalenko and Yarmolyuk 1995). The carbonatite and eruptive trachyte breccia subtype usually form large mineralized bodies. The magnetite-apatite deposit subtype occurs as dikes and stock-like bodies. The bastnaesite carbonatite deposit type forms mainly dike. Ore minerals represent as bastnaesite, carbonate, fluorite, celestite and barite, cerussite, magnetite, apatite, and monazite in all three subtypes. There are more than 200 dikes of carbonatites that range from 20 m to 2000 m in length and from 0.2 to 1 m in width within the Mushgai Khudag alkaline series. Quartz veins with celestite-fluorite-calcite and calcite-celestite-fluorite are also occurring besides calcite carbonatites. Magnetite-apatite rocks are associated with this carbonatite complex. Another example is the Lugiin Gol carbonatite deposit related to the Lugiin Gol alkaline complex. The complex is composed of nepheline syenite with ring and radial dikes of nepheline syenite, pulaskite, syenite-porphyry, and tinguaites. The main REE minerals at the Lugiin Gol deposit are synchysite, synchysite-parisite and synchysite-bastnaesite intergrowths, and Nd-enriched synchysite (Batbold 1997). Other carbonatite occurrences are those at Bayan Khoshuu, Ulgii, Tsogt Ovoo, and Khotgor (South Gobi). The carbonatites occur as dikes up to 1–2 m wide within plutons or host late Paleozoic volcanic-sedimentary rocks in these areas. The carbonatites are composed of calcite with mica, apatite, fluorite, bastnaesite, synchysite, rutile, zircon, and sulfides. Alterations are replacement of pre-existing minerals and recrystallization of pyroclastic rocks, formation of silica-jarosite, silica-fluorite, and silica-barite-fluorite metasomatic rocks.

6.2.2 Fe-REE in Carbonatite or Magnetite-Apatite

The Fe-REE carbonatite or magnetite-apatite deposit type consists of veins and small bodies with 1–14.5% REE (mostly LREE) that mainly occur in apatite, magnetite, phlogopite, and rare celestite in alkaline igneous rocks (Samoilov et al. 1984; Kovalenko and Koval 1984). In this type of deposit, magnetite is mostly altered to hematite. The magnetite and apatite content vary widely in stock-like bodies and in veins. Some bodies contain bonanza ore of massive, very coarse-grained magnetite or fine-and medium-grained apatite. This deposit type is genetically and spatially associated with REE-fluorspar carbonatite, REE-apatite carbonatite, and REE-strontium carbonatite deposits in alkaline igneous rocks and in explosive carbonatite breccia. Examples occur in the Mushgai Khudag district in southern Mongolia and in the Uvs Nuur district in northwestern Mongolia.

6.2.3 Peralkaline Granite-Related REE (Nb-Zr-REE)

The peralkaline granite-related REE (Nb-Zr-REE) deposit type examples are Khalzan Buregtei REE-Zr-Nb deposit in western Mongolia (Kovalenko et al. 1995a, b), the Khanbogd REE deposit in southern Mongolia (Vladykin et al. 1981), and several occurrences in western and southern Mongolia. These deposits mainly occur in the apical part of cupolas and are generally associated with highly fractionated magmatic phases including peralkaline pegmatite. The host granites are composed of potassium feldspar, quartz, albite, arfvedsonite, aegirine, fluorite, and various REE minerals, such as elpidite, gittingsite, zircon, pyrochlore, monazite, REE fluorcarbonate, polylitionite, and others. Replacement by epidote, orthoclase, and postmagmatic albite represent the alteration. The deposits are generally hosted in microcline-albite granite and albite metasomatic rocks composed of quartz, albite, pyroxene, and microcline. Quartz-epidote metasomatic rocks contain zircon, fergusonite, allanite, chevkinite, and titanite in vein-like zones. Fergusonite and zircon contain heavy REE (HREE) and Y (Kempe and Dandar 1995). Accessory minerals include amphibole, magnetite, zircon, epidote, ilmenite, fluorite, beryl, chevkinite, pyrite, and galena. REE pegmatites and quartz-fluorite veins may also occur (Getmanskaya et al. 1995).

Zr-REE mineralization shows magmatic origin and is concentrated within highly fractionated REE-bearing granites (Kovalenko et al. 1985). In contrast, other researchers suggest a metasomatic origin. For instance, some researchers explained that the Khalzan Buregtei deposit, and the Tsakhir, Shar Tolgoi, and Ulaan Tolgoi occurrences in the Mongolian Altai region are formed in the endocontact of alkaline plutons that consist of metasomatically altered alkaline syenite nordmarkite (Kempe et al. 1999; Andreev et al. 1994). These intrusions range from fine-grained granite to pegmatite and form metasomatically altered small bodies. Nordmarkite, dolerite, and syenite are also metasomatically altered. There are more than 70 ore minerals determined. The rock-forming minerals are potassium feldspar, albite, quartz, calcite, zircon, and Sr-fluorite. Pyrochlore, barite, rutile, titanite, epidote, REE carbonates, and fluorite occur as an accessory mineral. Metasomatic rocks contain quartz, epidote, zircon, titanite, fergusonite, pyroxene, chevkinite, and others (as at the Tsakhir occurrence).

Other examples of peralkaline granite-related REE (Nb-Zr-REE) mineralization are the Gurvan Uneet and Ulaan Uneet occurrences in Western Mongolia.

6.2.4 REE-Albite Type

The REE-albite deposit type forms in the endocontact of alkaline plutons which are mainly composed of metasomatically altered alkaline syenite (nordmarkite) (Andreev et al. 1994; Kempe et al. 1999). The main occurrences are Teeg Uul, Durvun Dert Uul in southern Mongolia. Other examples of this deposit type are the Khalzan Buregtei deposit and the Tsakhir, Shar Tolgoi, and Ulaan Tolgoi occurrences in the Mongolian Altai.

6.2.5 REE in Albite Nepheline Syenite

The REE-albite nepheline syenite deposit type occurs in peralkaline volcanic rocks such as comendite, pantellerite, peralkaline trachydacite, trachyrhyolite, and trachybasalt (Kovalenko and Yarmolyuk 1995). The main examples of this type of mineralization are the Boomyn Khyar and Zart Khudag occurrences in South Gobi. The ore minerals are represented by REE-Zr-Nb minerals.

6.2.6 REE in Pegmatite

The REE pegmatite deposit type contains the majority of potentially economic pegmatites in Mongolia. These deposits are related to late Paleozoic or Mesozoic REE-bearing pegmatites that are mainly associated with calc-alkaline and Li-F leucocratic granite. There are three main subtypes of REE pegmatites are defined (Kovalenko and Koval 1984): Li-mica, muscovite (muscovite-albite), and muscovite-microcline. The Li-mica subtype contains Ta-Nb minerals, cassiterite, Li-mica, quartz, albite, microcline, apatite, tourmaline, topaz, beryl, and other minerals. The main examples are at Khukh Del Uul and Unjuul. The muscovite-albite subtype is represented by Berkh, which consists of columbite, tantalite, quartz, albite, microcline, and muscovite. The muscovite-microcline subtype is hosted by subalkaline granite and includes cassiterite, wolframite, quartz, microcline, and muscovite. The main examples of this subtype are at Tumen Tsogt, Bayan Delger, Bayan Ovoo and Khalzan Uul. Many of the known REE pegmatite deposits occur in northeast Mongolia. The Khukh Del Uul, Unjuul, and Berkh REE pegmatite deposits and others occur as dike-like or lenticular bodies that range in size from few meters to hundreds of meters in length and from 1 to 10 m in width.

Li-Sn-Be pegmatites (Kovalenko et al. 1986; Kovalenko and Yarmolyuk 1995) consist of Li-mica, Ta, and Sn-W minerals hosted in post-accretionary intrusions that postdate the peak of batholiths emplacement. The associated granites are mainly calc-alkaline and Li-F leucogranites and their volcanic and subvolcanic equivalents (see Chap. 5, Rare Metals).

6.3 Description of REE Deposits and Occurrences

6.3.1 REE Carbonatite Deposits

The REE carbonatite deposits include REE-P-Sr-Ba-fluorite-Pb deposits and occurrences that are associated with late Mesozoic alkaline volcano-plutonic complexes (Samoilov and Kovalenko 1983). The distribution of these deposits is controlled by large-scale tectonic structures of the North Gobi rift zone (coinciding with the Main Mongolian Lineament), Gobi-Tien Shan fold belt, and Lake tectonic zone, as well as by the local geomorphology (Fig. 6.1). There are many carbonatite-bearing alkaline complexes in South Mongolia that contain REE, but except for three of them (the Mushgai Khudag, Khotgor, and Lugiin Gol), they are not presently of economic importance.

The deposits are controlled by tectonic structures of latitudinal direction and are associated with Early Cretaceous rifting (120–139 Ma). Associated REE fluorcarbonate mineralization in carbonatites and britholite in nelsonites at the Mushgai Khudag, Bayan Khoshuu, and Khotgor deposits hosts up to 12 wt.% RE_2O_3 (predominantly LREE).

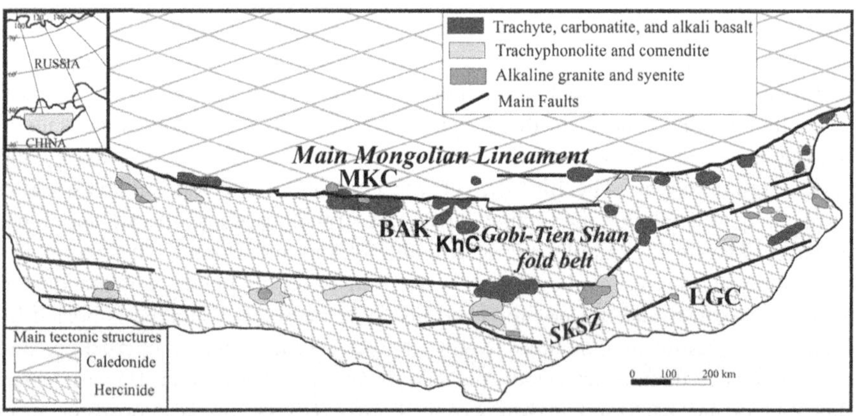

Fig. 6.1 Distribution of alkaline igneous complexes in South Mongolia (modified after Samoilov and Kovalenko 1983). MKC-Mushgai Khudag Complex, BAK-Bayan Khoshuu, KhC-Khotgor Complex, LGC-Lugiin Gol Complex, SKSZ-Sulinkheer suture zone

The carbonatites of the Gobi-Tien Shan fold belt are associated with Triassic rifting (240–244 Ma) along the southern side of the Gobi-Tien Shan fold belt, near the boundary with the Sulinkheer (Solonker) suture zone. The REE contents in Lugiin Gol are ten times higher than the global average for carbonatites (up to 15 wt.% RE_2O_3), and the rocks are significantly enriched in LREE.

6.3.1.1 Mushgai Khudag Deposit

The Mushgai Khudag deposit is located 100 km south of Dalanzadgad, a city of the Southgobi (Umnugovi) province. The Mushgai Khudag rare earth deposit was discovered in 1975, prospected in 1983–1984 by the Mongolian government, and explored in more detail between 1989 and 1994 by private companies. The most recent exploration campaign took place between 2007 and 2012 by the Mongol Gazar Company. Rare earth-metal-hosting alkaline magmatic rocks were discovered in the early 1970s by a Soviet-Mongolian scientific expedition. An investigation continued until mid-1980s.

Carbonatite-hosted deposits are associated with an alkaline syenite-trachyte volcanic-plutonic complex forming a ring structure of about 27 km diameter. Primary carbonatite is overlain by regolith. In an area of more than 100 km^2, Late Mesozoic (142 $\pm$ 3 Ma) nephelinite, melanephelinite, melaleucitite, trachyte, phonolite, trachydacite, trachyrhyodacite, syenite, nepheline syenite, shonkinite porphyry stocks and dikes, welded tuff, and lava flows form a volcanic-plutonic alkaline complex. The main plutonic host rocks are nepheline syenite, granosyenite, and shonkinite porphyry. Trachyte and latite are the dominant volcanic rock types. Apatite-magnetite and carbonatite diatremes or dikes cut these rocks. Carbonatites are composed of calcite and calcite-fluorite (Samoilov and Kovalenko 1983). Alkaline rocks are intruded in grabens that extend along the Main Mongolian Lineament, overlaying sedimentary-volcanogenic deposits, and cut Late Paleozoic granite (Fig. 6.2). Northwest and northeast trending faults control subvolcanic and volcanic pipe facies rocks (Samoilov and Kovalenko 1983). Trachytes are abundant and occur as dikes and plugs.

The lower part of the alkaline complex is composed of melanephelinite, phonolite, latite, and the upper part mainly of trachyte and trachylatite associated with syenites. Ore mineralization spatially and genetically related to the syenite intrusion.

The volcanic-plutonic complex is the K-rich with 0.4–1.5% P_2O_5, 0.2–2.35% F, 0.3–1.3% SrO, 0.2–0.8% BaO and with high content of REE (Baskina et al. 1978; Samoilov and Kovalenko 1983). Apatite-magnetite mineralization was discovered in 1974 and studied in detail by Samoilov and Kovalenko (1983) and Enkhtuvshin (1995). Presently, six ore zones (Tumurtei, Khurenkhad, Main ore, High-grade ore, Monazite, and Jonshit) have been discovered with tens of different ore bodies, which may be steeply dipping tabular or lenticular, breccia, stockworks, and veins or subhorizontal volcanic sheets (Fig. 6.2). The Mushgai Khudag Fe-REE carbonatite deposit mineralogy is complex. Four styles of mineralization have been reported: carbonate cemented mineralized breccia; mineralized carbonatite (mainly

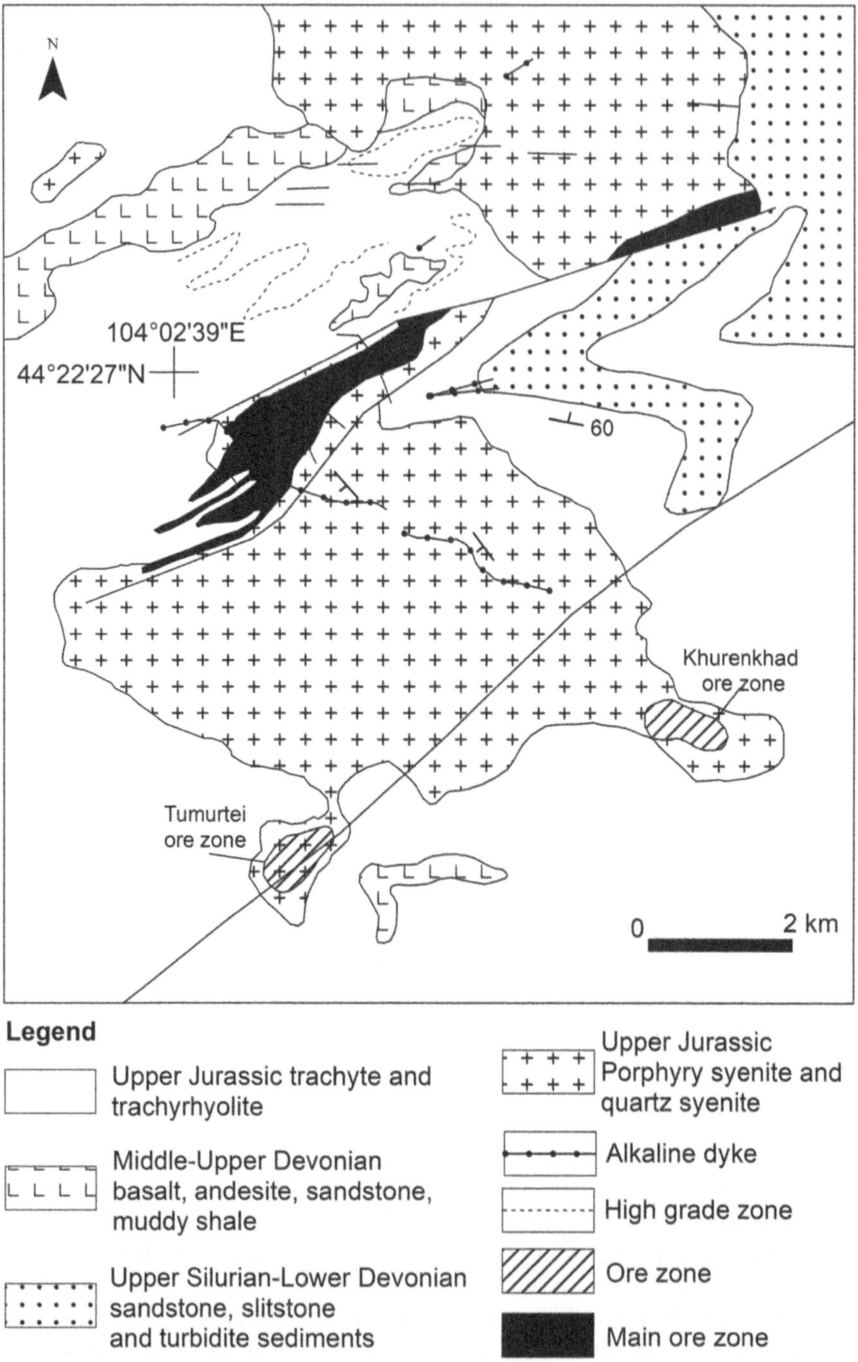

Fig. 6.2 Schematic geological map of Mushgai Khudag deposit

"apatite-bastnaesite ore"); magnetite-apatite ore; and apatite ore (mainly "phosphatic ore"). Regarding the composition, apatite-bastnaesite and apatite ores are the most important ones, and each of them has been subdivided into several specific types.

The main REE minerals are bastnaesite, apatite, perrierite, parisite, and synchysite; minor ore minerals are iron minerals, phosphate minerals, and monazite. Magmatic minerals are REE-bearing apatite, monazite, magnetite, potassium feldspar, fluorite, phlogopite, calcite, dolomite, celestite, and quartz. Accessory minerals are albite, alkaline amphibole, pyroxene-aegirine, garnet, ankerite, siderite, rodochrosite, barite, pyrite, hematite, ilmenite, and rarely occurring monazite, rutile, sphene, zircon, orthite, tourmaline, arsenopyrite, and chalcopyrite. Four stages are recognized (Ontoev et al. 1979): (1) magnetite-apatite; (2) fluorite-carbonate (REE content is low: 0.5–0.7%); (3) quartz-celestite-fluorite; and (4) carbonatite.

(1) Magnetite-apatite mineralization occurs in the contact of syenite-porphyry stocks and is represented by phlogopite-titanomagnetite-apatite or predominate apatite ore, or in lens-like bodies, composed mostly of titanomagnetite (Khuren Khad ore zone). Ore mineralization is accompanied by metasomatic alteration represented by microcline and albite zone forming magnetite-feldspar ore. Ore zones are cut by syenite-porphyry and trachyrhyolite dikes.
(2) Fluorite-celestite-magnetite-apatite mineralization is developed on magnetite-apatite ore forming quartz and carbonate-bearing fluorite-celestite mineralization (Main ore zone). The REE-containing mineral is apatite and fine-grained bastnaesite. Relatively low content of REE, but high content of Sr and F are characteristic features of this stage.
(3) The quartz-celestite-fluorite stage is developed later than the above mentioned syenite-porphyry and trachyrhyolite dikes and superimposed on the first stage mineralization forming dikes and lens-like bodies composed of quartz-celestite, fluorite, and calcite (Main ore zone). Ore mineralization is accompanied by silica, sericite, and fluorite alteration. REE content is 0.5–0.7%.
(4) Generally, high REE, F, P, and Fe contents and low Nb, Ta, and Zr characterize the geochemistry of the Mushgai Khudag deposit. In early stage, apatite and late stage, F-carbonate-bastnaesite are main REE-bearing minerals.

Interesting was the discovery of 2-m-thick and 50-m-long lens-like body elongated in a latitudinal direction with Pb-enriched cerussite-bearing volcanic glassy tuff-breccia. Enrichment in Pb and REE was due to hydrothermal-metasomatic alteration. The average content of Pb and REE reaches 15% Pb and 4.86% REE + Y in altered silica-jarosite rock.

Magnetite is replaced by martite, phlogopite, and feldspar composing magnetite-apatite ore. Quartz, albite, amphibole, rutile, and calcite are common. In this type, REE concentrated in apatite, P and FeO contents are high, but Ba and F contents are relatively low. Apatite is a main REE-bearing mineral in magnetite-apatite and fluorite-celestite-magnetite-apatite ore bodies. Apatite is formed in three generations and occurs as fine-grained aggregate. REE content in apatite is up to 13.5%. Magnetite is euhedral and its aggregates are from 0.2–0.5 cm to 1–2 cm large in magnetite-apatite and fluorite-celestite-apatite ore. Fluorite has a violet color, fine-grained, 0.02–0.2 mm in size. REE content in fluorite varies from 0.15% to 0.40%. Bastnaesite (Ce,La)CO_3F) occurs as fine-grained aggregate in quartz-carbonate-fluorite, carbonate-fluorite, and fluorite ore bodies. Bastnaesite forms needle-like crystals and aggregates that contain 7.68 RE_2O_3.

Seventeen ore bodies, spaced 100–1200 m apart of various shapes, structures, and compositions, are known. In particular: (1) steeply dipping lenticular and pillar-like pipes 900 m in extent and 200 m thick; (2) stock-like and irregularly shaped bodies from 10–30 to 20–70 m in size; (3) vein-like bodies up to 3 m thick and up to 100 m long; (4) stockwork up to 30 × 200 m in size; and (5) subhorizontal volcanic sheets 0.3 × 0.5 km^2 in size and up to several tens of m thick. They can be traced 50–150 m downdip without phase-out.

The Mushgai Khudag deposit has unique LREE mineralization due to various stages of magmatic and hydrothermal segregations. The fluorine, phosphate, and carbonate played a leading role in the accumulation of light lanthanides in the first stage of magmatism. LREE-bearing minerals are dominantly fluorapatite to hydroxyl-bearing fluorapatite and monazite. Fluorapatite to hydroxyl-bearing fluorapatite can be found in all mineralization zones, whereas monazite is distributed in a limited area. Sulfur and Sr-rich individual monazite grains have been found in the rim of fluorapatite and hydroxyl-bearing fluorapatite from the Monazite ore zone. LREEs have strong mobility in the last stage of metasomatism due to sulfate-rich fluid complexed with LREE (Enkhbayar et al. 2016).

The reserves and resources of the deposit are about 34 Mt. of RE_2O_3. The average rare earth oxide (REO) grade in Mushgai Khudag is 1.36% and even reaches 6.15% in the high-grade ore zone.

6.3.1.2 Khotgor Deposit

The deposit lies 68 km northeast of Dalanzadgad in the center of Southgobi (Umnugovi) province. The Khotgor deposit is located just south of the Main Mongolian lineament, within a Devonian-Carboniferous island arc. The Khotgor rare earth deposit was discovered in 1979–1982, during Soviet-Mongolia joint mapping project, and later the QGX LLC (Canadian 100% wholly-owned company) did detailed exploration activity during 2005–2009. The deposit is in a volcanic-breccia pipe with apatite-magnetite-phlogopite mineralization associated with nepheline syenite. Nepheline syenite has a tube-like body and is composed of potassium feldspar, nepheline, and some phlogopite, magnetite, hematite, fluorite, apatite, and celestite. The ore mineralogy is very complex, e.g., apatite-fluorite, magnetite-apatite, fluorite-celestite-magnetite-apatite, brecciated rocks with carbonate cement, and eruptive breccia of different contents. There are two main types of ore: REE minerals-bearing veins and breccia with REE mineralized cement. The majority of the wall rocks around and also within the Khotgor deposit are formed by metasediments of the Silurian Mandal Ovoo Formation. Mineralized syenite bodies and other rocks are altered and strongly brecciated. Ore bodies are pipe-shaped, vein-like, or in breccia form, steeply dipping. Approximately, 60° dipping and up to about

30-cm-thick vein-like ore bodies can be followed at the surface along its strike for about 30 to 40 m. Carbonatites are involved in metasomatic and hydrothermal processes. The width of individual ore bodies ranges from 1 to 30 m, and they extend lengthwise from 10 to 150 m along strike. Carbonatite breccia with REE-bearing minerals represents the second important ore type. Alteration-related nests of lilac-colored fluorite were rather frequent in the ore. Two apatite ore types are recognized: magnetite-apatite and fluorite-celestite-magnetite-apatite. The main REE minerals are bastnaesite, apatite, parisite, and synchysite.

The REE distribution patterns for the ore are rather similar to most REE-hosting carbonatites worldwide with slightly increased values for the heaviest rare earth. Average REO ore at Khotgor shows very similar REE distribution patterns when compared to the Mushgai Khudag deposit. Estimated resources and reserves are 40 Mt., and total rare earth oxide (TREO) + Y 1.22 wt.% (Elsner et al. 2011).

6.3.1.3 Bayan Khoshuu Occurrence

The Bayan Khoshuu occurrence is located 50 km eastward of the Mushgai Khudag deposit. The Bayan Khoshuu complex consists of Late Jurassic-Early Cretaceous trachytes, shonkinites, syenites, and monzonites hosted in Late Paleozoic granosyenite, granite, and Silurian volcanic-terrigenous rocks. The hosting rocks contain the veined body of dolomitic tuff. Syenite and monzonite occur in the core of the complex. The monzonite yields an Rb-Sr whole-rock-mineral isochron age of 130.6 ± 9.3 Ma (Munkhtsengel and Iizumi 1999a). The initial Sr ($^{87}Sr/^{86}Sr = 0.705$) ratio shows that the alkaline complex originated from enriched mantle source EMII and exhibit geochemical features characteristic for a post-collisional setting (Gerel et al. 2005) These features can also result in a low degree of melting of subduction modified by CO_2, H_2O and other volatile-rich fluid metasomatized lithospheric mantle that has produced rocks extremely enriched in LILE and LREE. Mineralization occurs as mineralized breccia and apatite-magnetite rocks, carbonatites and fluorite, fluorite-celestite, and barite-celestite veins. Carbon isotope values vary from −0.6 up to −0.9‰, and oxygen isotopes range from 17.9 up to 23.8‰. The tuff is marked by a heavier oxygen isotope as compared with carbonatite.

6.3.1.4 Lugiin Gol Deposit

The Lugiin Gol deposit is located 50 km south of Khatanbulag soum, Eastgobi (Dornogovi) province, and 100 km east of the Khanbogd alkaline pluton. Carbonatite mineralization is associated with the Lugiin Gol nepheline syenite pluton. Lugiin Gol pluton has been studied in the 1970s by researchers of the Mongolian-Russian joint expedition of the Academies of Sciences. The REE placer was discovered in 1984–1989 during the exploration work of the Mongolian-Polish joint expedition.

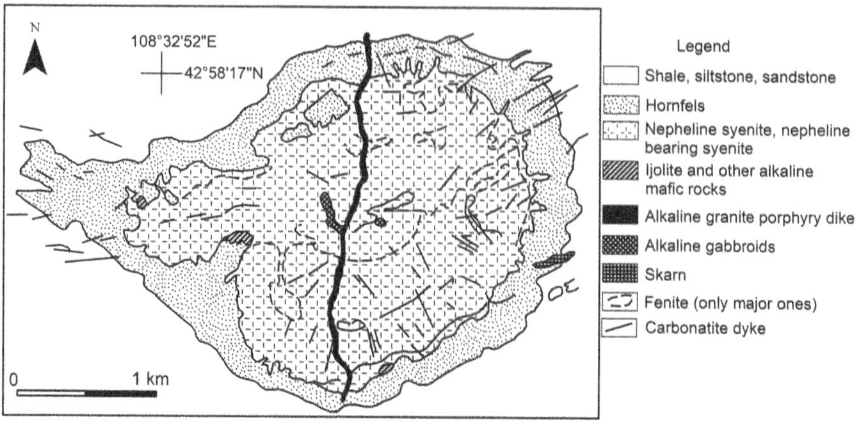

Fig. 6.3 Geological map of the Lugiin Gol pluton (Batbold 1997)

The Lugiin Gol deposit occurs on the eastern part of the South Gobi belt of alkaline rocks controlled by a near EW-trending regional fault. The Lugiin Gol pluton is 3 km in diameter forming a ring structure. Radial dikes in the internal contact of pluton are represented by syenite-porphyry, microsyenite, pseudoleucite syenite-porphyry, and tinguaite. Dikes trending mainly to E and NE are 20–100 m up to 1 km long (Fig. 6.3). The Permian shale-argillite sequence, with interbeds of siltstone and limestone, in contact with the pluton are converted into biotite-feldspathic, fenitized hornfels in a contact aureole. In the Paleozoic, South Mongolia was developed as an active continental margin, and at the end of the Permian collision, the Sulinkheer (Solonker) ocean closed. The Lugiin Gol Complex originated after this collision, and geochemical and isotopic data show a post-accretion signature (Munkhtsengel et al. 2013). The whole rock and mineral Rb-Sr age of nepheline syenite is 244.9 and 222 Ma, and the late granite-porphyry dike age is 210.3 Ma. Alkaline rocks could have originated from enriched mantle (EMII) (Munkhtsengel 2001; Munkhtsengel and Iizumi 1999b).

Mineralization is located in a linear or oval zone of eruptive breccia cemented by carbonatites, which are noted in the western part of the deposits. At least 17 carbonatite veins were traced within the deposit: eight veins occur within the pluton of nepheline syenites, seven veins within the host rocks, and two veins at the contact of pluton and host rocks. They are enriched in fluorite, feldspar, sericite, muscovite, hematite, and iron sulfides. The veins are arranged into an en-echelon pattern. They are from a few to 430 m long, the prevailing length being 200–259 m; their thickness ranges from 0.1 to 0.8 m, with 0.3–0.4 m values being dominant. Two types of carbonatite veins are recognized: calcite and calcite-fluorite. Their major component, synchysite, forms a finely divided dissemination. The RE_2O_3 content is 0.22–3.68%, the average being 3.23%; the content is somewhat higher on the surface due to leaching of calcite. The ore is enriched on account of stable synchysite. Rare earth belongs to the cerium subgroup. Of the total sum of rare earth, Ce comprises

50.77%, La 32.96%, Nd 9.14%, Pr 4.96%, and Sm 0.97%, the remaining minerals tens and hundreds fractions of a percent. Ore-bearing rocks are altered to muscovite, carbonate, and sulfides, the host rocks are clarified and pyritized. The alteration zone is not wide, on average 0.3 m but from 50 to 100 m long. Metasomatized carbonatite are enriched in Zr, REE, F, Pb, Zn, Li, Rb, Ta, and Nb. Batbold (1997) reported in an area of 8 km^2 more than 200 veins. Sixty of them have REE mineralization, including 24 with economic importance. The distance between veins varies from 50 to 800 m.

The alteration zone is limited to the depth and the oxidized zone is traceable to 20 m. Ore is coarse- to medium-grained and massive. REE mineralization is impregnated. Ore minerals are concentrated in the central part where selvage sulfide minerals dominated. Main ore minerals are synchysite, parisite, accessory: pyrite, galena, goethite, and limonite. Rare minerals are sphalerite, pyrolusite, psilomelane, monazite, hematite, ilemnorutile, apatite, celestite, barite, bismutite, zircon, bastnaesite, and strontianite. Gangue minerals, calcite; accessory minerals, dolomite, ankerite, fluorite, quartz, siderite, mica, and feldspar. Parisite is rich in REE and depleted in Th. Synchysite dominates in oxidized ore. Synchysite and parisite are fine-grained, and REE content varies from 0.1 to 27.0%, with average 3.2%. Co-product (Sr, Ba, Mn, Pb, Zn, Zr oxides, and fluorite) content is high. Th oxide content is 0.08%.

The carbonatites consist of primary calcite which is medium- to coarse-grained (0.1–10.0 mm), forming an euhedral mosaic of crystals (70–98%) with accessory amounts of primary automorphic apatite (0.01–2.00 mm) which is enriched in Na (0.18–1.35%), REE (0.26–2.13%), and SrO (0.33–2.87%). Lamellar and drop-like exolution textures of strontianite were observed under high-contrast BSE imaging in slightly altered centers of metasomatized calcite. REE-rich carbonatite from Lugiin Gol has some distinctive hexagonal or prismatic pseudomorph up to 1 cm in diameter and 3 cm in length. The pseudomorphs consist of an assemblage of secondary synchysite (Ce) and strontianite. Detailed mineralogical studies of nanomicroscopic relicts (0.1–25.8 μm) have indicated that the precursor mineral was burbankite $(Na,Ca)_3(Sr,Ca,REE,Ba)_3(CO_3)_5$ or calcioburbankite $(Na,Ca)_3(Ca,Sr,REE,Ba)_3(CO_3)_5$ (Kynicky 2003).

The primary mineral association is overprinteded by hydrothermal-metasomatic association of secondary fluorite–fluorocarbonates-strontianite cut by younger extension veins (thickness to 10–100 μm) barite + celestite + quartz. The carbonatites are practically penetrated by veins of fluorite with associated REE-fluorocarbonates. Under the polarizing microscope, strontianite and fluorocarbonate mineralization appear as small, dusty, and fibrous patches intergranular to the calcite and fluorite grains.

In the case of Lugiin Gol carbonatites, most commonly this takes the form of bastnaesite-parisite-synchisite syntaxes which attract the calcite crystals at the contact with or near the fluorite. Syntaxial intergrowths and replacement reactions between these fluorocarbonate phases are common. Hematite pigment is produced by metasomatic disintegration of calcite and occurs at the contact of fluorite and calcite or fluorocarbonates and calcite.

Reserves and resources are 3900 t graded at 1.22%, and 0.5 Mt. and total rare earth oxide (TREO) + Y 2.67% (Elsner et al. 2011).

6.3.2 Peralkaline Granite-Related REE Deposits

6.3.2.1 Khanbogd Alkaline Pluton

The Khanbogd pluton is located in approximately 600 km south of Ulaanbaatar in the Southgobi aimag. The first geological study of the Khanbogd pluton was made by Russian scientists, who conducted a geological survey of the area. Subsequently, several researchers (Vladykin et al. 1981; Kovalenko and Yarmolyuk 1995, Kovalenko et al. 2006b) studied geology, geochemistry and related mineralization. Vladykin et al. (1981) discovered here two new minerals, armstrongite and mongolite, and reported the first Rb-Sr whole-rock isochron age for the Khanbogd granite. New geochronological data were obtained within the framework of the Uudam Tal project by the Japan International Cooperation Agency and Metal Mining Agency of Japan (JICA and MMAJ 1992). Garamjav (2007) proposed a spiral-vortical structure for the Khanbogd pluton. The shape and internal structure of the complex can be readily seen from satellite imagery (Fig. 6.4). The Khanbogd alkaline granite pluton, one of the largest in the world, occupies the central part of the Late Paleozoic trough, in which the island arc calc-alkaline differentiated volcanic sequences are replaced by bimodal basalt-comendite-alkali granite associations. The inner ring structure is particularly typical of the western body and accentuated by ring dikes and roof pendants of the country island-arc complex. According to preliminary gravity measurements, the pluton is a flattened intrusive body (laccolith) with its base subsiding in a stepwise manner northwestward. The position of Khanbogd is controlled by the intersection of sublatitudinal Gobi-Tien Shan rift zone with the oblique transverse fault, controlling bimodal association. The pluton has a concentric structure (Kovalenko et al. 1995b) and consists of two circular bodies, western and eastern, and comes into sharp intrusive contact with island-arc rocks and tectonic contact with the bimodal complex. Granites host xenoliths of Carboniferous volcanic rocks. The Khanbogd pluton is unconformably overlain by thick Cretaceous sedimentary rocks with dinosaur fossils.

The rocks of the Khanbogd pluton were emplaced in the following order (Fig. 6.5): (1) phase-I granites (western body); (2) phase-II granites (eastern body); (3) ekerite and ekerite-porphyry dikes; (4) pantellerite dikes; (5) fine-grained syenite and monzonite dikes; and (6) an alkali granite porphyry dike (Vladykin et al. 1981). Based on the modal composition classification I and II phase rocks are classified as alkali-feldspar granite. U-Pb ages of accessory zircon and titanite (Kovalenko et al. 2006b; Vaglio 2007) show that the main-phase granites were emplaced at 290–292 Ma, nearly contemporaneously with the granite-hosted pegmatites and aplites (290.0 ± 0.4, 290.9 ± 0.3, 290.7 ± 0.9 Ma). The granite porphyry

Fig. 6.4 Satellite image of the Khanbogd complex. Landsat TM6

crosscutting the main-phase granites is distinctly younger (272.5 ± 1.2 Ma) and indicates protracted magmatic activity in the area. The older igneous rocks of the island-arc complex in the framework and roof pendants of the pluton are dated at 330 Ma (Amar-Amgalan 2004). The alkali granites of phase I are coarse-grained and show hypidiomorphic equigranular texture. The second phase granites are medium-grained equigranular with graphic intergrowth of quartz and potassium feldspar.

Granites are composed of potassium feldspar with a perthitic texture, quartz and albite (<2%), aegirine ($Ae_{86–97}$) and alkaline amphibole. Fluorite, apatite, zircon, titanite, allanite, and opaque minerals occur as an accessory phase. Fine-grained and massive or trachytic ekerite dikes are common in the Khanbogd complex. They are composed of alkali feldspar, quartz, albite, alkali amphibole, and aegirine. Granophyric and spherulitic pantellerite dikes are composed of potassium feldspar, quartz, and aegirine (rarely alkali amphibole). Phenocrysts are represented by quartz and potassium feldspar. The latter also forms anhedral grains commonly intergrown with quartz to form a granophyric texture. Spherulites consist of skeletal needles (dendrites) of riebeckite. The alkali granite porphyry dike is petrographically distinct from the rest of the intrusive rocks, including the presence of calcic amphibole, biotite, and abundant plagioclase.

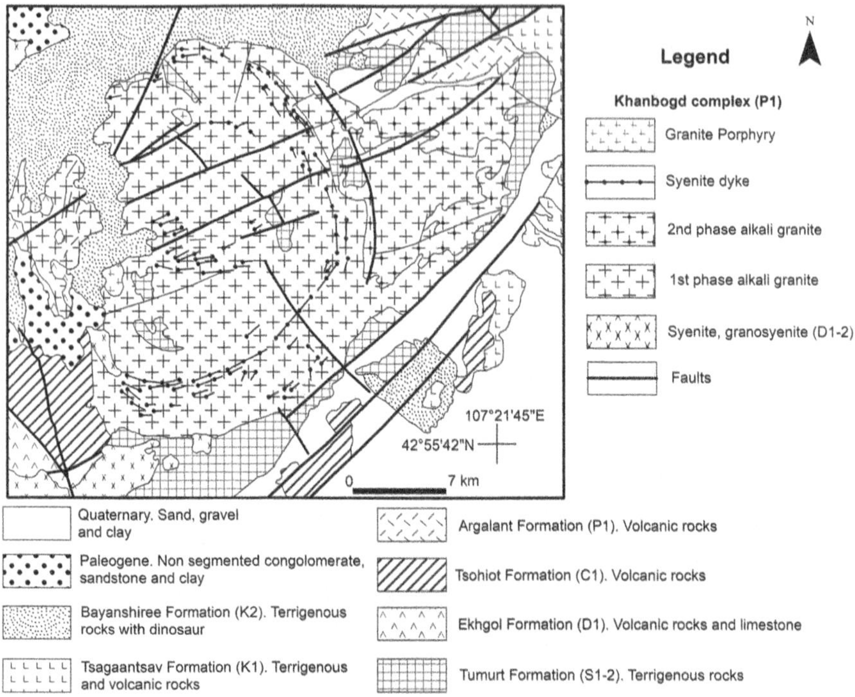

Fig. 6.5 Geological map of Khanbogd alkaline pluton

Orbicular granite was identified only in the southwestern part of the pluton. It is coarse-grained and contains quartz and large potassium feldspars crystal orbicules in the core and mafic minerals, such as arfvedsonite and aegirine, in the rim.

The Khanbogd granites are characterized by high alkali, Zr, Nb, Ga, Y, and F; low MgO, CaO, and Sr contents; and high FeO*/MgO and Ga/Al ratios. The rocks show characteristic of alkaline granite flat chondrite-normalized REE patterns with a strong Eu negative anomaly. The Khanbogd pluton is an example of the syn- or late-orogenic A-type granites. The Khanbogd granites show a low initial Sr isotope ratio $^{87}Sr/^{86}Sr$ and moderately high initial Nd value suggesting that the magma source was isotopically depleted.

The geodynamic model of the Khanbogd pluton formation suggests the activity of a mantle plume with two models proposed. According to the first model, the mantle plume, after collision with the margin of the North Asian paleocontinent, reworked the subducted lithosphere and formed a structure similar to an asthenospheric window, which served as a source of rift-related magmatism and the Khanbogd pluton proper. In the second model, the emergence of hot mantle plume resulted in flattening of the subducted plate; cessation of the island-arc magmatism; and probably in origin of a local convective system in the asthenosphere of the mantle wedge, which gave rise to the formation of a magma source. The huge body

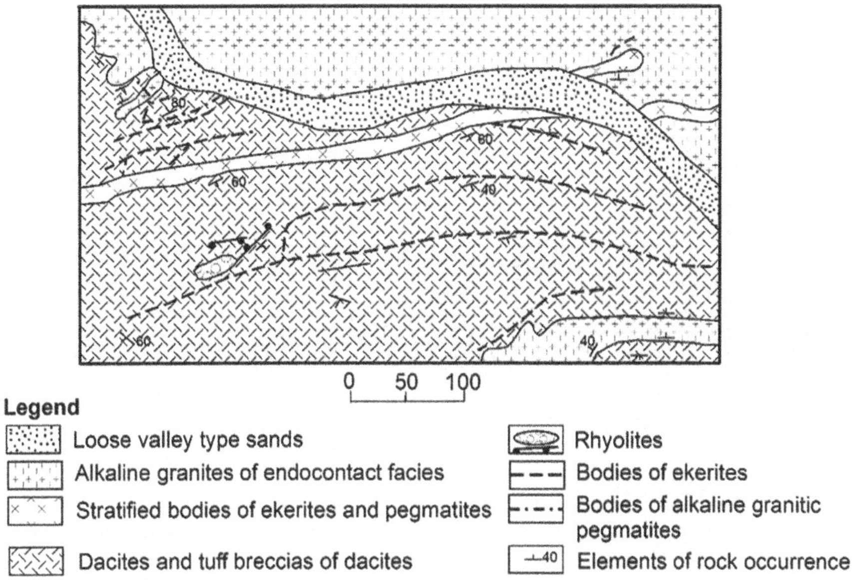

Fig. 6.6 Scheme of geological structure of Northern site (Vladykin et al. 1981)

of the Khanbogd alkali granite pluton and related volcanic rocks, as well as its ring structure, resulted from the caldera mechanism of the emplacement and evolution of magmatic melts.

REE mineralization. In Khanbogd pluton REE mineralization is concentrated in the second phase granites and pegmatites. REE content is not high: 0.8% RE_2O_3, 620 ppm Nb_2O_3, and 0.04% Hf_2O_3. Vladykin et al. (1981) reported very high concentration of Zr in pegmatites. Figure 6.6 displays the contribution of alkaline granites in the roof sagging. They commonly form the peaks of hypsometric elevations. They are underlain by the apical parts of the pluton, in places occurring as schlieren pegmatoids (stock sheider or divider) and ekerites. The roof sagging basically consists of dark gray hornfelsic volcanic rocks, rarely cutting them by subvolcanic rhyolites, similar to the rocks of the island arc association.

In the northern part of the pluton the roof sagging of biotite granites forming rounded fragments of biotite granites, as well as transitional amphibole-biotite granites, being the result of interaction of biotite and alkaline granites (Fig. 6.6). The evidence of assimilation with alkaline rocks of biotite granites is quite common. The presence of breccia of rounded segregations of biotite leucogranites cemented with alkaline granites is indicative of interaction of «cold» magmas of biotite leucogranites with "hot" magmas of alkaline granitoids when magma is intruded into magma. That is why it seems that magmas of these two granite types could occur at different times (biotite leucogranites are somewhat older, then alkaline granites).

Diverse dyke of alkaline granitoid, including their rare-metal varieties, is spatially located in the roof sagging in the pluton. In the eastern body, the roof sagging is

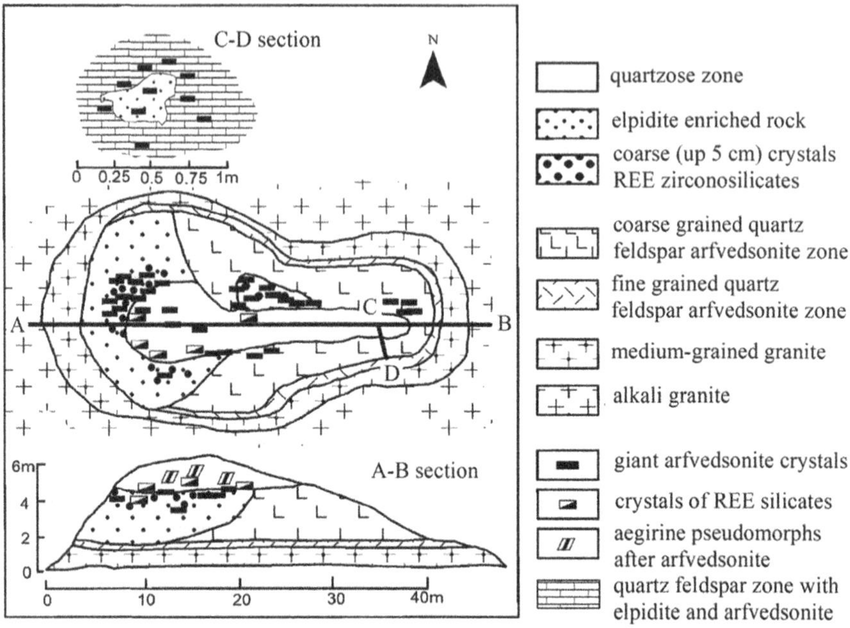

Fig. 6.7 Section of pegmatite body of the Khanbogd complex (Vladykin et al. 1981)

located in its western part, partly tracing the contact zone of the western and eastern bodies of alkaline granites. The hypsometric level of roof sagging shows that the depth of erosional cross-section of the massif is not in excess of a hundred meters. Location of the alkaline granite pluton in the upper part of the of the Upper Paleozoic island arc, which is overlapped by bimodal sequence is possibly due to a change of regime of tectonic compression in the formation of island arc sequences into tectonic extension in the formation of the bimodal association with its numerous dykes and volcanic units.

There are 12 mineral prospects in the Khanbogd pluton associated with granitic pegmatite (Vladykin et al. 1981). The pegmatites (Fig. 6.7) occur as zoned lenses 5–100 m in length or as layers alternating with aplite and ekerite dikes (Vaglio 2007). Pegmatites of both types are located in the cupola of the western granite body, commonly occurring near its contact with the wall rocks at the roof of the pluton. The mineralized pegmatites contain a quartz core 1–5 m across and large crystals of arfvedsonite, along with microcline, plagioclase, and quartz, and are commonly enriched in armstrongite ($CaZrSi_6O_{15}\cdot 3H_2O$) in the matrix, whose presence imparts a pinkish color to the rock. Large crystals of alkali zirconosilicates are absent and aegirine is rare. The content of Zr is ~100 ppm and varies among the individual zones of a pegmatite body. In pegmatites and aplites, large crystals of arfvedsonite are oriented with their growth vector pointing toward the pegmatite core. The texture of the mineralized pegmatites is poikilitic, and they composed of quartz, potassium feldspar, aegirine, arfvedsonite and a variety of alkali zirconosilicates, such as armstrongite, elpidite ($Na_2ZrSi_6O_{15}$ $\cdot 3H_2O$), and gittinsite ($CaZrS_{i2}O_7$). The "Dorozhny" ("Roadside") pegmatite was reported to contain large composite pseudomorphs of quartz and lilac kovalenkoite [$(Ca,Sr,K,Ba)_7Nb_{12}Si_8O_{52}(O,OH) \cdot 15H_2O$] after an unknown mineral and foliated aggregates of mongolite [$Ca_4Nb_6Si_5O_{24}(OH)_{10} \cdot nH_2O$] replacing the kovalenkoite (Vladykin et al. 1981; Minerals of Mongolia 2006). Other accessory minerals include titanite, zircon, fluorite, calcite, and fluorocarbonates of REE [bastnaesite-synchysite series: $REE_xCa_{ay}(CO3)_{x+y}(F, OH)_x$].

Aplites are similar to the alkali granites by their major-element geochemistry, as well as Zr/Hf, Th/U, Nb/Ta, Y/Ho, and $(La/Yb)_{CN}$ ratios. In the aplite–pegmatite series, the peralkalinity index increases from 1.4 in the aplite to 2.5 in the blocky zone of pegmatites, which reflects a drop in Al_2O_3 content toward the interior of pegmatite veins. This is accompanied by a drastic increase in Mn, Sr, Ba, Y, REE, Zr, Nb, Ta, Th, and U contents and a decrease in Na content, whereas the abundances of other elements do not follow a consistent pattern of variation. In the blocky zone of pegmatite, all key trace-element ratios greatly deviate from chondritic values, i.e., Zr/Hf = 1083, Th/U = 1.7, Nb/Ta = 70, Y/Ho = 93, Yb/Y = 0.04 and $(La/Yb)_{CN} = 11$ (Kynicky et al. 2011).

Lead isotope data indicate a common origin of the pegmatites and granites (Vaglio 2007). The two most notable characteristics of the pegmatites are their enrichment in light rare-earth elements, LREE [$(La/Yb)_{CN} = 11$] coupled with enrichment in Y relative to the similar-sized HREE (Y/Ho = 93). Although the enrichment of Zr other HFSE in granitic rocks at Khanbogd complex can be attributed mainly to magmatic differentiation, there is no doubt that these elements were remobilized during the postmagmatic evolution. Field and textural evidence indicates that postmagmatic alteration affected both main-phase and aplite-pegmatite units but is the most extensive in apical parts of the intrusion. The extent of alteration, probably reflecting variations in fluid-to-rock ratio, ranges from isolated thin veinlets of secondary zirconosilicates crosscutting the precursor elpidite to pseudomorphs lacking any relict material. The alteration appears to have been a two-step process, involving replacement of the primary elpidite by Ca-rich elpidite or armstrongite, and development of zircon at the expense of the elpidite and armstrongite. The late-stage zircon is associated with either calcite or gittinsite plus quartz. That most secondary zirconosilicates replacing primary elpidite contain significant Ca and are associated with calcite, fluorocarbonates and fluorite indicate that the fluid responsible for remobilization of Zr and other high-field-strength elements was enriched in Ca, CO_2, and F. Preferential partitioning of the LREE+Y and Zr into the fluid relative to HREE and Hf would explain the observed trend of increasing whole-rock $(La/Yb)_{CN}$, Y/Ho and Zr/Hf values from granites to pegmatites, as well as comparatively lower $(La/Yb)_{CN}$, Y/Ho and Zr/Hf values in primary elpidite from pegmatites, assuming that the pegmatite-forming melt reached elpidite saturation after the separation of the fluid. This interpretation is consistent with the published trace-element compositions for hydrothermal zircon and fluorite, which indicate that CO_2-bearing fluids facilitate Y–Ho and Zr–Hf decoupling in granites (Kynicky et al. 2011).

6.3.3 REE-Nb-Zr Deposits

The deposits occur in the Lake zone (terrane) composed of Cambrian age island arc volcanic rocks with fragments of ophiolites of Neoproterozoic-Early Cambrian age. The island arc rocks are represented by calc-alkaline basalt, andesite, rhyolite, tuff, volcanoclastic rocks, and granodiorite, granite, and granosyenite plutons. Post-accretionary assemblages include Silurian, Devonian and Mississippian, Permian, Jurassic, Cretaceous clastic sediments, Ordovician quartz diorite, granodiorite and granite of U-Pb zircon age of 451–464 Ma, Devonian alkali plutons of U-Pb age of 380–390 Ma, and Permian subalkaline plutons (Badarch et al. 2002; Kovalenko et al. 1995a, b). Several ore fields as Ulaan Tolgoi (Khyargas Lake area), Khalzan (Khar Us Lake area), and Maikhan Uul occupied more than 1500 km^2 area each (Fig. 6.8). REE mineralization of this region is spatially and genetically associated with the alkaline Khalzan Complex, and its distribution is controlled by Tsagaan Shiveet deep faults divided the Mongolian Altai orogen. REE-Nb-Zr deposits associated with alkaline granites and syenites in Mongolian Altai are characterized

Fig. 6.8 Location of REE-Nb-Zr deposits in Western Mongolia in satellite map

by high REE content and have complex mineralization. Unfortunately grade and tonnage of many occurrences are not known, and it is very difficult to do a quantitative resource assessment. Even deposits with calculated reserves need more detailed exploration. But in South and Northwestern Mongolia many alkaline complexes with REE and rare metal mineralization, including U-REE mineralization in metasomatic zone, could be of economic interest.

6.3.3.1 Khalzan Buregtei REE-Nb-Zr Deposit

The Khalzan Buregtei deposit is located in the east bank of Khovd River, in the Mongolian Altai Range, 40 km northwest of Khovd River mouth and 45 km northeast of the Khovd province center. There is a group of plutons oriented in a northwest (Altai system) direction (Fig. 6.8). One of them, the Khalzan Buregtei pluton has an oval shape 30 km long and 8 km wide. The pluton position is controlled by the Altan Khukhii regional fault, which divides early and middle Paleozoic structural zones in Western Mongolia. The deposit occurs in the Lake (Nuur) island arc terrane in the contact with the Khovd zone, interpreted as an accretionary wedge (Badarch et al. 2002).

The Lake zone is composed of Cambrian volcanic rocks of basalt-andesite-dacite series with fragments of ophiolite metamorphosed to greenschist facies. Country rocks are dominated by Neoproterozoic-Cambrian and lower-middle Devonian sedimentary-volcanic rocks, intruded by calc-alkaline Devonian-Carboniferous biotite-amphibole granite cut by Khalzan Buregtei pluton. Granite-porphyry and dolerite dikes form a dike series in a northern direction developed within Devonian-Carboniferous granites and cut both volcanic-sedimentary rocks and granites (Fig. 6.9). Alkali granite plutons are younger and cut all described rocks. They comprise seven types, from older to younger as follows: (1) nordmarkite and syngenetic dolerite; (2) alkali granite and syngenetic dolerite; (3) ekerite dikes, fine-grained alkali granite, and alkali-granite pegmatites; (4) pantellerite dikes; (5) rare-metal alkali granitoids; (6) alkaline mafic rocks and quartz syenites; and (7) miarolitic alkali rare-metal granites. All rocks are co-magmatic (Kovalenko et al. 2004; Kovalenko et al. 2006a, b). Two phases 5 and 7 are of economic interest. They are composed of K-feldspar (36%), 26.5% quartz, 0.5% arfvedsonite, 9% aegerine, and 25% of different minerals (more than 120) and fluorite, including Zr, Nb, Y, REE (synchysite, bastnaesite), U, Th, Li, and Be minerals. The U-Pb zircon age is 392.2 ± 2.3 and 390.8 ± 1.2 and for Tsakhir occurrence 395.0 ± 1.3 Ma (Kovalenko et al. 2006b). The Khalzan Buregtei and other plutons of alkaline rocks and mineralization originated in an intracontinental rift.

Rare metal mineralization occurs in two prospects: Khalzan Buregtei and Tsakhir and others that are alkali-granite pegmatoid and quartz-zircon-allanite-epidote albitites sited in a roof pendant north of the pluton. The Khalzan Buregtei pluton has a concentric zonal structure defined by dike arrangement. All plutons have a dike-like shape along a dike belt confined to Tsagaan Shiveet fault.

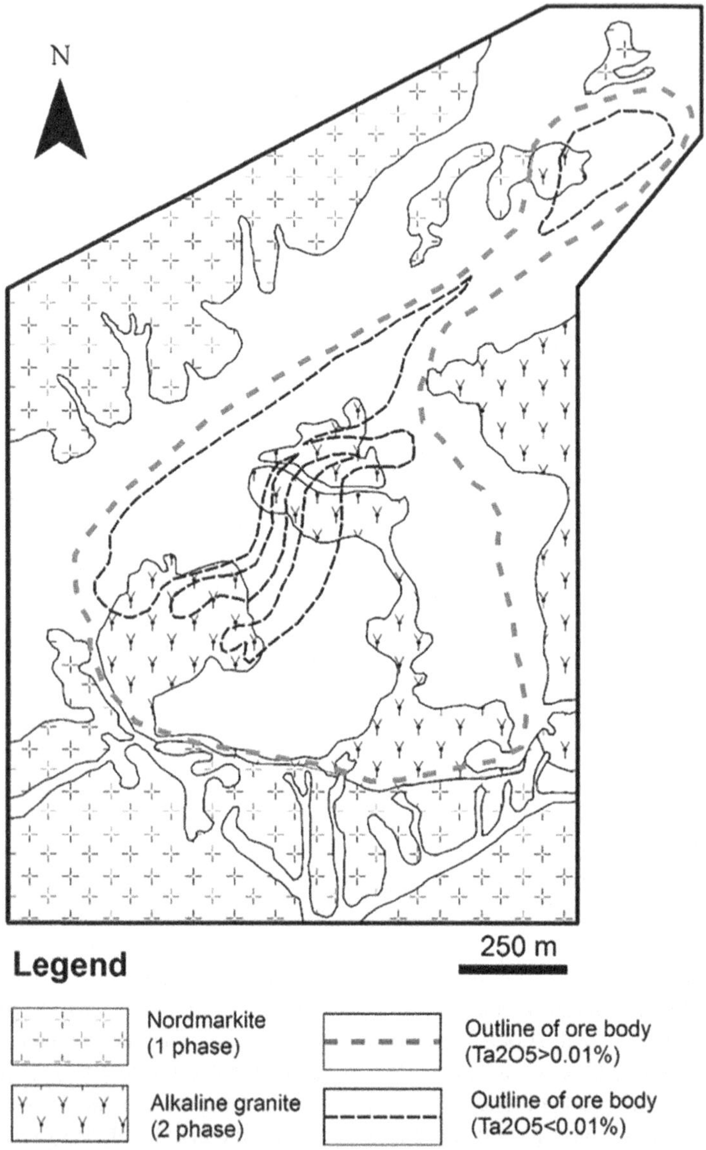

Fig. 6.9 Geological map of the Khalzan Buregtei deposit (Kovalenko et al. 2006b)

The area is characterized by extensive but low-grade mineralization of rare earths and rare metals in the central part of the mountain range. Within this low-grade mineralization, high-grade ore occurs as gently dipping bodies, stockworks in K-feldspar granite, and dissemination within alkali granites and syenites.

Mineralization is related to peralkaline granite and alkaline syenite rocks. Ore types and ore mineralogy is represented by the impregnation of zircon, pyrochlore etc. with light and heavy REE, Nb, Ta, Zr, Be, and Y. Major ore minerals are pyrochlore, columbite, elpidite-armstrongite, bastnaesite, zircon, amorphous REE minerals, zircon, zirconiumsilicates and hydro-silicates. Pyrochlore and columbite carry tantalum and niobium. The REE occur in many minerals, mainly in bastnaesite, monazite, synchysite, xenotime, and perhaps in fluorite and apatite. Mineralization is represented by elpidite (up to 17%), gittinsite, pyrochlore, polilithionite, REE fluorcarbonates, monazite, zircon, amorphic zircon-Nb-REE silicates, and fluorite.

Fluorite is present in all rock types. Microcline, albite and greisen alteration developed in the small Shar Tolgoi pluton of 250 × 75 m in size, composed of nordmarkite (Andreev et al. 1994). The main ore minerals are zircon and xenotime. In the Ulaan Tolgoi pluton albite alteration is associated with columbite and pyrochlore.

According to Kovalenko et al. (1985), Zr-REE mineralization is of magmatic origin and is concentrated within highly fractionated REE-bearing granites. In contrast, Kempe et al. (1999), Kempe et al. 2015, and Andreev et al. (1994) suggest a metasomatic origin. Their studies of the Khalzan Buregtei deposit, and the Tsakhir, Shar Tolgoi and Ulaan Tolgoi occurrences in the Mongolian Altai region show that these deposits are formed in endocontact of alkaline plutons that consist of metasomatically-altered alkaline syenite (nordmarkite).

Grade and Tonnage: Granite contains REE 0.33–0.4% (Kovalenko et al. 1985). Calculated resources and reserves are 49 Mt. of total rare earth oxide TREO (REE + Y) of 0.6 wt.% grade. Mineralization increases to the depth, and geological and geochemical features show economic importance.

6.3.3.2 Ulaan Del Zr-Nb-REE Deposit

The deposit occurs in the Lake zone (terrane) composed of Cambrian island arc volcanic rocks with fragments of ophiolites of Neoproterozoic-Early Cambrian age. The Ulaan Del deposit is hosted in Cambrian-Ordovician granodiorites of the Togtokhynshil Complex and is composed of various composition dikes of basalt-andesite, dacite, trachyte, trachyrhyolite, and breccia. Dikes are crosscutting each other and form a kind of stockwork. Dikes are 0.5–2 m in width and can be up to 500 m along strike. The thickness can be up to 3 m and more, as revealed by drilling. Dike colors are pinkish-reddish and reddish brown. The composition is not stable; trachyrhyolite changed to trachyte/syenite along strike. In addition, weak brecciated, tuffaceous, and greisenized segments can be observed often within the same dike. Dikes are strongly altered by quartz-albitite metasomatic rocks. Sericite, silica, carbonate, pyrite, fluorite, hematite and limonite are frequent alteration minerals. Potassic and late albite alteration with complete replacement of primary texture also observed. Common minerals are zircon-elpidite, xenotime, synchysite, apatite, monazite, REE-bearing fluorite, parisite, orthite, microlite, rutile, sphene, malachite, pyrite, ilmenite, magnetite, chalcopyrite, arsenopyrite, and pyrrhotite.

The dikes at the Ulaan Del area are altered at variable stages/levels and overprinted by hydrothermal-metasomatic alteration, breccia, and various veins and veinlets. The REE-bearing dykes well developed, 0.5–2 m wide, and form a "corridor" like in a 2 × 2 km area associated with alterations (together with brecciation) that may host REE-bearing fluorcarbonate minerals and above-mentioned hydrothermal alterations. REE mineralization is hosted predominantly in pinkish-reddish brown dykes and is characterized by zircon-xenotime-synchysite-apatite-carbonate-pyrite-goethite-magnetite mineral assemblages.

Geological resources at Ulaan Del deposit can be estimated as 6.3 Mt. Zr-REE-bearing ore to depth of 50 m.

6.4 Concluding Remarks

Rare earth mineral deposits in Mongolia are carbonatite and alkaline rock related. Carbonatite deposits formed in continental rift zones in South Mongolia and have complex mineralization, e.g., REE-P-Sr-Ba-F-Pb and Fe- REE. Deposits associated with alkali granite and syenite also have Zr-Nb-REE mineralization and formed in an active continental margin or rift environment in Western Mongolia. Other rare earth occurrences include Late Paleozoic and Mesozoic REE pegmatite associated with calc-alkaline and Li-F leucogranite, and REE-albite and REE-albite nepheline syenite deposit types. Although Mongolia is rich in rare earth occurrences, many of them are small size and low grades. Only Khalzan Buregtei, Mushgai Khudag and Khotgor are of economic potential. Another possible deposit type could be REE-(Ti–P–Nb) residual deposition in alkaline igneous and carbonatite complexes that now has not been studied.

References

Amar-Amgalan S (2004) Petrographical and geochemical studies on the Khanbogd alkaline complex in South Mongolia. Master thesis, Shimane University, Matsue, Japan

Andreev GV, Ripp GS, Sharakshinov AO, Minin AD (1994) Redkometal'naya minerlaizatsiya shchelochnych garnitoidov Zapadnoy Mongolii (Rare metal minerlaization of alakline granitoids in Western Mongolia). BNTs SB RAS, Ulan-Ude, p 137

Badarch G, Cunningham WD, Windley BF (2002) A new terrane subdivision for Mongolia: implications for the Phanerozoic crustal growth of Central Asia. J Asian Earth Sci 21:87–110

Baskina VA, Volchanskaya IK, Kovalenko (1978) Kalievie shelochnoi vulkano-plutonicheskye kompleks Mushugai-Khuduk na yuge MNR i sviyazannaya s nim mineralizatsiya (potassium alkaline volcano-plutonic complex Mushgai Khudag in southern Mongolia and related mineralization). Soviet Geol 4:86–99

Batbold D (1997) Mineralogy of carbonatites from the Lugiin Gol Alkaline pluton, south Mongolia: Master thesis, Shimane University, Japan

Elsner H, Buchholz P Schmitz M and Tamiraa A (2011): Industrial Minerals and Selected Rare Metals in Mongolia—An Investors' Guide MRAM, Ulaanbaatar, p322

Enkhbayar D, Seo J, Choi S-G, Lee YJ, Batmunkh E (2016) Mineral chemistry of REE-rich apatite and sulfur-rich monazite from the Mushgai Khudag, alkaline volcanic-plutonic complex, South Mongolia. Int J Geosci 7:20–31. https://doi.org/10.4236/ijg.2016.71003

Enkhtuvshin Kh (1995) A petrological study on the Late Mesozoic and Cenozoic Volcanic Rocks of the Mongolian Plateau Master thesis, Shimane University, Japan

Garamjav D (2007) Baigal yertents dekh khuilraltiin uzegdluud (Natural spiral-vortical phenomenon) Munkhiin Useg, Ulaanbaatar, p 422

Gerel O, Munkhtsengel B, Enkhtuvshin K, Iizumi S (2005) Mushgai Khudag and Bayan Khoshuu volcanic-plutonic alkaline complexes with REE±ta, Nb, Fe carbonatite mineralization. In: Seltmann R, Gerel O, Kirwin D (eds) , pp 215–221Geodynamics and Metallogeny of Mongolia with a special emphasis on copper and gold deposits, London

Getmanskaya TI, Dandar S, Kempe U (1995) Novye dannye po genezisu ferberitovyekh mestorojdenii Khovd-gol i Mushguu (Mongoliskii Altai) (New data about the genesis of ferberite deposits of the Khovd gol and Mushguu (Mongolian Altai): Gold-REE mineralization of Altan-Khokhii mountain) In: Problems of Mongolian Altai Geology, Ulaanbaatar, 3:43–47

JICA and MMAJ (1992) Report on the Mineral Exploration in Uudam-Tal Area, Mongolian People's Republic (Phase I)

Kempe U, Dandar S (1995) Nb-Zr-REE mineralization: of a possible source of HREE. In: Pasava J, Kribek T, Zak P (eds) Mineral deposits: from their origin to their environmental impact. Balkema, Rotterdam, pp 463–466

Kempe U, Gotze J, Dandar S, Habermann D (1999) Magmatic and metasomatic processes during formation of the Nb-Zr-REE deposits Khaldzan Buregte and Tsakhir (Mongolian Altai): indications from a combined CL-Sem study. Mineral Magazine 63(2):165–177

Kempe U, Möckel R, Graupner T, Kynicky J, Dombon E (2015) The genesis of Zr–Nb–REE mineralisation at Khalzan Buregte (Western Mongolia) reconsidered. Ore Geol Rev 64:602–625

Kovalenko VI, Goreglyad AV, and Tsareva GM (1985) Khalzan-Buregtegskii massiv-novoe proyavlenie redkometal'niikh schelochniikh granitoidov MNR Dokladii AN SSSR, 280, 4:954–959. (Khalzan-Buregtei massif: New occurrence of REE alkaline granitoids in Mongolia) Doklady AN SSSR, 280, 4:954–959

Kovalenko VI, and Koval PV (1984) Endogennye redkozemel'nye i redkometal'nye rudnye formatsye Mongolii (Endogenous rare-earth and rare-metal ore formations of Mongolia) In: Endogenic ore-bearing formations of Mongolia Nauka, M. 38: 50–75

Kovalenko VI, Koval PV, Yakimov VM, and Sherchan O (1986) Metallogeniya Mongol'skoi Narodnoi Respubliki (Volfram, olovo, redkie i redkozemel'nye elementii) Akademiya nauk SSSR, Akademiya Nauk MNR, Novosibirsk 1986 (Metallogeny of the Mongolian People's Republic (tungsten, tin, rare and rare-earth elements): USSR Academy of Sciences, Siberian Branch, 52p

Kovalenko VI, Kuzmin MI, Zoenshain LP, Nagibina MS, Gundsambuu T, Tseden T, Goreglyad AV (1971) Redkometal'nye granitoidy Mongolii (petrologiya, raspredeleniye redkych elemnetov i genezis) (Rare metal granitoids of Mongolia: petrology, distribution of rare elements and origin). M Nauka 5:239

Kovalenko VI, Yarmolyuk V, Bogatikov O (1995a) Magmatism, geodynamics and metallogeny of Central Asia MIKO, Commercial Herald Publishers

Kovalenko VI, Tsaryeva GM, Goreglyad AV, Yarmolyuk VV, Troitsky VA (1995b) The Peralkaline granite-related Khaldzan-Buregte rare metal (Zr, Nb, REE) deposit, Western Mongolia. Econ Geol 90:530–547

Kovalenko VI, Yarmolyuk VV (1995) Endogenous rare metal ore formations and rare metal Metallogeny of Mongolia. Econ Geol 90:520–529

Kovalenko VI, Yarmolyuk VV, Salnikova EB (2004) Khaldzan-Buregtey massif of alkaline and rare-metal magmatic rocks: structure, geochronology, and geodynamic setting in the Caledonides of Western Mongolia. Petrol 12(5):467–494

Kovalenko VI, Yarmolyuk VV, Andreeva IA et al (2006a) Magma type and their source in the Earth history Vol 2 IGEM RAN Moscow 280p

Kovalenko VI, Yarmoluyk VV, Sal'nikova EB (2006b) Geology, geochronology, and geodynamics of the Khan Bogd alkali granite pluton in southern Mongolia. Geotectonics 40(6):450–466. https://doi.org/10.1134/S0016852106060033

Kynicky J (2003) Primary and secondary minerals of carbonatites in South Mongolia. Mineral 11:57–61

Kynicky J, Chakhmouradian AR, Xu C, Krmicek L, Galiova M (2011) Distribution and evolution of zirconium mineralization in peralkaline granites and associated pegmatites of the khan Bogd complex, southern Mongolia. Can Mineral 49:947–965

Mineraly Mongolia (2006) Mineralogicheskii muzei im. AE. Fersmana Rossiiskaya Akedemiya Nauk, Moskva (Minerals of Mongolia) M.I. Novgorodova, Ed., Moscow, Russia, p 352

Munkhtsengel B (2001) Petrology and Geochemistry of Mesozoic potassic magmatism in South Mongolia (Lugiin Gol, Mushgai Khudag). Unpubl. PhD thesis, Ulaanbaatar, p 146

Munkhtsengel B, Gerel O, Kynicky J, Iizumi S, Comin-Chiaramonti P (2013) Some notes on the Lugiin Gol, Mushgai Khudag and Bayan Khoshuu Alkaline Complexes, Southern Mongolia. Int J Geosci 04(08):1200–1214. https://doi.org/10.4236/ijg201348114

Munkhtsengel B, Iizumi Sh (1999a) Rb-Sr geochronology and Sr-Nd isotope systematics of the Mushgai Khudag Syenite and Bayan Khoshuu Monzonite in South Mongolia. Mongolian Geoscientist. Special Issue International Geological Symposium on East Asia, 14:14–16

Munkhtsengel B and Sh Iizumi (1999b) Petrology and geochemistry of the Lugiin gol Nepheline Syenite Complex in the Gobi-Tien Shan fold belt, South Mongolia: a post collisional potassic magmatism Mongolian Geoscientist, Special Issue International Geological Symposium on East Asia, 14:12–13

Ontoev DO, Luvsandanzan B, Gundsambu Ts (1979) Geologicheskoe stroenie i endogennaya mineralizatsiya Mushugaiskogo qnkj-j-;[kz-mel'nogo mestorozhdeniya (MNR) (Geology and endogenic mineralization of the Mushgai Khudag fluor-rare earth deposit (MPR)). Geologiya rudnych mestorozhdenii 3:27–42

Samoilov VS, Ivanov VG, Kovalenko VI, and Smirnova EV (1984) Redkozemel'no-apatitovye i redkozemel'no-svintsovye rudy kompleksa Mushugai-Khuduk (Yujnaya Mongoliya) (Rare earth element-apatite and rare earth element-lead ores of the Mushgai Khudag complex (southern Mongolia) In: Endogennye rudnye formatsii Mongolii. Sovmestnaya Sovetsko-Mongoliskaya ekspeditsiya. Izd. Nauka, Moskva, Trudy, 38:180–199

Samoilov VS and Kovalenko VI (1983) Kompleksy shelochnykh porod I karbonantitov Mongolii (Alkaline and carbonatite rock complex of Mongolia) Sovmestnaya Sovetsko-Mongolskaya nauchno-issledovatelskaya geologicheskaya ekspeditsiya, Izd. Nauka, Moskva, Trudy, 35:200

Vaglio D (2007) Mineralogy, geochemestry and geochronology of pegmatites and associated alkaline granite rock of the Khan Bogd Complex, South Mongolia. Unpubl. MSc Thesis, Univ. Genève, p 97

Vladykin NV, Kovalenko VI, and Dorfman MD (1981) Mineralogicheskie i geochimicheskie osobennosti Khanbogdinskogo massiva schelochnyekh granitov (Mongolskaya Narodnaya Respublika) (Mineralogical and geochemical peculiarities of Khan Bogd massif of alkaline granites (Mongolian People's Republic) Izdatelstvo Nauka, Moskva, p 136)

Chapter 7
Lead-Zinc Deposits

Bayaraa Batkhishig

7.1 Introduction

Numerous researchers have conducted investigation on polymetallic and lead-zinc mineral systems in Mongolia. Significant researchers are V.A. Yakovlev, R. Kh. Bakhteev, I.A. Chijova, K.V. Podlessky, D.K. Aksuk, P.F. Vlasova, L.J. Naftali, V. Volkov, Ts. Konsulov, J. Lkhamsuren, Sh. Batjargal, D. Dorjgotov, Yu. B. Mironov, N.S. Soloviev, V.K. Lyvov, Yu. N. Pecherkin, B.M. Davidenko, N. Nirgev, D. Jargalsaihan, M. Kazmer, Z. Baras, D. Sanjaadorj, D. Garamjav, V. Kovachev, A.E. Fomenko, G.S. Ripp, V.I Sudakov, W.J. Nokleberg, O. Gerel, G. Dejidmaa, Kh. Gantumur, F.J. Dahlkamp, S.H. Jiang, F.J. Nie, S.K. Hu, H.Q. Yan and M. Ye.

In this chapter a selected of polymetallic deposits are presented, including skarn Tumurtiin Ovoo; hydrothermal vein and stockwork Tsav, Dulaankhar, Mungun-Undur, Ulaan, Mukhar, Uulbayan and Nominy Am; and epithermal vein systems Khartolgoi, Kharmorit, Borts Uul, Burgedtas, and Bayan-Uul deposits.

7.2 Metallogeny and Distribution of Polymetallic Deposits in Mongolia

Polymetallic (Pb-Zn $\pm$ Ag, Cu) deposits and occurrences in Mongolia are different in terms of their geology, mineralogical characteristics, timing, and genetic processes. Therefore, researchers have suggested different classifications for the

B. Batkhishig (✉)
Department of Geology and Hydrogeology, School of Geology and Mining, Mongolian University of Science and Technology, Ulaanbaatar, Mongolia
e-mail: bkhishig@must.edu.mn

O. Gerel et al. (eds.), *Mineral Resources of Mongolia*, Modern Approaches in Solid Earth Sciences 19, https://doi.org/10.1007/978-981-15-5943-3_7

Table 7.1 Comparison on classification of polymetallic mineral deposits[a] by different researchers

Yakovlev (1977)	Garamjav et al. (1989)	Dorjgotov (1996)
Lead-zinc skarn	Zinc-lead skarn	Iron-polymetallic skarn
Arsenic-lead ore formation	Carbonate and silicified rock hosted polymetallic	Silver-lead epithermal
Lead carbonate-quartz vein		Zinc-lead hydrothermal
Polymetallic-quartz vein		Carbonate hosted zinc-lead
		Polymetallic massive Sulfide

[a]Ore formation is a group of ore deposits of similar mineral composition, formed in similar geological and physicochemical conditions near the surface or in deeper levels

Mongolian polymetallic deposits. For example, Yakovlev (1977) developed the first outline of the polymetallic deposit classification with copper ore formations; Garamjav et al. (1989) proposed two types of polymetallic formations; and Dorjgotov (1996) classified five polymetallic ore systems (Table 7.1; Fig. 7.1) (Dorjgotov 1996; Dorjgotov and Batjargal 2001; Dorjgotov 2002, 2009).

Dorjgotov's research was conducted in order to classify the Pb-Zn deposit models and types, based on mineral association, rocks related to mineralization, host-rock lithology, alteration, and ore structure and relative to the deposit models developed by American geologists (Cox and Singer 1986). According to Dorjgotov (1996, 2002, 2009), Pb-Zn deposits and occurrences in Mongolia are classified into four deposit types. Each type is further subdivided into 1–4 deposit models (Table 7.2).

7.3 Lead-Zinc Deposit Types

The following three types of deposits form the major metallic industry in Mongolia: (1) Fe-Zn skarn (contact metasomatic) related to subalkaline granite; (2) Ag-bearing Pb-Zn hydrothermal vein related to subalkaline polyphase intrusion; and (3) Ag-Pb epithermal vein related to subalkaline subvolcanic rocks. Especially skarn-type Fe-Zn-Pb and hydrothermal vein-type Ag-Pb-Zn deposits represent more than 90% of the total Pb-Zn resources of the country.

7.3.1 Skarn: Contact Metasomatic (Fe-Zn)

Most of the Fe-Zn skarn deposits occur within the uplifted blocks of Late Paleozoic and Early Mesozoic rocks located in the Eastern Mongolian intercontinental volcanic belt (Dejidmaa et al. 2005; Podlessky et al. 1984, 1988). The uplifted blocks are

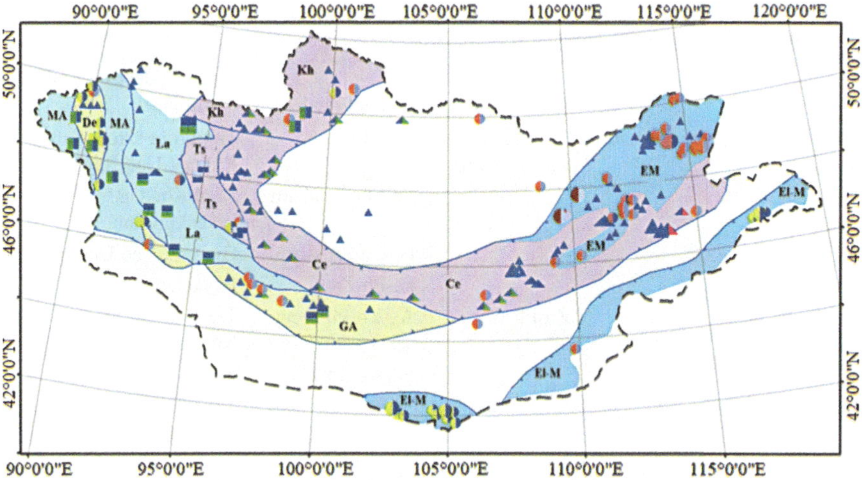

	Eras	Legend	Metallogenic Belt	Deposit mineralization type
1	Neoproterozoic		Khuvsgul /Khu/	Calcic skarn, Massive sulfide ore in metamorphic rocks, Carbonate hosted lead and zinc
			Tsagaan-Olom /TsO/	
2	Early Paleozoic		Mongolian Altai /MA/	Massive sulfide ore in metamorphic rocks
			Lake /La/	Massive sulfide ore in volcanogenic rocks
3	Middle Paleozoic		Deluun /De/	Silver-lead mineralzation, Gold-silver-polymetallic mineralization
			Gobi-Altai /GA/	Polymetallic massive sulfide ore /Cu dominant/, Silver-lead mineralization
4	Late Paleozoic-Mesozoic		Central /Ce/	Polymetallic calcic skarn
5	Late Mesozoic		East Mongolian /EM/	Silver-polymetallic vein
			Elgen Uul-Modon /El-M/	Silver-lead mineralization

Fig. 7.1 Metallogenic map and distribution of polymetallic deposits and occurrences in Mongolia (After Ukhnaa and Batkhishig 2018)

surrounded by Late Mesozoic volcanic grabens with Permian and Triassic granitoids that intrude Proterozoic and Paleozoic metamorphosed sedimentary rocks.

Skarn mineralizations generally occur along contacts and faults between Proterozoic and Devonian limestone and shale sequences and Triassic subalkaline granites and leucocratic granites. Skarns form lenses-like bodies (Bayandun, Tumurtei, Kharaat, Salkhit, and other occurrences), or layers (Tumurtiin Ovoo). Skarn bodies' sizes are ranging from 10 to 100 m in thickness and are n*100 m along strike. The intrusive rocks are rarely altered. The main skarn minerals are andradite, hedenbergite, and grossular associated with sphalerite. Zonation is typically

Table 7.2 Classification of Pb-Zn deposits related to subalkaline magmatic rocks in Mongolia (Dejidmaa et al. 1999; 2005; Dorjgotov 1996, 2002, 2009)

Deposit type	Ore formation	Deposit model	Example deposits and occurrences
Skarn—Contact metasomatic (Fe-Zn)	Carbonaceous Zn-Pb skarn	Fe-Zn skarn	Tumurtiin Ovoo, Salkhit Kharaat-Uul, Bayandun
		Fe-Mo-Zn skarn	Khol Khudag, Ervenkhoshuu
		Fe-Cu-Zn skarn	Oyut, Erdene-Undur
		Fe-Pb-Zn	Bayanjargalan
Hydrothermal vein and stockwork (Ag-bearing Pb-Zn)	Zn-Pb skarn	Ag-Zn-Pb (vein, stockwork)	Tsav, Baits-Ovoo, Khavtsgait, Dulaankhar
		Sn-Ag-Zn-Pb vein	Mungun-Undur, Tugalgatain nuruu
		Au-Ag-Zn-Pb (vein, stockwork)	Bayan-Uul
Epithermal vein (Ag-Pb)	Cu-Pb-Zn vein	Ag-Pb-Zn (vein, pipes)	Ulaan, Mukhar
		Cu-Zn-Pb	Zaanshiree, Berkh-Uul
	Ag-Pb vein	Sb-Ag-Pb	Khartolgoi group occurrence
		Ag-Pb	Modon group occurrence, Salaagiin Ulaan
Telethermal	Carbonate hosted Zn-Pb	Zn-Pb stratiform	Nukhet group occurrence

developed with epidote-feldspar, epidote-andradite, andradite-magnetite, andradite-pyroxene-magnetite, and pyroxene-magnetite zones. Characteristic retrograde minerals are actinolite, quartz, calcite, and chlorite (Dejidmaa et al. 2005).

Sphalerite and magnetite are the major ore minerals, and chalcopyrite, hematite, bismuthinite, molybdenite, pyrite, and galena may also occur. Fe-Zn mineralization is associated mostly with garnet and garnet-pyroxene skarn and occurs as disseminated mineralization. Pb/Zn/Cu ratios are 0.2/4.5/0.1. The deposits have four stages of mineralization: garnet-pyroxene skarn, andradite-magnetite exoskarn, sulfide, and quartz-carbonate (Dejidmaa et al. 1999, 2005).

7.3.1.1 Tumurtiin Ovoo Polymetallic Fe-Zn Skarn Deposit

The Tumurtiin Ovoo polymetallic deposit is located 18 km from the Baruun-Urt City, Sukhbaatar Province. This deposit (Yakovlev 1977; Podlessky et al. 1988; Dorjgotov 1996) is related to a calcic skarn that occurs along the contact between Devonian limestone and Mesozoic subalkaline granite (Fig. 7.2). Mesozoic leucogranite is cut by fine-grained granite, diorite porphyry, and gabbro-diabase dykes. The volcanogenic-sedimentary rock outcrop is $0.6 \times 0.8\ km^2$. It is subdivided

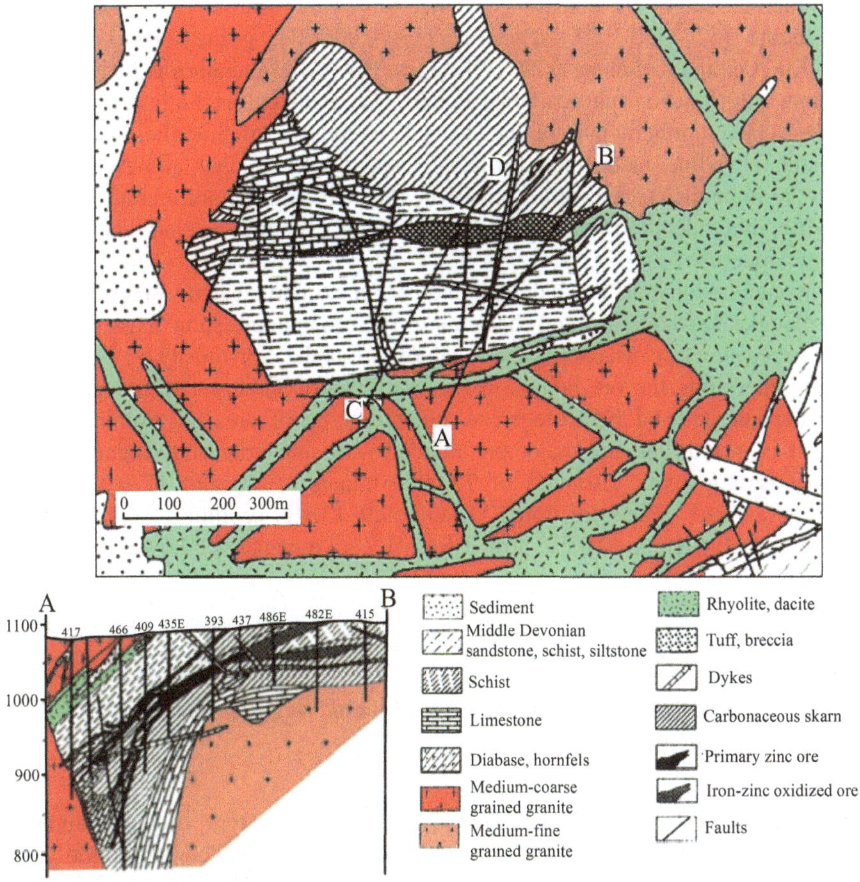

Fig. 7.2 Geological map and cross section of the Tumurtiin Ovoo polymetallic deposit (From Ukhnaa and Batkhishig 2018)

into three members of diabase, limestone, and shale. Volcanogenic sedimentary rocks are strongly altered by metasomatic and metamorphic processes. Diabase rocks are altered to hornfels, limestone is altered to marble, and shale is altered to plagioclase-pyroxenite and hornfels. Polymetallic skarn mineralization is formed at the contact between volcanogenic-sedimentary rocks and granitic intrusion and is covered by shale and hornfels. This shale and hornfels cover plays an important role for the accumulation of high-grade zinc ore.

The skarn is elongated to the NW and dips concordantly with host rock to the SW, forming lenticular bodies with different sizes. The main ore body is trending to NW, dipping to S and SW 10–20° near the surface. In the central part, it extends for about 800 m along strike, 480 m down dip, and 200–230 m down dip on the eastern and western flanks. The average thickness is 14 m. The major skarn minerals are andradite, hedenbergite, grossular, epidote, quartz, and wollastonite. Skarn and hydrothermal stages are determined to be wollastonite+garnet (>570 °C), garnet +magnetite (527–330 °C), sphalerite (330–200 °C), pyrite+quartz+carbonate (170–90 °C), and carbonate (<90 °C) stages. Host rock alteration is mainly silicification, chloritization, and carbonatization. The oxidized ore body occurs in the northeast and northern part of the deposit; it consists of limonite, smithsonite, hematite, covelline, malachite, azurite, cerussite, anglesite, opal, and gypsum. The main ore minerals are magnetite, sphalerite, hematite, and minor amounts of chalcopyrite, molybdenite, pyrite, marcasite, chalcocite, bornite, galena, scheelite, and opaque minerals.

The gangue minerals are andradite, wollastonite, quartz, calcite, chlorite, pyroxene, plagioclase, rhodonite, and chalcedony. The ore minerals show massive texture and layered structure. Magnetite and sphalerite mineralization mostly developed into granite; therefore, the ore body contact is not clear. Garnet-magnetite-sphalerite, garnet-sphalerite, and hematite-pyrite-sphalerite-type of ores are dominant. The deposit is zoned, and the major ore minerals are sphalerite and magnetite. The deposit is large with resources of 750,000 tonnes of Zn and 1770 tonnes of Cd. The average grades are iron 17%, zinc 9.9–13.1%, cadmium 0.02%, indium 0.002%, molybdenum 0.02%, iron 26.8–41.7%, manganese 4.5%, and cobalt 0.01%.

7.3.2 Hydrothermal Vein and Stockwork (Ag-Pb-Zn; Cu-Pb-Zn) Deposits

The hydrothermal vein and stockwork Cu-Pb-Zn deposit type occurs in Jurassic intrusive along faults. Numerous quartz-sulfide and quartz-carbonate-sulfide veins are developed within these intrusives in alteration zones. The veins are related to small bodies of diorite, granodiorite, and granite that intrude into metamorphosed sedimentary rocks. Host rocks undergo metasomatic alteration, consisting of widespread propylitic zones and narrow alteration zones composed of quartz, sericite, carbonate, and pyrite. The alteration shows zonation that is with quartz, sericite, and pyrite commonly occurring in the center, whereas quartz and pyrite commonly occur in the periphery, along with an increase of carbonate and chlorite. The mineralization characteristic of this type of deposits is complex, but similar between deposits. The main ore minerals are galena, sphalerite, pyrite, and Ag minerals, sometimes with chalcopyrite, pyrrhotite, tetrahedrite, boulangerite, and native gold. The ore minerals are irregularly distributed and exhibit streaky, massive, and brecciated structures, whereas barren zones are characterized by disseminated to linear disseminations and occasional coated and nested structures. The mineralization is subdivided into three ore forming stages: quartz-sulfide, quartz-carbonate-sulfide, and carbonate-fluorite.

Ulaan and Mukhar deposits are the main examples of the stockwork-type deposits; they occur in explosion breccia pipes that are formed along steeply dipping faults. The ore-bearing pipes have trapezoid or wedge shapes and cut through metamorphosed Late Mesozoic sedimentary and volcanic rocks. The breccias are hydrothermally altered intensively with a well-defined mineralogical zoning. For instance, in the central part and deep horizons of the breccia pipes, the cement varies from epidote-actinolite-sulfide to epidote-actinolite. In other cases the cement is quartz-fluorite-sulfide. Polymetallic mineralization is developed mainly in the pipes and occasionally in fractures near the breccia pipes. The ore bodies occur as veins to stockworks, striking for several n*100 m, with thicknesses of *n**10 m, and extend down dip for up to 700 m. The major ore minerals of these deposits are galena, sphalerite, chalcopyrite, and pyrite. Also arsenopyrite, pyrrhotite, magnetite, tennantite, bornite, native silver, and gold are occurring. Gangue minerals are quartz, feldspar, fluorite, epidote, actinolite, carbonate, and chlorite. The ore minerals are distributed as disseminated ore-pocket, disseminated, and breccia structures. Deposits exhibit five stages of mineralization: diopside-magnetite, epidote-actinolite, quartz-sulfide, quartz-fluorite sulfide, and quartz-fluorite (Dejidmaa et al. 1999, 2005).

7.3.2.1 Tsav Polymetallic Pb-Zn ± Cu (±Ag, Au) Deposit

The Tsav deposit is one of the largest Ag-Pb-Zn in northeast Asia (Gantumur et al. 2005; Gerel 1998; Yang et al. 2009; Nie et al. 2010, 2014). It is located in the Kherlen soum, Dornod Province, NE Mongolia, 146 km northeast of the city of Choibalsan.

The Tsav deposit was first discovered in 1975, by a Russian geological mapping expedition which was working in Mongolia. 1986–1992 pre-exploration work has been done, and the resource was estimated by the Dornod expedition geologists Ya. Dolgor, O. Chuluun, V.I. Melnikov, and D. Tserenpuntsag. The deposit is situated in the Ereendavaa metamorphic terrain (Badarch et al. 2002). Proterozoic gneiss, leptynite, schist, and Tsav granitoid intrusions outcropped in the N and E parts of the Tsav district. Jurassic shoshonite, trachyandesite, trachydacite, dacite, rhyolite, ignimbrite, tuff, and tuffaceous rock are predominant in the region (Jargalsaihan et al. 1996; Baljinnyam et al. 1989; Dorjgotov 1996; Bat-Ulzii 1996; Nie et al. 2015). Jurassic porphyritic intrusions, including dykes and stocks of diorite, monzodiorite, and monzonite, intrude into the volcanic rocks (Fig. 7.3). Hornblende from monzodiorite dykes in the region shows a K-Ar isotopic age of 135 ± 5 Ma (Nie et al. 2010, 2014). The chemical analyses reveal that rock compositions are ranging from high-K calc-alkaline to shoshonitic (Gantumur et al. 2005; Nie et al. 2010). Silicification, carbonatization, and quartz-sericite-pyrite alterations are present in the host rock (Nie et al. 2015). Host rock alteration is closely related to the mineralization, and intensity of the alteration increases with increasing ore grade. The alteration zone is 10–20 m wide. Central part of the alteration zone contains an assemblage of quartz-sericite-pyrite, and a peripheral zone has less pyrite, chlorite, sericite, and carbonate.

Most of the ore bodies are hosted by the Jurassic volcanic-sedimentary rocks and show a close relation to Mesozoic monzonite, monzodiorite, monzogranite, and

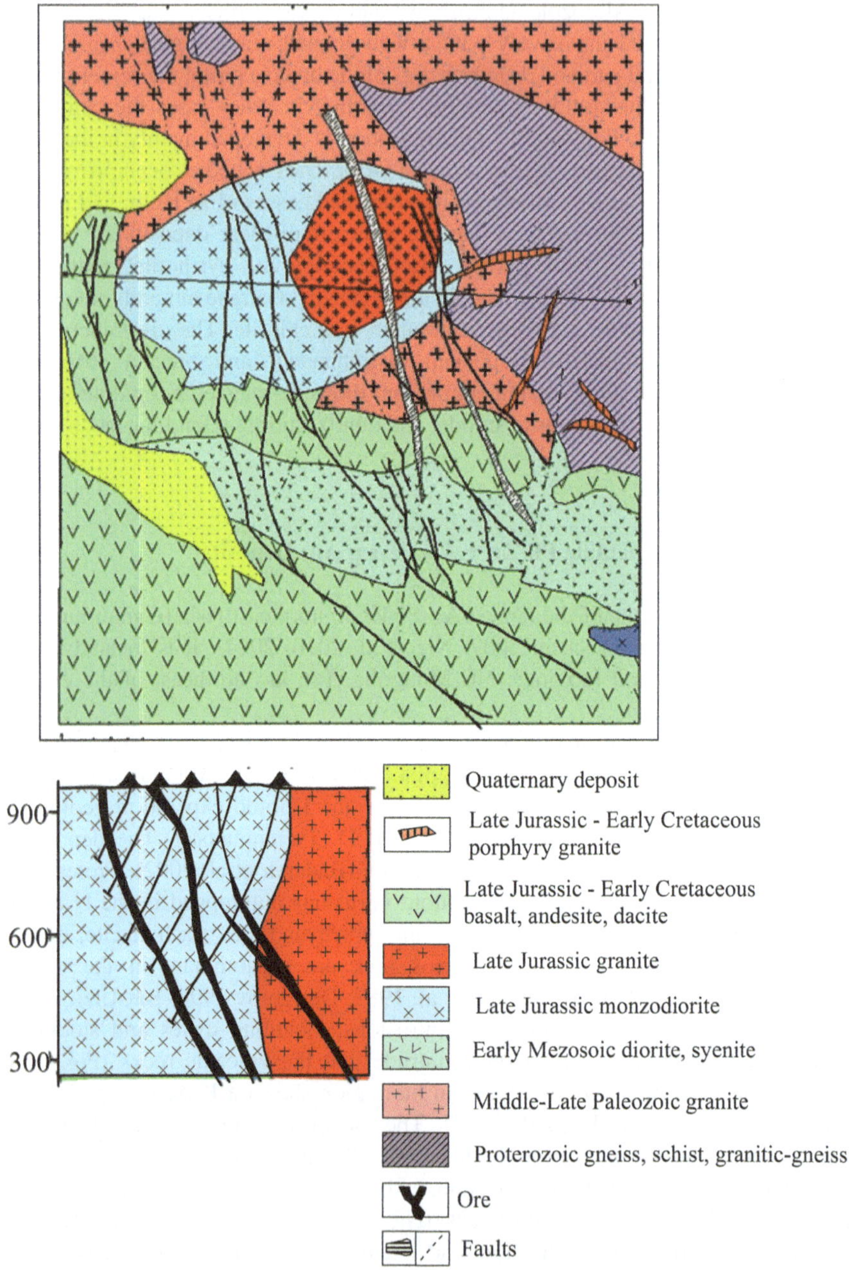

Fig. 7.3 Geological map and cross section of the Tsav polymetallic deposit (From Ukhnaa and Batkhishig 2018)

syenite dykes. Ag-Pb-Zn mineralization occurs within the Jurassic volcanic rocks, forming veins, veinlet, stockworks, and disseminations that constitute a number of irregular pods, pockets, and/or lenses interconnected by joint fillings and stringers (Nie et al. 2015). Overall, 21 ore zones have been defined in the Tsav mineralized district which covers an area of 5.4 km^2; some of the ore zones contain several ore bodies.

Ore bodies are 350–2200 m long, 0.5–15 m thick, and 125–550 m down to dip. Two types of Ag-Pb-Zn ores are classified: early quartz-dominated vein-stockworks are superimposed by later carbonate-dominated vein-stockworks (Nie et al. 2015).

Both quartz and carbonate lodes show colloform, crustiform, and cockade textures. The main ore minerals are sphalerite, galena, magnetite, pyrite, chalcopyrite, pyrrhotite, arsenopyrite, and Ag minerals with small amount of sulfoantimonite, polybasite, cleiophane, bismuthite, tetrahedrite, altaite, electrum, freibergite, argentite, hessite, chlorargyrite, calaverite, pyrargyrite, native gold, and native silver. Gangue minerals are quartz, sericite, muscovite, eligonite, rhodochrosite, chlorite, and calcite, as well as oligonite, Mn-siderite, epidote, smectite, muscovite, chlorite, illite, and kaolinite. Average quartz lode ore grades are Ag 141 g/t, Pb 6.48–7.21%, Zn 3.42–3.53%, and Cu 0.23% and for the carbonate lode ore are Ag 372 g/t, Pb 3.61%, Zn 2.66%, and Cu 0.08%. The deposit has an ore resource of 420,000 tonnes, which belongs to a medium-sized deposit (Nie et al. 2010; 2014; Ukhnaa and Batkhishig 2018).

The $\delta^{34}S$ value of various sulfides (pyrite, pyrrhotite, arsenopyrite, and galena) in the Tsav deposit is similar to that of typical magmatic hydrothermal deposits (Gantumur et al. 2005). Also, the isotopic analysis of fluid inclusions and H-O isotopes shows that the ore-forming fluid was mixed with magmatic water (Gantumur et al. 2005; Yang et al. 2009; Nie et al. 2010; Zhai et al. 2013). Ore mineralogy, textures, and alteration characteristics of the Tsav deposit are indicated to be similar to the intermediate sulfidation epithermal deposit (Hedenquist et al. 2000; Sillitoe and Hedenquist 2003; Jiang et al. 2004). This suggested origin is an important signature; it may have potential for porphyry-style mineralization at deeper levels in the Tsav mineralized district and neighboring area (Nie et al. 2015).

7.3.2.2 Dulaankhar Uul Ag-Pb Epithermal Vein Deposit

The Dulaankhar Uul deposit is tectonically situated in the Khovd subterrane, northeast part of the Mongolian Altai terrane (Tomurtogoo et al. 1999; Byamba and Dejidmaa 1999; Badarch and Tomurtogoo 2001). The Dulaankhar Uul polymetallic deposit is located on the SE margin of the Dulaankhar volcano-tectonic structure which is along the Dulaankhar Uul tectonic fault and associated with the Achit Nuur granitic intrusion. This deposit occurs as layers and sheets like bodies within the Early to Middle Devonian Dulaankhar Formation. The Formation consists of mainly rhyolite tuff, flows, tuffaceous sandstone, rhyolite and sometimes dacite porphyry of subvolcanic bodies, dykes, and diabase dykes (Fig. 7.4).

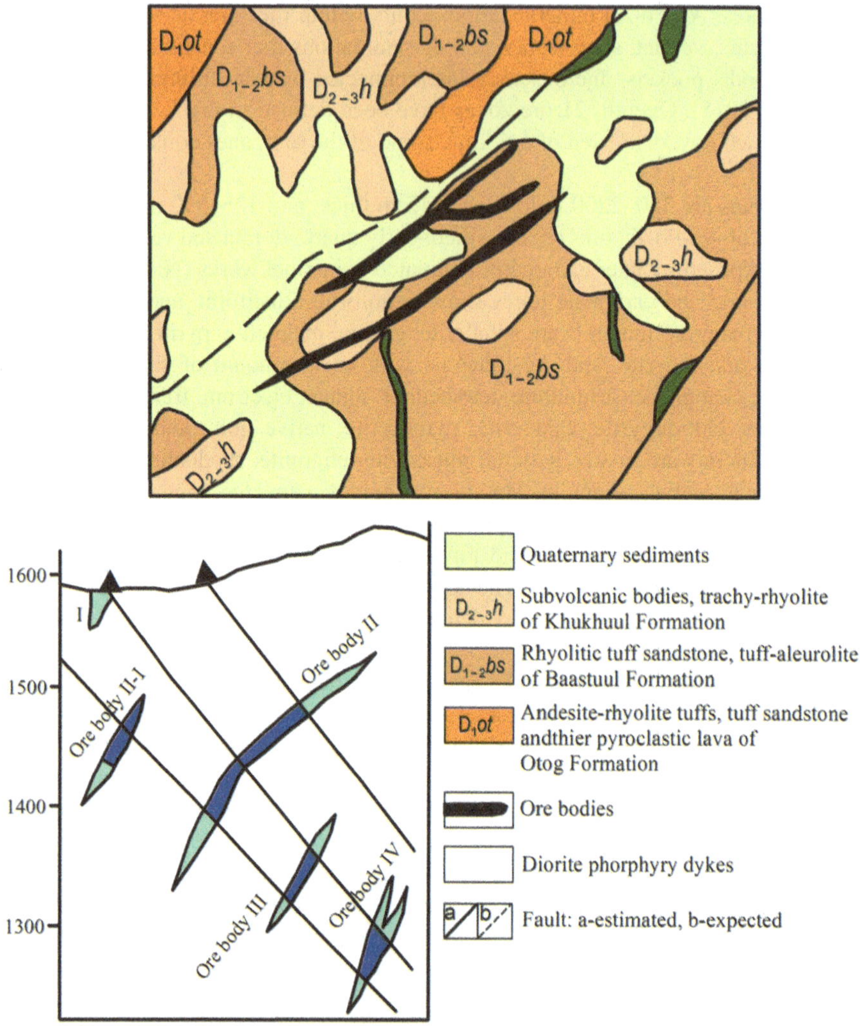

Fig. 7.4 Geological map and cross section of the Dulaankhar Uul polymetallic deposit (From Ukhnaa and Batkhishig 2018)

Overall, five ore bodies are recognized, and they are 200–700 m long and 10–40 m wide. Two of them outcrop to the surface, and the others are tested by drilling. Sulfides occur in altered chlorite-sericite-quartz tuff and at the intersection of the Dulaankhar fault and dykes. This zone is about 600 m wide, and the sulfide bodies are 0.7–20 m thick, occurring in layers, lenses, and veins; they extend more than 100 m on the surface and down dip to a depth of 400 to 500 m. A significant SP anomaly in the NE part of the deposit is suggesting new sulfide bodies. The oxidized parts of the deposit contain galena, wulfenite, cerussite, calamine, barite, fluorite, calcite, and malachite. There are five mineralization stages that have been identified as follows.

1. Pre-mineralization siliceous stage—K-feldspar, muscovite, quartz, biotite.
2. Metasomatic quartz-sericite-biotite stage—quartz, feldspar, muscovite, biotite, pyrite, sphalerite, and minor amount of arsenopyrite, marcasite, kaolinite, chlorite, and hematite.
3. Quartz-polymetallic stage—quartz, muscovite, calcite, sphalerite, galena, chalcopyrite, pyrite, and minor amount of arsenopyrite and siderite. This association constitutes 90–95% of the ore mass.
4. Post-mineralization stage—quartz, carbonate, pyrite. It is mainly composed of quartz, calcite, pyrite, and minor amounts of muscovite, siderite, and sphalerite.
5. Supergene stage—cerussite, anglesite, goethite, and hydrogoethite. This mineral assemblage constitutes 0.5–1% of the ore mass.

The Dulaankhar Uul deposit is classified as a volcanogenic, hydrothermal vein-type deposit, based on these mineralization sequences and mineral assemblages. The ore grades are Pb 1%, Zn 1–10%, Cu 0.5–1%, Ag up to 2000 g/t, Ba up to 0.2%, and Au 0.2–4 g/t. The eastern and central parts of the deposit are hosted in tuff breccia. However, in the southwest part of the deposit, 700 × 10–20 m-sized silica cap occurs, and it contains Cu oxides and hematite. These areas contain anomalous Pb, Zn, Cu, Mo, and Co. The average grade is Zn 2.05%, Pb 0.1–1.76%, Au 1.1 g/t, Ag 1–40 g/t, and Cu 0.1–0.3%. The deposit is large with resources of 665,000 tonnes of Zn, 430,000 tonnes of Pb, and 16 tonnes of Au.

7.3.2.3 Mungun-Undur Sn-Pb-Zn Vein Deposit

The Mungun-Undur Ag polymetallic deposit was first discovered and prospected in the early 1980s by geologist O. Gurjav. The deposit type was defined by G.S Ripp and V.I. Sudakov as polymetallic vein (Dejidmaa et al. 1999). Further prospecting works were conducted for silver mineralization by Mongolrostsvetmet in the 1990s. The Mungun-Undur Pb-Zn-Ag deposit is located in the Umnudelger soum, Khentii Province, 70 km northwest of the city of Chinggis in NE Mongolia. Tectonically the deposit is situated in the SE margin of Undurtsagaan intrusive dome, eastern part of Khentii uplift. Geology of the deposit area is characterized by Lower Devonian Tsarigiingol Formation terrigeneous flyschoid rocks that consist of chlorite-sericite slate and quartz-muscovite slate with layers of sandstone and siltstone and intruded during the Devonian by gabbro and diorite-porphyry dykes (Fig. 7.5; Dorjgotov 2009). Stockworks in fine-grained granite and granite porphyry rocks were identified in the NW part of the Mungun-Undur deposit (Kovachev et al. 1990). Granitoids were strongly altered to greisen. The deposit is hosted within Tsarigiingol Formation, which is consisting of mainly siltstone-sandstone layers of terrigeneous rocks. Mineralization may have formed by metasomatic processes developed into the terrigeneous rocks through faults and fractures, due to the intrusion of the gabbroic

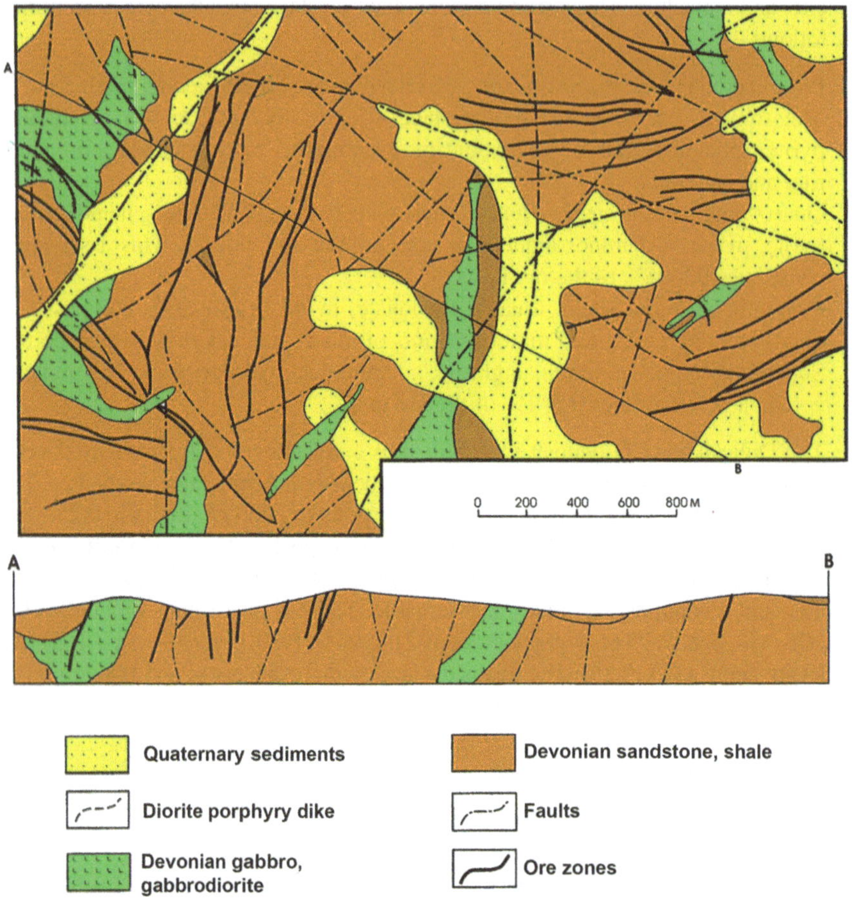

Fig. 7.5 Geological map of the Mungun-Undur polymetallic deposit (From Dorjgotov 2009)

body. Ore mineralization occurred in the veins along the fault zones (Dorjgotov 2009).

G.S. Ripp and V.I. Sudakov studied host rock alteration. Hydrothermal alterations of chloritization, silicification, pyritization, and carbonatization are developed in the host rocks. The following alteration stages are recognized: quartz-amphibole-chlorite-muscovite-feldspar replacement zones with polymetallic and Sn minerals with thicknesses of 1 m to 6 m; → quartz chlorite-muscovite-carbonate and rhodochrosite replacement associated with polymetallic sulfide mineralization; → silicified and argillized zones with Ag and Sb minerals that are often overprinted on the second type of replacement stockwork associated with Ag-polymetallic mineralization, and → late stage carbonate alteration (Dejidmaa et al. 1999, 2005). An oxidization zone was developed up to 3–5 m depth from the surface. The oxidation zone consists of hematitization and limonitization with oxidized ore of Zn >0.5% and Pb >0.3%. The Mungun-Undur deposit is classified as medium-temperature hydrothermal vein-type Ag polymetallic deposit based on the geology, alteration of the host rock, and the physicochemical condition of the ore-forming minerals. Overall 24 mineralized zones were mapped in the deposit field, but the more significant three zones are named Central, Northeast, and Southeast. The Central mineralized zone is situated along the fault zone, composed of 6–7 ore bodies. The Southeast mineralized zone is composed of 12 ore veins. Ore mineralized zones vary from 750 m to 1200 m in length along its strike, up to 300 m in length along its dip, and up to 60 m in width. The ore bodies are subvertical dipped vein type with 1–4 m thickness on average.

The main ore minerals are sphalerite, galena, cassiterite, pyrite, arsenopyrite, pyrrhotite, and marcasite. The main metals are Ag, Sn, Pb, and Zn, and the potential by-products are Cu, Cd, and In. Second-order ore minerals are freibergite, chalcopyrite, stannite, antimony, goethite, manganite, plumbojarosite, and minor amount of wolframite, scheelite, pyrargyrite, psilomelane, native silver, gold, bismuthite, ilmenite, and rutile (Kovachev et al. 1990; Dorjgotov 2009). The oxidized ore is composed of goethite, pyrolusite, psilomelane, anglesite, plumbojarosite, jarosite, kaolinite, and silica. Ore styles are mostly vein and veinlets-like, disseminated and rarely brecciated, massive, and lenticular structures. The ore minerals are formed by hydrothermal processes through following the stages and mineral assemblages (Naftali et al. 1984; Dorjgotov 2009).

- I stage: Quartz-pyrite association developed along the host rock and cut by all of the later stage veins.
- II stage: Quartz-arsenopyrite-pyrrhotite mineralization mainly occurs in the north part of the deposit. Pyrrhotite is the dominant mineral of the ore in this stage.
- III stage: Quartz-sphalerite-freibergite. It mainly occurs in the central part of the deposit; additionally galena, pyrite, and chalcopyrite are present.
- IV stage: Rhodochrosite-galena-pyrargyrite. This mainly occurs in the central and east part of the deposit. It forms matrix which was brecciated during the previous ore stages. The pyrargyrite-sphalerite-galena ores of the III–IV stages are the main mineable ore formation.
- V stage: The quartz-stibnite association is formed in the upper part of the ore deposit, forming individual small-sized veins and veinlets. They are crosscutting the sphalerite-galena veins and replacing freibergite.

These assemblages result in geochemical zones of Mo, (Cu-Mo) W, Bi-As, Fe-Sn (Sn)-Pb, Zn-Ag-Pb, and Ag (Naftali et al. 1984). The main ore assemblage of the Mungun-Undur deposit is quartz-sphalerite-freibergite and rhodochrosite-galena-pyrargyrite. Supergene processes took place near-surface and generated goethite, psilomelane, anglesite, and plumbojarosite minerals. The average grade of the deposit is 1.3–1.89% Zn, 1.67–2.18% Pb, 81.85–188 g/t Ag, and 0.14% Sn. Expected resources of the metals were estimated as 135.4 thousand tonnes Zn, 155.1 thousand tonnes Pb, and 700 tonnes Ag (Ukhnaa and Batkhishig 2018). The K-Ar age (175 ± 9 Ma) of the sulfide mineralization was determined by Murao et al. (1998).

7.3.2.4 Ulaan Ag-Pb-Zn Deposit

The Ulaan deposit was discovered in 1973, by a Russian geological mapping expedition. Exploration work was carried out by V.Ya. Kiselev, Yu.B. Mironov, N.S. Soloviev, and G.N. Gubkin in 1980–1986, and more recent works are conducted by O. Chuluun, Yu. B. Mironov, and others. Since 2004, Chinese Shin Shin Co. Ltd. started mining (Ukhnaa and Batkhishig 2018).

The Ulaan deposit is located 110 km NW from the Choibalsan City. It is situated in the central part of the Mesozoic volcano-sedimentary basin, which unconformably overlies or is in fault contact with the Ereendavaa metamorphic terrain (Badarch et al. 2002). The Ereendavaa terrain in NE Mongolia is dominated by Proterozoic quartzite, gneiss, amphibolites, black schist, and marble, intruded by Early Paleozoic granodiorite and diorite intrusions. The geology of the overlying Mesozoic basin is dominated by volcanic rocks, including basalt, andesite, dacite, oligophyric rhyolite, shoshonite, latite, trachyandesite, trachydacite, pyroclastics containing vitrophyric felsic ash tuff, ignimbrite, and tuffaceous rocks (Fig. 7.6). Coal-bearing sedimentary rocks are also distributed in the district. Ore-bearing volcanic rocks can be classified

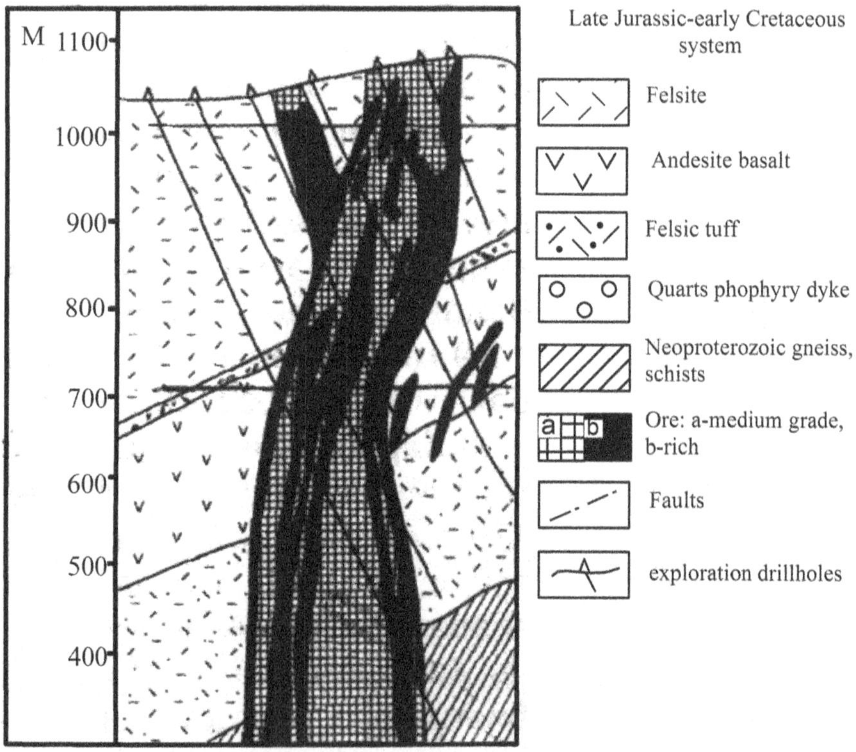

Fig. 7.6 Exploration cross section of the Ulaan polymetallic deposit (From Ukhnaa and Batkhishig 2018)

into two categories: (1) alkaline rocks and (2) high-K volcanic rocks transitional to shoshonite (Nie et al. 2010, 2014; Dahlkamp 2009).

The deposit situated at the intersection of NE trending Mukhar Gol fault and sublatitudinal Gurvanbulag fault is in explosive breccia pipes. The ore-controlling fault zone extends 10 km long and 2 km wide. In the Ulaan Ag-Pb-Zn district, two parallel crypto-explosive pipes have been identified. Breccia pipes which host the Ag-Pb-Zn ore bodies penetrate the Ereendavaa metamorphic unit and the overlying volcanic strata (Nie et al. 2015). The breccias contain clasts of basalt, andesite, dacite, rhyolite, trachyandesite, and shoshonite, vary in size from 23 cm to 40 m, and are cemented by tuffaceous material containing sulfides, carbonate, and fluorite. Associated wall-rock alteration includes silicification, carbonatization, and chloritization (Nie et al. 2015).

Two types of mineralization at the Ulaan deposit have been classified: tabular skarn-type (at deeper levels) and vein-stockwork lodes (at a shallow depth) (Nie et al. 2015).

The skarn-type mineralization occurs as sulfide-rich lenses and layers, with high-grade pods, along the contact zone of Ereendavaa marble and Cretaceous monzogranite and monzodiorite porphyry dykes. The dykes show a LA-ICP-MS zircon ages between 157 and 150 Ma (Hu et al. 1998; Nie et al. 2014). The ore minerals are marcasite, magnetite, pyrargyrite, bismuthite, argentite, tetrahedrite, native Au, bornite, bismuthinite, bismuthine, chalcocite, chlorargyrite, pyrite, chalcopyrite, galena, sphalerite, pyrrhotite, and arsenopyrite. Precious metal-bearing phases include electrum, argentite-acanthite, argentobismuthite, tintinaite, jalpaite, native silver, native gold, and bornite. The gangue minerals are primarily K-feldspar, actinolite, diopside, chlorite, epidote, garnet with minor quartz, fluorite, sericite, albite, dolomite, quartz, calcite, and ankerite. The average grades of the main ore bodies are Ag 46–86 g/t, Pb 2.2–3.6%, Zn 6–12%, Cu 0.56%, and Au 3.5 g/t (Nie et al. 2015). Resources estimated at present are over 3000 tonnes of Ag and 2 Mt. of combined Pb-Zn in the deposits (Nie et al. 2014; Dahlkamp 2009; Ukhnaa and Batkhishig 2018).

7.3.2.5 Mukhar Zn-Pb-Ag-Cu-Cd Deposit

The Mukhar deposit is located just 850 m southeast of the Ulaan deposit along the East Mukhar fault zone in NE Mongolia. The geology, mineralization, and alteration of the Mukhar deposit is the same as the Ulaan deposit. The ore minerals are the same as in the Ulaan deposit except for the lack of gold and higher content of chalcopyrite. In the deposit 1108 t Ag with grade of 113 g/t were estimated along with 332.5 thousand tonnes of Zn, 61.8 thousand tonnes of Pb, 11.9 thousand tonnes of Cu, and 1079 tonnes of Cd.

7.3.2.6 Uulbayan Zn-Pb Vein Deposit

The Uulbayan deposit is located in the Uulbayan soum, Sukhbaatar Province. It is situated in the western part of the Gobi-Khayngan structure-formation zone, Khentii uplift. The deposit belongs to the island arc terrain of the Gobi Altai-Sukhbaatar (Tomurtogoo 2012). The geology of the deposit is characterized by the Permian Erdenetsogt Volcanogenic-Sedimentary Formation. It is mainly composed of andesite, trachydacite, dacite, rhyolite, and their tuff and tuff breccias. The Erdenetsogt Volcanogenic-Sedimentary Formation originated in an active continental margin geodynamic setting. These volcanogenic rocks are cut by intermediate subvolcanic stocks and felsic porphyritic dykes. Dykes and stocks occur along the fault zones, which are associated with the mineralization. There are also sedimentary rocks of the Cretaceous Tsagaantsav Formation and Late Jurassic subvolcanic rocks in the deposit area (Fig. 7.7).

The mineralization occurs along the shear zone, within the quartz vein, near the exocontact of dacite porphyry bodies. Layered chalcedony-like quartz veins have high content of sulfide minerals of As, Sb, and Ag. Minor amount of disseminated sphalerite and galena are present. Host rock alterations are propylitic, chloritization, carbonatization, pyritization, and epidotization. Sericitization is dominant in porphyritic dacite. The main ore minerals are sphalerite, galena, pyrite, and chalcopyrite. The mineralization occurs as disseminations, brecciated, fracture filling, vein, and veinlets within the andesite porphyry, diorite, and rarely sedimentary rocks of conglomerate and sandstone. Overall nine parallel veinlike ore bodies were distinguished, and they continued up to 80–240 m in depth. The ore bodies' thicknesses vary from 3 to 31.5 m. The deposit has a resource of 3.2 thousand tonnes of ore and 110 thousand tonnes of metal Zn. The metallic ore consists of zinc, lead, and silver, and equivalent Zn grade is 3.4%.

7.3.2.7 Nominy Am Pb-Zn ± Cu (±Ag, Au) Vein and Stockwork Deposit

The Nominy Am deposit is hosted in the Early to Middle Devonian Otogiin Formation that consists of dacite and andesite, tuff, tuff breccia, and quartz-calcite sandstone. The cataclastic host rock is intensely altered to silica and limonite. Ore mineralization consists of sulfides in quartz and quartz-barite veinlets in a breccia zone. The breccia zone strikes NW and is 60–100 m wide and 300 to 350 m long in size and occurs at the intersection of NW and NE-striking faults at the contact between volcanic and terrigenous rocks. The quartz and quartz-barite veins vary from 0.1 to 1.5 m in thickness and extend for 60–70 m along the strike.. The main ore minerals are chalcopyrite, pyrite, galena, arsenopyrite, malachite, and azurite. The largest vein is 1–1.5 m by 70 m in size, and grades are Cu 0.3–1%, Pb 0.0015–0.7%, Zn 0.007–0.5%, As up to 1%, Ba up to 1%, Au 0.003–0.015 g/t, and Ag 1.5–500 g/t.

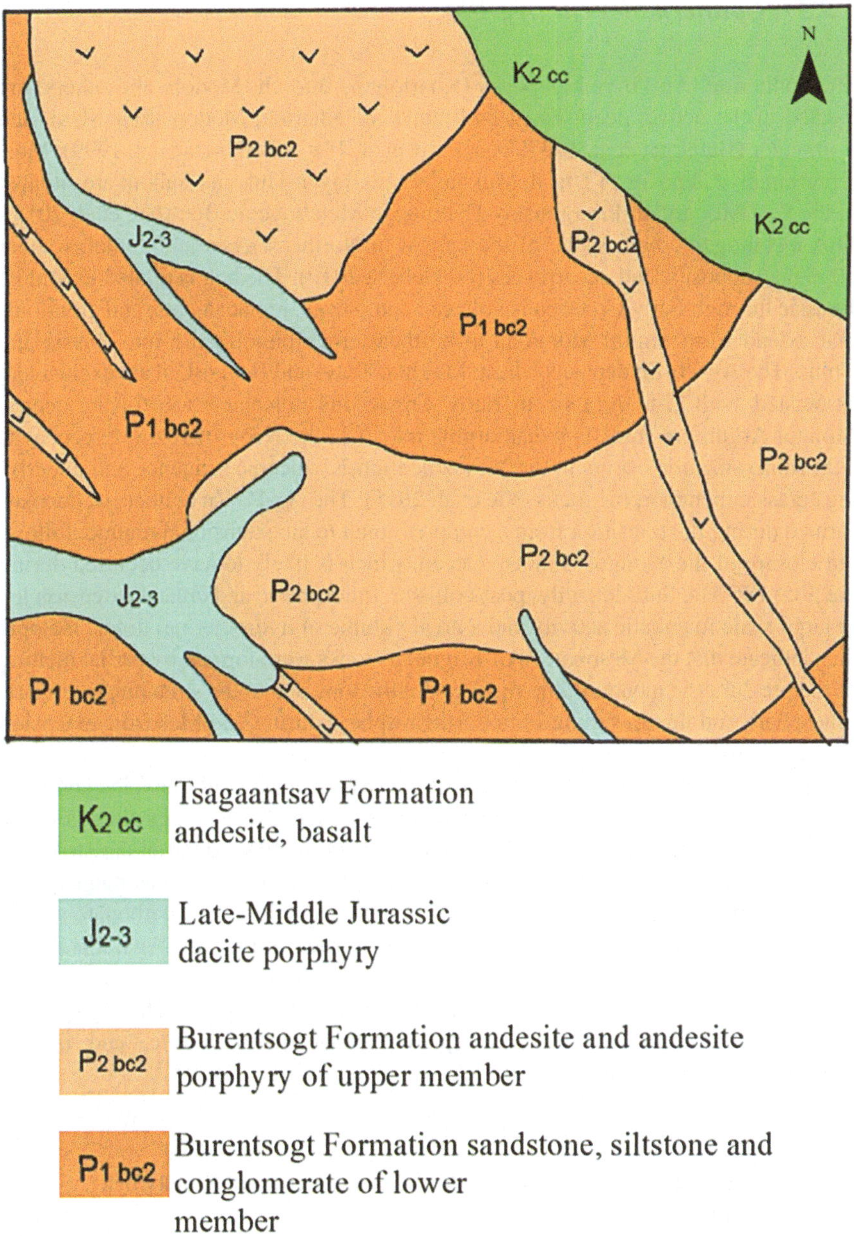

Fig. 7.7 Geological map of the Uulbayan polymetallic deposit (From Ukhnaa and Batkhishig 2018)

7.3.3 *Epithermal Vein (Ag-Pb)*

The epithermal Ag-Pb vein deposits (Khartolgoi, Boorch, Modon, and others) are located in the active, deep-seated fault zones of Kherlen, Modon, Elgen-Uul, and Tolbonuur (Batjargal et al. 1997; Dorjgotov et al. 1997; Dejidmaa et al. 1999). Most representative deposits of Ulaan, Mukhar, Tsav, Bayan-Uul, and Salkhit are situated in the East Mongolian-Priargunskiy-Deerbugan Metallogenic Belt (Nie et al. 2015). This metallogenic belt is one of the largest in Northeast Asia and contains about 80 mines, deposits, and occurrences (Nokleberg 2010). The belt is related to Middle Jurassic through Early Cretaceous volcanic complexes, replacements, and granitoids that extend from central Mongolia to northeastern Mongolia and into Russia and China. The Ag-Pb-Zn deposits Ulaan, Mukhar, Tsav, and Bayan-Uul are exclusively associated with Late Jurassic to Early Cretaceous volcanic rocks in the central Mongol-Argun pre-Jurassic metamorphic massif of the Neoproterozoic Ereendavaa belt. Mineralization occurs in the Mesozoic high-K volcanic sequence and underlying Jurassic metamorphic rocks (Nie et al. 2015). The Ag-Pb-Zn epithermal deposits formed during the transition from a compressional to an extensional regime, following closure of the Mongol-Okhotsk Ocean, which is likely to have occurred during the 205–185 Ma. Subsequently, post-collision lithospheric and crustal extension led to large-scale magmatic activity and a great volume of hydrothermal fluids. Isotopic data indicate that the Mesozoic high-K igneous rocks were formed by partial melting of the pre-Jurassic metamorphic rocks and intrusions due to the stretching of mantle lithosphere and the upwelling of hot asthenosphere during late Mesozoic extension (Nie et al. 2015).

Most of the epithermal Ag-Pb deposit consists of quartz-sulfide veins and mineralized zones in various rocks intruded by mafic dykes. The host rock alterations are quartz, chalcedony, kaolinite, and chlorite. The ore bodies extend along strike for $n*100$ m and down dip up to 300 m. The thickness of the ore bodies ranges up to $n*10$ m. The ore bodies consist of quartz-carbonate-sulfide and carbonate-sulfide types. The major ore minerals are galena, arsenopyrite, stibnite, and Ag minerals and subordinate chalcopyrite, sphalerite, cinnabar, and pyrite. Gangue minerals are quartz, siderite, chalcedony, kaolin, calcite, barite, and fluorite. The deposit has three major mineralization stages: quartz-galena, quartz-fluorite, and quartz-carbonate.

7.3.3.1 Khartolgoi Pb-Zn ± Cu (±Ag, Au, As) Vein and Stockwork Deposit

The Khartolgoi deposit is located 170 km S from Dalanzadgad City in South Mongolia. The deposit is situated in the Khartolgoi-Sulinheer Metallogenic Belt (Nokleberg 2010), hosted in Neoproterozoic metamorphic rock that is thrust over Carboniferous plutonic rocks composed of gabbro, diorite, granodiorite, granosyenite, and granite and Permian terrigeneous sedimentary rocks, and they

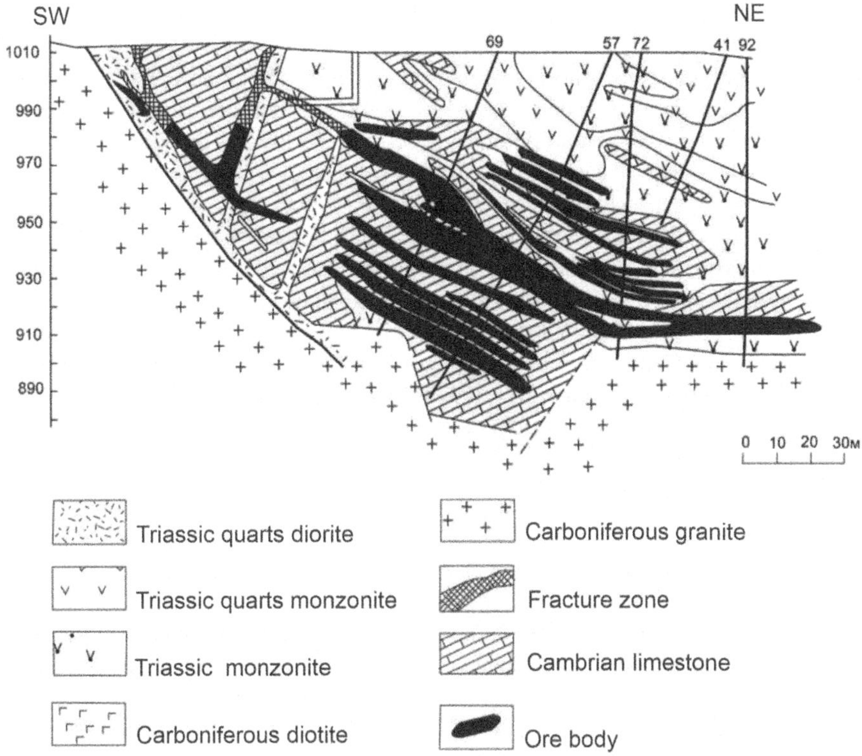

Fig. 7.8 Geological cross section of the Khartolgoi polymetallic deposit (Nokleberg 2010)

are intruded by Late Triassic and Early Jurassic subvolcanic dykes of monzonite, quartz monzonite, and quartz-porphyry (Yakovlev 1977; Marinov et al. 1977; Jargalsaihan et al. 1996; Dorjgotov 1996). The Khartolgoi deposit is mainly hosted in marble and sometimes in monzonite and quartz porphyry. Individual bodies occur along a major N-NW-striking, N-dipping thrust and parallel faults and also in NW-striking, steeply dipping faults associated with the major thrust. Pb-Ag veins and lenticular ore bodies have dimensions of 100–400 m long, 2–29 m wide, and down to 120 m, mainly in thrust faults and veins in steep-dipping faults. Wallrock hydrothermal alteration consists of carbonate and silica alteration (Fig. 7.8).

Ore mineral assemblages are siderite-sulfide, massive sulfide, and sulfide-magnetite. Main sulfides are galena, pyrite, stibnite, arsenopyrite, and Ag minerals. In addition to high grade Pb, high contents of Ag, Sb, As, Cd, Cu, and Zn are existing. Major gangue minerals are chalcedony, siderite, quartz, and calcite. The deposit is intensely oxidized and leached at the surface and at a shallow level. The deposit has average grades of 1.8–13.4% Pb, 126–145 g/t Ag, and 2.6% Sb and a resource of 60,000 tonnes of Pb, 30,000 tonnes of Sb, 138 tonnes of Ag, and 20,000 tonnes of As.

7.3.3.2 Kharmorit Pb-Zn ± Cu (±Ag, Au) Vein and Stockwork Deposit

Kharmorit polymetallic deposit (Marinov et al. 1977; Yakovlev 1977) is hosted along a fault in Permian sandstone and shale that are intruded by a Late Triassic through Early Jurassic granite porphyry stock. It also hosts a Sn vein and greisen deposit. Diabase and diabase porphyry dykes are closely related to the polymetallic fluorite greisens that occur in the granite porphyry stock and in adjacent, contact-metamorphosed Permian sandstone and shale. Polymetallic veins are altered to sericite, chlorite, and limonite and contain pyrite, arsenopyrite, galena, and chalcopyrite. Gangue minerals are quartz, feldspar, and calcite. The Pb grade ranges from 0.1% to 11% at the surface. Yakovlev (1977) interpreted that the polymetallic veins are closely related to the granite-porphyry stock. Sn vein and greisen deposits formed in separate stages. However, polymetallic veins may have related to diabase and diabase porphyry dikes and are younger than granite porphyry stock and associated Sn deposits.

7.3.3.3 Borts Uul Volcanogenic Cu-Zn Massive Sulfide Deposit

The Borts Uul deposit consists of sulfide-rich lenses and tabular bodies in volcanic rock at the intersection of the Khangai and Zavkhan faults. The deposit contains three parts named Northern, Central, and Pyrite. The Northern part is hosted in andesite, dacite, basalt tuff, and volcanic breccia. The host rock is altered to chlorite and epidote alterations. Ore bodies are forming tabular and layers-like and lens-like bodies. Sulfide ore bodies range from 1 to 17 m thick and extend up to 1.4 km long. Ore minerals occur in irregular masses, disseminations, stringers, and nests. Average ore grades in sulfide bodies are Cu 0.5–0.6%, Ag up to 60.0 g/t, and Au up to 0.4 g/t. The ore minerals are chalcopyrite, chalcocite, bornite, cuprite, covellite, and copper oxides. The Central part consists of sheets and lenses of andesite, basalt, dacite tuff, tuff, and tuff breccia, which strikes NW and extends for 0.5 km. The Northern and the Central ore zones range from 2 to 15 m thick and contain sulfide lenses or tabular ore bodies that range from 0.2 to 2.0 m thick and dip steeply. Other features are similar to the Northern part. The average grades are Cu 1.3% and Ag 5.0 g/t. The Pyrite part occurs 1.5 km E of the Central part and is hosted in dacite porphyry and tuff. Finely disseminated pyrite occurs in a zone of 100 × 250 m. Pyrite is intensively oxidized and limonite is well developed. Cu minerals are rare. The grades are Cu up to 0.1%, Au up to 0.4 g/t, and Ag 5.9 up to g/t.

The Borts Uul deposit has an overall average grade of 0.6–1.3% Cu with a cutoff grade of 0.1% Cu. The reserves in the Northern part are 28,200 tonnes of Cu with average grade of 1.0–1.5% Cu to a depth of 100 m.

7.3.3.4 Burgedtas Pb, Zn ± Cu, Ba, Ag, Au Volcanic Hosted Deposit

The Burgedtas deposit (Dorjgotov 1996) is hosted in Devonian felsic volcanic and sedimentary rock of altered sandstone and siltstone. The host rocks are hydrothermally altered into silicification, sericitization, and limonitization. The sulfide ore zones are up to 3.2 km long and n * 100 m wide in size. The main ore minerals are pyrite, sphalerite, chalcopyrite, galena, arsenopyrite, and iron oxides. The major gangue minerals are quartz, sericite, kaolinite, and chlorite. The average grade is Cu 0.1–1.0%, Zn 0.2%, and Pb 0.2–1%.

7.3.3.5 Bayan-Uul Zn-Pb-Cu-Ag-Cd Deposit

The Bayan-Uul deposit is located 130 km northeast of Choibalsan City in northeastern Mongolia. The Bayan-Uul deposit was discovered in 1975 and it was explored in 1986 and 1990. The deposit is located in a Late Mesozoic volcano-plutonic belt, which consists mainly of monzodiorite, alkaline granite, and dykes of fine-grained granite and diorite. Neoproterozoic gneiss, orthogneiss, and marbles are exposed as elongated small bodies bounded by faults in the central and marginal parts of the deposit district. Ore mineralization including silver mineralization occurred in the silica and sericite alteration zone along the Bayan-Uul fault zone, i.e., Mo, Au-Ag-Zn-Pb, sulfide-Au, and quartz-fluorite-Mo. Native silver and tetrahedrite are identified as minor ore minerals in the deposit. About 358 tonnes of Ag were estimated along with 23.5 thousand tonnes of Zn, 10.6 thousand tonnes of Pb, 204 tonnes of Cd, and 4.9 thousand tonnes of Cu in the Bayan-Uul deposit.

7.4 Concluding Remarks

Polymetallic lead and zinc deposits in Mongolia occur in calcic skarn (Zn-Pb-Fe), hydrothermal vein (Cu-Pb-Zn), and epithermal vein (Ag-Pb-Zn) types, related to the Mesozoic volcano-plutonic rocks. Representative mineable deposits are located in Eastern Mongolia and may become one of the main economic products of Mongolian exports. The Tumurtiin Ovoo and Ulaan deposits are under active mining, the Tsav, Dulaankhar Uul, and Uulbayan deposits are due to be mined, whereas the Mungun-Undur, Salkhit, Kharaat Uul, Bayandun, and Tugalgatain nuruu deposits are expected to have a future economic potential.

References

Badarch G, Dickson CW, Windley BF et al (2002) A new terrane subdivision for Mongolia: implications for the Phanerozoic crustal growth of Central Asia. J Asian Earth Sci 21:87–110

Badarch G, Tomurtogoo O (2001) Tectonostratigraphic Terranes of Mongolia. Gondwana Res 2:143–144. https://doi.org/10.1016/S1342-937X(05)70667-5

Baljinnyam V, Batjargal S, Batulzii D, Lhamsuren J (1989) Tsavskii vulkano-plutonicheskii kompleks granitoidov vostochnoi Mongolii. (The volcanic and plutonic rocks of Tsav in Eastern Mongolia In: Gerel O, Ivanov AN (eds) Problems Geology and Metallogeny Mongolia) 3:34–44

Batjargal S, Lkhamsuren J, Dorjgotov D (1997) Lead-zinc ore deposits in Mongolia. Mong Geosci 2:2–14

Bat-Ulzii D (1996) Zuun khoit Mongoliin latitiin magmatizmiiin petrologi ba geikhimi (Petrology and geochemistry of latitic magmatism in notheast Mongolia). Dissertation, Mongolia University of Science and Technology

Byamba J, Dejidmaa G (1999) Terranes of Mongol Altai area. Mong Geosci 14:16–19

Cox DP, Singer DA (eds) (1986) Mineral deposit models. US Geol Surv Bull 1693:379

Dahlkamp FJ (ed) (2009) Uranium deposits of the world. Springer, Heidelberg

Dejidmaa G, Dorjgotov D, Gerel O, Gotovsuren A (1999) Preliminary description of mineral deposit models (types) for Mongolia. MINDEP (5):99–165. https://pubs.usgs.gov/of/1999/of99-165/data/MINDEP/MINDEP5.PDF

Dejidmaa G, Dorjgotov D, Gerel O, Gotovsuren A (2005) Preliminary description of mineral deposit models (types) for Mongolia. In: Seltmann R, Gerel O, Kirwin DJ (eds) Geodynamics and metallogeny of Mongolia with a special emphasis on copper and gold deposits. IAGOD guidebook Series 11: CERCAMS/HNM London, pp 31–52

Dorjgotov D (1996) Dornod Mongoliin mezozoin vulkan-pluton busiin tsair, khar tugalganii khuderjilt (Govi-Ugtaal-Bayanjargalan, Dornodiin khudriin duurguudiin jisheen deer: hkudriin formats, burdel, markhaltiin ontslog. (Lead, zinc mineralization of the Eastern Mongolia volcano-plutonic bel (Gobi-Ugtaal-Bayanjargalan, Eastern Mongolian ore districts: ore formation, complex and distribution feature) Dissertation, Mongolian Technical University, Institute of Geology, Mongolian Academy of Sciences

Dorjgotov D (1998) Distribution map of polymetallic deposits and occurrences in Mongolia. Scale 1:3000000

Dorjgotov D (2002) Mongoliin tsair, khar tugalganii ordiin angilal (classification of lead, zinc deposits of Mongolia) problems of. Geology 5:94–105

Dorjgotov D (2009) Khartugalga, tsairiin ore deposits (Lead and zinc deposits in Mongolia). In: Byamba J, Lkamsuren J (eds) Mongolian geology and mineral resources, book VI, pp 85–108

Dorjgotov D, Batjargal Sh (2001) Zakanomernosti razmesheniya svintsogo-zinkivi orudneniya Vostochno-Mongoligkogo vulkano-plutonicheskogo poyasa (Distribution characteristics of lead-zinc mineralization of Eastern Mongolian volcano-plutonic belt) Problems of Geology Mongolia (3–4):418–429

Dorjgotov D, Murao S, Nakajima T, Batjargal S (1997) Genetic types of Mesozoic lead and zinc deposits in the Dornod Metallogenic zone (DMZ) of Mongolia. Mong Geosci 2:15–21

Gantumur H, Batulzii D, Wang LJ, Zhu HP (2005) Tsav: a shoshonite-hosted intermediate sulfidation epithermal Ag–Pb–Zn deposits, eastern Mongolia. In: Mao JW, Bierlein FB (eds) Mineral deposit research: meeting the global challenge. Proc 8th SGA Meeting Beijing, China, pp 389–392

Garamjav D, Zaryanov YuP, Dorjgotov D, Ganbaatar D, Gantumur Kh (1989) Rudnie formatsii n mineralinie tipi mestorojdenii medi, svintsa i tsinka Mongolii i ikh prognoznaya otsenka (Lead, zinc and copper mineralization types and ore formation of Mongolia and their evaluation on perceptivity. In: Geology and Mineral Resources of Mongolian Peoples' Republic, Book II, Nedra, Moscow), 37–141

Gerel O (1998) Phanerozoic felsic magmatism and related mineralization in Mongolia. Bull Geol Surv Jpn 49:239–248

Hedenquist JW, Arribas AJ, Gonzalez UE (2000) Exploration for epithermal gold deposits. Rev Econ Geol 13:245–277

Hu SK, Yan HQ, Ye M, Xiang WD (1998) Metallogenic focus-area and superlarge mineral deposits in the bordering zones between China, Russia and Mongolia. Sci China (Series D) 41:28–41

Jargalsaihan D, Kazmer M, Baras Z, Sanjaadorj D (1996) Guide to the geology and mineral resources of Mongolia, p 330

Jiang SH, Nie FJ, Zhang Y, Hu P (2004) The latest advances in the research of epithermal gold deposits. Earth Sci Front 11(2):401–411

Kovachev B, Strashimirov S, Bakirdjiev S (1990) Mineralinaya n temperaturnaya zonalinosti Undrtsaganskogo rudnogo polya. (Mineralization and temperature zonation of the Undurtsagan ore field. In: Geology and Mineral Resources of Mongolian Peoples' Republic, Book III, Nedra, Moscow) 123–130

Marinov NA, Khasin RA, Khurts Ch (1977) Marinov NA (ed). Geologiya Mongoliskoi Narodnoi Respubliki TOM III, Poleznie iskopaemie (Geology and Mineral Resources of Mongolian Peoples' Republic, Book Fig. III, Nedra, Moscow)

Murao S, Dorjgotov D, Tseden T (1998) K-Ar dating of granitoids and hydrothermal micas from the northern part of Kherlen Depression, Mongolia. Bull Geol Surv Jpn 49(6):249–255

Naftali LJ, Volkov B, Konsulov T (1984) Geology and mineral resource of MPR (book II), pp 102–109. Нафтали Л.Ж, Волков В., Консулов Ц., 1984. Raionov s geokhimihceskaya zonalinosti Undur-Tsaganskogo rudnogo polya po vtorichhim oreolom rassieniya (The geochemical zonation of the Undurtsagaan ore field by secondary geochemical halo. In: Geology and Mineral Resources of Mongolian Peoples' Republic, Book II, Nedra, Moscow) 102–109

Nie FJ, Jiang SH, Bai DM, Hou WR, Liu YF (2010) Types, temporal and spatial distribution of metallic ore deposits in southern Mongolia and its neighboring area. Acta Geosci Sin 31 (3):267–288

Nie FJ, Li QF, Liu CH, Ding CW (2015) Geology and origin of Ag–Pb–Zn deposits occurring in the Ulaan-Jiawula metallogenic province, Northeast Asia. J Asian Earth Sci 97(B):424–441

Nie FJ, Liu YF, Li QF, Jiang Z, Wu KR, Zhang XK, Ding CW, Cao Y (2014) Mesozoic Multiple Magmatic Activities and Ag–Pb–Zn and U Mineralization Occurring within the Wulaan-Jiawula Metallogenic Province, Conjunction Area of China, Mongolia and Russia. In; No.16 Internal Research Report of Institute of Mineral Resources, Chinese Academy of Geological Sciences 1–258

Nokleberg WJ (ed) (2010) Metallogenesis and Tectonics of Northeast Asia: U.S. Geol Surv Prof Paper 1765

Podlessky KV, Aksuk DK, Vlasova PF (1984) Rudonosnie skarni tsentralinoi I vostochnoi Mongolii (mineralized skarns of central and eastern Mongolia). Kovalenko VI, Jamsran M (eds) Endogenic Ore-bearing formations of Mongolia, transactions, Moscow, (38): 124–143

Podlessky KV, Vlasova DK, Kudrya PF (1988) Skarni I svyazannoe c nimi orudenenie Mongolii (Skarns and connected ores of Mongolia, Joint Soviet-Mongolian Scientific-Research Geological Expedition, Transactions, Moscow, (45):149)

Sillitoe R, Hedenquist JW (2003) Linkages between volcanotectonic settings, ore fluid compositions, and epithermal precious-metal deposits. Soc Econ Geol Geochem Soc Spec Publ 10:285–313

Tomurtogoo O (2012) Mongol ornii orogeny mujuudiin tektonik mujlalt (Classification of orogenic belt of Mongolia. Institute of Geology and Mineral resources of Mongolia. Scientific transaction (21):5–25)

Tomurtogoo O, Badarch G, Orolmaa D, Byamba J (1999) Terranes and accretionary tectonics of Mongolia. Mong Geosci (14):5–10

Ukhnaa G, Batkhishig B (2018) Tsair khar tugalga: Ordiin geologi, erel khaiguul (Lead and zinc deposits: Geology, prospecting and exploration. Ulaanbaatar) 338

Yakovlev VA (1977) Medi, svinets i zink (Copper, lead and zinc. In: Geology and Mineral Resources of Mongolian Peoples' Republic, Book III, Nedra, Moscow) 141–146

Yang ZL, Zhang DQ, Li JW (2009) Ore-forming types, metallogenic zoning and potential prospecting areas in southwestern sector of Deerbugan metallogenic belt. Miner Depos 28 (1):53–62

Zhai DG, Liu JJ, Wang JP, Yao MJ, Liu ZJ, Wu ZJ, Wu SH, Fu C, Wang SG, Li YX (2013) A study of stable isotope geochemistry of the Jiawula large Pb–Zn–Ag ore deposit, Inner Mongolia. Earth Sci Front 20(2):213–225

Chapter 8
Iron, Manganese, Chromium, Titanium and Vanadium Deposits

Bayaraa Batkhishig

8.1 Introduction

Iron and manganese deposits of Mongolia have been described by Marinov et al. (1977); Filippova and Vydrin (1977); Bakhteev and Chijova (1984); Podlessky et al. (1984); Podlessky et al. (1988); Tsend-Ayuush and Chebotarev (1989); Chebotarev et al. (1990); and Dejidmaa (1998); and those previous works have been summarized in Dejidmaa et al. (1999); United Nations Publication (1999); Ochirbat et al. (2008); and Ukhnaa and Baasan (2016). The iron and manganese deposit types of Mongolia discussed here are derived from Eckstrand (1984), Cox and Singer (1986), and Dejidmaa et al. (1999).

8.2 Ferrous Metal (Fe-Mn, ±Cr, Ti, V) Metallogeny of Mongolia

Based on geological studies of more than 100 deposits of iron occurrences found in Mongolia, Ministry of Geology and Mining, Mongolian People's Republic, and Ministry of Geology Geological Survey of USSR works have compiled map of the distribution of Fe-Mn mineralization in Mongolia (Marinov et al. 1977). In this research, they have established three metallogenic provinces and further subdivided into belts, ore zones, ore districts, and mineralized areas (Fig. 8.1; Table 8.1).

Ferrous metal (Fe, Mn, Ti, Cr, and V) deposits and occurrences of Mongolia are shown in a 1:3,200,000 scale distribution map compiled by Dejidmaa (Dejidmaa

B. Batkhishig (✉)
Department of Geology and Hydrogeology, School of Geology and Mining, Mongolian University of Science and Technology, Ulaanbaatar, Mongolia
e-mail: bkhishig@must.edu.mn

O. Gerel et al. (eds.), *Mineral Resources of Mongolia*, Modern Approaches in Solid Earth Sciences 19, https://doi.org/10.1007/978-981-15-5943-3_8

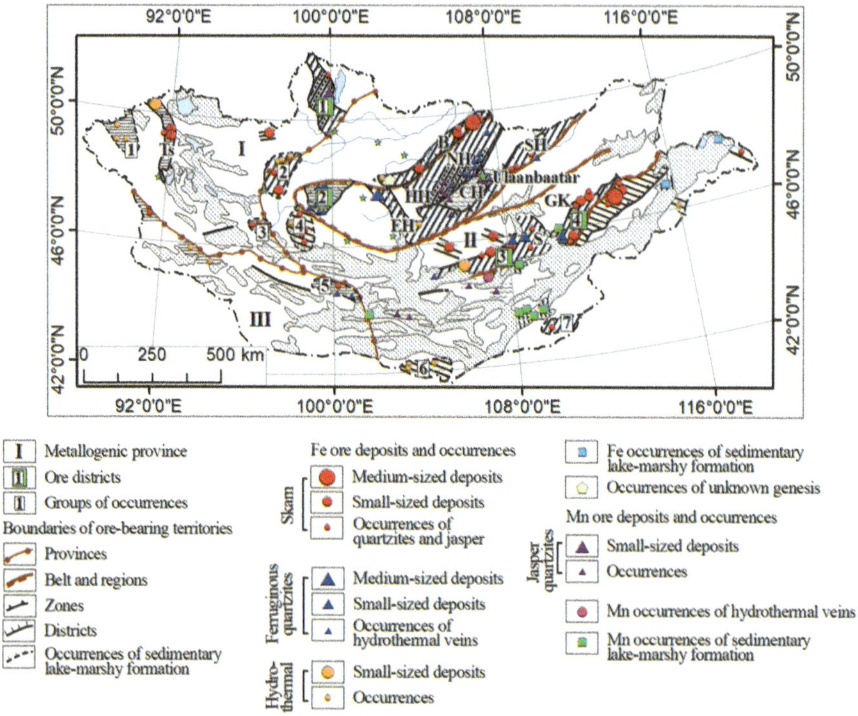

Fig. 8.1 Schematic metallogenic districts/provinces for iron and manganese deposits and occurrences of Mongolia (Marinov et al. 1977). Metallogenic provinces: I-Northern Mongolian, II-Eastern Mongolian, III-Southern Mongolian. Metallogenic belts: *Iron bearing* (GK, Gobi Kherlen) and *manganese-iron ore regions* (HH, Khangai-Khentii). Zones: *Iron ore bearing* (TS, Tsagaan Shiveet; B, Bayan Gol; EH, East Khangai; SH, South Khentii; S, Southern), *manganese-iron ore bearing* (NH, North Khentii; CH, Central Khentii; SH, South Khentii; S, Southern), *manganese-iron ore bearing* (NH, North Khentii; CH, Central Khentii). Ore districts: 1-Prikhubsgul, 2-West Khangai, 3-Middle Gobi, 4-South Kherlen. Groups of occurrences: 1-Deluun Sagsai, 2-Ider, 3-Zavkhan, 4-Buutsagaan, 5-Gobi Altai, 6-South Gobi, 7-Central Toto Shan

1998). Iron ore deposits and occurrences are mostly distributed in the eastern Mongolian Province and to a lesser extent in the northern and southern Mongolian Provinces.

8.2.1 *Eastern Mongolian Metallogenic Province*

The northern part of the Eastern Mongolian Metallogenic Province hosts most of the iron ore deposits and occurrences of the country. The deposits have been grouped southwards into the Bayan Gol iron ore zone, Khangai-Khentii manganese-iron ore

Table 8.1 Metallogeny classification of Mongolia (Filippova and Vydrin 1977)

Metallogenic provinces	Metallogenic belts	Ore zones	Ore districts	Mineralized areas	Representative deposits and occurrences
Northern Mongolian Province		Tsagaan Shiveet			Khairsan, Kharganat, Shiveen Gol, Tamir Gol
Eastern Mon-golian Province	Gobi-Kherlen		South Kherlen		Khiid, Chandmani Uul, Khar Undur, Munkh Khan
			Middle Gobi		Bayanjargalan, Ereen, Khongor, Durvuljin
				Khangai Khentii	Tamir Gol
		Bayan Gol			Bayan Gol, Tumur Tolgoi, Khudag Baishin, Ikh Bumbiin
		Delger			Tumurtiin Ovoo, Tsookhor Ovoo, Khol Khudag
Southern Mongolian Province					Tayan Nuur

area, Gobi-Kherlen iron metallogenic belt, and Delgeriin iron ore zone. The iron mineralization in this metallogenic province is of metamorphic-sedimentary, skarn (contact-metasomatic), hydrothermal, and sedimentary genetic types.

Gobi-Kherlen belt occupies the middle part of the Central Mongolian Fold System trending north-east, 50–100 km wide and 600 km long in size. Iron mineralization is related to the Neoproterozoic metamorphic and carbonaceous-terrigenous rocks. In lower part of the geological section, Neoproterozoic metamorphic rocks host ferriferous iron quartzite-type mineralization, while in upper part of the section, younger formations host skarn-type mineralization. The Gobi-Kherlen belt is subdivided by the Khar-Airag fault into the Middle Gobi and South Kherlen ore districts.

Middle Gobi ore district hosts 15 iron ore deposits, and occurrences of skarn, ferruginous quartzite (ironstone), and hydrothermal vein types proved to be of no economic interest due to a low iron content and high concentration of impurities. There is, however, a potential for discovering larger deposits of iron ore within this district due to a wide distribution of potentially iron ore-bearing formations of Precambrian age. Geologically, the district situated in Eastern Mongolian volcanic belt is characterized by Precambrian-Upper Cambrian metamorphic and carbonaceous schists and upper Paleozoic to Mesozoic volcanic rocks, intruded by lower Paleozoic and Permian granites. The main structure is represented by the Delger fault. Ore controlling structure of Middle Gobi zone is the Khar Airag fault. Airborne magnetic survey results show that the Middle Gobi ore district is controlled by a linear structure, with weak negative magnetic anomalies. However, along the main structure direction, weakly positive anomalies appear, whereas Khar Airag fault zone shows complex anomalies of (−300)–(+500–1000) nT. Most prominent deposits in this district are the Bayan-Dzargalant and Durvuljin skarn-type deposits and Ereen and Khongor occurrences.

South Kherlen ore district hosts 17 deposits and occurrences of metamorphic ferruginous quartzite and skarns. The deposits of both types in this district are of no economic value due to a low iron content and insignificant reserve. Khiid, Chandmana Uul, Khar Undur, and Munkh Khan deposits and occurrences are in the South Kherlen ore zone.

Bayan Gol iron ore zone lies on the right bank of Orkhon River within the boundaries of Selenge and Tuv provinces. Tectonically this zone is located in Bayan Gol island arc terrain. Basement rocks of this zone are characterized by terrigenous-carbonaceous rocks of Upper Precambrian-Lower Cambrian overlain by Devonian volcanic rocks and Lower-Middle Carboniferous marine sediments. They are intruded by mostly Cambrian and to a lesser extent by the Devonian and Mesozoic granitoids. Middle-Late Devonian granodiorite-granite intrusions form a chain, along the Bayan Gol fault; locally also small bodies of gabbro occur. Devonian subalkaline granite and syenites are co-magmatic with the Devonian volcanics, and Carboniferous-Early Mesozoic gabbro-diorite and granites are present. All of these rock formations are folded and structurally extending sub-parallel to the Bayan Gol deep fault zone. Aerial magnetic surveys indicate that the Bayan Gol ore zone is controlled by distinct positive magnetic anomalies of an n100–n1000 nT and forms large normal fault structure. The Bayan Gol fault zone is perpendicularly cut by later faults, and the mineralization mostly lies on the crosscutting junctions of the faults. According to the magnetic surveys, significant mineralization is associated with contact zone of Cambrian intrusions with carbonaceous-terrigeneous rocks. Next, in contact between the Devonian subalkaline intrusions and carbonaceous rocks, iron mineralization is present (Filippova and Vydrin 1977). Bayan Gol iron ore zone extends for 350 km being up to 60 km wide. This zone hosts 13 iron ore deposits and occurrences being concentrated in the Tumurtei, Tumurtolgoi, Khudagbaishin, and Ikhbumbiin area at distances of 50–75 km from each other. Almost all the deposits and occurrences of this zone belong to skarn iron ore type. The most prominent deposits of this zone are Bayan Gol and Tumur Tolgoi. Bayan Gol deposit is described in Sect. 7.3.1.

Delger iron ore zone lies in the territory of Sukhbaatar and East Gobi provinces and extends for 300 km in a northeast direction. Delger ore zone is located at the southeast end of the Mandal Ovoo island arc terrane, characterized by Lower Devonian iron ore bearing carbonate, carbonaceous-terrigeous rocks. Devonian formations are intruded by Carboniferous-Permian intrusions and overlain by Permian, Late Jurassic-Early Triassic volcanic formations. Magnetite-hematite skarn-type mineralization formed along the contact between Devonian carbonaceous formations and those intrusions. The skarn-type mineralization may have economic potential for further mineral exploration. This zone hosts two deposits and four occurrences of skarn-hematite-magnetite type, including the large Tumurtiin Ovoo deposit and Tsookhor Ovoo and Khol Khudag occurrences. Tumurtiin Ovoo deposit is discussed in Sect. 7.3.1.

Khangai-Khentii manganese-iron mineralized area of the Eastern Mongolian Metallogenic Province occupies the Khangai and Khentii uplifts and structurally coincides with the Mongol-Transbaikal fold system. The iron ore deposits and occurrences of the Khangai plateau occur in the eastern Khangai iron ore zone and western Khangai iron ore region. Khangai-Khentii manganese-iron mineralized area of the Eastern Mongolian Metallogenic Province occupies the Khangai and Khentii uplifts and structurally coincides with the Mongol-Transbaikal fold system. The iron ore mineralization of the Khangai-Khentii area has many common features in terms of host environment, geological structure and shape of deposits, and their genetic type and mineralogy. The host environment is mostly the Upper Precambrian-Lower Cambrian and Carboniferous volcano-terrigenous formations, and major depositional genetic types are ferruginous quartzites, jasperoidal quartzites, and skarns. The most prominent and typical deposit of this area is the Tamir Gol deposit which deserves further evaluation and exploration for establishing a steel industry in the country. The iron ore deposits and occurrences of the Khangai-Khentii area are less prominent in terms of geological and economic prospects.

The known parameters of iron deposits and occurrences of the Eastern Mongolian Metallogenic Province are summarized in Table 8.2.

8.2.2 Southern Mongolian Metallogenic Province

Earlier studies designate that the Southern Mongolian Metallogenic Province contains only few small iron ore occurrences of various genetic types, located along the southern slope of the Mongol Altai and Gobi-Altai ranges. The skarn magnetite and quartz-magnetite hydrothermal vein type mineralization in these localities has no economic potential for further mineral exploration (United Nations Publication 1999).

However, in this province Tayan Nuur mineralization was first discovered by D. Togtokh during a field survey in 1991; subsequently trenches were completed on the west mineralized body which uncovered skarn-type alteration with magnetite mineralization. In 2006, the first drilling exploration and operation of the Altai Khuder Tayannuur mine began. The Tayan Nuur deposit is described in Sect. 8.3.1.

8.2.3 Northern Mongolian Metallogenic Province

Iron ore mineralization of this province is represented by small occurrences and deposits of no economic interest, distributed in Mongol Altai, northern Mongolia, and southwestern Khangay ranges. Mongol Altai incorporates the Tsagaan Shiveet iron ore zone and Deluun Sagsai group of iron ore occurrences.

Table 8.2 Selected features of iron ore deposits and occurrences in the Eastern Mongolian Metallogenic Province

Name and number on shown on map	Locality	Mineralization	Major mineral and content	Accessory minerals and content	Genetic type	Reserves
Bagdain Arshaan deposit (17)	60 km E from Uliastai city	17 orebodies of 30–360 m long and 3–15 m thick	Magnetite 48.18–55.23% Fe_2O_3	Cassiterite 0.1–0.05% Manganese 0.7% Cyrtolite, hematite, thorite, azurite	Skarn	60 Mt. to a depth 150 m
Sartantiin Ulaa occurrence (13, 14, 15)	8–15 km NW from Telmin and Toson Tsengel soums	Several lenses of 30 m long and 2–10 m thick	Siderite 10% Fe_2O_3	n.a	Hydrothermal metasomatic or sedimentary	n.a.
Zagastyin occurrence (16)	14 km from Zagastyin soum	n.a.	Magnetite	n.a.	Skarn	n.a.
Zavkhan Gol (18), Shar Bulag (19) occurrences	11 km E from Gobi-Altai, and 22 km W, SW from Tsagaan Olom soum, Zavkhan	Three lens-like bodies of 100–200 m extent and 3–8 m up to 100 m wide	Magnetite	Malachite, garnet	Skarn	5000–7000 tonnes
Buutsagaan occurrence (24)	10 km from Buu Tsagaan soum	14 magnetite bodies of 0.3–0.45 m thick and 6–300 m long	Magnetite up to 48.7% Fe_2O_3	Chalcopyrite, malachite, azurite 0.01–0.83%	Skarn	n.a.
Asgat Uuliin occurrence (20)	32 km E, NE of Bayan Bulag soum	One vein of 1 m thick	Magnetite 56.38% Fe_2O_3	n.a.	n.a.	n.a.
Tumurtei deposit (37)	95 km E, NE of Darkhan city	Three orebodies concordant with host rocks. 150–500 m long and 50–100 m thick	Magnetite 59–66.78% Fe_2O_3	Martite, hematite, limonite sulfur 0.06–0.28%, phosphate 0.06–0.08%	Skarn	20 Mt. to a depth of 150 m
Khudag Baishin deposit (27)	3 km S of Mt. Khudag Baishin	15 lenses of 2–6 m thick and 10–20 m long	Magnetite-hematite up to 66.7% Fe_2O_3	Sulfur, phosphate 0.04%	Skarn	60,000 tonnes to a depth of 10 m

Tsakhtsag Uul occurrence (28)	3.5 km SW from Khudag Baishin deposit	Eight lenses of 20–50 m long and 0.5–2 m thick	Magnetite up to 68.5%	n.a.	Skarn	20,000 tonnes
Ikh Bumbyin occurrence (25)	11 km W of Ulziit soum	n.a.	Magnetite up to 62.44% Fe_2O_3	Manganese 0.3%, phosphate 0.006%	Skarn	n.a.
Ulziit occurrence (26)	Right bank of Orkhon River, 4 km NE of Ulziit soum	One layer of 1.5 m thick and from 150 to 2000 m long	Magnetite 48.38–8.2% Fe_2O_3	Zinc 0.13%, sulfur 0.01%, phosphate 0.12%	n.a.	n.a.
Ungut deposit (46)	10 km NW of bayan Undur soum	Small lenses of hematite controlled by granite-aplite dyke	Hematite	n.a.	Skarn	15,000–20,000 tonnes
Zag deposit (39)	13.5 km SE of Khangai soum	Elongated lens of 800 m long and 15–20 m thick in jasper sequence	Magnetite-hematite 36.18–42.2% Fe_2O_3	Psilomelane, pyrolusite, malachite, azurite, sulfur phosphate	Metasomatic	21–25 Mt. of ore by possible reserve category
Muhetyin occurrence (43)	23 km W, SW of Jargalant soum, Bayankhongor	Inliers of hematite 0.35–0.5 m thick and up to 75 m extent in jasper sequence	Hematite 55–60% Fe_2O_3	Manganese 9.54–16.86%, phosphate 0.79–0.89%	Metasomatic	n.a.
Navchatiin Gol (64)	30 km SE of Khentii soum	Two orebodies of jasperous quartzites of 2500 and 1800 m long and 20 and 30 m thick	Hematite 32–33% Fe_2O_3	Minor	Metasomatic	23.2 Mt. by indicated and 35 Mt. by possible reserve categories
Ar-Nuruut Gol occurrence (63)	In river valley of Ar-Nuruut Gol	Lens of magnetite of 30 m long and 12 m thick	Magnetite up to 60% Fe_2O_3	High	Skarn	n.a.
Bayan-Jargalant deposit (72)	10 km NE of Bayan-Jargalant soum	Lens of 20 km^2 in size	Hematite 63.32% Fe_2O_3	Magnetite, limonite sulfur 0.62%, phosphate 0.06%	Skarn	2.52 Mt. by possible reserve categories
Nukhet deposit (69)	20 km NE of Gurvansaikhan soum	Eight orebodies of 50–70 m^2 in size	Magnetite 57.24% Fe_2O_3	Hematite, quartz	Skarn	20,000 tonnes by possible

(continued)

Table 8.2 (continued)

Name and number on shown on map	Locality	Mineralization	Major mineral and content	Accessory minerals and content	Genetic type	Reserves
						reserve categories
Tegiin Uul deposit (67)	n.a.	Several quartz-hematite veins and irregular shape bodies of 7 m thick and up to 30 m long each. Total extent of ore-bearing zone is 2 km	Hematite	Quartz	Hydrothermal veins	20,000–30,000 tonnes
Even deposit (73)	16 km SW of even soum	Six orebodies of 750–950 m extent, 150–280 m width, and 100 m depth	Magnetite and hematite-magnetite quartzite 33–39% Fe_2O_3	Hydrohematite, limonite, epodite, chlorite, quartz, manganese 1.09%, phosphate 0.02–1.18%	Metamorphic-sedimentary	158.3 Mt. by indicated reserve categories
Khongor deposit (74)	10 km SE of Khongor soum	17 orebodies of lens or layer shape of 57 m average thickness and 110 m maximum extent. Total size of deposit 2 km^2	Magnetite-hematite, quartzites, 35–52% Fe_2O_3	Quartz sulfur 0.5%, phosphate 0.07–0.35%	Metamorphic-sedimentary	13 Mt. ore
Durvuljin deposit (76)	55 km N from Khar-Airag railway station, Khentii	Three elongated orebodies of 40–85 m long and 15–35 m thick and one orebody of 380 m long 40–100 m thick	Magnetite 40–60%, Fe_2O_3 average 42.5% Fe_2O_3	Zinc 0.7–2.5%, cobalt, nickel, titanium, vanadium, chromite, germanium	n.a.	n.a.

Khiid deposit (79)	48 km west of Delgiin Khiid	Six concordant layers of hematite-magnetite lenses grouped in zone of 1500 m extent and 100–150 m thick	Hematite 32–35% Fe_2O_3	Silver 2–5 g/t	Ferruginous quartzite	420,000 tonnes by indicated reserve categories
Chandmani-Uul	15 km SW of Delgiin Khiid	Small hematite-magnetite lenses grouped in zone of 1500 m extent and 100–150 m	Hematite/magnetite 52.32% Fe_2O_3	Copper 0.06%, zinc 0.19–0.54%, phosphate 4.16%	Skarn	
Khar-Undur deposit (86)	43 km SW of Mt. Munkh-Khanui	Four orebodies of 50–140 m long and 30–60 m thick	Magnetite 58.7% Fe_2O_3	Manganese 2.35%, sulfur 0.35%	Skarn	
Oyut-Ovoo deposit (66)	5–7 km of Tsagaan Ovoo soum, middle Gobi	Orebody of irregular shape and series of quartz-magnetite veins of 600 m extent and 15–120 m thick	Magnetite 55.8% Fe_2O_3	Malachite, azurite 5.1%	Skarn	Possible reserves to 100 m depth 10 Mt

Tsagaan Shiveet iron ore zone is associated with a N-S-trending fault of the same name separating Late and Early Caledonian structures of the Mongol Altai and Northern Mongolian fold structures. The zone extends for 300 km being 15–20 km wide. Tectonically the zone is dominantly belonging to Lake Island arc terrane; however, to the NW a small section of the zone is located in Tsagaan Shivee terrain, which is characterized by a metamorphic rock. The region is characterized by Ediacaran-Lower Cambrian volcanogenic-sedimentary rocks; Ordovician, Silurian carbonaceous-terrigenous rocks; Lower-Middle Devonian volcanogenic rocks; and Carboniferous and Jurassic molasse formations. Iron mineralization is related to Ediacaran-Lower Cambrian and Devonian rocks. The Tsagaan Shiveet zone is mainly controlled by N- and NW-trending Tsagaan Shiveet fault zone and locally intersected by vertical shear zones. Small intrusive bodies of Devonian granite and Late Devonian-Early Carboniferous granites and gabbro are emplaced in the cross-cutting junctions of the faults and shear zones. Several contact metasomatic and hydrothermal metasomatic iron mineralized deposits Khairsan, Kharganat, Shiveen Gol, and Tamir Gol and other occurrences were observed in the zone.

8.3 Ferrous Metal (Fe, Mn) Deposit Types of Mongolia

The iron and manganese deposit types of Mongolia are discussed by Bakhteev and Chijova (1984); Eckstrand (1984); Cox and Singer (1986); and Dejidmaa et al. (1999).

Earlier, Baktheev and Chijova did numerical modeling based on geological information of 250 deposits and occurrences (Bakhteev and Chijova 1984). They have classified six iron-manganese ore formations (deposit types) within the 11 metallogenic zones (Fig. 8.2).

Ferrous metal deposits and occurrences of Mongolia were shown in 1:3,200,000 scale map of Fe, Mn, Ti, Cr, and V deposits and occurrences (Dejidmaa 1998). The ferrous deposit types in this map consider that both descriptive and genetic information are systematically arranged to describe the essential properties of class of mineral deposits. The classification of ferrous deposits was derived principally from the mineral deposit types of Eckstrand (1984) and Cox and Singer (1986) and data of the Mongolian authors. Iron ore deposits are formed in magmatic, metamorphic, and sedimentary rocks, during different metallogenic times. They are eight main deposit types, based on their host rock lithology, origin, mineral assemblages, and orebody morphology.

Deposits related to marine volcanogenic rocks:

Volcanogenic Fe (±Mn, Au, Ag)

Stratiform deposits in marine fine-grained clastic, siliceous sedimentary rocks:

Volcanogenic—sedimentary Fe and Mn

Deposits in continental slope and shelf carbonate, chemical sedimentary rocks:

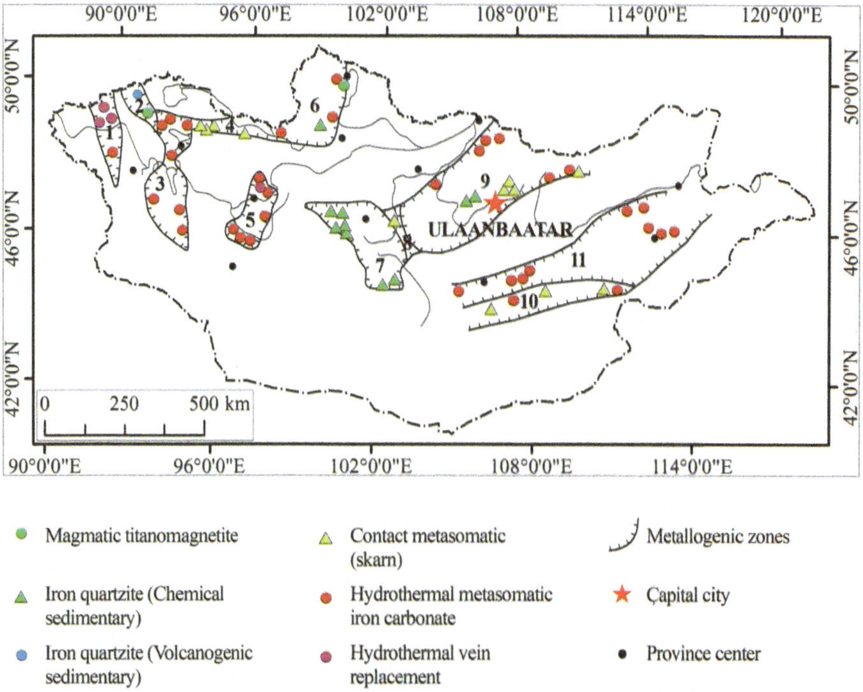

Fig. 8.2 Metallogenic zones and deposit types of Mongolia, by Bakhteev and Chijova 1984. Metallogenic zones: 1- Deluun Yustid, 2- Tsagaan Shuvuut, 3- Lake, 4- Tes River, 5- Bogdiin Rashaan, 6- Khuvsgul, 7- Khangai, 8- Khangai-Khentii, 9- Khentii, 10- Middle Gobi, 11- Kherlen

Chemical—sedimentary Fe, Mn, and Fe-Ma
Chemical—sedimentary V

Deposits related to subaerial volcanics and volcanic-sedimentary rocks:

Volcanogenic sedimentary Fe
Volcanogenic sedimentary Mn

Deposits related to continental basin sedimentary rocks:

Chemical—sedimentary Fe
Chemical—sedimentary Mn

Deposits related to calk-alkaline and alkaline granitic intrusions:

Veins and replacements
Hydrothermal vein and stockwork (Au-magnetite, tourmaline-quartz)

Skarns:

Skarn Fe, Fe-Cu, Fe-Cu-Au, Fe-Zn, and Fe-Sn

Porphyry intrusion-related deposits:

Breccia pipe (Au-magnetite, sulfide-quartz)

Deposits related to mafic and ultramafic rocks:

Podiform Cr
Zoned mafic-ultramafic-related Fe-Ti and apatite-Fe-Ti

Deposits related to regionally metamorphosed rocks:

Banded iron formation
Ilmenite-bearing metamorphic rocks

Deposits related to carbonatite:

Magnetite-rare earth element-carbonatite

Deposits related to surficial process:

Placer ilmenite

Economically important and widespread deposit types of Mongolia and their representative deposits are described in the following sections.

8.3.1 Skarn Deposits

Skarn-type deposits can be subdivided by Fe-Zn Skarn and Fe and Fe-sulfide skarn deposit types (Dejidmaa et al. 1999).

The skarn-type Fe-Zn deposit occurs mainly in uplifted blocks of Late Paleozoic and Early Mesozoic rocks located on the periphery of the Late Mesozoic Eastern Mongolian intercontinental volcanic belt (Bakhteev and Chijova 1984; Podlessky et al. 1984, 1988). The uplifted blocks (horsts) are surrounded by Late Mesozoic volcanic grabens with Permian and Triassic granitoids that intrude Proterozoic and Paleozoic metamorphosed sedimentary rocks. The skarns mainly occur along contacts and faults between Proterozoic and Devonian limestone and shale sequences and Triassic subalkaline granite and leucogranites. Skarns usually form lens-like (Bayandun, Tumurtei, Kharaat, Salkhit deposits, and other occurrences) or layer-like (Tumurtiin Ovoo deposit) bodies. Their size is ranging from tens to hundreds of meters in thickness and several hundred meters along the strike. The main skarn minerals are hedenbergite, andradite, and grossularite associated with abundant sphalerite. Zonation is typically developed with zones of epidote-feldspar, epidote-andradite, andradite-magnetite, andradite-pyroxene-magnetite, and pyroxene-magnetite. Typical retrograde minerals are actinolite, quartz, calcite, and chlorite. Main ore minerals are sphalerite and magnetite, and chalcopyrite, hematite, bismuthinite, molybdenite, pyrite, and galena may also occur. The deposits contain Fe-Zn mineralization which is irregular in distribution and occurs mostly in garnet and garnet-pyroxene skarn and minor amount of Pb and Zn. Pb/Zn/Cu ratios are 0.2/

4.5/0.1. Deposits have four stages of mineralization: garnet-pyroxene skarn, andradite-magnetite external skarn, sulfide, and quartz-carbonate.

Representative magnetite skarn deposits are Tumurtiin Ovoo, Tayan Nuur, Kharganat, Shiveengol, Bargilt, Khudag Baishin, Ikh Bumbiin, Ar Khuruut, Bayanjargalant, Baruuntsogt, Bayandun, Tumurtei, Kharaat, and Salkhit.

8.3.1.1 Tumurtiin Ovoo Deposit

The Tumurtiin Ovoo deposit is located in Sukhbaatar province, 18 km west from Baruu-Urt city. Deposit is hosted in Devonian sedimentary rocks enclosed in a polyphase granite batholith of Triassic-Jurassic age (Fig. 8.3). The large skarn orebody consists of two parts with a lower part rich in quartz and poor in manganese and zinc and an upper part with the opposite relationship. Magnetite occurs in the whole skarn body. In the lower part, magnetite forms massive lenses of up to 20 m in thickness. Jacobsite is restricted to the upper part. Sphalerite forms locally a low-grade mineralization at the top of the lower part, related to quartz-rich layers. Sphalerite is concentrated in three stratiform zones in the upper part. Sphalerite coexisting with jacobsite contains more Mn than sphalerite coexisting with magnetite, indicating that the minerals crystallized in chemical equilibrium. The zinc mineralization does not extend into the sedimentary host rocks and the surrounding granite. The strata-bound position of the zinc ores, the stratified structures of the skarn, and the conformable alternation of marble and skarn beds indicate

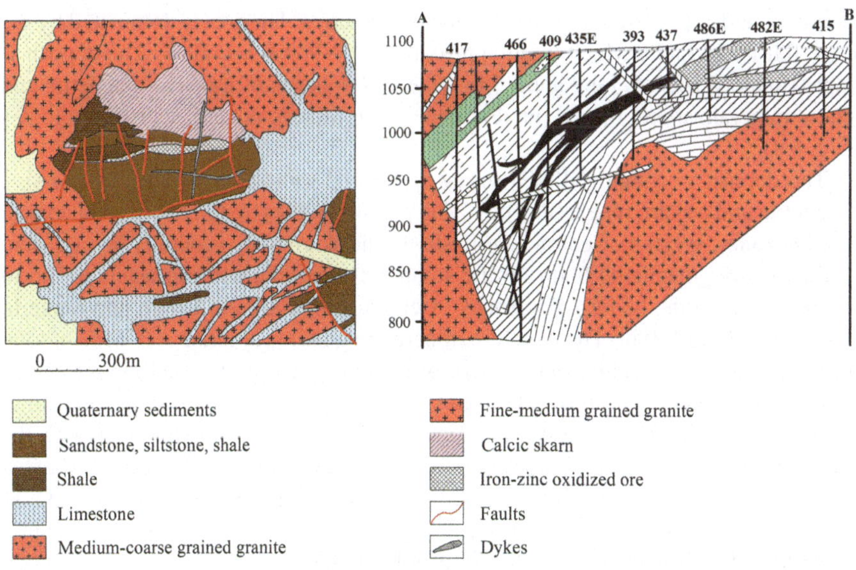

Fig. 8.3 Geological map (left) and cross section (right) of the Tumurtiin Ovoo deposit

sedimentary deposition for the formation of skarn and ore. Therefore, the origin of the deposit is classified as hydrothermal sedimentary (Gottesmann and Kampe 2013).

The deposit iron content is 63%, and total resources of iron ore are estimated at 217 Mt.

8.3.1.2 Tayan Nuur Iron Deposit

The Tayan Nuur iron deposit is located in Tseel soun, 162 km south of the Altai city, Gobi-Altai Province. Tayan Nuur mine produces ~60% Fe grade iron ore concentrates and exports to steel mills in China. The deposit is located at the southeastern end of the Mongolian Altai Fold Belt. The series of metamorphic and deformation events have occurred in the region. Ordovician to Silurian stratified sediments occupy some 80% of the region and have been classified into three concordant members; these include:

Ordovician-Lower Silurian unclassified sediment occurs predominately in an east-west-trending synclinal structure with strata typically dipping toward the southwest at 50° to 70. The Ordovician-Silurian metamorphosed sediment was subsequently intruded by upper Silurian gneissic granite, granite diorite, diorite, and gabbro diorite of the Khovd complex and lower Devonian pegmatite plagiogranite and associated dykes. Sediments show evidence of successive alteration, such as chloritization and epidotization. In addition there is a regional metamorphic overprint. Secondary metamorphic overprinting includes silicification and sulfidic and argillic alteration. Quartz veining and fracturing are associated with hydrothermal alteration and copper mineralization in the upper series. The area is dominated by the large Ikh Bogd faults which trend approximately NW/SE and E-W. The deposit is approximately 3600 m long and 75 m wide and consists of massive magnetite and skarn mineralization, hosted predominately in limestone and schist lithologies (Fig. 8.4).

The deposit is comprised of a series of parallel mineralized bodies: a massive magnetite and a skarn magnetite body. The parallel bodies trend NW/SE and consist of three zones, Western, Central, and Eastern. The skarn body generally lies adjacent to the massive magnetite body on the south/western side, with occasionally up to 20 m separating the bodies. The zones extend up to 1200 m in length and range in width from 31 to 160 m. The main ore mineral magnetite comprises 60–65% of total Fe with minerals including pyrite, pyrrhotite, and chalcopyrite combining to comprise the remainder. The host lithologies are amphibole-pyroxene schist and limestone unit with mineralization a product of metasomatic processes related to the Khovd intrusion of the Tseel limestone formations. Statement of Mineral Resources on May 2011 reported that Inferred, Indicated and Measured Resources of 134.31 Mt. grading 38.50% TFE (total iron) and 28.30% mFe (metallic iron) containing 51.68 Mt. of TFE metal and 38.02 Mt. of mFe metal.

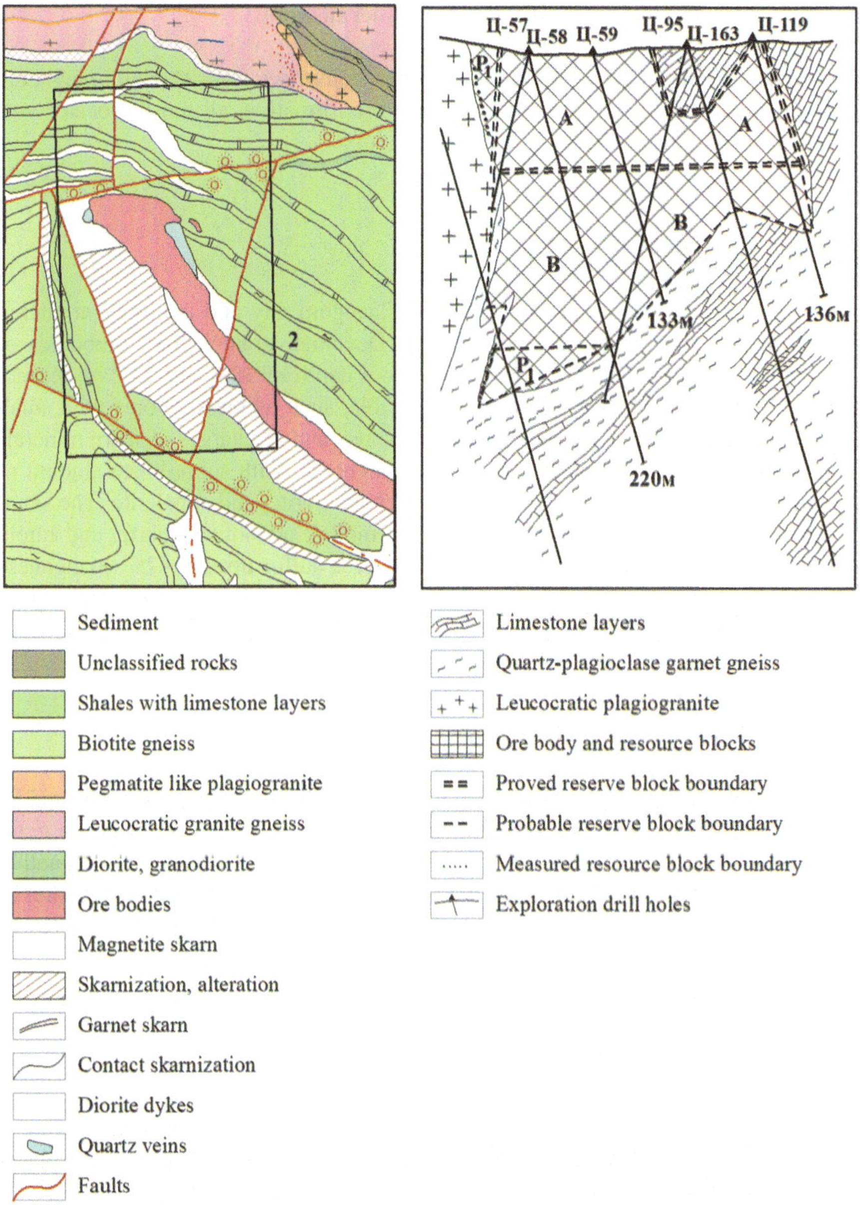

Fig. 8.4 Geological map of Tayan Nuur iron ore deposit (left) and cross section of the deposit (right) (Chuluunbat and Byambaa 2010)

8.3.1.3 Kharganat Group of Iron Ore Mineralization

Kharganat group of iron ore mineralization comprises two hematite-magnetite deposits and three occurrences of skarn iron ore formation, all mostly located east of the Tsagaan Shiveet fault zone.

8.3.1.4 Kharganat Deposit

Kharganat deposit occurs in 45 km SE of Ulaangom city and is hosted in altered limestone and epidote-garnet endoskarn of Devonian granite. Two orebodies of 430 and 750 m in length have been traced to a depth of 38.6 and 27.2 m, respectively. Three types of iron ores are recognizable, such as massive ores with small impregnation of skarn minerals, disseminated ores, and discharges of skarn minerals and massive ores. The main ore mineral is magnetite, with subordinate amount of hematite, martite and rare pyrite, pyrrhotite, chalcopyrite, and sphalerite. The skarn minerals include epidote, garnet, chlorite, tremolite, actinolite, quartz, and rarely apatite. The iron content varies from 48% to 67% sulfur up to 0.12%, manganese 0.07–0.38%, phosphorus 0.75%, and vanadium up to 0.04%. Total resources of iron ore are estimated at 3.9 Mt.

8.3.1.5 Shiveen Gol Iron Ore Deposit

Shiveen Gol iron ore deposit is located 67 km S of Ulaangom city, being hosted by Lower-Middle Devonian volcano-sedimentary formation near a contact with granitoid intrusions. This contact-metasomatic zone contains 15 lenses of magnetite-hematite ores of 0.8–1.5 m thick and 10–25 m long and three elongated orebodies of up to 5 m thick and 100 m long. The iron content varies from 32% to 59.6%, and possible iron ore reserves are estimated at 3–3.5 Mt.

8.3.1.6 Fe and Fe-Sulfide Skarn Deposit Type

Iron and iron sulfide skarn deposit type consists of magnetite and magnetite-hematite that are associated with calc-alkaline granitic rocks and contact metasomatic replacements (Podlessky et al. 1984, 1988). The skarns generally occur along the contacts between the early Paleozoic granodiorite and granite intrusions and host dolomite marble and rarely along contacts between dolomite and gneiss. There are two types of skarns which can be classified as 1- magnesium-silicate skarn and 2- calcium-silicate skarn. The main skarn minerals are fassaite, spinel, forsterite, diopside, phlogopite, and serpentine.

Magnetite-bearing magnesium-silicate skarn has following idealized zonations (Podlessky et al. 1984, 1988): (1) granitoid rock; (2) pyroxene-plagioclase rock;

(3) spinel-pyroxene skarn; (4) pyroxene-forsterite or spinel-forsterite skarn; (5) forsterite calciphyre; and (6) dolomitic marble. Magnetite occurs mostly in the periphery of the skarn. Main ore mineral magnetite forms massive, coarse to fine disseminations, vein like, lens like, and irregularly shaped bodies with an insignificant amount of sulfide minerals that are classified as Fe skarn. Most deposits and occurrences of the Fe skarn deposit occur in association with Proterozoic metamorphic basement rocks that are intruded by Early to Late Paleozoic calc-alkaline to subalkaline granitic rocks.

Magnetite-bearing calcium-silicate skarn is genetically related to post-magmatic (retrograde) granite rocks that intrude marble, limestone, dolomite, and calcareous-clastic sedimentary rocks. Podlessky et al. (1984, 1988) describe the following zonation: (1) granite; (2) potassium feldspar or amphibole-epidote; (3) epidote monomineralic zone; (4) epidote-garnet endoskarn; (5) garnet exoskarn; and (6) carbonate rock. The main ore mineral assemblages are magnetite, magnetite-hematite, and quartz-hematite. Ore mineralization is mainly related to garnet and rarely in epidote and epidote-garnet skarn. Ore mineral formation occurs in a characteristic sequence consisting of garnet, amphibole, chlorite, calcite, and quartz with anomalous Zn, Cu, and rare Sn.

Complex sulfide-magnetite minerals may occur in the deep levels of this type of deposits. Representative deposits of this type are Bayan Gol and Tumur Tolgoi. For example, the Tumur Tolgoi deposit of the Bayan Gol iron ore zone contains sulfide-magnetite ore at depth. Sulfide minerals are mostly insignificant in magnesium-silicate skarn but are significant in calcium-silicate skarn. The Fe-sulfide skarn type contains significant sulfides such as pyrrhotite, pyrite, chalcopyrite, sphalerite, and others.

8.3.1.7 Bayan Gol Iron Ore Deposit

The Bayan Gol iron ore deposit was located by aerial magmatic survey 75 km N-E from Darkhan city in the area of the Bayan Gol fault zone. Two stages of structural units can be classified in the Bayan Gol deposit area. Upper structural stage is represented by Neoproterozoic Darkhan Formation which is affected by NE- and E-W-trending linear folding. The lower structural stage is characterized by green schist formation rocks, which was subjected to NW-trending complex, steep folding. Main structural control is the Bayan Gol fault, and along this fault the Permian Selenge intrusive complex was intruded (Fig. 8.5). Due to the tectonic activities, deformations, metamorphic and hydrothermal-metasomatic alterations, and pre- and post-mineralization fracture zones are developed. Pre-mineralization fracture zone is mostly characterized by NE 45–50° striking and NW 75–85° dipping faults and fractures. The main mineralization is mostly related to the NE 14–20° trending, steeply dipping normal faults. This tectonic structural characteristic is same for almost all iron deposits and occurrences in the Bayan Gol fault zone (Ukhnaa and Baasan 2016).

Fig. 8.5 Geological map of Bayan Gol iron ore deposit. Map compilers: Y.P. Ryzhkov and V.N. Philatov in Marinov et al. (1977)

Mineral exploration confirmed the existence of 15 orebodies with lens or layer shapes of variable (150–2800 m) extent and (10–80 m) thickness. The orebodies have been traced up to a depth of 200 m with sub-latitudinal strike and from 75–85° southern dip. The mineralogical composition, structure, and size of orebodies vary within three distinguished areas of the deposit; however, in most cases, magnetite mineralization is being controlled by diorite or dioritic porphyrite dikes. The iron content is high, varying from 40% to 50% in southwest of the deposit through a range of 15–50% in its central part to a range of 50–52% in the western part of the deposit. The average iron ore content is 51%. The mineralization also contains from 0.09% to 1.5% sulfur and from 0.016% to 0.14% phosphate. Total resources of iron ore estimated to a depth of 200 m are 100 Mt.

8.3.1.8 Tumur Tolgoi Iron Ore Deposit

The Tumur Tolgoi iron ore deposit occurs in the NE part of the Bayan Gol zone, from 25 to 30 km south from Darkhan city and 45 km north from Baruun Kharaa soum. The deposit consists of seven orebodies within a NE-trending arcuate zone of 1000 m long controlled by the major fault. Four major orebodies from 6 to 150 m thick extend for 50 m and have been traced to a depth up to 220 m. The biggest western orebody contains more than 70% of iron ore. The orebody is overlain by

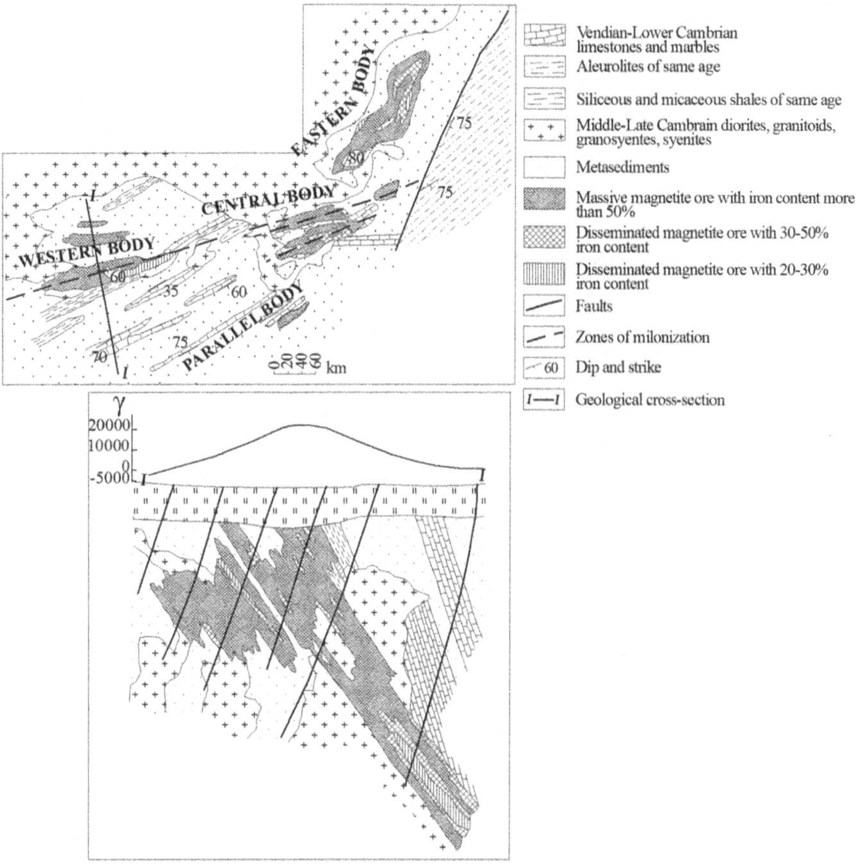

Fig. 8.6 Geological map of Tumur Tolgoi iron ore deposit. Map compilers: V.V. Taranin and K.A. Agamelyan in Marinov et al. (1977)

Quaternary sediments from 30 to 50 m thick (Fig. 8.6). The geological structure and shape of all orebodies of the deposit vary to a great extent; however mineralogical and textural parameters are similar. The massive magnetite ores are prevalent; subordinate minerals include martite, hematite, pyrrhotite, and pyrite. The highest iron content from 50 to 65% was assayed in massive ores and lesser contents from 20% to 50% in impregnated ores. The average iron content in the deposit varies from 22.37% to 58.68%, sulfur content from 0.08% to 3.16%, and phosphate content from 0.04% to 0.01%. Indicated reserves of 25.05 Mt. of iron ore are amenable for open pit mining to a level of +800 m to a depth up to 200 m.

8.3.2 *Metamorphic Deposits*

Metamorphosed iron and manganese deposit types are subdivided into volcanogenic sedimentary Fe-Mn; ironstone (clastic magnetite in sedimentary rocks); and chemical-sedimentary iron or ironstone (Fe-Fe, Mn-Mn) (Dejidmaa et al. 1999).

8.3.2.1 Volcanogenic-Sedimentary Fe-Mn Deposit Type

The volcanogenic-sedimentary deposit type consists of stratiform massive and disseminated magnetite-hematite ore that occurs in sheet-like, lens-like tabular bodies that are interbedded with marine volcanic-sedimentary and sedimentary rocks. Filippova and Vydrin (1977) named this deposit type jasper-quartzite Fe-Mn Formation. The host rocks are mostly chert, quartzite, sandstone, siltstone, shale, siliceous rocks, jasper, and jasperoid. Ore minerals of the deposit type are complex. According to mineralogy and Fe and Mn grade, this type is divided into three subtypes: (1) volcanogenic sedimentary Fe; (2) volcanogenic-sedimentary Fe-Mn; and (3) volcanogenic-sedimentary Mn. The ore minerals are magnetite, hematite, pyrolusite, hausmannite, and braunite.

The depositional environment consists of marine epicratonal and intracratonic basins, back arc basins, and volcanic terrigenous units of ophiolite complexes. Deposits and occurrences of this deposit type are widespread in Mongolia. They are located in the Northern Mongolian Metallogenic Province: Khangai metallogenic belt related to the Devonian to Carboniferous volcanogenic-terrigenous rocks of Khangai Group, Govi-Altai metallogenic belt in relation to Early Paleozoic Tugrug volcanogenic Formation and Ukhiinnuruu volcanic-terrigenous Formations, Urgamal metallogenic belt related to Neoproterozoic siliceous carbonate Shuvuun Formation, and other metallogenic belts.

8.3.2.2 Chemical-Sedimentary Fe or Ironstone (Fe-Fe, Mn-Mn) Deposit

Generally this deposit type is hosted in sedimentary and clastic carbonate rocks, including limestone and dolomite. The ore-bearing layers consist of sedimentary chert, quartzite, quartz-sericite-chlorite schist, and clastic carbonate. Filippova and Vydrin (1977) referred to this deposit type as iron-quartzite formation. Its ore forms sheet-like and lens-like layers containing massive and disseminated Fe and Mn oxide and carbonate minerals. Based on mineralogy, and Fe and Mn grade, the deposit type is divided into Fe, Fe-Mn, and Mn subtypes. The main ore minerals are magnetite, hematite, siderite, pyrolusite, hausmannite, braunite, and rhodochrosite.

In Mongolia, this type of deposit exists in early Paleozoic chemical-sedimentary carbonate rocks formed in basins overlying Precambrian granitic and metamorphic complexes and in Late Cretaceous and Neogenic intracratonic lake basins. Main example is the Khovsgol carbonate basin located in northern Mongolia where Fe,

Fe-Mn, and Mn occurrences are associated with sedimentary phosphate, allunite, and vanadium deposits.

Fe subtype occurrences are located in southern and southeastern Mongolia and are associated with sedimentary gypsum, alunite, and rare barite and bauxite deposits and occurrences.

8.3.2.3 Khairsan Group of Deposit and Occurrences

The Khairsan group of deposit and occurrences includes the Khairsan iron ore deposit and Khavtsal small occurrence of hydrothermal-metasomatic genesis being hosted by the Middle Devonian carbonate-terrigenous rocks. The Khairsan iron ore deposit occurs from 6 to 7 km NE of Uureg Nuur lake on the eastern slope of a mountain of the same name. The siderite orebody is hosted by the Middle Devonian limestone bed of 300 m thick and 2.5 km long. The orebody is intensively dislocated and intruded by stocks of the Devonian gabbro. The associated minerals are malachite and azurite. The iron content by visual estimate does not exceed 15%. The deposit is of no economic interest.

8.3.2.4 Tamir Gol Iron Ore Deposit

The Tamir Gol iron ore deposit is located 57 km east-northeast from Tsetserleg city and represented by ferruginous quartzites being hosted by metamorphic schists and quartzites of Upper Precambrian-Lower Cambrian. The deposit contains 11 outcrops of magnetite-hematite orebodies intercalated by ferruginous quartzites. The thickness of orebodies varies from 40 to 68 m, and the iron content is 48%, manganese from 1.18% to 5.86%, and phosphate from 0.07% to 0.4% with traces of sulfur. Accessory minerals are pyrite and sulfides of copper and lead. Possible iron ore reserves are estimated from 150 to 200 Mt.

8.3.2.5 Deluun Sagsai Group of Iron Occurrence

The Deluun Sagsai group of iron occurrence is located in the north of Mongol Altai near the border with the Russian Federation (RF) and represents the southeastern extension of the Altai group of hydrothermal iron ore deposits of the RF. Three hydrothermal quartz-hematite occurrences are hosted by the Lower Devonian effusive rocks of intermediate and acid composition occurring in the Deluun-Sagsai regenerated trough. The Tsagaan Uul iron ore occurrence is located 25 km west-northwest of Tsagaan Uul soum in a fault zone of the trough. The Oroku Gol occurrence is situated in the mouth of Oroku Gol River, and the Bayan Ulgii occurrence was found 18 km north of Bayan Ulgii soum in the eastern wing of the trough. The Tsagaan Uul occurrence is represented by strongly brecciated silicified rock enriched by hematite. The iron ore has localized in hematite ore lenses from 0.2

to 0.3 m thick and from 2.5 to 3 m long. The thickness of iron ore mineralization is up to 10 m, and extent is from 300 to 350 m. Hematite content by visual estimate does not exceed 30%. Two other occurrences are similar to that of the Tsagaan Uul occurrence.

8.3.3 Magmatic Deposits

Magmatic deposits are firstly classified by A.E. Izokh in Mongolia, as mafic-ultramafic-related Ti-Fe and apatite-Ti-Fe, carbonatite-related Fe-REE, and magnetite-apatite deposit types (Dejidmaa et al. 1999).

Few occurrences of zoned mafic-ultramafic Ti-Fe (and apatite Ti-Fe) deposits occur in differentiated, concentrically zoned gabbro, gabbro-anorthosite, and gabbro-pyroxenite plutons and in rare diabase sills. Based on mineralogy and host rock composition, three subtypes are classified as follows: (1) ilmenite-titanomagnetite and apatite-ilmenite-titanomagnetite related to gabbro-anorthosite; (2) Ti-Fe titanomagnetite related to clinopyroxenite-gabbro; and (3) Ti-Fe mineralization related to diabase sills. In all subtypes, ore minerals formed in two main stages: (1) early magmatic stage ore occurs as layers and lenses concordant to layering in the host mafic-ultramafic intrusions. The ores occur in pyroxenite and peridotite layers as lenses and consist of massive and disseminated ore minerals; (2) post-magmatic stage ore that are mostly occurring as veins.

Only a few occurrences are known in Mongolia. The ilmenite-titanomagnetite and apatite-ilmenite-titanomagnetite ores are hosted in Proterozoic gabbro-anorthosite that is part of an Archean and Early Proterozoic metamorphic basement. The host rocks are mainly high alkaline, high ferruginous, and high titaniferous clinopyroxene-gabbro intrusions of Must Uul, Khojuuliingol, and Olon Khudag that consist of gabbro, amphibole gabbro, clinopyroxenite, and ore-bearing clinopyroxenite. Ore-bearing pyroxenite and ilmenite-titanomagnetite and apatite-ilmenite-titanomagnetite layers, lenses, and veins are occurring in the abovementioned plutons. The main ore mineral is titanomagnetite; other subsidiary minerals are ilmenite, apatite, rare pyrrhotite, pyrite, hematite, and spinel. The Ti-Fe subtype is hosted in Upper Neoproterozoic layered clinopyroxene-gabbro plutons in northern Mongolia in the Khuvsgul, Urgamal, and Lake terranes. Examples of this subtype are the Khoshimgol occurrence and the Uet-Undur occurrence. The Fe-Ti subtype also occurs in diabase sills in Late Paleozoic to Early Mesozoic volcanic-plutonic belt in northern Mongolia. The diabase sills contain disseminated titanomagnetite which are intruding the Permian volcanic and volcanogenic-sedimentary rocks of the Khanui Group.

8.3.3.1 Khoshim Gol Ti Occurrence

The Khoshim Gol Ti occurrence is located 163 km N from Murun soum, Khuvsgul province. It is associated with a massif of the early Paleozoic gabbro-diorites. The magnetite and titanomagnetite mineralization was found in thin fissures of 0.1–2 cm thick. It contains Fe_2O_3 57.1% and Ti 8.4%. Indicated reserves of titanomagnetite ore have been estimated at 1.5 Mt. This occurrence belongs to a magmatic titanomagnetite type of deposit, proved to be of economic significance. Therefore motivating the further mineral exploration for similar type of larger deposits within Mongolia is important.

8.3.3.2 Fe-REE Carbonatite or Magnetite-Apatite Deposit

This deposit type consists of magnetite-apatite veins and bodies in alkaline plutonic and coeval volcanic rocks and in explosive carbonatite breccia. The deposit is genetically and spatially associated with REE-fluorspar carbonatite, REE-apatite carbonatite, and REE strontium carbonatite deposits. Examples of the deposit type occur in the Mushgai Khudag district in southern Mongolia and in the Uvs Nuur district in northwestern Mongolia.

Ore contains 1–14.5% REE (mostly LREE) that occur in apatite, magnetite, phlogopite, and rare celestine (Dejidmaa et al. 1999). Magnetite is mostly altered to hematite. The amount of apatite and magnetite varies strongly in stock-like bodies and in veins. Some bodies contain rich ore consisting of massive, very coarse-grained magnetite or fine- to medium-grained apatite.

8.4 Manganese Deposits

Manganese mineralization is closely associated with iron ore and mostly distributed in the eastern Mongolian Metallogenic Province (United Nations Publications 1999). The Eastern Mongolian Metallogenic Province hosts around 40 small manganese occurrences, of which 28 are distributed in the Khangai Khentii manganese-iron ore area and 12 in the south of the province. None of the manganese occurrences has so far attained economic interest, due to a low MnO_2 content and limited mineralization. The Northern Mongolian Metallogenic Province hosts only the Saikhan Gol manganese deposit and Bayan Tsagaan Gol occurrence. Major genetic types are metamorphic sedimentary, hydrothermal, and lake marsh (peat).

8.4.1 *Saikhan Gol Manganese Deposit*

The Saikhan Gol manganese deposit is located near to the Saikhan River, being hosted by the Ediacarian-Lower/Middle Cambrian siliceous rocks. The siliceous host rock is 20 m thick and contains a series of lenses and layers of manganese ore which have been traced over 2 km. The average thickness of ore lenses is from 1.5 to 2 m. The major ore mineral of the deposit is pyrolusite of massive and laminated texture with a variable MnO2 content ranging from 9.26 to 36.72%. Inferred reserves estimated for 9.26 average MnO_2 content and 150 m extensional depth are 450,000 tonnes.

8.5 Chromium Deposits

Nine chromium occurrences of chromium-spinel are known to exist all localized in ultramafic rocks in eight extended ophiolite belts along deep-seated fault systems (Fig. 8.7). The ultramafic rocks are represented by lenses and pods of serpentinite,

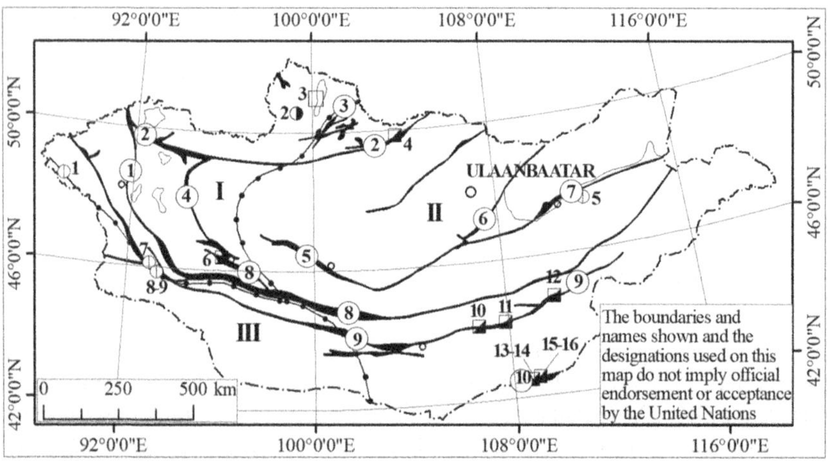

LEGEND

- Metallogenic provinces: I - North Mongolian; II - East Mongolian; III - South Mongolian
- Major ophiolite belts
- Occurrences of magmatic chrome-spinel formation
- Occurrences of hydrothermal veins vanadium formation
- Occurrences of magmatic titanium-magnetite formations
- Occurrences of hydrothermal veins titanium-magnetite formations
- Faults

OPHIOLITE BELTS

1. Tsagaan Shiveet
2. Khankhukhii
3. Arguingol
4. Zavkhan
5. Bayankhongor
6. North Gobi
7. Kherlen
8. Ikh Bogd
9. Gobi
10. Sulinkheer

Fig. 8.7 Schematic distribution map of chromium, titanium, and vanadium occurrences in Mongolia (Marinov et al. 1977) peridotite, and rarely dunite and pyroxenite. Relatively large chromium-bearing rocks have been found so far within the Gobi and Sulinkheer ophiolite belts, where the Manlai and Sulinkheer groups of chromium occurrences have been distinguished. Two chromium occurrences have been found in the Northern Mongolian fold system (United Nations Publication 1999).

8.5.1 Manlai Group of Podiform Chromite Occurrences

The Manlai group of podiform chromite occurrences incorporates the Ulkhid and two other occurrences (10, 11) located in Dornogovi province. The chromite mineralization is associated with peridotite, pyroxenite, and serpentinite intrusions that can be up to 20 km long and 2.5 km wide. Ulkhid occurrence has been found 60 km EW of the Manlai soum, in an intrusion of serpentinized peridotite. The mineralization is traceable within an area of 6x12 m as nest-like pods from 10 to 20 cm wide and veins of 1 m long and 3–10 m wide. The Cr_2O_3 content is 29.25%, Ni 0.01%, and Co 0.006%.

8.5.2 Sulinkheer Group of Chromite Occurrences

Sulinkheer group of chromite occurrences has been located in Dornogovi province, near to the border area with China. This group includes the Sulinkheer Ovoo (14), Ontsgoi Tolgoi (13), and Baga Shar Khad (15) occurrences and several small chromite showings (16). The chromite mineralizations occur in rocks of the Sulinkheer ultramafic belt associated with the NE, E-W striking Sulinkheer fault.

8.5.3 Sulinkheer Ovoo Chromite Occurrence

Sulinkheer Ovoo chromite occurrence is located 64 km SW of Sulinkheer soum. Two silicified and serpentinized peridotite and dunite lithologies from 8 to 28 km long and from 2.5 to 7 km wide are in the area (Fig. 8.8). The mineralization has been found in eastern and western areas being separated from each other by 800 km. The eastern area occupies 0.5 km^2 and contains 15 elongated orebodies of lens-like shape striking in E, NE direction. The mineralized bodies crop out at the surface 3–25 m wide, extend for up to 100 m, and dip 70–80°NW. The texture of the ore is massive, with eight having disseminations and laminated styles.

The content of Cr_2O_3 in lenses of massive ores (up to 1 m thick) varies from 37 to 45% and in other types of ore from 10% to 26%. The western area of 1.5 km^2 size includes two lenses of chromite ore up to 45 m long from 3.65 to 3.8 m thick and with around 33% Cr_2O_3 content. The average Cr_2O_3 content in the eastern and

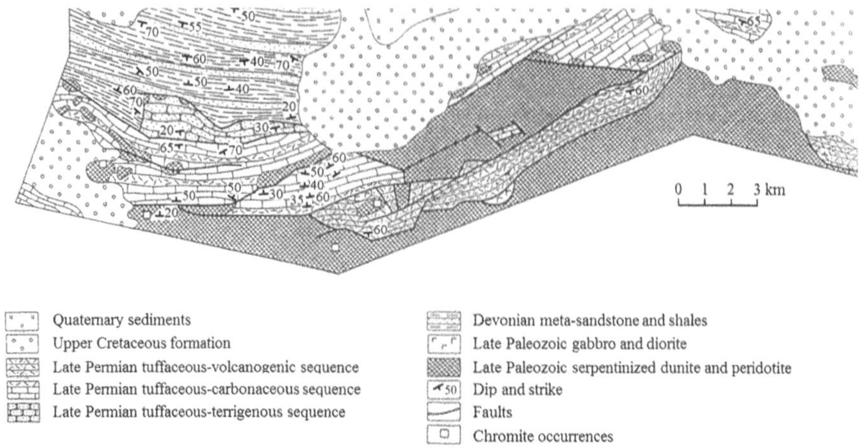

Fig. 8.8 Geological map of Sulinkheer Ovoo chromite occurrence (Marinov et al. 1977)

western area is 18–19% and 20%, respectively. The major ore mineral is chromite; it is closely associated with olivine. The Cr_2O_3 content in chromite ore reaches a range of 55–60% after flotation. All ore concentrates are conditioned for ferrochromite production.

8.5.4 Azargyin Chromium Occurrence

Azargyin chromium occurrence (3) has been located 77 km NW of Buren soum, Selenge province. The mineralization is associated with small bodies of serpentinized peridotite hosted by limestone and schists rocks of the Ediacarian-Lower Cambrian. Alluvial samples assayed 60% Cr_2O_3.

8.5.5 Nogoon Tolgoi Chromite Occurrence

The Nogoon Tolgoi chromite occurrence (9) is located 15 km SW of Yusunbulag soum, Gobi-Altai province. The occurrence is associated with the Early Caledonian Ikh Bogd ophiolite belt, which hosts numerous ultramafic rocks. The chromium mineralization occurs in an elongated (300 km^2 in size) serpentinite outcrop and boulders. The chromium massive ore shows good quality, having minor iron content, Cr_2O_3/Fe_2O_3 ratio of 4.6 and 53.19% of Cr_2O_3 content.

8.6 Titanium and Vanadium

Rare occurrences of titanium and vanadium are known in western Pre-Khuvsgul, Mongol Altai, and eastern Mongolia (United Nations Publication 1999; Ochirbat et al. 2008).

Vanadium occurrences in Mongolia are commonly belonging to the chemical sedimentary deposit type (Dejidmaa et al. 1999) and consist of V-bearing chert-carbonaceous shale and cherty, sheet-like layers. The chert-carbonaceous shale layers consist of interbedded siliceous shale, carbonaceous-calcareous shale, clay-shale, siltstone, chert, and rare limestone. The chert layers consist of chert and sedimentary quartzite. Main example of this type is vanadium occurrences in the Lower Paleozoic Khuvsgul Shelf basin.

Altogether six titanium occurrences are known in northern Mongolia and Mongol Altai. Except of Khoshim Gol occurrence, all others belong to quartz-magnetite-ilmenite and quartz-ilmenite mineralization type. The occurrences are represented by hydrothermal quartz veins containing magnetite and ilmenite or only ilmenite. The quartz veins are hosted by Carboniferous gabbro-diorites in the northern part of the Central Mongolian fold system, Cambrian sandstone and schists of Caledonian Mongol Altai, and Devonian granite-gneisses in the southern part of Mongol Altai. The grade of V ranges from 0.01% to 1.0%. These V-bearing horizons vary from 20.0 to 70.0 m wide and from 0.5 to 2.7 km long. The mineralogy of these occurrences is not studied.

The most prospective Tsohio vanadium occurrence has been located 65 km NW of Khatgal soum. It is hosted by Neoproterozoic-Lower Cambrian limestone and schist formation. Hydrothermal quartz and quartz calcite veins occur in shear zone (100–150 $\times$ 10 m) of the above formation. The hydrothermal veins are 0.1–0.3 m thick and 25–90 m long, containing sulvanite (Cu_3VS_4) and sulfides of lead and copper. The vanadium oxide content varies from 0.2% to 3.6% and copper from 0.8% to 25.1%.

8.7 Concluding Remarks

Ferrous deposits in Mongolia are dominantly characterized by iron deposits. Magmatic apatite-Ti-Fe, titanomagnetite, and Fe-REE occurrences existed in relation to mafic and ultramafic and alkaline rocks.

According to the announcement in 2012, from the Mineral Resource Authority of Mongolia, Mongolia has 1 billion tonnes of iron ore reserve and 5.5 billion tonnes of geological resource. The economic iron deposits of Mongolia are considered as Fe, Fe-Zn skarn type, and few metamorphosed ironstone type. Tumurtiin Ovoo, Tayan Nuur, Bayan Gol, and Tumur Tolgoi deposits are under mining, while the Bargilt and Kharganat iron deposits are expected to have future economic potential.

However, the manganese, chromium, titanium, and vanadium mineralization is mostly represented by occurrences, in relation to mafic and ultramafic rocks in ophiolite zones along paleo-subduction suture zones. Mn, Cr, Ti, and V occurrences do not show economic interest at the present technology.

References

Bakhteev RKh, Chijova IA (1984) Zhelezorudnoye formatsii Mongolii i zakanomernosti ikh razmeshcheniya (Iron-ore formations of Mongolia and regularities in its spatial distribution) Endogenic ore-bearing formations of Mongolia. Joint Soviet-Mongolian Scientific-Research Geological Expedition): Transactions, Moscow, 38:115–123

Chebotarev MV, Chebotareva NA, Tsend-Ayush J (1990) Mestorozhdeniya magnetitovykh skarnov poyasa Bayan-Gol: geologiya i mineral'nyye resursy Mongol'skoy Narodnoy Respubliki (Magnetite skarn deposits of the Bayan Gol Belt: Geology and Mineral Resources of the Mongolian People's Republic), International Geological Expedition in Mongolia), Transactions, Moscow, 3:145–151

Chuluunbat D, Byambaa B (2010) Gobi Altai aimgiin Tseel sumiin nutagt orshikh Tayan Nuur tumriin ordiin tuv bietiig ashiglah tehnik ediin zasgiin undeslel (Feasibility study for Tayan Nuur iron ore deposit located in Tseel soum of Altai aimag). Geological Information Center Mongolia, Report #6311, p 53

Cox DP, Singer DA (ed) (1986) Mineral deposit models. US Geol Surv Bull 1693:379

Dejidmaa G (1998) Mongolian Fe, Mn, Ti, Cr and V deposit types. Mong Geosci 10:32

Dejidmaa G, Dorjgotov D, Gerel O, Gotovsuren A (1999) Preliminary description of mineral deposit models (types) for Mongolia. MINDEP (5):99–165. https://pubs.usgs.gov/of/1999/of99-165/data/MINDEP/MINDEP5.PDF

Eckstrand OR (1984) Canadian mineral deposit types: a geological synopsis: geological survey of Canada. Econ Geol Rep 36:86

Filippova IB, Vydrin VN (1977) Chornyye metally; Jelezo (Black metals: Iron Geology and Mineral Resources of Mongolian Peoples' Republic, Book III), Nedra, Moscow, 90–140

Gottesmann W, Kampe A (2013) The Tumurtiin-ovoo Fe-Mn-Zn skarn deposit, Mongolia: sedimentary features, distribution of main minerals and the whole-rock chemical composition. J Mineral Geochem 191(1):1–32

Marinov NA, Khasin RA, Khurts Ch (1977) Marinov NA (ed) Geologiya Mongoliskoi Narodnoi Respubliki TOM III, Poleznie iskopaemie (Geology and Mineral Resources of Mongolian Peoples' Republic, Book III), Nedra, Moscow

Ochirbat P, Purev L, Lkhansuren J, Chuluun D, Sun YS, Gerel O, Dandar S, Minjin Ch, Jargalan S (2008) Mongol ornuu geologi uul uurkhain nom (Geology and Mining Book of Mongolia), Masterprint, p 666

Podlessky KV, Aksuk DK, Vlasova PF (1984) Rudonosnie skarni tsentralinoi I vostochnoi Mongolii (mineralized skarns of central and eastern Mongolia). In: Kovalenko VI, Jamsran M (eds) (Endogenic ore-bearing formations of Mongolia) transactions, Moscow, vol 38, pp 124–143

Podlessky KV, Vlasova DK, Kudrya PF (1988) Skarni I svyazannoe c nimi orudenenie Mongolii (Skarns and connected ores of Mongolia) Joint Soviet-Mongolian Scientific-Research Geological Expedition, Transactions, Moscow, (45):149

Tsend-Ayuush J, Chebotarev MV (1989) Zhelezorudnyye mestorozhdeniya MNR i voprosy ikh osvoyeniya (Iron ore-formations of Mongolian People's Republic and their potential for mining) Geology and exploration of the territory of Mongolian People's Republic, International Science Conference for 50th year jubilee of Geological Survey of Mongolian Peoples' Republic), Ulaanbaatar, 105–107

Ukhnaa G, Baasan B (2016) Tumur: Ordiin geologi, erel khaiguul (iron deposit: geology and mineral exploration methods). Gecom Press, Ulaanbaatar, p 283

United Nations Publication (1999) Geology and mineral resources of Mongolia. Atlas of mineral resources of the Escap region, United Nations, vol 14, New York, p 192

Chapter 9
Silver Deposits

Otgonkhuu Javkhlan and Baatar Munkhtsengel

9.1 Introduction

Silver occurrences and deposits are known to occur mainly in the northeastern and western Mongolian metallogenic province and less so in the southern Mongolian metallogenic provinces (Fig. 9.1). Commonly, silver is a minor metal of gold-sulfide-quartz and polymetallic deposits and rarely occurs in small silver occurrences. There are a few numbers of silver ore deposits and a number of silver-bearing base-metal deposits in Mongolia; we present an overview of different types of Mongolian silver deposits and discuss their geological characteristics and economic significance.

9.2 Descriptions of Silver and Silver-Bearing Deposits

9.2.1 Asgat Ag-Sb-Cu-Bi Deposit

The Asgat silver and base-metal deposit is located within the Deluun-Yustid basin in the Mongolian Altai fold belt in NW Mongolia. The basin is characterized by a strongly deformed and folded Devonian black shale formation. Middle-Upper Devonian dark-gray and blackish sandstone and black argillite of the Khatuugol Formation and medium to coarse-grained, porphyritic granite of the Carboniferous Yustid Complex and related aplite dykes, fine-grained granite and pegmatites, and Jurassic dyke complexes are distributed in the area (Fig. 9.2; Lkhamsuren 2009).

O. Javkhlan (✉) · B. Munkhtsengel
Department of Geology and Hydrogeology, School of Geology and Mining, Mongolian University of Science and Technology, Ulaanbaatar, Mongolia
e-mail: o.javkhlan@must.edu.mn; tsengel@must.edu.mn

O. Gerel et al. (eds.), *Mineral Resources of Mongolia*, Modern Approaches in Solid Earth Sciences 19, https://doi.org/10.1007/978-981-15-5943-3_9

Fig. 9.1 Silver and silver-bearing deposits in Mongolia. Small and bigger circles are Pb-Zn deposits and occurrences. Metallogenic belts: (I) Deluun, (II) northwestern Khangai, (III) Eastern Mongolia, (IV) Govi—Nukhetdavaa. Major districts: (1) Zavkhanmandal, (2) Tsagaan Uul, (3) Dulaan Khar Uul, (4) Tsav, (5) Khartolgoi, (6) Tumurtiin Ovoo, (7) Ulaan (modified after Batjargal et al. 1997)

The Khatuugol Formation hosts economic silver-sulfosalt ore mineralization. Hornfels and hornfelsed rocks occur along the contact between the Turgen Gol and Yustid plutons. The folded structure of the deposit has been intersected by a number of faults with various ages and origins including the lake zone fault which controls silver mineralization. Along a total length of 45 km 15 ore zones and a large number of small-sized and parallel bands with mineralization signs are distinguished within the deposit. The deposit structure is like a "horse tail," and its thickness decreases in a SW direction and branches into a NE direction.

The first ore zone is comparatively bigger and well-studied and contains the main reserve of silver-base metal ore. The contacts of the ore zones are sharp (?) and elongated from NE to SW. The contact of the ore zones is vertical, stretches from the north east to the west, and is a few meters in thickness. These ore zones are composed of siderite veins, breccias, and stockworks and contain sulfides and sulfosalt minerals. The surface is dark and dark-brown and covered with iron oxide. The main ore mineral is native silver; chalcostibite, pyrite, and arsenopyrite are minor and siderite; and calcite and mica occur as gangue minerals (Lkhamsuren 2009).

Tetrahedrite is a widespread mineral and constitutes the main portion of the silver and ore components (Lkhamsuren 2009). The mineral occurs as porphyritic, veinlets, nest-like, cemented grains, and occasionally crustification crystals. Tetrahedrite grains are 0.001 mm–3 cm, most of them 0.1–1.0 cm in size and show hypidiomorphic and allotriomorphic texture. There are three generations of tetrahedrite and generation II is dominant.

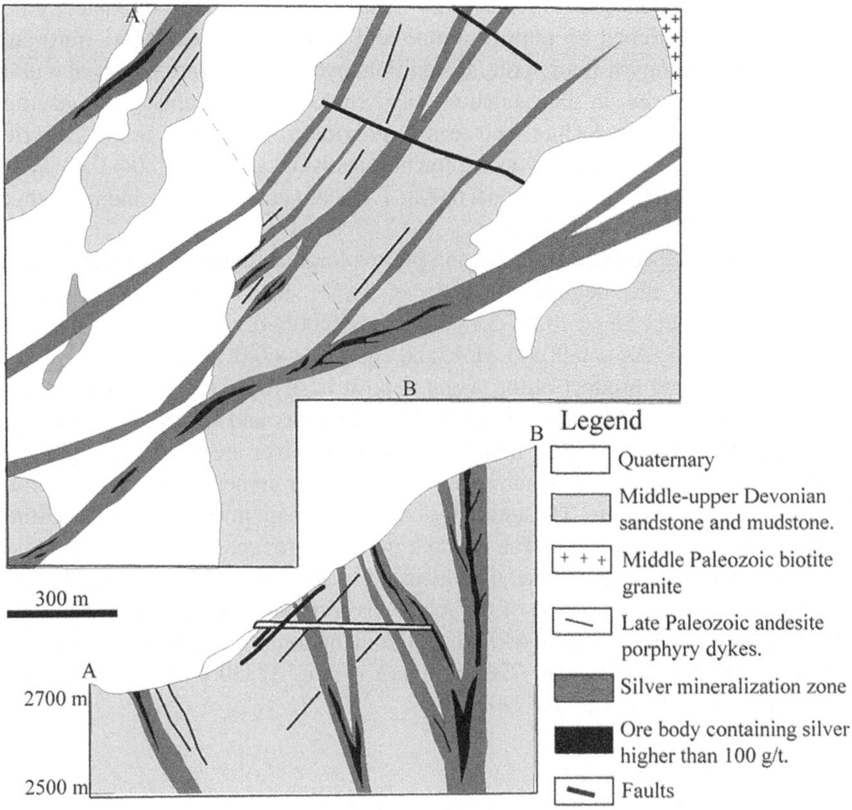

Fig. 9.2 Sketch geological map of the Asgat Ag-Sb-Cu-Bi deposit, western Mongolia (Lkhamsuren 2009)

There are fine inclusions (up to 0.01 mm in size) of pyrite and pyrrhotite associated with quartz within the tetrahedrite. Early formed pyrrhotite, burnonite, chalcostibite, and bismutite are observed. Chalcostibite occurs as a relict, individual, and nest-like grains in tetrahedrite. Silver content in tetrahedrite is 780–22,000 g/t; Bi, 0.13–4.08%; Zn, 0.01–0.38%; Pb, 0.01–2.18%; As, 0.015–2.61%; Au, 30 g/t; and Hg, 0.006–0.08%. Chalcopyrite is the next dominant mineral after tetrahedrite by its distribution and can be classified into four stages. It occurs as veinlets, porphyritic, and nest-like grains of a few mm to 1 cm in siderite. Chalcopyrite associates with early formed minerals in fault zone and replaces tetrahedrite grains, and its fine veinlets (sometimes together with sphalerite) cut the tetrahedrite.

The relationship of tetrahedrite and chalcopyrite is variable and has different ore textures. The boundary between chalcopyrite and tetrahedrite in porphyritic mineralization is normal, and chalcopyrite contains small grains of sphalerite and few pure bismutite surrounded by tetrahedrite. In the nest-like and veinlet ore, tetrahedrite grains are sometimes seen within the chalcopyrite shells with a thickness of up to

1 mm or more. Tetrahedrite is replaced by metasomatite alteration of chalcopyrite. Chalcopyrite is enriched by pure bismutite and very fine grains (0.001 mm) and fragments of arsenopyrite and pyrite. In addition, pyrite veinlets are observed within cracks in chalcopyrite. In the oxidation zone, chalcopyrite is replaced by azurite, covellite, chalcocite, malachite, and iron hydroxides. Ag content in chalcopyrite ranges from 74 to 15,350 g/t by mono fraction analysis, and Pb is 0.04–0.43%, Zn 0.02–2.05%, As 0.035–3.22%, and Bi 0.008–1.1%. Au content in one mono fraction analysis reaches up to 30 g/t.

During the prospecting evaluation and pre-exploration studies in the deposit, the following results are determined such as Zn < 0.1%, Cu < 0.01–3.56%, Ag < 2–1990 g/t, Sb < 0.01–2.14%, Bi < 0.005–0.31%, S < 0.1–4.24%, Fe = 4.83-46.56%, As = 0.005–1.81%, and Hg = 0.3–60.0%.

The main valued product of the Asgat deposit is silver, and the byproducts are copper, antimony, and bismuth (Ochirbat 1999). Mercury and arsenic are considered as harmful elements. The main minerals containing silver are tetrahedrite, chalcopyrite, chalcostibite, and cubanite. The main copper minerals are chalcopyrite, tetrahedrite, and malachite. The antimony-containing main minerals are tetrahedrite, chalcostibite, and gudmundite. The bismuth minerals are represented by bismuthite, native bismuth, and chalcostibite. The main mineral of arsenic is arsenopyrite. All ore reserves for the exploitation of the Asgat silver deposit were calculated in 1996 and in C + C_2 category were 6402.6 thousand of tonnes. Metal reserves are as follows: 2247.8 tonnes of Ag, 7264.6 tonnes of Cu, 31830.9 tonnes of Sb, and 3319.8 tonnes of Bi (Ochirbat 1999).

9.2.2 Ulaan Ag-Pb-Zn Deposit

The Ulaan deposit is located 110 km northwest of Choibalsan, one of the major Ag-Pb-Zn deposit in NE Mongolia. It exposed in the central region of the Mesozoic volcano-sedimentary basin, which unconformably overlies or is in fault contact with the Ereendavaa metamorphic terrane (Badarch et al. 2002). The Ereendavaa terrane in NE Mongolia is dominated by Proterozoic quartzite, black schist, marble, gneissic rock, and amphibolite, intruded by Early Paleozoic plutonic rocks. The Mesozoic basin is dominated mainly by mafic, intermediate, and felsic volcanic rocks with variable alkaline in compositions and pyroclastics (Fig. 9.3). Coal-bearing sedimentary formations are also distributed sporadically in the mineral district. Volcanic rocks of alkaline, high-K volcanic rocks transitional to shoshonite are host in the ore deposits (MITM 2002; Nie et al. 2010, 2014; Dahlkamp 2009).

Two parallel crypto-explosive pipes, about 20 m apart, have been recognized in the Ulaan Ag-Pb-Zn deposit area. Both breccia pipes developed in the Ereendavaa belt metamorphic unit and the overlying Mesozoic volcanic layers. These breccia pipes represent the major host of the Ag-Pb-Zn ore bodies (Nie et al. 2015). The breccias comprise clastic materials of basalt, andesite, dacite, rhyolite, trachyandesite, and shoshonite. Clastic materials vary in size from ~20 to

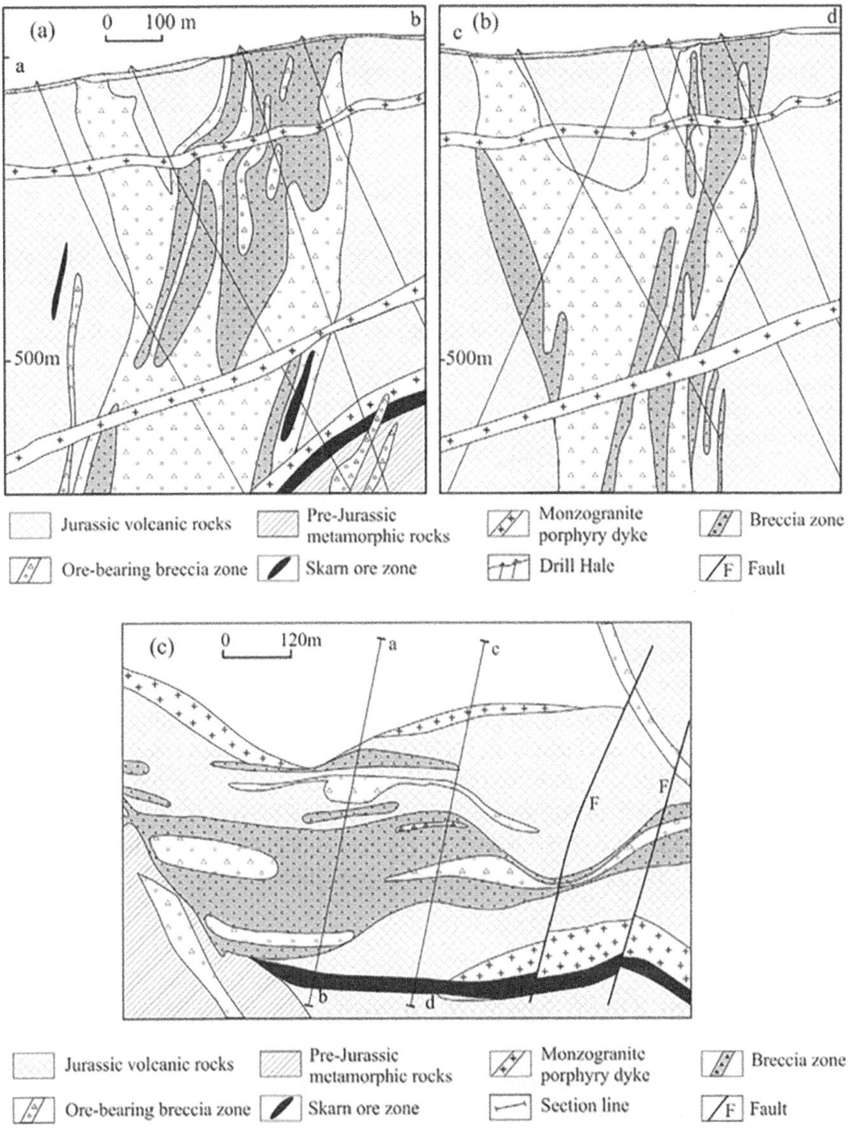

Fig. 9.3 Geological map of the Ulaan Ag-Pb-Zn deposit, NE Mongolia (Nie et al. 2015)

~4000 cm and are cemented by tuffaceous material including carbonate, sulfides, and fluorite. Two kinds of mineralization at Ulaan deposit district have been identified: a tabular skarn-type at deeper levels and vein-stockwork lodes at shallower depths (Nie et al. 2015). The skarn-type mineralization characterized as sulfide rich strata and tabular form bodies with high-grade pods is occurred along the contact zone of Ereendavaa marble and Cretaceous monzogranite and monzodiorite porphyry dykes. LA-ICP-MS zircon ages (c. 157–150 Ma) of dykes are obtained (Yan et al. 2000; Hu et al. 1998; Nie et al. 2014).

The main, precious metal-bearing and gangue minerals are listed in following table. The average grades of the major ore bodies are 86 g/t Ag, 3.6% Pb, and 12% Zn (Nie et al. 2015).

Main ore minerals	Precious metal-bearing ore minerals	Gangue minerals
Marcasite, magnetite, pyrargyrite, bismuthite, argentite, tetrahedrite, native Au, bornite, bismuthinite, bismuthite, chalcocite, chlorargyrite, pyrite, chalcopyrite, galena, sphalerite, pyrrhotite, and arsenopyrite	Electrum, argentite-acanthite, argentobismutite, tintinaite, jalpaite, native silver, native gold, and bornite	K-feldspar, actinolite, diopside, chlorite, epidote, garnet with minor quartz, fluorite, sericite, albite, calcite, and ankerite

Vein-stockwork-type Ag-Pb-Zn ore mineralization is controlled by two parallel crypto-explosive breccias pipes. Sulfide-bearing quartz veins, vein networks, and as disseminated to massive bodies are occurred as lenticular or more continuous podiform shapes. Most veins and ore bodies trend NE and dip to NW with obvious shrinkage, extension, and branching along-strike and down-dip. The main metallic minerals of the ore-bearing stockwork-veins composed mainly of sphalerite, galena, pyrite, and a minor (trace) amount of pyrrhotite, marcasite, magnetite, chalcopyrite, arsenopyrite, and electrum. The main gangue minerals consist mainly of dolomite, quartz, and calcite. Associated wall rock alterations are characterized by silicification, carbonatization, and chloritization (Nie et al. 2015). The average grades of the main ore bodies are 3.5 g/t Au, 46 g/t Ag, 2.2% Pb, 6%Zn, and Cu 0.56%. According to Nie et al. (2015), over 3000 tonnes of Ag and 2 Mt. of combined Pb-Zn in the deposits (MITM 2002; Nie et al. 2014; Dahlkamp 2009).

9.2.3 Mukhar Zn-Pb-Ag-cu-cd Deposit

The Mukhar deposit is located just 850 m southeast of the Ulaan deposit along the East Mukhar fault zone in NE Mongolia. The geology, mineralization, and alteration of the Mukhar deposit are the same as the Ulaan deposit. The ore minerals are similar to those in the Ulaan deposit, except for the lack of gold, and they are richer in chalcopyrite. In the Mukhar deposit, the reserves of 1108 t Ag with grade of 113 g/t were estimated besides 332.5 thousand tonnes of Zn, 61.8 thousand tonnes of Pb, 11.9 thousand tonnes of Cu, and 1079 tonnes of Cd.

9.2.4 Tsav Ag-Pb-Zn Deposit

The Tsav deposit is located in NE Mongolia, 146 km northeast of the city of Choibalsan. It is one of the larger Ag-Pb-Zn deposits in NE Asia (Gantumur et al. 2005; Gerel 1998; MITM 2002; Yang et al. 2009; Nie et al. 2010, 2014).

The geology of the Tsav district is characterized as gneiss, leptynite, and schists of the Ereendavaa terrane; intruded by granitoids in the northern and eastern parts; and covered by Jurassic shoshonite, trachyandesite, and trachydacite, dacite, rhyolite, ignimbrite, tuff, and tuffaceous rocks (Nie et al. 2015). These volcanic rocks are also intruded by Jurassic hypabyssal porphyritic dykes and stocks of diorite, monzodiorite, and monzonite (Fig. 9.4). The K-Ar isotopic dating on the hornblende from monzodiorite dykes yields 135 $\pm$ 5 Ma (Nie et al. 2010, 2014). Compositions of intrusive rocks showed high-K calc-alkaline to shoshonitic (Gantumur et al. 2005; Nie et al. 2010).

Twenty one ore zones have been explored in the Tsav mineralized district. Some of the ore zones include several ore bodies. Ore zones are characterized as (a) early quartz-dominated vein-stockworks and (b) later carbonate-dominated vein-stockworks (Nie et al. 2015). Ag-Pb-Zn mineralization occurs as veins, veinlet, stockwork, and disseminated zones that constitute a number of irregular pods, pockets, and/or lenses interconnected by joint fillings, stringers, and veinlets of sub-ore-grade materials. These mineralized zones are hosted by the Jurassic volcano-plutonic complex (Nie et al. 2015) and display a close spatial relation to Mesozoic monzonite, monzodiorite, monzogranite, and syenite dykes. Ore bodies vary from 350 to 2200 m in length and 0.5 to 15 m in width, with depths from 200 to 550 m. The shape of the ore bodies varies greatly along-strike and down-dip and display complex pinch-and-swell morphology, as well as branching and dilation characteristics. Ore bodies become thicker with depth by drilling data (Nie et al. 2014). Both quartz and carbonate lode ores display crustiform, colloform, and cockade textures. The major ore minerals are pyrite, arsenopyrite, galena, sphalerite, chalcopyrite, and marcasite with small amounts of sulfoantimonite, polybasite, tetrahedrite, cleiophane, altaite, bismuthite, freibergite, electrum, argentite, chlorargyrite, hessite, pyrargyrite, calaverite, native gold, and native silver. Gangue minerals are oligonite, calcite, rhodochrosite, calcite, chlorite oligonite, Mn-siderite, rhodochrosite, epidote, muscovite, smectite, illite, chlorite, and kaolinite.

Silicification and carbonatization are regarded as main wall rock alteration which is accompanying mineralization, and the degree of alteration increases with increasing grade. Also chlorite, sericite, and epidote are conspicuous (Nie et al. 2015). The average grade of the quartz lode ore is 141 g/t Ag, 3.42% Zn, 7.21% Pb, and 0.23% Cu, and for the carbonate lode ore, 372 g/t Ag 3.61% Pb, 0.08% Cu, and2.66% Zn, (Nie et al. 2010, 2014; MITM 2002).

The distribution pattern of $\delta^{34}S$ value of various sulfides of the Tsav deposit is similar to that of typical magmatic hydrothermal deposits (Gantumur et al. 2005; Corbett and Leach 1998). Isotopic data suggest ore-forming fluid is a mixed solution dominated by magmatic water (Gantumur et al. 2005; Yang et al. 2009; Nie et al.

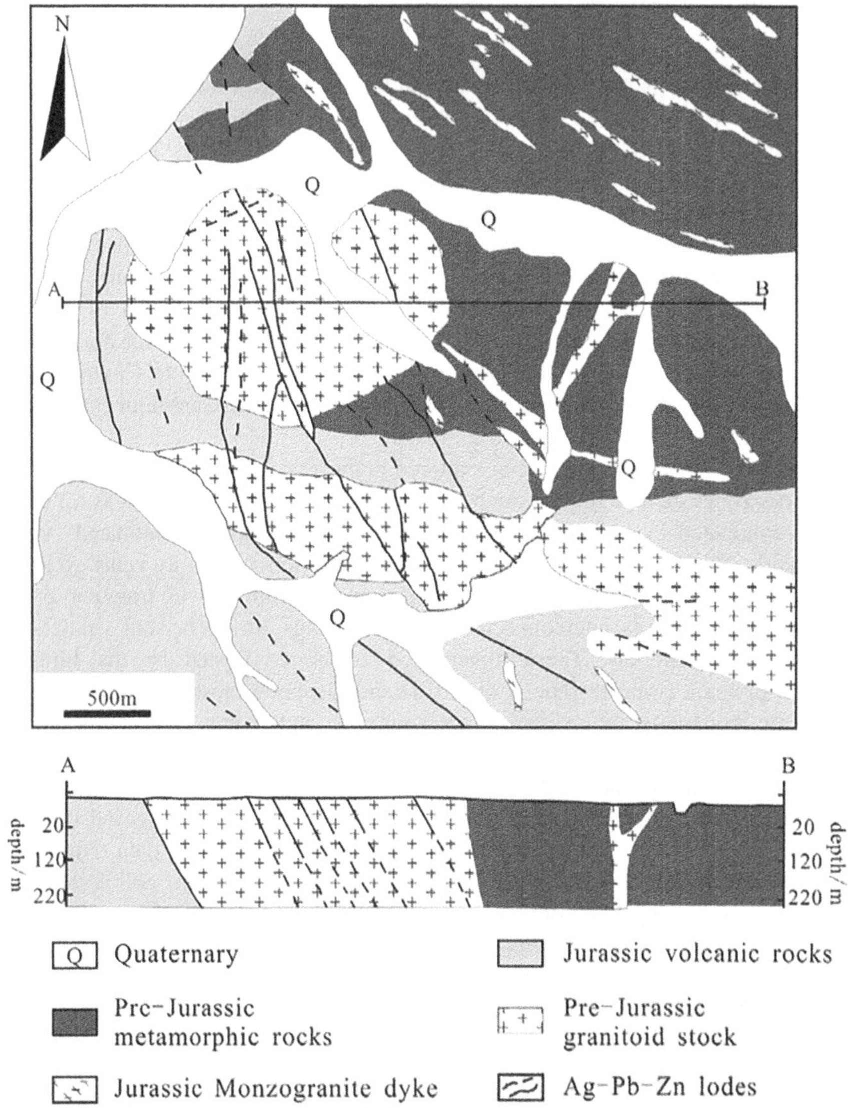

Fig. 9.4 Geological sketch map of the Tsav Ag-Pb-Zn deposit, northeastern Mongolia (Nie et al. 2015)

2010; Zhai et al. 2013). The Tsav deposit are distinguished as intermediate sulfidation epithermal deposits (c.f. Hedenquist et al. 2000, Sillitoe and Hedenquist 2003, Jiang et al. 2004) based on the textures, mineralogy, and alteration. Furthermore, Nie et al. (2015) assumed that such an origin is a potential for porphyry style mineralization at deeper levels in the Tsav mineralized district and adjacent areas.

9.2.5 *Bayan-Uul Zn-Pb-Cu-Ag-Cd Deposit*

The Bayan-Uul deposit is located 130 km northeast of Choibalsan city in NE Mongolia. The Bayan-Uul deposit was discovered in 1975, with prospecting works being carried out in 1986 and 1990. The deposit is hosted by a Late Mesozoic volcano-plutonic belt consisting mainly of monzodiorite, alkaline granite, and dykes of fine-grained granite and diorite. Neoproterozoic gneiss, orthogneiss, and marbles are exposed as stretched small bodies bounded by faults in the central and marginal parts of the deposit district (Fig. 9.5).

Ore mineralization including silver mineralization occurred in the silica and sericite alteration zone along the Bayan-Uul fault zone, i.e., Mo, Au-Ag-Zn-Pb, sulfide-Au, and quartz-fluorite-Mo. Native silver and tetrahedrite are identified as minor ore minerals in the deposit. The reserves of about 358 t Ag were estimated besides 23.5 thousand tonnes Zn, 10.6 thousand tonnes Pb, 204 tonnes Cd, and 4.9 thousand tonnes Cu in the Bayan-Uul deposit.

9.2.6 *Mungun-Undur Pb-Zn-Ag Deposit*

The Mungun-Undur Pb-Zn-Ag deposit is located in NE Mongolia, 70 km northwest of the city of Chinggis. The deposit was discovered and prospected in early 1980s by geologist O. Gurjav. Further prospecting works were conducted for silver mineralization by Mongolrostsvetmet in the 1990s. The Mungun-Undur deposit hosted by the Lower Devonian Tsarigyngol Formation consists of chlorite-sericite schist and quartz-muscovite schist with layers of sandstone and mudstone, intruded by gabbro

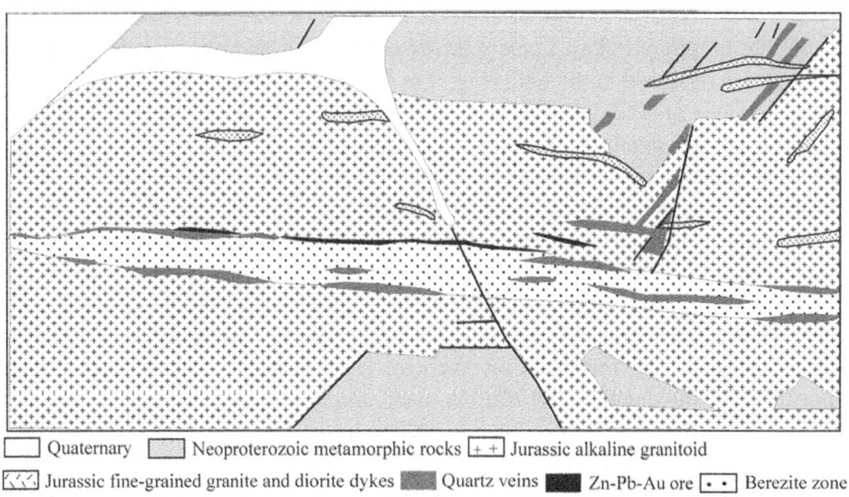

Fig. 9.5 Geological sketch map of the Bayan-Uul Zn-Pb-Cu-Ag-Cd deposit, northeastern Mongolia (Dorjgotov 2009)

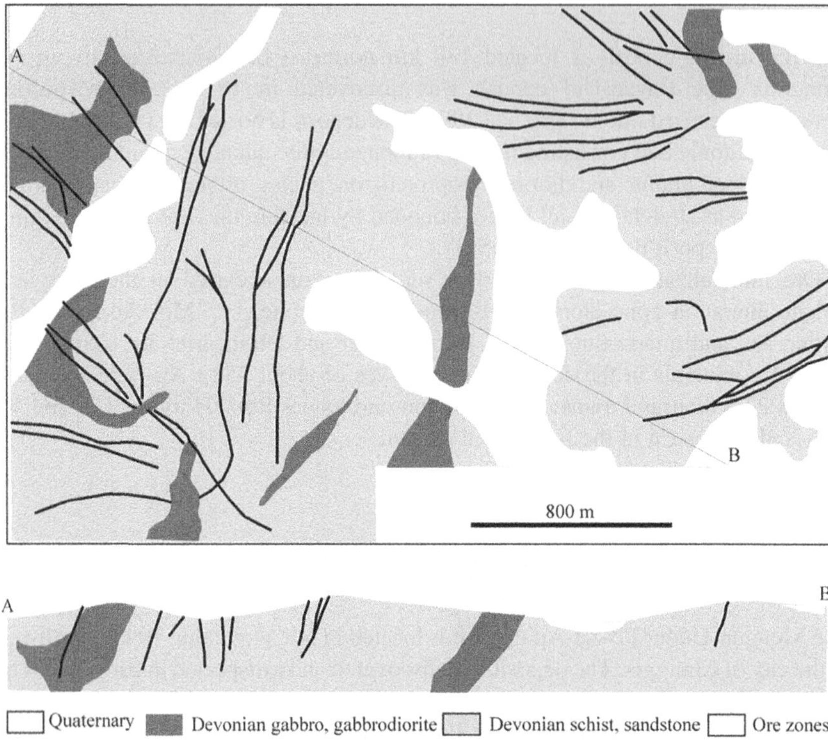

Fig. 9.6 Geological sketch map of the Mungun-Undur Pb-Zn-Ag deposit, northeastern Mongolia (Dorjgotov 2009)

and diorite-porphyry dykes. Locally, lavas and tuffs of andesite and rhyolite compositions are observed in the field (Fig. 9.6; Dorjgotov 2009).

Fine-grained granitoid stocks were identified under the Undurtsagaan Be-W-Mo deposit at a 450 m depth by a borehole northwest of the Mungun-Undur deposit. The granitoids are strongly altered to greisen. The Mungun-Undur deposit is hosted within Tsarigyngol schist-sandstone layers underlying on the greisenized granitoid stock. Ore mineralization occurred in the arc- or radiated-like fault zone and in the veins (Dorjgotov 2009). Twenty-four mineralized zones trending NE were identified. Ore zones vary from 750 to 1200 m in length along its strike, up to 300 m in length along its dip, and up to 60 m in width. Ore bodies are subvertical vein type with from 1 to 4 m thickness on average. Hydrothermal alteration consists of chloritization, silicification, pyritization, and carbonatization developed in the wall rocks. About 40 minerals have been identified in the Mungun-Undur deposit. The most important minerals are sphalerite, pyrite, galena, arsenopyrite, pyrrothine, marcasite, quartz, and rhodochrosite. Native silver and Ag-bearing minerals are relatively common.

Brecciated, disseminated, vein type, and massive textures were observed. Several assemblages of pyrrothine-pyrite, sphalerite-galena, carbonate-cassiterite, and carbonate-Ag minerals were distinguished. These assemblages formed geochemical zones of Mo, (Cu-Mo) W, Bi-As, Fe-Sn (Sn)-Pb, Zn-Ag-Pb, and Ag (Naftali et al. 1984). The main ore assemblage of the Mungun-Undur deposit is quartz-sphalerite-freibergite and rhodochrosite-galena-pyrargyrite. The average grade of the deposit is 1.3–1.89% Zn, 1.82–2.18% Pb, 146.8–188 g/t Ag, and 0.14% Sn. Expected resources of metals were estimated as 135.4 thousand tonnes of Zn, 155.1 thousand tonnes of Pb, and 700 tonnes of Ag. The K-Ar age (175 $\pm$ 9 Ma) of sulfide mineralization was determined by Murao et al. (1998).

9.2.7 Salkhit Ag-Au Deposit

The Salkhit deposit is located in the Gurvansaikhan soum, Dundgovi Province in the SE Mongolia. The deposit is situated in the Dundgovi volcano-plutonic belt (Badarch et al. 2002) and situated in the Gurvansaikhan ore district of the Dundgovi-Buyant metallogenic belt. The occurrence of Ag-Au was first documented in the report of geological mapping at a scale of 200,000 by V. Zabotkin in 1982. Mineral exploration was conducted in the Salkhit deposit by GPF Co., Ltd. in 2013. Ore bodies of the deposit are controlled by the Dundgovi Fault Zone (DFZ). Lower Permian mostly felsic volcanic lava formed along the DFZ. The lithologies consist of rhyolite, trachyrhyolite, ignimbrite, andesite, dacite, and tuffs. Most of the ore bodies of the Salkhit deposit are hosted by trachyrhyolite (Batzorig et al. 2013).

Trachyrhyolite consists mainly of plagioclase, K-feldspar, quartz, and biotite. Trachyte texture is developed in the rock. Silica (jasperoid), pyrite, and sericite alterations also developed in the rock. Trachyrhyolite is crosscut by micro veins of mica, quartz, and iron oxide. Related to these micro veins, the rock is enriched sharply by elements of Ti, Zr, and Rb.

The main silver mineralization took place in the quartz veins and quartz breccia. Quartz veins are defined as an epithermal type (Batzorig et al. 2013). The oxidation zone is identified as up to as 100 m in depth. Ore minerals are arsenopyrite, pyrite, chalcopyrite, tetrahedrite, galena, sphalerite, magnetite, and gold. Secondary oxidation minerals are malachite, goethite, and covellite. It is concluded that the main mineralization of silver is related to tetrahedrite and galena. Whereas, the gold mineralization is defined that it is related to arsenopyrite. Most of galena contains the argentite component in their interior part (Batzorig et al. 2013). Ore grade of Ag in the ore is 781 g/t, and Au is 0.81 g/t. All ore reserve for exploitation of the Salkhit silver deposit is calculated in 2012 in B + C category is 2,805,933 tonnes. Metal reserves are as follows: 858.46 tonnes of Ag and 1 tonnes of Au (Batzorig et al. 2013).

9.2.8 *Khartolgoi Pb-Ag-Sb-As Deposit*

The Khartolgoi deposit is located 170 km south of Dalanzadgad City in southern Mongolia. In the deposit area, Neoproterozoic carbonate rocks, Permian terrigeneous sediment, and Carboniferous plutonic rocks are exposed, and they are intruded by Late Mesozoic dykes (Fig. 9.7).

Ore bodies occur as vein and lenses formed within limestone intruded by a granitoid pluton. Mineralization occurred in the NW directed fault zone containing subvolcanic bodies. Ore minerals are dominated by siderite-sulfide, massive sulfide types. Ore bodies are oxidized in the shallow level, forming an iron cup. In the boundary zone between granitoid and limestone, non-mineralized assemblages of garnet, garnet-epidote, and garnet-pyroxene are determined. About 138 tonnes of Ag with grade from 126 g/t to 145 g/t were estimated. Other main metals were estimated as 59.6 thousand tonnes of Pb, 30.3 thousand tonnes of Sb, and 19.5 thousand tonnes of As.

9.3 Concluding Remarks

Silver mineralization in Mongolia can be classified in terms of age and host rocks. Age-wise, the silver mineralization may have taken place from Late Paleozoic to Early Mesozoic magmatism. The host rocks of silver occurrences are terrigeneous sedimentary rocks (Asgat, Mungun-Undur deposits), volcanic materials, plutonic rocks, and carbonate rocks. The Asgat deposit is situated within the Early Devonian-Early Carboniferous Deluun-Sagsai metallogenic belt in the Mongolian Altai mountain range which is related to granitoids and the Deluun sedimentary-volcanic-plutonic belt, according to the classification of Nokleberg (2010). The metallogenic belt is considered as one of two large metallogenic belts that occur in a large metallogenic aureole associated with Devonian Andean-type calc-alkaline igneous rocks in north and northwestern Mongolia (Berzin et al. 1994; Kovalenko et al. 1995). The metallogenic belts and host Deluun volcanic-plutonic belt later are overprinted by Middle and Late Devonian calc-alkaline granitoids. This overprinting complex (Byamba and Dejidmaa 1999) is developed across the Mongol Altai and Hovd terranes. The Deluun-Sagsai metallogenic belt is considered to formed during accretion event (Dandar et al. 2001).

The Ulaan, Mukhar, Tsav, Bayan-Uul, and Salkhit deposits are situated in the Middle Jurassic-Early Cretaceous East Mongolian-Priargunskiy-Deerbugan metallogenic belt. This metallogenic belt is closely associated with Jurassic-Cretaceous veins, volcanic complexes, replacements, and granitoids; extends from central Mongolia to northeastern Mongolia and into Russia and China; is one of the largest in Northeast Asia; and comprises about 80 mines, deposits, or occurrences (Nokleberg 2010). The Ag-Pb-Zn deposits Ulaan, Mukhar, Tsav, and Bayan-Uul are solely related with Late Jurassic to Early Cretaceous volcanic belt covering central

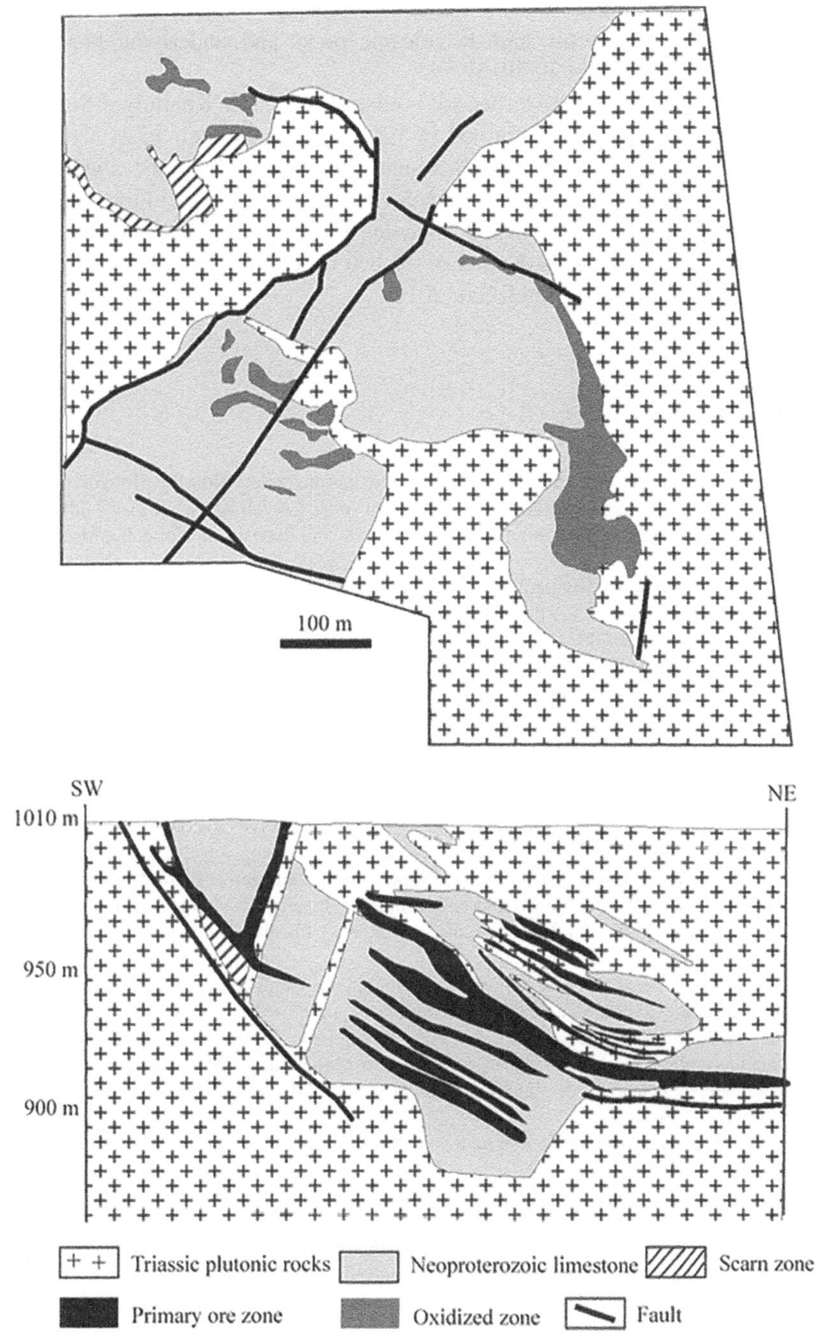

Fig. 9.7 Geological sketch map of the Khartolgoi Pb-Ag-Sb-As deposit in southern Mongolia (Dorjgotov 2009)

Mongol-Argun pre-Jurassic metamorphic massif (Neoproterozoic Ereendavaa belt). Ore is hosted by Mesozoic high-K volcanic rocks and underlying pre-Jurassic metamorphic rocks (Nie et al. 2015).

The Khartolgoi Pb-Ag-Sb-As deposit is characteristic into Khartolgoi-Sulinkheer metallogenic belt which was formed in Late Jurassic through Early Cretaceous (Nokleberg 2010). The belt extends roughly 1200 km along the southern and southeastern border zone between Mongolia and China. The part of the belt located in northwestern China is considered as having formed during the intrusion of post accretionary granite, whereas the part of the belt in southern Mongolia is formed as during back arc extension (Nokleberg 2010).

References

Badarch G, Dickson CW, Windley BF (2002) A new terrane subdivision for Mongolia: implications for the Phanerozoic crustal growth of Central Asia. J Asian Earth Sci 21:87–110

Batjargal S, Lkhamsuren J, Dorjgotov D (1997) Lead-zinc ore deposits in Mongolia. Mong Geosci 2:2–14

Batzorig B, Enkhbayar B, Kherlenbaatar M (2013) Brief introduction of exploration result of Salkhit Ag-Au deposit in the Zuun area, Gurvansaikhan soum, Dundgovi province, Unpublished report of GPF Co., Ltd, p 39

Berzin AN, Coleman RG, Dobrevtsov NL, Zonenshain LP, Xiao X, Chang EZ (1994) Geodinamicheskaya karta paleoaziacheskogo okeana. Geologiya I geofizika. Geodynamic map of the Paleoasian Ocean. Geol Geophys 35:8–28

Byamba J, Dejidmaa G (1999) Terranes of Mongol Altai area. Mong Geosci 14:16–19

Corbett GJ, Leach TM (1998) Southwest Pacific rim gold–copper systems: structure, alteration and mineralization. Econ Geol Spec Publ 6:1–238

Dahlkamp FJ (2009) Mongolia. In: Dahlkamp FJ (ed) Uranium deposits of the world. Springer, Berlin, Heidelberg, pp 285–312

Dandar S, Dejidmaa G, Enhbaatar Sh (2001) Mongol Altain nuruunii colliziin ba colliziin umnuh ueiin metallogeny. Pre-collisional and collisional metallogeny of Mongol Altai area. In: Problems of geology: Mongolia, Magazine of Earth Science Faculty, National University of Mongolia, 368–399

Dorjgotov D (2009) Lead and zinc deposits in Mongolia. In: Gerel O (ed) Mongolian geology and mineral resources (book VI), Ulaanbaatar, Mongolia, pp 85–107

Gantumur H, Batulzii D, Wang LJ, Zhu HP (2005) Tsav: a shoshonite-hosted intermediate sulfidation epithermal Ag–Pb–Zn deposits, eastern Mongolia. In: Mao JW, Bierlein FB (eds) Mineral deposit research: meeting the global challenge. Proc 8th SGA Meeting Beijing, China, pp 389–392

Gerel O (1998) Phanerozoic felsic magmatism and related mineralization in Mongolia. Bull Geol Surv Jpn 49:239–248

Hedenquist JW, Arribas AJ, Gonzalez UE (2000) Exploration for epithermal gold deposits. Rev Econ Geol 13:245–277

Hu SK, Yan HQ, Ye M, Xiang WD (1998) Metallogenic focus-area and superlarge mineral deposits in the bordering zones between China, Russia and Mongolia. Sci China (Series D) 41:28–41

Jiang SH, Nie FJ, Zhang Y, Hu P (2004) The latest advances in the research of epithermal gold deposits. Earth Sci Front 11(2):401–411

Kovalenko VI, Tsaryeva GM, Goreglyad AV, Yarmolyuk VV, Troitsky VA, Hervig RL, Farmer GL (1995) The peralkaline granite-related Khaldzan-Buregtey rare metal (Zr, Nb, REE) deposit, Western Mongolia. Econ Geol 90:530–547

Lkhamsuren J (2009) Silver deposit in Mongolia. In: Gerel O (ed) Mongolian geology and mineral resources (book VI), Ulaanbaatar, Mongolia, pp 301–306

Ministry of Industry and Trade, Mongolia (MITM) (2002) Investors' Forum. Internal Mining and Oil Industry Information Bulletin, 1–350

Murao S, Dorjgotov D, Tseden T (1998) K-Ar dating of granitoids and hydrothermal micas from the northern part of Kherlen depression, Mongolia. Bull Geol Surv Japan 49:249–255

Naftali et al (1984) The geochemical zonation of the Undurtsagaan mineral district. Geology and mineral resource of MPR (book II), 102–109

Nie F, Li Q, Liu C, Ding C (2015) Geology and origin of Ag–Pb–Zn deposits occurring in the Ulaan-Jiawula metallogenic province, Northeast Asia. J Asian Earth Sci 97:424–441

Nie FJ, Jiang SH, Bai DM, Hou WR, Liu YF (2010) Types, temporal and spatial distribution of metallic ore deposits in southern Mongolia and its neighboring area. Acta Geosci Sin 31 (3):267–288

Nie FJ, Liu YF, Li QF, Jiang Z, Wu KR, Zhang XK, Ding CW, Cao Y (2014) Mesozoic Multiple Magmatic Activities and Ag–Pb–Zn and U Mineralization Occurring within the Wulaan-Jiawula Metallogenic Province, Conjunction Area of China, Mongolia and Russia. No. 16 Internal Research Report of Institute of Mineral Resources, Chinese Academy of Geological Sciences, 1–258

Nokleberg WJ (2010) Metallogenesis and Tectonics of Northeast Asia. U.S. Geological Survey Professional Paper 1765:624

Ochirbat P (1999) Strategy and ecology of the precious metal (gold, silver, diamond) development, Book, Ulaanbaatar, 347–376

Sillitoe R, Hedenquist JW (2003) Linkages between volcanotectonic settings, ore fluid compositions, and epithermal precious-metal deposits. Soc Econ Geol Geochem Soc Spec Publ 10:285–313

Yan HQ, Hu SK, Ye M, Xiang W (2000) Mutielement deposit prospecting area in the western slope of the Great Hinggan Mts. In: Tu GZ (ed) Superlarge ore deposits of China. Science Publishing House, pp 273–292

Yang ZL, Zhang DQ, Li JW (2009) Ore-forming types, metallogenic zoning and potential prospecting areas in southwestern sector of Deerbugan metallogenic belt. Mineral Deposits 28 (1):53–62

Zhai DG, Liu JJ, Wang JP, Yao MJ, Liu ZJ, Wu ZJ, Wu SH, Fu C, Wang SG, Li YX (2013) A study of stable isotope geochemistry of the Jiawula large Pb–Zn–Ag ore deposit, Inner Mongolia. Earth Sci Front 20(2):213–225

Chapter 10
Platinum Group Elements Mineralization

Baasandorj Altanzul and Ochir Gerel

10.1 Introduction

Platinum group element (PGE) mineralization in Mongolia is associated with ultramafic and mafic intrusions. There are three major types of PGE mineral systems known in Mongolia. The most common are (1) PGE mineralization related to ophiolites, which includes podiform chromite, (2) PGE in layered gabbro intrusions, and (3) PGE-Au placer deposits. Other types of unconventional (i.e., amagmatic) sources of Pt and Pd, for example, black shale-hosted Au deposits (Ermolayev et al. 1992), later included as orogenic Au-Pt (Pd) deposits in carbonaceous and metasedimentary rocks, and deposits associated with near-surface enrichment of ophiolitic ultramafic rocks have not been sufficiently studied (Johan 2002, 2006).

Investigation of the PGE mineralization in Mongolia began with a research team of the Mongolian-Russian joint geological expedition in the 1960s (Zonenshain et al. 1990). This project included the study of ultramafic and mafic rocks, their structural position, petrography, and associated chromite and asbestos mineralization (Pinus et al. 1955; Dandar et al. 2003). A specialized study was conducted in 1970–1980s by Pinus et al. (1984) and later Agafonov et al. (1991) and Izokh et al. (1990). In the 1980s Mongolian-Russian teams carried out reconnaissance work in several occurrences with PGE mineralization, which led to the accomplishment of detailed research on the Naran mafic-ultramafic ophiolite, a part of Khantaishir ophiolite complex, located in the Govi-Altai Province (Krivenko et al. 1985; Prichard et al. 1994). In the 1990s Jamsrandorj and Sharkhuukhen carried out an exploration project on the Naran ophiolite, Bayan Nuur (Uvs Province) placer occurrence and

B. Altanzul (✉) · O. Gerel
Department of Geology and Hydrogeology, School of Geology and Mining, Mongolian University of Science and Technology, Ulaanbaatar, Mongolia
e-mail: altanzul@must.edu.mn; gerel@must.edu.mn

O. Gerel et al. (eds.), *Mineral Resources of Mongolia*, Modern Approaches in Solid Earth Sciences 19, https://doi.org/10.1007/978-981-15-5943-3_10

on Khantaishir (Govi-Altai Province), and Altan Uul (Umnugovi Province) (Byamba and Bat-Erdene 1999).

In 1995, a 1:1,000,000 scale geological map was compiled by (Jamsrandorj et al. 1995) showing the distribution of ultramafic and mafic rocks. In Mongolia there are about 120 ultramafic and about 100 mafic complexes (Jamsrandorj et al. 1995). Some of them contain PGE mineralization. The most common type is ophiolite-related PGE mineralization (Fig. 10.1). The majority of ophiolite complexes are supra-subduction formed in fore-arc, back-arc, or volcanic arc (Jian et al. 2014; Furnes and Safonova 2019).

The gabbroid intrusions are not well studied for PGE mineralization. There are several gabbro plutons: Nomgon, Khairhan, Ontsuul, Oortsog, and Dulaan. The Nomgon pluton, which occurs 15 km to NE of the confluence of Orkhon and Tuul rivers, is a layered troctolite-anorthosite-leucogabbro intrusion. Sulfide mineralization is represented by chalcopyrite and bornite. The sulfide-bearing zone contains Pt and Pd with As and Sb, sperrylite, Au, and Ag. Sperrylite has rhodium up to 12.21%. Mao et al. (2018), studied the mafic-ultramafic intrusions of Oortsog, Dulaan, and Nomgon for Ni-Cu mineralization. The distribution patterns of platinum group elements of the Dulaan and Oortsog intrusions record slight Rh, Pt, and Pd (PPGE) enrichment relative to Os, Ir, and Rh (IPGE). But only the Nomgon intrusion may feature a PGE mineralization potential (Mao et al. 2018).

Placer occurrences formed by the erosion of the platinum group minerals (PGM) bearing ultramafic rocks occur in Bayan Nuur (Uvs Province), Zost (Umnugobi Province), and were studied by Banzragch (2000) and Sharkhuukhen and Gerel (2001). Most of PGM nuggets in Ulziit (Bayankhongor Province) (Tomurtogoo et al. 1999), Ulaannachin, Burgastai (Uvs Province), and Zost (Umnugovi Province) were discovered as accessory mineralization during the mining and exploration of gold placer deposits, and their mineralogy was studied in detail (Amitan and Dandar 1999; Altanzul and Kitakaze 2005; Altanzul and Dandar 2010; Altanzul et al. 2015).

Few occurrences with PGE mineralization were discovered during 1:50,000 scale geological mapping, e.g., the Godroi ultramafic complex (Govi-Altai Province). According to Erdenebayar et al. (2017), it might be thought that this complex is similar to PGE-poor ophiolite complexes because of the thin layers of ultramafic and low content of ore minerals. PGE minerals were observed in serpentinite and chromite samples (Erdenebayar et al. 2017). The link between chromitite and the PGE mineralization in the original deposit is still debated. Most authors agree that the chromite mineralization is magmatic, but PGE mineralization seems to have undergone a complex post-emplacement evolution. In Mongolia, no economically important deposits have yet been discovered, but several occurrences of PGE mineralization are known in western Mongolia (Lkhamsuren 2011).

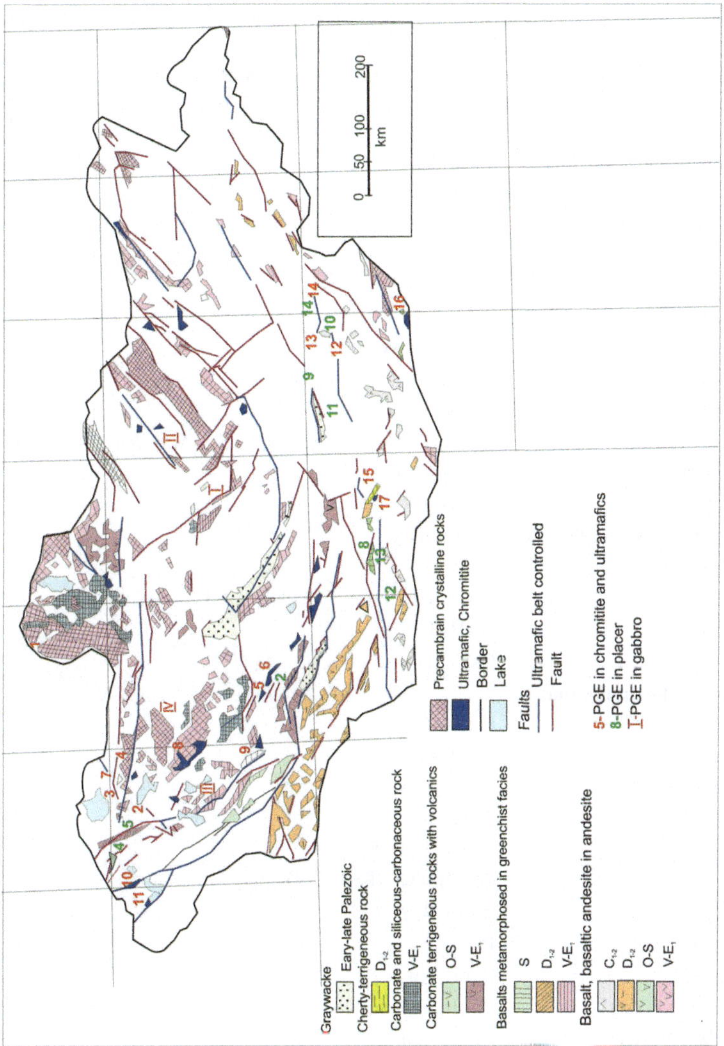

Fig. 10.1 Distribution on mafic-ultramafic intrusion and PGE mineralization in Mongolia (modified after Agafonov et al. 2005); explanation of locality numbers is given in Table 10.1

Table 10.1 Location and composition of PGE mineralization in Mongolia

No	Name of PGE deposits and occurrence	Composition of PGE minerals
Related to ophiolite complexes (numbers in red color in the map)		
1	Shishged	Pt, $IrAs_2$, RuOsIr, (Os, Ir, Ru, Ni)
2	Icheetgol	RuS_2, Os_2, RuAsS, (Os, Ru)AsS, RuIrOs
3	Dumberel	RuS_2, Os_2, FeNiPt, Pt_3Fe, RuIrOs, (Os, Ir), (Ir, Os), RuAs, RuAsOs, IrAsS
4	Jargalant	(Ir, Os), (Os, Ir), (OsRu)AsS, RuAsS, OsS_2
5, 6	Naran (chromitite), Naran (vebsterite)	RuIrOs, RuS_2, IrAsS, (Os, Ir), Pt_2FeNi, $(Ni, Ir)_3As$, (Rh, Ni, As)
7	Bayannuur (conglomerate)	Pt, (Ir, Os)Ru
8	Khutul	IrAsS, RhAsS
9	Bayannuur (Darvi ridge)	RuS_2, OsS_2, IrAsS, (Os, Ru), $(Ir, Rh, Ni, Cu)_2S_3$, RuS_2
10	Godroi	(Pt, RuIrOs), (RuIrOs), (Ru, Os), RuS_2, OsS_2, RuS_2, (Ir, Os), (Os, Ru)AsS, IrAsS,
11	Pridorozhny	Os, Ir
12	Kharaat Uul	FeNiPt, $PtAs_2$
13	Ulaankhiid	(Ru, Ir, Os), (Os, Ir), (Ir, Os)
14	Khalga	Os, Ir
15	Gurvansaikhan	Pt_3Fe, (Pt, Os, Ir), (Ir, Os), RuS_2, OsS_2, Ru, (Ru, Ir, Os)
16	Sulinkheer	Os, (Ir, Os)
17	Altan Uul	Pt (Os, Ir)
Related to ultramafic-mafic intrusive complex		
I	Oortsog, Dulaan Uul	Pt, Pd
II	Nomgon	$PtAs_2$, $Pd_8(Sb, As)_3$, PdNiAs, Pd_3Sb, Pd_3As
III	Khairkhan	Pd_3HgTe
IV	Ontsuul	OsS_2, (Os, Ru)AsS
Placer (green color in the map)		
1'	Bayannuur	Pt, Pd, Os, Ir
2'	Naran, Jong gol	(Ru-Ir-Os)
3'	Ongon Ulaan Uul	Pt, (Ir, Os)
4'	Ulaannachin	Pt_3Fe (Ir, Os, Ru)
5'	Burgastai	Os-Ir-Ru, Pt-Fe
6'	Zost	(Ru, Ir, Os), (Fe_3Pt), (Os, Ir)
7'	Ulziit	(Os-Ir-Ru)
8'	Kharaanen	Pt
9'	Bayanleg	Ir, Os, Ru
10'	Khashaat Khudag	Pt, (Ir, Os)
11'	Ikh Ulziit Nuruu	Pt, (Ir, Os)
12'	Bayan Bor Nuruu	Ir, Os, Ru

10.2 Geology

Ultramafic and mafic intrusions in Mongolia are known in Precambrian structures, where they occur in the southern margin of the Siberian Craton. The ultramafic bodies are fragments of ophiolite sequences and are fault-controlled. The ultramafic bodies form small lenses of tens of meters long. Precambrian rocks are metamorphosed and comprise gneisses, amphibolites, schists, and carbonaceous rocks. Ultramafic intrusions in Precambrian rocks are serpentinized, but relatively fresh rocks are not rare. Very common is the association with rock-forming minerals: antigorite ± olivine ± tremolite ± anthophyllite ± talc ± carbonate ± chlorite. Accessory minerals are represented by magnetite and chrome spinels (Agafonov et al. 1991).

Tectonically, Mongolia is subdivided into a northern and southern domain separated by the Main Mongolian Lineament, an approximate regional topographic and structural boundary separating dominantly Precambrian and Lower Paleozoic rocks to the north, from dominantly Lower to Upper Paleozoic rocks to the south (Marinov et al. 1973).

The northern domain (megablock) contains outcrops of Precambrian and Lower Paleozoic metamorphic rocks, Neoproterozoic to Early Cambrian ophiolites, Lower Paleozoic island arc volcanics and associated volcaniclastic sediments, granitic rock of various ages, Devonian to Carboniferous sediments, and Permian volcanic-plutonic belts. The southern domain (megablock) is dominated by Lower to Middle Paleozoic arc-related volcanic and volcaniclastic rocks with fragments of ophiolites and serpentinite mélanges, Late Devonian to Carboniferous granitoid rocks, and Permian-Mesozoic alkaline rocks. The eastern parts of both domains are intruded by numerous Mesozoic rare metal granite plutons and are overlapped by Upper Jurassic to Cretaceous nonmarine volcanic and sedimentary rocks and Quaternary alkali basalts.

In the northern domain, ophiolite sequences consist of ultramafic and mafic bodies, basalts and basaltic andesites, and carbonaceous-siliceous and volcaniclastic rocks, metamorphosed in greenschist facies. Ultramafic rocks are also serpentinized, but talc and antigorite are not present (Pinus et al. 1984). Among ultramafic rocks, dunites and harzburgites are dominant: pyroxenites, vehrlites, and lherzolites are also present. The largest body of this age is the Naran ophiolitic sequence in southwest Mongolia, a part of the Khantaishir ophiolite of 570 Ma (Janousek et al. 2018). In western Mongolia, ultramafic bodies occur in Mongolian Altai ultramafic belts. Host rocks are represented by terrigenous rocks, rarely basalts. Ultramafic bodies are totally serpentinized with quartz-carbonate (listvenite) alteration. Listvenite is a hydrothermally metasomatic quartz-carbonate rock, which contains pyrite, fuchsite, chlorite, talc, serpentine, sericite, rutile, and actinolite (Lkhamsuren 2011).

In the southern domain, along the faults of Main Mongolian Lineament, a number of ultramafic bodies such as Ikh Bogd, Undurshil, and others are present. All ultramafic belts here have a latitudinal orientation. Host rocks are metamorphosed to greenschist facies. Ultramafic rocks are intensely serpentinized with a common quartz-carbonate alteration. Occasionally, serpentine is replaced by silica, and rocks look like jasper with relics of serpentinite (Zoolen and Gurvansaikhan ophiolite complexes). The quartz-carbonate alteration is cut by hydrothermal quartz veins. Dykes of gabbro, altered gabbro and diabase are spatially associated with ultramafic rocks. All dykes are replaced by rodingite. In the southern megablock, weathering crust is developed on the ultramafic rocks. The weathering crust, developed after dunite, is composed of birbirite, massive siliceous rock. Magnesite and dolomite are cut by quartz veinlets. Some of ultramafic bodies were uplifted and the weathering crust removed by erosion (Nemegt ophiolite complex), but within overlapping upper Cretaceous coarse-grained terrigenous rocks, birbirite and magnesite fragments of weathering crust are found. This allows estimating the uppermost time of weathering crust formation as the end of Cretaceous. Chromitites here are found as blocks, often deformed and rounded. Some ultramafic bodies are relatively well preserved with the undisturbed layering of small lens-like bodies of pod-like chromitites (Ulaankhiid, Kharaat Uul, and Suujiin Nuruu). Based on Agafonov et al. (1991), all ultramafic bodies occur along large regional faults and are steeply dipping and well defined by the gravimetric and geomagnetic surveys. Gently dipping bodies are rare (Agafonov et al. 1991). These authors distinguished 25 ultramafic belts (Fig. 10.1). In early Paleozoic structures 17 are known and five ultramafic belts in Mid-Late Paleozoic. In western Mongolia, Stupakov et al. (1994) distinguished five other belts: Sagsai (Terek), Tolbonuur, Khovd, Bairam, and Tsagaanshiveet. In early Paleozoic structures, the following occur: East Sayan, Khuvsgul-Sangilen, Onon, Khankhukhii, Armakgol, Egiingol, Zalaatuul, Yeroogol, Zavkhan, Bayankhongor, North Gobi, Kherlen, Delger, in Devonian-Early Carboniferous: Tolbonuur, Khovd, Bairam, and Sagsai. In Late Paleozoic structures, there are Bayanleg, Manlai, Modon Ovoo, South Gobi, and Sulinkheer. All these belts have similar or close ages. Some of belts, e.g., Tsagaanshiveet, Ikh Bogd, and Undurshil, border different age complexes.

There are about 150 placers and heavy mineral concentrates around and near the outcrop of ultramafic rocks.

10.3 PGE Mineralization Types and Classification

The PGE mineral systems in Mongolia can be divided into three types (Table 10.1):

1. Magmatic podiform chromitites related to ophiolite complexes: Naran ophiolite at Khantaishir ophiolite, Altan Uul, Godroi ophiolite.
2. Mafic-ultramafic intrusion related type: Oortsog, Dulaan, Nomgon intrusion.
3. Placer (with Au placer): Bayannuur, Zost, Ulaannachin, and Ulziit.

10.3.1 Podiform Chromitites with PGE Mineralization Related to Ophiolite Complex

Detailed studies of the ophiolite-related PGE have been performed by Leblanc (1991), Naldrett and Gruenewaldt (1989), and Naldrett (1993). The ophiolitic rocks were classified as PGE mineralized and non-mineralized and their petrochemical, mineralogical, and geochemical differences determined as well as their geodynamic origin (Prichard et al. 1990, 1996). Throughout Mongolia, ophiolite complexes have Neoproterozoic-early Cambrian age in western Mongolia and Lower-Mid-Paleozoic to Late Paleozoic in southern Mongolia. Based on geochemical classification, the dominant type of Mongolian ophiolites is subduction-related type (supra-subduction), including back-arc, fore-arc, and island arc subtypes (Gerel and Munkhtsengel 2015; Furnes and Safonova 2019), and only the Bayankhongor ophiolite is related to mid-ocean ridge type (MOR). The PGE associated with ophiolite is of two types (Leblanc 1991).

The first type is PGE in chromite within dunite and harzburgite, a part of the lower mantle (Naldrett and Gruenewaldt 1989). This mineralization is of magmatic origin and chromite precipitated from the magma. Some researchers believe that PGE-poor chromites occur within slow-spreading ridge ophiolite, such as Semail, Oman, and Troodos, Cyprus, whereas PGE-rich chromites occur within fast-spreading ridge ophiolite (Boudier and Nicolas 1985). Typical Pt and Pd poor sequences occur both in northern Semail, Oman, and in Troodos, Cyprus. The Unst complex of the Shetland Isles is a Pt- and Pd-rich ophiolite complex. Chromite minerals containing PGE as in the Urals also belong to this type. The best example of disseminated chromite mineralization is the Naran ophiolite massif of the Khantaishir ophiolite complex (Byamba and Bat-Erdene 1999; Bat-Erdene 2001).

The Nuur (Lake) island arc terrane contains Taishir, Dush mountain, Dumberel, Khets, Onts mountain and Godroi ophiolite sequences. Mineralization associated with the first type in Mongolia is characterized by sulfide disseminations in chromitites similar to those of the global PGE mineralization. They could not have been the cause of large deposits but are more important as the sources of placer deposits.

10.3.1.1 Neoproterozoic Ultramafic Rocks

Shishged Ophiolite Complex (1)[1]

Few ultramafic bodies occur in Precambrian rocks. One example is 800 Ma the Shishged ophiolite (Fig. 10.1) that occurs in the northernmost part of Mongolia and hosted in Neoproterozoic rocks, metamorphosed to greenschist and

[1]*Numbers as cited in* Table 10.1

epidote-amphibolite facies (Kuzmichev et al. 2005). Metamorphosed marbles, dolomites, and gneisses are also present. They are superimposed by an oncolith-bearing sequence of the Ediacaran-Cambrian age. The Shishged ophiolite is 200 km^2 in size and elongated in N-S direction to 55–60 km. This ophiolite is composed of serpentinized dunites and harzburgites and rarely wehrlites. In the contacts with host rocks, ultramafic rocks are metamorphosed forming olivine-enstatite-talc-antigorite rocks, locally with carbonate, anthophyllite, and tremolite. Ultramafic rocks are intruded by gabbro (Agafonov and Stupakov 1983). A large gabbro body occurs in the core of ophiolite system. Here in the contact of gabbro and serpentinites, nephrites were found. Chromitite bodies are seen only in the dumps, not in outcrops, because of very difficult alpine relief and intense tectonic activity. Other ultramafic plutons occur in the Russian territory. PGE mineralization is recognized only in two samples of chromitites and is represented by iridarsenide and rutheniridosmine and contains Ni, Au, and Ag.

The mineralogy is characterized by sulfides, arsenides, and sulfoarsenides of Pt, Os, Ru, Ir, and Rh. The Os-Ru-Ir alloys are limited with predominated cubic solid solutions. Also, sulfides and arsenides of Ni are present with awaruite, native Ni, and native platinum.

Naran Ophiolite (5, 6)

The Naran ophiolite is the lower part of the Khantaishir ophiolite complex of 570 Ma (Janousek et al. 2018), and its outcropping rocks are usually harzburgite and serpentinite formed by alteration of harzburgite. Among these, up to 5 m thick lherzolith and dunite bodies occur (Fig. 10.2). Serpentinite formed after pyroxenite is composed of relatively large prisms of bastite, reaching sizes of 0.07–2.6 mm. Ore minerals such as magnetite and chromite occur as accessories. In harzburgite, orthopyroxene content is 10–35 vol%. Dunites with chrome spinel particles are rare. Lherzolite consists of olivine, orthopyroxene, and clinopyroxene and occurs within harzburgite as small lenses. The northern boundary of the Naran, ophiolite extends from the southeast to the south to serpentinite body, with the formation of serpentine-carbonate and quartz-carbonate rocks. This body consists of 20–60 vol% calcite and 20–25 vol% chlorite.

The northern part of the Naran ophiolite contains a 10x3 km chromitite zone from Taliin Bulag to Jongiin Khoid Am. A secondary type of chromitite is widely spread from northwest to southeast on the north part of the ophiolite sequence. Chromitite is characterized by a relatively high content of PGE. Mineral compositions of chromitite: 70.3% chromite, 22.8% olivine, 3.3–2.1% amphibole, 1.1% magnetite, 0.2% hematite, 0.1–0.15% Ni, pyrite, pyrrhotite particles, gold, and reddish iridosmin grain, were identified. Rutheniridosmine in the chromitite ore is found in one sample with iridium, ruthenium, and osmium composite sulfide, while another sample contains laurite, irarsite, rhodium, nickel, and arsenic. According to Sharkhuukhen and Gerel (2001) in Naran chromitite Ir, Os and Ru are common.

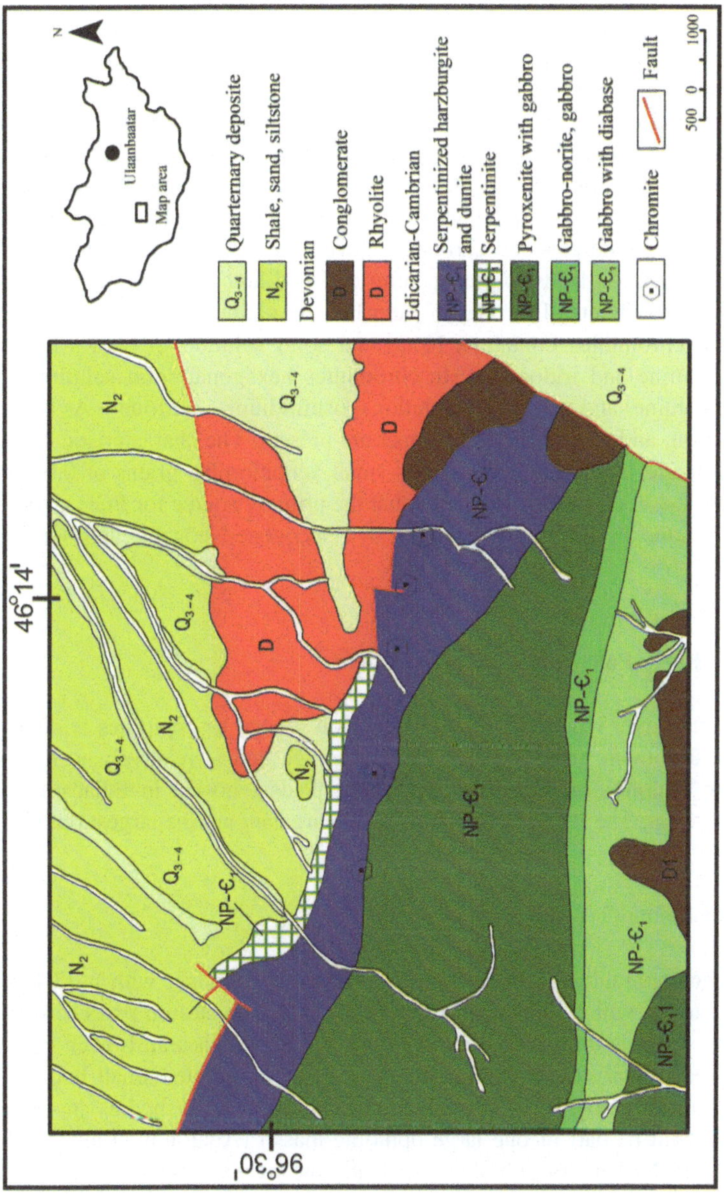

Fig. 10.2 Schematic geological map of Naran ophiolite massif (Sharkhuukhen and Gerel 2001)

Conglomerates of Bayan Nuur Lake (7)

Conglomerates occur in the Agardag fault zone. The Agardag ophiolite a part of the Agardag ophiolite complex of 570 Ma is situated in the territory of Tuva, but the Agardag fault zone extends to Mongolian territory. Bayan Nuur Lake conglomerates (7) contain fragments of different rocks and minerals, including chromitite fragments up to 0.3 m in size from large ophiolitic sequences (Sharkhuukhen and Jamsrandorj 1992). The PGM are concentrated in chromitites. Conglomerates are displaying alteration and intercalation with carbonate-clay rocks. The matrix also has gravel-carbonate-clay composition. Besides chromitites, conglomerates contain fragments of gabbro, serpentinites, serpentinized peridotites, dunites, slates, magnesites, and vebsterite. Conglomerates and carbonate-clayey rocks are intermediate collectors, and their destruction leads to the opening of PGM and the supply of Quaternary loose coastal sediments of the Bayan Nuur Lake (Agafonov et al. 1991).

Massive chromitites contain a hexagonal solid solution of Os-Ru-Ir system: rutheniridosmine and iridosmine. In chromitites hexagonal solid solution of Os-Ru-Ir (iridosmine) and cubic solid solution—osmiridium—are found. As admixture rhodium, iron, and also nickel and copper are present. The characteristic of PGE is high Pt content. Isoferroplatinum forms small xenomorphic grains of 0.01 mm in size. Agafonov et al. (2005) suggested that the primary source for PGE could be an ultramafic nonexposed body, covered by Paleogene-Neogene and Quaternary sediments.

10.3.1.2 Early Paleozoic Ultramafic Rocks

The most common type of PGE mineralization related to ophiolites is distributed widely in ultramafic rocks of early Paleozoic age and is relatively well studied. Ultramafic plutons occur in the Khankhukhii Ridge, hosted in Ediacaran-Lower Cambrian rocks. The Ichitgol, Dumberel, and Jargalant are the largest ophiolites.

Ichitgol Ophiolite (2)

The Ichitgol ophiolite is situated in the Khankhukhii Ridge, with a 27 km long latitudinal trend, with widths from 10 m up to 3 km. The eastern part is covered by Devonian volcanic-terrigenous rocks, dissected by Ediacaran-Lower Cambrian volcanic-sedimentary deposits. Serpentinites occur as isolated small bodies. Geophysical survey recognized that these isolated serpentinite bodies in Devonian sequence could be part of one large ophiolite massif (Agafonov et al. 1990). The major part of the Ichitgol ophiolite, especially the western part is represented by a melange composed of volcanic-terrigenous and intrusive rocks of different ages. This melange zone is hosted by Ediacaran-Lower Cambrian carbonaceous-terrigenous and volcanic-terrigenous rocks. Ultramafic rocks of the mélange zone are represented by serpentinites with dominant antigorite-serpentine-lizardite.

Petrographic studies have shown that primary rocks were harzburgites. Serpentinites are altered to quartz-carbonate rocks (listvenites). Cataclastic serpentinites display strong foliation with a polished surface. Several dykes of granite, basaltic andesite, and diabase cut the serpentinites.

Mélange includes black siliceous rocks, locally with an admixture of carbonate blocks of different size. Terrigenous rocks varying from siltstones to conglomerates in places chromitite bodies occur in the mélange zone. Chromitites contain sulfides, sulfoarsenides, and hexagonal solid solution of Os-Ru-Ir system. Only three PGE-bearing minerals were found. One is laurite-erlichmanite, another is ruarcite-osarsite, and the third one is rutheniridosmine. Sulfide and sulfoarsenide of Ru and Os contain Ir component.

Dumberel Ophiolite (3)

This ophiolite is 13 × 9 km in size. Around this ophiolite, listvenites and serpentinites controlled by subsidiary faults occur as xenoliths in younger Devonian granite and as small bodies in the carbonate-terrigenous rocks. Host rocks are represented by Ediacaran-Cambrian basaltic andesites intercalated with schists. The upper part of the section is composed of recrystallized limestone with siltstone, sandstone, and clastic rocks. The Dumberel ophiolite complex is surrounded in all sides besides its western contact by Devonian pink granite and tectonic wedge in the western contact (Agafonov et al. 1991). The serpentine-dominated mélange zone is developed in the southern contact. Foliated serpentinites, serpentinized harzburgites, and dunites occur on the margin of the ophiolite. Gabbroid bodies are occasionally found in association with pyroxenites. Relatively fresh ultramafic rocks occur mainly in the western part of the ophiolite. Harzburgites are dominant in the eastern part and dunites in the western. Alteration is represented by serpentine, talc, and magnesite replacements. Common are mafic dykes of fine-grained texture, sometimes replaced by rodingites. Nephrite veins are also present. Chrysotile asbestos occurrence as shown by geophysical work was found in the core part of the massif. There are two occurrences of chromitites. An eastern one occurs as disintegrated blocks, and the western one, is elongated in the longitudinal direction up to 250 m and 15 m wide zone. Here chromitites occur as veins, schlieren, and disseminated varying from poor to massive. The maximum vein thickness is up to 0.5 m. Sometimes chromitites mantle dunites, showing that chromitites were formed after dunites. Chromitites also contain dunite xenoliths.

The PGE mineralization only occurs within chromitites from a western occurrence. Total PGE content is about 1 g/t, and Ir has a high content of Pt and Ru. Platinum group minerals are concentrated in chromitites, in ore-bearing dunite, only one example of erlichmanite, Ag, minerals of cobaltite-gersdorffite, and Ni sulfide—maucherite—are present. Barren dunite contains only maucherite. Together with PGE, native Cu and Ni sulfides occur. Pt-Fe alloys are common in the Dumberel pluton.

Jargalant Ophiolite (4)

In the Khankhukhii Ridge, this is one of the largest bodies together with the Dumberel ophiolite and Ichitgol melange zone. The Jargalant ophiolite is 17 km long and from 0.5 to 3 km wide (Agafonov et al. 1990) and consists of dunite-vehrlite-clinopyroxene gabbro intrusion. The intrusion is located in the central part of the Khankhukhii Ridge and is related to a deep fault zone, which separates highly metamorphosed Precambrian rocks from Ediacaran-Cambrian terrigenous-volcanic rocks. Fault zone is E-W oriented and complicated by numerous dislocations along a longitudinal direction, which break up the pluton into different size blocks. Ultramafic and mafic rocks have tectonic contacts.

In the northern contact area, the Archean rocks, composed of gneisses, migmatites, amphibolites, and crystalline schists, are intruded by mid-Paleozoic granitoids. The quartz-sericite-chlorite, chlorite-epidote-actinolite schists, basaltic andesites and their tuffs, sandstones, marbled limestones, and quartzites of Ediacaran-Cambrian age occur in the southern contact. Small subvolcanic bodies and dykes of gabbro-diabase cut this sequence.

Ultramafic rocks are represented by serpentinized dunites, vehrlites, and serpentinites. Very characteristic is banding, shown by the alternation of gabbroid, pyroxenite, vehrlite, and dunite. All rocks are altered, and dunites are strongly serpentinized. Additional banding is related to alternation of leucocratic and melanocratic layers, composed of gabbro, clinopyroxenite, and vehrlite.

There are three chromitite occurrences, two of them occur in the western part of the ophiolite unit forming disintegrated blocks and one in the eastern part. The eastern occurrence is represented by massive and disseminated chromitite. Mineralization occurs in a 15 m thick layer of serpentinites after dunite, alternating with altered olivine diopside. Chromitites are poor, sparse, disseminated, and massive. Massive and highly disseminated chromitites form lenses up to 10 cm thick, forming a chain-like sequence in a banding structure with serpentinized dunites. Poorly disseminated chromitites form bands up to 1 m thick, which contain schlieren up to 0.2 m in thickness. These chromitites contain PGM, represented by sulfoarsenides and Os-Ru-Ir alloys. Besides it, there are erlichmanite, silver, heazlewoodite, maucherite, and minerals of cobaltite-gersdorffite. In heavy concentrates sperrylite and gold were found (Agafonov and Stupakov 1983).

Khutel Ophiolite (8)

This ophiolite occurs in the western part of Zavkhan ultramafic belt, 20 km south of Urgamal soum. It is one of the largest ophiolites (95 km^2) localized in the early Paleozoic structures. The Khutel ophiolite is located in Zavkhan deep fault zone, which separates Proterozoic metamorphic rocks from Ediacaran-Lower Cambrian volcanic-sedimentary rocks. The ophiolite has a tetragonal shape, broken down by faults of different trends into blocks, dissected by fault wedges that are composed of Precambrian metamorphic rocks metamorphosed in epidote-amphibolite facies.

Near the eastern contact, blue quartzites together with amphibolites overlap ultramafic rocks.

The Khutel ophiolite f is composed of serpentinites; sometimes relics of serpentinized harzburgites and dunites are preserved with quartz-carbonate replacement. Along the western contact antigorite serpentinites are common; in other sections chrysotile serpentinites with chrysotile asbestos and asbestos-bearing veins occur.

Granites and gabbro intruded the ultramafic rocks. Three occurrences of densely disseminated chromitites are present. Chromitites are highly affected by serpentine-carbonate alteration, and chrome spinel is replaced by magnetite. Few grains of sulfoarsenides approximately 0.03 mm in size occur and represented by irarsite-hollingwortite with Pt, Os, Fe, and Ni admixture. The most important is a ruthenium admixture.

Bayannuur Ophiolite (9)

The Bayannuur ophiolite occurs within Ikh Darvi Ridge, comprised of Precambrian, Ediacaran-Lower Cambrian, and Middle Paleozoic terrigenous-volcanic rocks (Pinus et al. 1984). Within the Ikh Darvi Ridge, two ultramafic bodies occur. One is the Alaguul is composed of serpentinized dunites, harzburgites, and lherzolites with known chrysotile-asbestos and chromite mineralization. The PGE mineralization in chromitite and in the mélange zone in the southwest part of the ridge was not found. The second ultramafic body, named Bayannuur (Izokh et al. 1990), is a layered dunite-wehrlite-clinopyroxenite and a gabbro intrusion. On the top of sequence, gabbro cycles with wehrlite, pyroxenite, and leucocratic gabbro. Contacts between rocks within one cycle are gradual. In meso- and melanocratic gabbro, titanomagnetite and sulfide disseminations occur. Bottom of layered intrusion is represented by serpentinized dunites changed to wehrlites, olivine clinopyroxenites and clinopyroxenites. The PGM mineralization is in chromitites which form disintegrated blocks in serpentinized dunites.

PGM in chromitites are mostly laurite-erlichmanite. Some ore minerals are sulfoarsenides: irarsite-osarsite-ruarsite alloys. PGM minerals are about 0.05 mm in size. There are intergrowths of erlichmanite-laurite with laurite, and osarsite-ruarsite-irarsite with ruarsite-osarsite-irarsite. In general, PGE minerals in Darvi Ridge are represented by iridium, which forms mineral kashinite $(Ir, Rh)_2S_3$. Sulfides and sulfoarsenides contain Rh, Fe, and Ni from 0.06% to 1.79%, and Co content is low.

Godroi Ophiolite (10)

The Godroi ophiolite is situated within the Khovd ultramafic belt and comprised of isolated lens-like bodies forming up to 16 km NW-trending sequence. The ophiolite is in the Godroi Uul mountain and is 5 km long and 1.5 km wide. Host rocks are represented by Lower Cambrian volcanic-siliceous rocks. All contacts with ultramafic rocks are tectonic. In the northwestern part, pluton is covered by red terrigenous Devonian rocks (Stupakov et al. 1994).

The ophiolite is composed of serpentinites formed after dunites and harzburgites. They are replaced by carbonate and talc forming listvenites and talc-carbonate alteration. One primary occurrence of massive chromitite is 15 m long and 0.5 m wide elongated in a sub-longitudinal direction. The other two, occur as diluvial disintegrated blocks 2 km north of the primary chromitite occurrence, one in the eastern contact within listvenites and another broke-up block in the southern part. Dykes of gabbro-diabase and diabase dykes cut serpentinites.

PGE mineralization was discovered in chromitite. PGM grains are 0.1–0.2 mm large; one is 1 mm in massive chromitite. The PGM mineralization occurs as sulfides and As-poor sulfoarsenides of the Os-Ir-Ru system in a proportion of 60 to 40%. One grain was ruthenosmirid with iridium and platinum and osmium and ruthenium. Additionally, small grains of silver (0.05 mm) were recognized in chromitites.

Pridorozhny Ophiolite (11)

Pridorozhny ophiolite (Agafonov et al. 2005) consists of contiguous lens-like bodies along the left slope of the Khovdgol river valley. In the western part, the largest body reaches up to 6.5 km and an E-W extension. To the east lens-like bodies trend from NNE up to a northerly direction. The host rocks are represented by Ordovician sericite-chlorite slate, siltstone, sandstone, conglomerate, and diabase. All contacts with ultramafic rocks are tectonic. Ultramafic rocks are composed of serpentinites after harzburgites. Serpentinized dunites are relatively rare. Along the contacts, ultramafic rocks and host rocks display strong foliation with host rock enclosures in the ultramafic rocks. Near the contact, serpentinites are replaced by quartz-carbonate and talc-carbonate rocks (listvenites), especially in the largest western body, where altered gabbro is present. Gabbro is medium-grained with low-grade sulfide disseminations. Wehrlites and clinopyroxenites are locally associated with gabbro.

One occurrence was found with broken diluvial fragments of massive and disseminated chromitites. A small PGM of 0.05 mm in size was recognized. Hexagonal Os-Ru-Ir alloys are quite common, besides rare sulfoarsenides of irarsite-osarsite composition and silver. All PGM contain Pt, Rh, rhodium, Cu, and Fe and sulfoarsenides Ru and Ni.

10.3.1.3 Mid-Late Paleozoic Ultramafic Rocks

In South Mongolia, PGE mineralization is associated with Mid-Late Paleozoic ultramafic belts. Major ophiolites sequences are Kharaat Uul, Ulaakhiid, Khalga, and ultramafic bodies in the Gurvansaikhan Ridge. Ophiolites are hosted in mafic volcanic rocks, intercalated with sandstones, siltstones, and limestone of Devonian and Early Carboniferous ages. These ophiolites are quite small varying from 1–2 × 0.3 km up to 10–25 × 3 km. Ultramafic rocks are represented by dunites or serpentinized dunites and harzburgites and also wehrlites and lherzolites. Ultramafic rocks are replaced by quartz-carbonate and jasper-like silica rocks. Chromitites form zones up to a width of 5 m and from 0.1 to 1 m wide subvertical lens-like bodies. PGE mineralization is represented by Ir-Os-Ru-Fe alloys. In addition, native Ag and awaruite are present. Os-Ru-Ir hexagonal alloys are the most common, enriched by Pt; sperrylite and ferronickel platinum are rare.

Kharaat Uul Ophiolite (12)

The Kharaat Uul ophiolite is located in South Mongolia within folded Paleozoic structures near the Ulaan Khiid ultramafic ophiolite. In chromitites from the Kharaat Uul ophiolite Os-Ru-Ir-Fe alloys are recognized (Agafonov and Khmel'nikova 1991). These two ophiolitic sequences: Kharaat Uul and Ulaan Khiid, are dissected by a 7 km wide valley, filled by Quaternary loose sediments. Some authors (Pinus et al. 1984) described these two ophiolites s as one body, but this is not confirmed by geophysical data. Distinct PGE mineralization in chromitites from these two ophiolites confirms that they are two separate ophiolite bodies. The Kharaat Uul ophiolite is hosted in mafic volcanic rocks alternated with sandstones, siltstones, and limestones of Devonian-Early Carboniferous age. This ophiolite is 2 × 0.3 km in size and composed of serpentinites after dunites and serpentinized dunites. Pegmatoidal websterites, enriched in orthopyroxene, and gabbro-pegmatites occur in the eastern part of the ophiolite body. Weathering crust developed here, and also in the western edge of the ophiolite, birbirites with veins and networks of dolomites and magnesites are present. Quartz-carbonate alteration is relatively rare.

Six chromitite occurrences are identified within a zone up to 5 m wide, which contains thin (from 0.1 to 1.0 mm) subvertical chromitite bodies. These bodies are covered by loose sediments. In the core of ophiolite body, massive chromitites occur as thin (about 0.2 m) veins extending for up to 100 m. Chromitites are massive or disseminated. Carbonate alteration is in places very strong. The PGM are characterized by Pt, Ni, Fe, Cu alloys, diarsenide Pt-sperrylite and Ni sulfide – hizlewoodite. Mineral grains are about 0.1 mm in size. Ferronickelplatinum grain's shape is irregular, but sperrylite always has a regular shape. Platinum content in PGM of Pt-Fe-Ni-Cu system varies from 32.88 to 54.08 wt.%, Ni from 22.30 to 47.54 wt.%, and Cu up to 4.92%. Sperrylite contains an admixture of rhodium, sulfur and As.

Ulaan Khiid Ophiolite (13)

The Ulaan Khiid ophiolite is in Late Paleozoic folded belt and hosted in Devonian-Early Carboniferous mafic volcanic rocks and clastic sediments. This ophiolite has a lens-like shape of 3x10 km in size and is oriented in the latitudinal direction. The Ulaan Khiid ophiolite is composed of serpentinized dunites and peridotites (harzburgites are dominant, lherzolites and wehrlites are subordinate). Fine- and medium-grained gabbro and granitoids cut ultramafic rocks. Veins of a giant- and coarse-grained orthopyroxenites are very common. Old weathering crust up to 10 m thick, composed of carbonate-silica rocks with alternations of magnesites, dolomites, and birbirites, is preserved in some part of the ophiolite (Bannikov 1986). Within the weathering crust in erosion windows along dry ephemeral riverbed, dunites can be seen. Within dunites, many lens-like chromitite bodies varying from few cm to 0.5 m thick and up to 50 m long have been discovered. Chromitite bodies exhibit a latitudinal trend and steeply dip to north.

Chromitites are disseminated, massive, and nodule-like types. Chromspinelides in dunites and chromitites are similar. The PGM discovered only in very few samples occur as small grains of 0.1 mm in size, and only one reaches 0.25 mm. PGM (Ir-Os-Ru-Fe alloys) are represented by rutheniridosmine, iridosmine, and osmiridium. Platinum is found in some grains, and in all PGM, Rh content is up to 4 wt. %, and Ni and Ag are present.

Khalga Ophiolite (14)

This ophiolite is situated 20 km south of Sainshand town. It is a small ultramafic body about 1 km long and 0.3 km wide. The ophiolite has a northeast orientation, different from others, which have a mainly latitudinal direction parallel to the major structures in South Mongolia. Geophysical anomalies show similar orientation, and it is possible that this ophiolite is associated with northeast structures of Carboniferous age (Agafonov et al. 2005).

The Khalga ophiolite comprises foliated serpentinites and serpentine-quartz-carbonate composition (listvenites) rocks. The major part of the ophiolite is covered by Quaternary sediments. In its southern part, the ophiolite is covered by Jurassic terrigenous-carbonate deposits, in the middle part by Carboniferous terrigenous rocks with relics of Jurassic basalts. Eroded products of ultramafic and mafic rocks (conglomerate, clastic rocks, and sandstone) cover the ultramafic rocks. The presence of altered mafic rocks and chromitites can be recognized in rounded fragments of terrigenous rocks. Small gabbroid bodies cut ultramafic rocks. In the northern part of the ophiolite, small lenses (5x1 m) of strongly cataclastic and altered chromitite occur.

PGM occur only within altered chromitites that contain Cr-garnet-uvarovite. PGM are rare and represented by iridosmine and native Os. Iridium and ruthenium are common admixture; Fe, Ni, and Cu are rare. Iridosmine contains Pt.

Gurvansaikhan Ridge (15)

A large number of ultramafic bodies are mapped in the Gurvansaikhan Ridge. These bodies are connected to each other or isolated by overlying Quaternary loose sediments. There are two main areas, north and south. The first one is in the southern belt of ultramafic rocks, and the second is the Ulaankhudag ophiolite in the northern part of ultramafic rocks.

The Ulaankhudag ophiolite is situated in the central part of the Gurvansaikhan Ridge, extends in a latitudinal direction for up to 25 km, is up to 2.5 km in width, and has tectonic contacts with vertical dips. This ophiolite is hosted in sedimentary-volcanic rocks: siltstones, silicic and carbonate rocks, diabase, basaltic andesite, and pyroclastic rocks. Ultramafic rocks contain fragments of host rocks, as well as gabbro, listvenites, and rodingites.

Ultramafic rocks are intensively serpentinized without relics of primary minerals, besides chromspinelides. Serpentinites are replaced by quartz-carbonate and silica (birbirites). Clinopyroxenites, wehrlites, amphibolites, and rodingites are rare. Chromitites occur as boudins or as diluvial fragments. PGM occurs as densely disseminated and massive chromitites with individual grains of 0.05 mm in size. PGM is found as Os-Ru-Ir alloys and laurite-erlichmanite. In the Gurvansaikhan Ridge, PGM is represented by dominated rutheniridosmine, iridosmine, and Rh-iridium. Os-Ru-Ir alloys contain Pt from 6.12 to 11.88 wt. %. Cu and Ni are rare.

10.3.1.4 Other Ophiolite Systems in Paleozoic Structures

Very few mineral grains of PGM, Au, and Ag were found in some ophiolites: Murun, Khukhmorit, and Sulinkheer. Some of them: Sharnuur, Bargilt, Dashuul, and Khets, are controlled by the Tolbonuur deep fault zone and Dolon and Khavtsal Gol by the Khovd fault zone.

Murun ophiolite consist of a series of serpentinite bodies within the Kherlen fault zone (Agafonov and Stupakov 1983) 20 km west of the Murun sum. Serpentinite bodies vary from the few hundred meters to 2 km and are from 0.1 to 1.0 km in width. Host rocks are represented by Ediacaran-Lower Cambrian terrigenous-volcanic rocks. Massive and disseminated chromitites are hosted in serpentinites as broken blocks. PGE mineralization is absent, but gold is present.

Khukhmorit ophiolite is situated on the left side of Zavkhan Gol River, 40–50 km northwest of Khukhmorit soum. Most of this ophiolite body is covered by Quaternary sediments and was mapped only by a magnetic survey. Host rocks are represented by Ediacaran-Lower Cambrian rocks. Ultramafic rocks are totally serpentinized. To the east of the ophiolite, a number of ultramafic bodies occur. Broken blocks of chromitites contain mineral grains of Os.

Sulinkheer Ophiolite (16)

The Sulinkheer ophiolite complex (253 Ma) is located within Late Paleozoic structures on the border with China and comprises most of the Suujiin Nuruu ridge. The major part of the ophiolite is situated in Chinese territory, where chromitite deposit occurs. The ophiolite is strongly weathered, and ultramafic rocks are replaced by silicic rocks, magnesites, and serpentinites preserved only in the deep erosion channels. Three occurrences of chromitites are known. They form lenses up to 3 m long and 0.5 m wide. Fifteen chromitite bodies were recognized during the 1:50,000 scale mapping project in 1959. Chromitites were not analyzed for PGE. There are a number of other occurrences of chromitites within serpentinized dunites near the Mongolian-Chinese border. Very few grains of PGM were found in chromitites of Bargilt, Khets, and Dushuul ophiolite massifs.

Altan Uul Occurrence (17)

The Altan Uul occurrence is located in southern Mongolia, where island arc and continental arc terranes were accreted during the Paleozoic (Badarch and Orolmaa 1999). The island arc terranes are composed of Devonian to Carboniferous tholeiitic and calc-alkaline pillow lavas, volcanoclastic rocks and contain Silurian-Devonian ophiolites. The geology of this region is dominated by Paleozoic volcanic and sedimentary rocks that are intruded by Late Devonian and Carboniferous granitoids and overlain by Mesozoic sedimentary rocks. The regional structure is characterized by east-west trending strike-slip faults. The geology of the Altan Uul area consists of Late Silurian and Early Devonian units that contain sedimentary rocks and intermediate-composition volcanic rocks of the Nomgon Formation (Fig. 10.3). The lower sequence of the Nomgon Formation consists mainly of intermediate volcanic rocks (andesite and dacite), and the upper sequence consists of sandstone and shale.

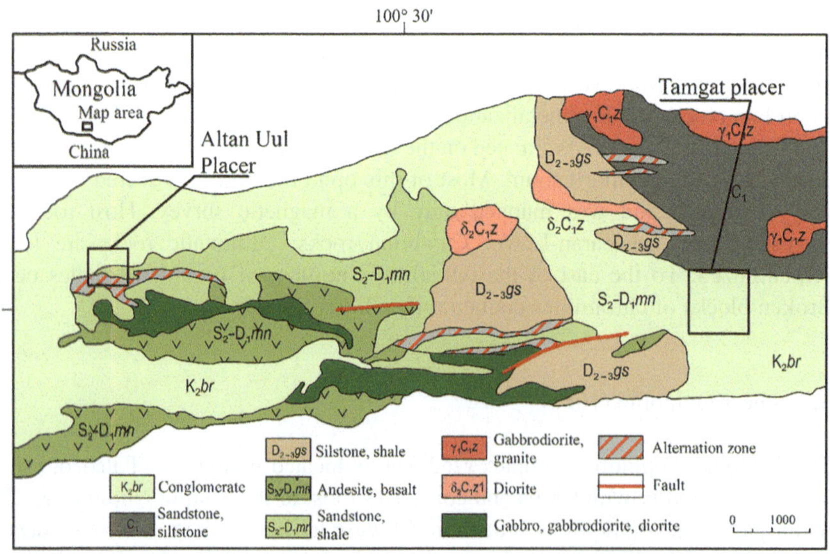

Fig. 10.3 Schematic geological map of Altan Uul area (Gerel et al. 2002)

Table 10.2 Chemical composition of PGE minerals

Sample number	Ir	Os	Ru	Fe	Mineral
Alt-99/5	38.00	33.70	27.80	0.10	Rutheniridosmine
Alt-99/6	44.10	50.20	5.20	0.10	Iridosmine
Tam-99/7	78.60	17.20	3.8	0.10	Iridosmine
Tam-99/8	38.80	40.50	20.10	0.20	Rutheniridosmine

The lower sequence exhibits silica alteration with pyrite and cut by east-west trending faults along with ophiolites consisting mainly of serpentinized harzburgite and dunite. The lower sequence is also intruded by gabbrodiorite and diorite plutons that are calc-alkaline, rarely tholeiitic, low and medium K, and metaluminous. The plutonic rocks exhibit characteristics of I-type granitoids. The upper sequence of the Nomgon Formation is composed of coarse-grained sandstone and shale that extend up 4–5 km long and 50–100 m wide in outcrop and contain quartzite bodies, quartz veins, and small serpentinite bodies. Overlying the Late Silurian is Early Devonian represented by a sequence of up to 7300 m in thickness consisting of thin conglomerate layers and shale, sandstone, siltstone, and gravel and contains silica altered carbonate layers with crinoids. These island arc rocks are overlain by a Carboniferous clastic sequence mostly conglomerate and sandstone. The conglomerate layer is from 1 to 2 m thick and up to 300 m long, containing quartzite pebbles. The conglomerate is crosscut by a zone with chromite serpentinite with pyrite and limestone. These rocks are intruded by Early Carboniferous gabbrodiorite-diorite-granite suite that is calc-alkaline, medium to high K and show geochemical characteristics of a continental arc. A widespread unit of the Late Cretaceous Baruungoyot Formation overlies the Late Silurian and Early Devonian sequence and forms topographic outliers on hilltops. The formation consists mainly of coarse- and medium-size pebbles, which range up to 2–3 m thick, is poorly sorted and locally contains large ultramafic boulders up to 1 m in diameter. The conglomerate also contains pebbles of all older units in the area, including Early Carboniferous conglomerate, granite, sandstone, shale, and rare ultramafic rocks. Baruungoyot Formation has two sequences. The lower sequence contains mainly a red sandstone matrix-supported conglomerate, and the upper sequence consists of a green pebble conglomerate with smaller clasts that range up to 10–20 m in thickness. Gold and PGE are concentrated at the base of the conglomerate, whereas the upper sequence is relatively poor in gold and PGE. The PGE placer deposits are interpreted as genetically related to the lode PGE occurrence.

PGE occurs in lenses or veins within an ophiolite sequence of ultramafic rocks that crop out as hydrothermally altered bodies that occur along east-west trending faults. Ultramafic bodies are completely serpentinized, and MgO/SiO_2 ratios vary from 1.01 to 1.05 suggesting that the primary rocks are dunite and harzburgite (Gerel et al. 2002). The serpentinized ultramafic rocks contain 0–0.1 ppm Pt, from 0.002 to 0.005 ppm Pd, 0.002 ppm Rh, 0.1 g/t Ir, 0.005 ppm Ru, 1500 ppm Ni, 500 ppm Co, and 3000 ppm Cr (Table 10.2). The placer PGE deposits contain Os, Ir and Ru alloys and chrome spinel. These relations indicate a genetic link between the ultramafic rocks and the PGE-bearing Cretaceous conglomerate. A complex geochemical anomaly that is defined by high values of Ni, Co, Cr, Cu, and Zn coincides with the ultramafic and diorite bodies and contains characteristic high values of Cr. The Ir-Os alloys are predominant and contain more Os than Ir and Pt, and platinum arsenides and sulfides are accessories. The Ir-Os alloys occur as fine grains and rarely as nuggets in the conglomerate.

Sharnuur Ophiolite

Sharnuur ophiolite is hosted in Devonian red beds, composed of conglomerates, sandstones, siltstones, and lesser felsic and intermediate volcanics. Ultramafic rocks are comprised of serpentinites replaced by quartz-carbonate alteration. Gold, cinnabar, cobaltite-gersdorffite, orselite, and pentlandite have been found in these quartz-carbonate rocks (listvenites). PGM have not been recorded.

Bargilt Ophiolite

Bargilt ophiolite is about 3 km long and hosted in Devonian red beds in the western contact and in Ordovician rocks in the eastern contact. The ophiolite is related to a fault, which divides it into two sequences. The Bargilt ophiolite is composed of serpentinites, listvenites, and rodingites. The ophiolite contains fragmentary blocks of chromitites with Os sulfide erlichmanite.

Dashuul Intrusion

Dashuul intrusion is elongated in the northwest direction up to 6 km and occurs in the western part of a large body of serpentinite. The largest fragments occur near the Khovd Gol (25×25 m). Ultramafics are represented by serpentinites, locally with quartz-carbonate replacement. Pyroxenites and rodingite boudins are also present. The pluton is situated in a fault zone and is hosted in volcanic-siliceous-shale rocks. Chromitites contain few grains of erlichmanite and silver.

Khets Pluton

Khets pluton is divided by Khovd River into two parts. The right side is represented by listvenites, which are elongated in a northwest direction up to 5 km and is named the Ulaan massif, on the left riverside is named the Khets massif. The Ulaan massif is hosted in early Cambrian rocks, while the Khets massif is in Devonian red beds. Ultramafic rocks are represented by melanged serpentinites, which contain boudins of rodingites, chromitites, and host rocks. One chromitite contains very few grains of Rh, Os, and Ag.

Dolon Pluton

Dolon pluton is an extension of the Godroi pluton. Here a small occurrence of Cu is known (chalcopyrite and siderite). Ultramafic rocks form small lenses of strongly melanged and foliated serpentinite. Chromitite block contains only some gold, and Ag sulfide – acantite.

Khavtsalgol pluton is located in the Khovd fault zone and composed of melanged serpentinites, within which rodingites and listvenites occur. This has the largest chromitite occurrence in Mongolia (60x10 m), which is situated in the northern part of the intrusion (Agafonov and Stupakov 1983). PGE mineralization is not found; only Ni sulfide heazlewoodite is present.

10.3.2 Mafic-Ultramafic Intrusions: Associated PGE Mineralization

Almost 90% of the world PGM mining is mined from mafic-ultramafic intrusions. Most mafic-ultramafic intrusions include chromite and titanomagnetite, as well as copper and nickel sulfide mineralization. In Mongolia, the main mafic-ultramafic intrusions are Oortsog and Dulaan (Arkhangai Province), Nomgon (Selenge Province) and Oyu Tolgoi (Zavkhan Province).

10.3.2.1 Oortsog and Dulaan Mafic-Ultramafic Layered Intrusions (I)

The Oortsog and Dulaan intrusions are located in the northern margin of the Khangai-Khentii Basin, close to the southern margin of the Permian-Triassic igneous belt (Mao et al. 2018). Zircon U-Pb dating indicates that the emplacement age of the Oortsog and Dulaan intrusions is ~270 Ma (Shelepaev et al. 2015). The Oortsog and Dulaan plutons intruded into Precambrian crystalline schists, gneisses, and amphibolites. The Oortsog intrusion consists of ultramafic unit (lherzolite) and mafic units (olivine-gabbronorite and gabbronorite) (Izokh et al. 1990). The lherzolite unit occurs as an irregular, sheet-like body within the mafic unit and has been severely altered into serpentine, which shows a poikilitic texture (Mao et al. 2018). The Dulaan intrusion is composed of a bulk mafic unit and some small, irregular ultramafic unit. The ultramafic units comprise plagioclase lherzolite and plagioclase olivine websterite. The main mafic unit is composed of olivine gabbronorite and gabbronorite and hornblende gabbro (Mao et al. 2018). The sulfide minerals consist dominantly of pyrrhotite, pentlandite, and mackewnite, with only fewer amounts of chalcopyrite and pyrite, and also chromite, magnetite, and ilmenite (Altanzul and Dandar 2008). The Cu-Ni-PGE mineralization related to the layered intrusion has been described by Sharkhuukhen and Baldorj (1993). According to this study, several primary zones of Cu-Ni-Co with average contents of Cu

0.005–0.03%, Ni 0.003–0.15%, and Co 0.003–0.01% have been distinguished. Also, a primary Cr zone with content of Cr 0.002–0.05% was detected. In pyroxene gabbro the Pt content is 10 ppb, Pd 30 ppb, Ir 2 ppb, Rh 5 ppb, and Ru 1 ppb; gabbro generally contains lower amounts of PGE (Pt 6 ppb, Pd 20 ppb, Ir 3 ppb, Rh 3 ppb, and Ru 1 ppb) than pyroxene gabbro. Serpentinite contains Pt 1 ppb, Pd 2 ppb, Ir 2 ppb, Rh 5 ppb, and Ru 1 ppb. Sharkhuukhen and Baldorj (1993) suggested that PGM content is not economic, but the lower part of the layered complex with Cu-Ni disseminated mineralization with PGE could be of economic interest. Mao et al. (2018) determined PGE, Ni, Cu, and S concentrations in whole rock as Os 0.04–0.22 ppb, Ir 0.02–0.26 ppb, Ru 0.03–0.21 ppb, Rh 0.08–0.4 ppb, Pt 0.27–5.89, and Pd 0.37–5.83.

10.3.2.2 Nomgon Mafic-Ultramafic Intrusion (II)

The 252 M Nomgon intrusion a (Shelepaev et al. 2015) is located at the southern margin of the Permian-Triassic volcanic-plutonic belt. Nomgon intrusion consists of gabbro, gabbro norite, troctolite, and gabbro with sulfide mineralization (Fig. 10.4). In the northern block in transition zone, gabbro enriched in sulfides has been distinguished. Based on Izokh et al. (1990), sulfide mineralization is primary magmatic and represented by bornite and chalcopyrite, lesser pentlandite, and pyrrhotite. Secondary minerals are Fe and Cu oxides. Disseminated sulfide occurs form a 200–250 m long and 20–25 m wide zone in the northern block. Similar zone is fixed in the southern block of layered series. Poor sulfide mineralization checked in gabbro is up to 100 m long and 10–15 m wide. Within gabbronorites of the northern block of layered intrusion, zone of sulfide mineralization up to 40–50 m long and up to 10 m in width has been discovered.

Similar rocks with visible sulfides are distributed in many zones within the intrusion; sulfide assemblage is dominated by chalcopyrite, whereas pyrrhotite and pentlandite are rare (Mao et al. 2018). Copper contents are 0.22–2.38 wt. %, PGE subordinately with Os of 0.02–0.71 ppb, Ir 0.01–1.03 ppb, Ru 0.01–0.84 ppb, 0.02–4.54 ppb, Pt 2.65–100.2 ppb, and Pd 0.94–282.7 ppb. Mao et al. (2018) concluded that parental magma of the Nomgon intrusion is severely depleted in Ni but significantly enriched in Cu and PGE. Sperrylite with elevated rhodium and sulfur, mertieite, isomertieite, and hollingworthite is developed within troctolites. Stibiopalladinite, arsenopalladinite, mayakite, and compounds of Pd with Bi and Te and Au and Ag are rare. In the upper zone of the sulfide-bearing anorthosites, PGM composition is different. The most characteristic is sperrylite, Pd forms compounds with Bi, Te, and rarely with Pb, Ag with Bi, and Te. Compounds of Pd with As and Sb are missing. Sperrylite is enriched in Rh up to 12.21 wt %. Sperrylite was found also in heavy concentrate.

Fig. 10.4 Geological map of the Nomgon troctolite-anorthosite pluton (Gerel et al. 2002)

10.3.2.3 Khairkhan Layered Intrusion (III)

This mafic-ultramafic is situated in western Mongolia in the southern part of Lake Valley in Bumbat Khairkhan Ridge with an area of about 80 km^2. It is a dunite-troctolite-gabbronorite layered intrusion. Copper occurrence (chalcopyrite-pyrrhotite) is located in the margin of pluton with a Cu grade of 1%. In the intrusion, the gabbro layer contains 1% Cu and about 1 g/t Pd. The Pd mineralization (temagamite Pd_3HgTe_3) is associated with bornite and chalcopyrite (Gertner et al. 2003). The PGM is represented by sperrylite, osarsite, and stibiopalladinite.

10.3.2.4 Ontsuul Mafic-Ultramafic Intrusion (IV)

The Ontsuul intrusion occurs in the early Paleozoic structures in western Mongolia. It is a dunite-vehrlite-clinopyroxenite-gabbro layered intrusion (Agafonov et al. 2005). Poorly disseminated sulfide mineralization is associated with plagioclase peridotite with a high content of platinum up to few g/t, as analyzed by atomic-emission method, but the mineralogical study did not support these analyses. Small grains of erlichmanite and osarsite were identified under an SEM. Besides PGM sulfides, arsenides and sulfoarsenides of Ni, Co, and Fe: pentlandite, pyrrhotite, maucherite, nickeline, and minerals of the cobaltite-gersdorffite group, have been recognized. Co content is 17.97–28.33 wt. %, Ni 5.85–14.82 wt. %, and Fe 2.33–10.40 wt. %. Pentlandite is contained Co from 1.57 to 1.88 wt. %. Ore minerals in the peridotite are represented by ilmenite, low chromium chromspinel, and magnetite. Gabbro-pegmatites are rare and contain few grains of Au, Ag, and Cu.

10.3.3 PGM $\pm$ Au Placers and Occurrences

10.3.3.1 PGM Placers

PGM placers are the main source of production and most important for economic mining. The largest deposits are Goodnews Bay, USA; Tulameen River, Canada; and Central Urals, Russia. There is only one PGM placer in Mongolia in Bayan Nuur, Uvs Province, but it is not economic (Jamsrandorj et al. 1995; 2002).

Several PGM minerals from Quaternary sediments were identified as accessory during exploration of many gold placer deposits including Ulaannachin, South Burgastai, and North Burgastai gold placer deposits Turgen region northwestern Mongolia (Amitan and Dandar 1999); Ulaanbulag, Duvunt, Ulziit gold placer deposits from Bayankhongor area in southwest Mongolia (Dandar and Altanzul 2006; Altanzul and Kitakaze 2007); and Zost gold placer deposit South Mongolia (Banzragch 2000; Sharkhuukhen and Gerel 2001).

Bayan Nuur Lake (1′)

The Bayan Nuur Lake is situated near the northern border of Mongolia to the east of the Uvs Nuur Lake. The lake is northeast oriented and together with Baga Nuur Lake underlies the Agardag fault zone. The major part of this zone is situated in the Tuva territory (Russia), where a number of ultramafic ophiolites are known (Pinus et al. 1955). The southwestern part of this zone is defined by the Khan Khuhei Ridge, where ultramafic bodies occur. The largest one is the Dumberel ophiolite described above. The Agardag zone in Mongolia is covered by loose sediments. In the area of Bayan Nuur Lake, gabbro, diorites, and red beds of Devonian age, conglomerates, and fine-grained sediments occur, which contain pebbles of ultramafic and mafic rocks. The age of conglomerates and fine-grained carbonaceous-clayey sediments is not clear but could be late Neogene. They form at least two sequences with alternating coarse-grained and fine-grained sediments up to 3 m in thickness. Carbonaceous-clayey rocks are strongly weathered in the supergene zone forming brown-reddish ochres. Terrigenous rocks containing fragments of ultramafic and mafic rocks occur in a relatively limited 1 km area. Coarse-grained clastic rocks contain fragments of serpentinites, serpentinized dunites, peridotites, gabbroid and chromitites, and magnesites and birbirites. Weathering of ultramafic rocks is very common in Mongolia. Although weathering crusts are preserved only in South Mongolia, the bottom of weathering crust is found everywhere, including the Khankhukhii Ridge. Supergene weathering results in the formation of siliceous profiles with magnesites and iron-silica rocks (birbirites). The presence of ultramafic and mafic fragments is favorable for Pt prospecting because weathering preserves Pt minerals. Fragments are cemented by fine-grained clayey-carbonate matrix, composed of calcite, dolomite, clay minerals (montmorillonite), quartz grains, chromspinelides, feldspar, mica, serpentine, talc, and artinite. Coastal lake sediments contain a large volume of chromspinelides. Study of pan concentrates of chromspinelides shows the presence of isoferroplatinum and erlichmanite. In the pan concentrates from the river bank and from ephemeral channels show the presence of rhuteniridosmine and Pt rhuteniridosmine, isoferroplatinum, and iridosmine. Primary source occurs close to placers and is covered by breakdown products. The presence of PGE in chromitites from fragments of conglomerates is indicative of a spatial and genetic relation between a primary source and pan concentrates.

PGE in heavy concentrates near the Naran ophiolite (2′). PGE is represented by Os-Ru-Ir alloys, rare isoferroplatinum, and sperrylite. Gold was also found. Comparison with primary PGE mineralization in the Naran pluton and PGE in heavy concentrate shows similarity in mineralogy. The Naran ophiolite is crossed by a number of small rivers and is covered by the Quaternary sediments. The thickness of Quaternary sediments is not very high, and near the ophiolite, the Neogene-Paleogene rocks cemented by carbonate occur. The Neogene-Paleogene sediments are not sorted and form an alluvial fan, where PGE and gold are not concentrated but are dispersed. PGM grains were found in 1984 from alluvium of Bideriin Gol, and sperrylite and awaruite were found from small John Gol. In the dispersion zone, Ru-Ir-Os alloys predominate with isoferroplatinum and tulameenite. Pure gold and native copper were also found. Comparison between PGE from a primary source and alluvial PGE shows predominant Os-Ru-Ir alloys. The primary source contains iridosmine and native Ru. In alluvial materials sulfide, arsenide, and sulfoarsenide of PGE are absent. In heavy concentrates rutheniridosmine rich in Ru and chromitites poor in Ru rutheniridosmine are characteristic. Tulameenite, isoferoplatinum, and ferronickelplatinum were not found in chromitites. All these differences suggest at least two different sources of PGE.

Heavy Concentrate Zones of Ongon Ulaan Uul Ultramafic Rocks (3′)

The Ongon Ulaan Uul is situated in the western end of elongated South Gobi ultramafic belt of latitudinal trend, which includes in the Gurvansaikhan Ridge from east to west number of ultramafic bodies such as Zuulun, Servei Uul, Altan Uul, Ongon Ulaan Uul, and Nogoon Tsav. Another unnamed one up to 13 km long is situated 50–60 km to west of the Ongon Ulaan Uul. All these bodies are remnants of ophiolites composed of serpentinites, replaced by listvenites with relics of weathering crust and silicification. Mafic dykes cut serpentinites. Dykes are more resistant to weathering and are preserved as slopes. Within this belt, ultramafics form positive topography (hills and ridges) between thick upper Cretaceous sediments, which occur as a detection cone. In the rewashed upper Cretaceous, sediments of the Ongon Ulaan Uul PGE minerals varying from 0.5 to 2.0 in size are present. One of the minerals is represented by isoferroplatinum; others are Os-Ru-Ir alloys.

10.3.3.2 PGM in Gold Placers

Ulaannachin Gold Placer Deposit with PGE Nugget (4′)

Platinum group minerals were discovered in the Ulaannachin and North Burgastai river of Turgen area, western Mongolia. Mineral exploration carried out by Amitan and Dandar (1999) detected significant concentrations of gold and minor PGM mineral phases in the Quaternary alluvial sediments. The PGM nuggets from Ulaannachin show a silver-gray color with a metallic luster and an average 1 mm in size. The shape of nuggets varies considerably (Altanzul and Kitakaze 2007). The majority occurs as anhedral grains; some are well rounded. Other grains are knobby with many rounded projections and cavities. Some grains have a smooth and shiny surface; others are with a dull and pitted surface and irregular brown ferruginous crusts. Analytical data show significant contents of Ir, Rh, and minor Os. The Pd and Ru concentrations in isoferroplatinum remain low (0.05 wt% and 0.01 wt%). Grains of Pt-Fe alloy are optically homogeneous under reflected light microscopy.

The concentration of Fe varies from 8.27 to 8.89 wt% and Pt from 89.1 to 90.62 wt%. According to the experimental work of Kessel et al. (2001), the mole fraction of iron (X_{Fe}) in (Pt, Fe) alloys rises with decreasing f O_2. At 1200 °C concentrated X_{Fe} between 0.20 and 0.36, which is characteristic of isoferroplatinum type alloys from Nizhni Tagil (Johan 2006). The large variability in the chemical composition of Pt and Fe alloys from Nizhni Tagil might be due to some significations, contents of Ir, Rh, and Os accompanied by a minor.

Burgastai Gold Placer Deposit with PGE Nuggets (5′)

The PGE mineral grains were discovered in 1998 during exploration of a gold placer deposit. Geochemical results of PGM from the South Burgastai river placer occurrence were published by Agafonov et al. (2005). The PGM mineral phases are subdivided into two groups based on their compositions: (1) predominant of Os-Ir-Ru alloys and (2) subordinate of Pt-Fe alloy. PGE nuggets are composed mainly of Os (41.38–52.88 wt%), Ir concentrations in native osmium vary between 36.4–39.23 wt%, and the Ru content varies between 8 and 14.89 wt%. Only one iridium nugget was observed, and Ir composition is increasing up to 71.94 wt%, containing 12.84 wt% Pt, 6.47 wt% Ru, and 5.2 wt% Os. This ruthenium composition is considered to indicate a very primitive ratio of PGE in the ore-forming system, before the separation of Ru as laurite by sulfidization process. The Rh and Pt concentrations are comparable to those found in native iridium. Grains of Pt-Fe alloy are generally rounded and up to 5 mm in size and have a bright metallic luster. The surfaces of Pt-Fe alloy grains show scratches and cavities and occur as flat flakes. Compositions of PGE are highly variable in all of the measured PGM mineral phases. Pt-Fe alloy is dominant. The concentration of Fe in the Pt-Fe alloy stability, ranging from 9.48 to 9.53 wt% and Pt content of alloy, ranges from 77.62 wt% to 79.17 wt%; such Pt-Fe alloy corresponds to isoferroplatinum Pt_3Fe. The contents of the minor PGE in isoferroplatinum are relatively high; Ir concentrations are as high as 7.07 wt%. The Rh content of Pt-Fe alloy in the South Burgastai river placer also is relatively high as much as 2.67 wt%. The concentration of other elements (Pd, Cu, Ni, and Co) is more restricted.

Zost Gold Placer Deposit with PGE Nuggets (6′)

In 1947, a Russian-Mongolian team searched and found gold deposits in Toromkhon, Zost, Tamgat, Alag Shand, and Khuurai Sair areas. These gold deposits belong to the Altan Uul region of the Nemegt-Altan Uul zone of arc terrane in South Mongolia. Also, PGM were discovered in gold deposits in 1988–1989. Banzragch (2000) and his team proved that they have good potential to be considered of economic value. The geology of the Altan Uul-Tamgat area (Fig. 10.3) consists of Late Silurian and Early Devonian units comprising sedimentary rocks and intermediate volcanic rocks of the Nomgon Formation. The lower sequence of the Nomgon Formation consists mainly of mafic and intermediate volcanic rocks (andesite and dacite), whereas the upper sequence comprises sandstone and shale. The lower sequence exhibits silica alteration with pyrite and is cut by east-west trending faults along which occur ophiolites that consist mainly of serpentinized harzburgite and dunite. The lower sequence is also intruded by gabbrodiorite and diorite plutons that are calc-alkaline, rarely tholeiitic, low- and medium K, and metaluminous. The plutonic rocks exhibit characteristics of I-type granitoids. The upper sequence of the Nomgon Formation is composed of coarse-grained sandstone and shale that extend up to 4–5 km long and 50–100 m wide in outcrop and also contains quartzite bodies and quartz veins. This sequence also contains small serpentinite bodies. Overlying the Late Silurian and Early Devonian units is the Middle and Late Devonian sequence that consists of thin conglomerate layers and shale, sandstone, siltstone, and gravels that ranges up to 7300 in thickness, overlain by a

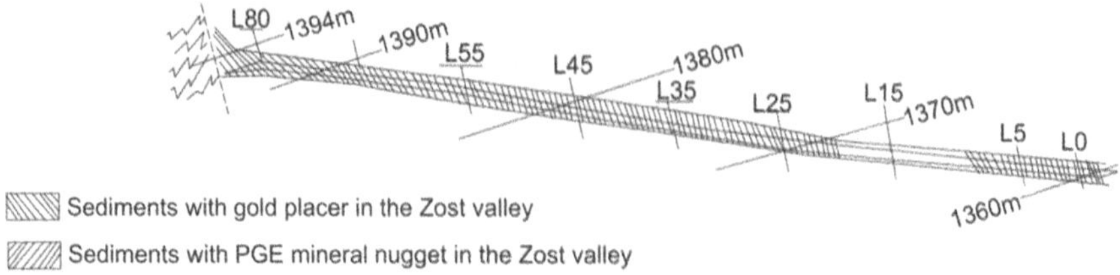

Fig. 10.5 Plan of PGE occurrence in Zost valley (Banzragch 2000)

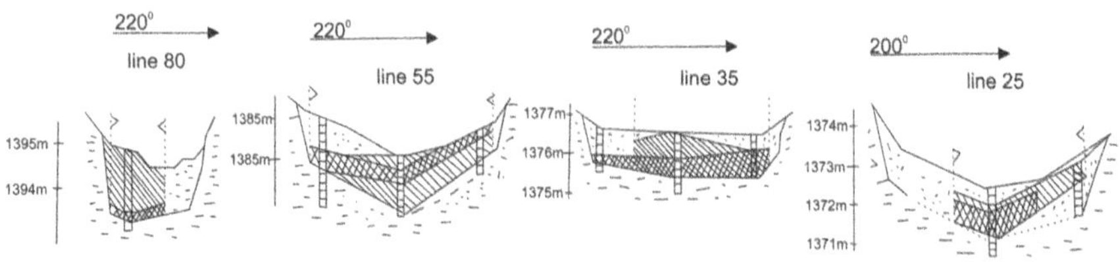

Fig. 10.6 Gross section by exploration lines (Banzragch 2000)

Carboniferous clastic sequence dominated by conglomerate and sandstone. Early Carboniferous sequence of gabbrodiorite-diorite-granite that is calc-alkaline varies from medium to high K and shows characteristics of forming in a continental-margin arc.

The Zost is one of the dry river beds covered by the Altan Uul. It is a 3.5 km long, 40–50 m, sometimes up to 100 m wide. It has alluvial-proluvial origin, which contains upper Silurian-Lower Devonian sedimentary-metamorphic and the lower boundary of the Upper Cretaceous terrigenous sediments (Fig. 10.5). The base of the proluvial sediments is the lower Cretaceous green-colored siltstone. The thickness of the proluvial sediments is 0.3–3.0 m, usually consisting of moderate and poorly sorted gravels, gravels, sand, sandy loops, and occasionally country rocks. The bed is located at V-notched valley. The thickness of the precious metal placer layer is 0.3–1.1 m (Banzragch 2000). The PGM mineralization in the ophiolite and placers is uniform, most of which are represented by rutheniridosmine of 1.5–2.0 mm in size. Rutheniridosmine grains are by 0.3–0.4 mm in size, glossy white, glossy, hexagon-like, and partly granular. This mineral composition is Ru 17.11–16.13%, Ir 33.29–32.79%, and Os 46.2–47.1%. Isoferroplatinum is a gray-colored, metallic, 0.1–0.2 mm in size, semi-glabrous submetallic luster, and oval-shaped. The chemical composition is, Pt 81–86%, Fe 8.1–8.6%. Osmiridium is light-colored and irregular, Ir 65.93–66.38%, Os 16.89%, and Ru 5.73%.

A comparison of the distribution of metals in plan and section shows that PGM distribution is almost overlapping with gold deposits horizontally (Fig. 10.5). In the section, PGM is distributed in the upper, middle, and bottom parts. In the upper and bottom parts of the deposit, the PGM distribution coincides with the upper and middle gold-bearing layer (Fig. 10.6). This can be explained by the similar specific gravity of the PGM and gold (18.6–22.7). Heavy minerals include gold, limonite, titanite, tourmaline, rutile, apatite, zircon, barite, monazite, and kyanite.

Ulziit Gold Placer Deposit with PGM Nuggets (7′)

PGM grains (nugget) are accessory minerals from area near the Bayankhongor ophiolite complex in several gold placer deposits such as Ulaanbulag, Ulziit, and Duvunt in Southwest Mongolia. The Ulziit gold placer deposit is located in Ulziit River valley in the western part of the Bayankhongor ophiolite complex (Jamsrandorj et al. 2002).

The PGM grains are well-shaped crystals or have rounded and flattened irregular shape and predominantly up to 1 mm in size (Altanzul and Kitakaze 2007). The surfaces of PGM grains often show scratches as the result of transport. Grains have a silver and gray color with a metallic luster, flattened hexagonal crystals, and some roundness and reach up to 3 mm in diameter (Fig. 10.7a, b).

The Os-Ir-Ru assemblage is dominant, and grains of osmium are mostly homogeneous and contain up to 1.8 wt% Pt and 0.19 wt% Fe. Compositions range from Os 78.53 to 49.87 wt%, Ir 43.02 to 19.27 wt%, and Ru 8.78 to 30.95 wt% (Table 10.3). Two grains of Os-Ir-Ru alloys display different color and drop-like

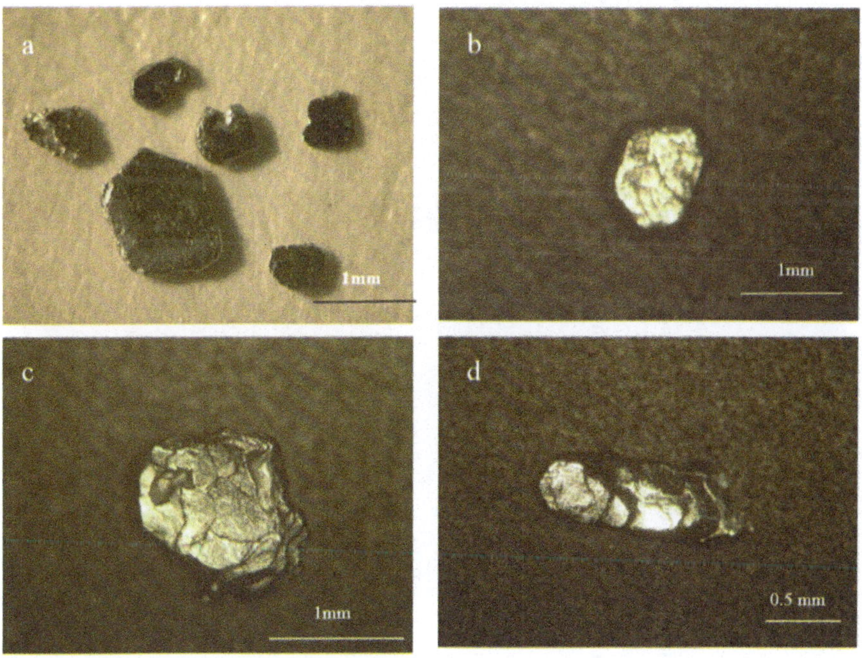

Fig. 10.7 (**a**) Photomicrographs of PGE alloys from Ulziit River. (**b**) Group 1 rounded and euhedral grains, (**c**) Group 2 subhedral grains, and (**d**) Isometric euhedral grains

Table. 10.3 Chemical composition determined by EPMA both of Os-Ir-Ru and Pt-Fe alloys, from Ulziit River

Os-Ir-Ru alloy									Pt-Fe alloy	
Element	bkh-1	bkh-2	bkh-3	bkh-4	bkh-5	bkh-6	bkh-7	bkh-8	bkh-9	bkh-11	bkh-101
Weight %											
Os	67.71	55.75	61.51	49.87	76.14	49.62	78.53	73.96	56.33	0.87	5.33
Ir	26.78	39.17	31.94	40.44	22.70	39.82	19.27	21.89	43.02	2.46	5.62
Pt	1.62	0.35	1.28	1.80	0.08	0.80	0.05	0.39	0.14	87.71	81.45
Ru	3.01	4.36	4.93	7.08	1.00	8.78	2.20	3.37	0.95	0.10	0.29
Fe	0.00	0.06	0.09	0.11	0.07	0.31	0.09	0.04	0.19	7.60	8.00
Total	99.12	99.7	99.74	99.29	99.99	99.32	100.12	99.65	100.63	98.74	100.69
Atomic ratio											
Os	66.74	53.99	59.18	47.35	75.52	46.22	76.93	72.18	55.52	0.76	4.51
Ir	26.12	37.54	30.42	37.99	22.28	36.70	18.68	21.14	41.96	2.12	4.71
Pt	1.55	0.33	1.20	1.67	0.08	0.73	0.04	0.37	0.14	74.43	67.26
Ru	5.58	7.95	8.92	12.64	1.87	15.38	4.06	6.18	1.75	0.16	0.47
Fe	0.01	0.19	0.28	0.35	0.25	0.97	0.44	0.12	0.63	22.54	23.05

intergrowth, which is typical of Pt-Fe alloys. The grains of Pt-Fe alloy intergrowths in osmium contain up to Pt 87.71 wt% and Fe 8.00 wt%. One grain of osmium, which contains of Pt-Fe inclusion, is the only one Pt-Fe intergrowth (Fig. 10.7d), while another one has many inclusions of Pt-Fe intergrowths (Fig. 10.7c).

Kharaanen Fault Zone (8′)

This zone contains ultramafic and mafic rocks, mainly cataclastic and serpentinized, and altered to listvenites. Gabbro associated with ultramafics is altered to rodingite. In the eastern part of the zone is the large Ulziit Tolgoi ultramafic ophiolite complex. PGM occur near this gabbro exposed in a narrow valley. The PGM and gold are very well rounded and identified as isoferroplatinum up to 0.1 mm in size. The Os-Ru-Ir alloys characteristic of ophiolites were not found, and the interpretation of PGM origin from ophiolite complex is problematic. Bayanleg ultramafic belt (9′). PGE mineralization was found in 2001 within chromitites in serpentinite with gabbro. In the Bayan Khovriin Bulag, Ir and Os component bearing laurite was identified. Another site is in the Shorvog Nuruu the western continuation of the Bayanleg ultramafic belt, which was affected by a transform fault to the north. There are many small bodies of strongly foliated serpentinites associated with gabbro. The serpentinites are mainly replaced by listvenites and gabbro by rodingites. In heavy concentrates, PGM was found up to 0.2 mm in size of rhuteniridosmine with a high content of Fe. Others like the Khashaat Khudag and Ikh Ulziit Nuruu occur in the eastern end of the Manlai ultramafic belt and the Bayan Bor Nuruu in the western end of the Manlai belt.

Khashaat Khudag (10′)

It is situated in the Argal Nuruu ridge 40 km southwest of the Sainshand. The geology of the ridge is dominated by sedimentary-metamorphic rocks, which alternate with diabase and diorite in the north. Quartz veins cut all rocks. In the heavy concentrate from the valley, PGM is identified as isoferroplatinum. Isoferroplatinum contains Ir, Os, Rh, Pd, and Cu. The primary source is not known but could be some ultramafic bodies that occur in the ridge, and some of them were totally eroded.

Ikh Ulziit Nuruu (11′)

It is situated 15–20 km south of the Saikhandulaan soum. Ulziit Nuruu ridge in the area of Khashaat Khudag is comprised of metamorphosed siltstones, sandstones, clastic rocks, and conglomerates with fauna. Heavy concentrates from the valley contain isoferroplatinum with Ir admixture.

Bayan Bor Nuruu (12′)

It is situated in Hercynian structures on the extension of the Manlai ultramafic belt. Manlai ridge comprises sedimentary-metamorphic rocks and diabase, gabbro, and gabbro-diabase. Heavy concentrates from valley contain one PGM grain identified as rhuteniridosmine.

10.4 Concluding Remarks

It can be concluded that economically important PGE deposits and occurrences are not yet discovered in Mongolia. The PGE mineralization is known from three deposit types: podiform chromite associated with the ultramafic part of ophiolite complexes, PGE associated with layered gabbro intrusions, and PGE-Au placer. The major type is ophiolite-related podiform chromite. The link between chromitite and the PGE mineralization in the original deposit is still debated. Most authors agree that the chromite mineralization is magmatic, but the PGE mineralization seems to have undergone a complex post-emplacement evolution. Chromitites from ultramafic rocks contain mainly Os-Ru-Ir alloys, and gabbro associated PGE mineralization contains Pt and Pd. The PGE mineralization in PGE-Au-placers shows a genetic relation of PGE with ultramafic rocks of ophiolite complexes. Because only a limited number of ultramafic rocks with PGE mineralization were studied, future detailed study will be important. The presence of PGM in heavy concentrates is a positive signature for detailed study.

References

Agafonov LV, Baikova VS, Stupakov SI (1991) Metamorfichskiye ultraosnovnyye i osnovnyye porody Mongoliskogo Altay (Metamorphic ultramafic and mafic rocks of Mongolian Altai, MPR). In: Hyperbasite associations of folded belts. 6. Mineralogy and geochemistry. Novosibirsk 9–27

Agafonov LV, Khmel'nikova OS (1991) Samarodnyye elementy v khromititakh Ulankhidskogo massiva MNR. (native elements in the Ulaakhiid ophiolite massif MPR). Geol Geophy 1:128–132

Agafonov LV, Lkhamsuren J, Kujuget KC, Oidup ChK (2005) Platinonosnost' ul'tramafit-mafitov Mongolii i Tuvy. (Platinum mineralization of ultramafic, mafic rocks in Mongolia and Tuva). Ulaanbaatar, p 224

Agafonov LV, Stupakov SI (1983) Petrologiya i mineralogiya bazitovykh, ultrabazitohykh i metamorphicheskikh porod Kerulenskhoi zoni razlemov. (Petrology and mineralogy of mafic, ultramafic and metamorphic rocks of the Kherlen fault zone). In: Ultramafics associations of the folded belts. Novosibirsk, IGG SB USSR Academy of Sciences, 2:71–104

Agafonov LV, Stupakov SI Sheremet OG (1990) Morphologiya i usloviya zaleganiya giperbazitovykh massivov Mongolii. (Morphology and environment of the ultramafic ophiolite massifs of Mongolia). In: Petrology of ultramafics and mafics. Novosibirsk, p 1–41

Altanzul B, Dandar S (2008) Khet suurilag nairlagatai Oorthog Dulaan Uul burdliin geologi mineralogiin sudalgaa. (Geological and Mineralogical study of ultramafic intrusion Oortsog, Dulaan Uul complex). Guideline, p 5–9

Altanzul B, Dandar S (2010). Shoroon ord dahi tsagaan altnii erdsiin nairlaga. Pt mineral composition in placer deposit. Mongolian University of Science and Technology, Geology 22:255–263

Altanzul B, Kitakaze A (2005) Petrography of the granitoids in Bayankhongor area, central Mongolia. Geological and Mineralogical Symposium in Japan, p 262

Altanzul B, Kitakaze A (2007) Mineralogical characteristics of platinum group minerals from some gold placer deposits, Mongolia. Mong Geosci 31:57–60

Altanzul B, Ukhnaa G, Erdenebayar J (2015) Development of mineral resource in Mongolia, A case study of PGE mineralization. Recent Advances in Resource and Science and Technology to Sustain Japan, p. 133–147

Amitan N, Dandar D (1999) Bairamiin gunii hagarliin tsagaan altnii erdesjilt. (First Platinum mineralization Mountain Bairam). Probl Geol Metallog Mong 12:55–56

Badarch G, Orolmaa D (1999) Accreted terranes of southern Mongolia and adjacent areas. Mong Geosci 14:10–11

Bannikov OL (1986) Kora vyvetrivaniya ul'traosnovnykh porod Ulakhidskogo massiva Mongolii. Weathering core of ultramafic rocks of the Ulaakhiid ultramafic pluton (MPR). In: Hyperbasite associations of folded regions. Petrochemistry, mineralogy, geochemistry. Novosibirsk, 3:76–91

Banzragch B (2000) Altan Uuliin Tsagaan altnii erdesjilt. (Platinum group mineralization in Altan uul). Mong Geol 1:82–83

Bat-Erdene G (2001) Mongoliin Tsagaan altnii buleg metal. (Platinum group metal in Mongolia). Mong Geol 2–3:249–255

Boudier F, Nicolas A (1985) Harzburgite and lherzolite subtypes in ophiolitic and oceanic environment. Earth Planet Sci Lett 76:84–92

Byamba J, Bat-Erdene G (1999) Khantashiir ophiolite complexiin unet metal. (precious metal of Khantaishir ophiolite complex). Probl Geol Metallog Mong 13:97–99

Dandar S, Altanzul B (2006) PGE mineralization in the Western Mongolia. In Symposium Environment and Transforming (Society Multidisciplinary Approaches) p 26

Dandar S, Ukhnaa G, Altanzul B, Oyunchimeg B (2003) Types and criteria of platinum mineralization of Western Mongolia. Mong Geosci 19:79

Erdenebayar J, Ogata T, Adachi T, Altanzul B, Baatar A (2017) Mineralogical characteristics of platinum group elements mineralization in Mongolia: case study on the Bayannuur and the Ulaannachin occurrences in Western Mongolia. p

Ermolayev NP, Sozinov NA, Filtsyan VA, Chinenov VA, Khoroshilov VL (1992) Novyye vidy rud dragotsennykh metallov i redkozemel'nykh elementov v uglerodistykh slantsakh. (New types of precious metal and rare earth element ores in carbonaceous slates) Nauka, Moscow

Furnes H, Safonova I (2019) Ophiolites of the central Asian Orogenic Belt: geochemical and petrological characterization and tectonic settings. Geosci Front 10(4):1255–1284

Gerel O, Munkhtsengel B (2015) Ophiolites in Mongolia: an overview. Acta Geologica Sinica. IGCP Workshop on: Ophiolites and related high-pressure rocks in Qilian Mountains. 89(2):72

Gerel O, Sharhuuhen D, Badarch G (2002) PGE-Au Occurrence at Altan uul Tamgat in South Gobi, Mongolia. Mineral deposits at the Beginning of the 21st Century. (Ed. Adam Piestrzynski et al.) Balkema. p 1105–1108

Gertner IF, Tishin PA, Stupakov SI, Agafonov LV (2003) Khairhan gabbro ophiolite massif as potential source for precious metals in Western Mongolia. Tomsk p 225–227

Izokh AE, Polyakov GV, Krivenko AP, Boginov VI, Bayarbileg L (1990) Gabbroidnye formatsi Zapadnoi Mongolii. (Ultrabasic complexes in North Mongolia) p 269

Jamsrandorj G, Dandar S, Kempe U, Wolf D, Baast G (2002) Buyankhongor busiin Ulaanbulag altnii shoroon ord dahi Tsagaan altnii erdes. (Pt mineralization in gold placer in Ulaanbulag in Bayankhongor zone). Problems of Geology, Ulaanbaatar, 5: 145–156

Jamsrandorj G, Tumurtogoo O, Bayarbileg L (1995) Mongol ornii PGM- huderjilt, prognoziin 1:1000000 mashtabtai zurag shemiin tailbar. (Report on the Prediction of PGE mineralization mapping scale 1:1000000, in Mongolia). Open file report, State Geological Fund of Mongolia, Ulaanbaatar Mongolia 48 p

Janousek V, Jiang Y, Buriánek D, Schuimann K, Hanzl P, Soejono I, Kröner A, Altanbaatar B, Erban V, Lexa O, Ganchuluun T, Kosler J (2018) Cambrian-Ordovician magmatism of the Ikh-Mongol arc system exemplified by the Khantaishir magmatic complex (Lake zone, south-Central Mongolia). Gondwana Res 54:122–149

Jian P, Ktoner A, Jahn B, Windley BF, Shi Y, Zhang W, Zhang F, Miao L, Tomurkhuu D, Liu D (2014) Zircon dating of Neoproterozoic and Cambrian ophiolites in West Mongolia and implication for the timing of orogenic processes in the central part of the Central Asian Orogenic Belt. Earth-Sci Rev 133:62–93

Johan Z (2002) Alaskan type complexes and their platinum group element mineralization. The Geology, Geochemistry, Mineralogy and Mineral Beneficiation of Platinum Group Elements. Canadian Institute of Mining, Metallurgy and Petroleum, 54:699–719

Johan Z (2006) Platinum group minerals from placer related to the Nizhni Tagil (middle Urals, Russia) Uralian-Alaskan type ultramafic complex: ore mineralogy study silicate inclusion in (Pt-Fe) alloys. Mineral Petrol 87:1–30

Kessel R, Beckett JR, Stolper EM (2001) Thermodynamic properties of the Pt-Fe system. Am Mineral 86:1003–1014

Krivenko AP, Sidorov EG, Izokh AE (1985) First Pt mineralization of Mongolia. Formation, Genesis and Metallogeny Ultramafic Rocks, Sverdlovsk, p 37–38

Kuzmichev AB, Kröner A, Hegner E, Dunyi L, Yusheng W (2005) The Shishkhid ophiolite, northern Mongolia: a key to the reconstruction of a Neoproterozoic island-arc system in Central Asia. Precambrian Res 138:125–150

Leblanc M (1991) Platinum group elements and gold in ophiolite complexes: distribution and fractionation from mantle to ocean floor. Kluwer Academic Publishers, pp 231–260

Lkhamsuren J (2011) Tsagaan altnii huderjilt. (Platinum mineralization). In: Metallic mineral deposits. Soyombo Printing, Ulaanbaatar, pp 319–332

Mao YJ, Bat-Ulzii D, Qin KZ, Bujinlkham B, Tang DM (2018) Comparisons among the Oortsog, Dulaan, and Nomgon mafic-ultramafic intrusions in Central Mongolia and Ni-Cu deposits in NW China: implications for economic Ni-Cu-PGE ore exploration in Central Mongolia. Russ Geol Geophys 59:1–18

Marinov NA, Zonenshain LP, Blagonravov VA (eds) (1973) Geologiya Mongol'skoy Narodnoy Respubliki. (geology of the Mongolian People's Republic). TOM II. Magmatism, metamorphism, tectonics. Nedra, Moscow, p 752

Naldrett AJ (1993) Models for the formation of strata bound concentrations of platinum group elements in layered intrusion. Geological Association of Canada. Special paper 40:373–387

Naldrett AJ, Gruenewaldt G (1989) Association of platinum group elements with chromitite in layered intrusions and ophiolite complexes. Econ Geol 84:180–187

Pinus GV, Agafonov LV, Lesnov PF (1984) Alipinotipnie gyperbaziti Mongolii. Alpine type ultramafic complexes in Mongolia. Moscow, Science p 200

Pinus GV, Kuznetsov VA, Volokhov IM (1955) Gyperbaziti Tuvy (Hyperbasites of Tuva) Publishing of Academy of Sciences of the USSR, p 135

Prichard HM, Ixer RA, Lord RA, Maynard J, Williams N (1994) Assemblages of platinum group minerals and sulphides in silicate lithology and chromite rich rocks within the Shetland ophiolite. Can Mineral 32:271–294

Prichard HM, Lord RA, Neary CR (1990) Platinum and palladium in the Troodos ophiolite complex, Cyprus. Can Mineral 28:607–617

Prichard HM, Lord RA, Neary CR (1996) A model to explain the occurrence of platinum and palladium rich ophiolite complexes. J Geol Soc Lond 153:323–328

Sharkhuukhen D, Baldorj B (1993) Oortsog Dulaan Uuliin Tsagaan altnii nudeerjiltend yavuulsan erliin ajliin tailan. (Exploration of PGE mineralization in Oortsog Dulaan and Zurkh Khairkhan area). Open file report, State Geological Fund of Mongolia, Ulaanbaatar

Sharkhuukhen D, Gerel O (2001) Platinum mineralization in mountain of Altan-Uul and Naran. Mongolian University of Science and Technology. Geology 12:257–264

Sharkhuukhen D, Jamsrandorj G (1992) Bayan nuur, Narangiin talbaid hiisen Tsagaan altnii metaliin huderjiltiin erliin ajliin tailan. (Report for exploration study of PGE mineralization in Bayannuur, Naran). Open file report, State Geological Fund of Mongolia, Ulaanbaatar

Shelepaev R, Vishnevsky A, Izokh A, Egorova V, Shelepov Y (2015) Permian mafic complexes of the Khangai region (Western Mongolia). In: First China-Russia international meeting on the central Asian Orogenic Belt and IGCP 592 workshop. Institute of Geology, Chinese Academy of Geological Sciences, Beijing, pp 78–79

Stupakov SI, Lesnov FP, Agafonov LV, Demin AN (1994) Giperbazity pozdnykh kaledonid Zapadnoy Mongoliiv (Hyperbasites of late Caledonids of Western Mongolia). Geol Geofiz 3:29–39

Tomurtogoo O, Nakajima T, Takahashi Y, Minjin Ch, Ichinnorov N, Oyungerel S (1999) Geological map L-47-XYI 1:200000. Mineral Resources authority of Mongolia

Zonenshain LP, Kuzmin MI, Natapov LM (1990) Geology of the USSR: A Plate-Tectonic Synthesis, American Geophysical Union, Geodynamic Series. 21:242

Part II
Non-metallic Mineral Resources

Chapter 11
Fluorite Deposits

Jargal Lkhamsuren and Yondon Majigsuren

11.1 Introduction

In Mongolia, fluorite is one of the major economic minerals, along with copper, molybdenum, gold, and coal. The production of fluorite of Mongolia is about 15% of the world's production (JICA MMAJ 1993). In Mongolia, fluorite was previously used to prevent decay of milk, and nowadays it is mainly used for the production of aluminum, iron, hydrofluoric acid, and specialized glass and optical glass.

The first fluorite deposit in Mongolia was found in 1933, and the first economical fluorite deposit was Dojiru discovered in 1939. After World War II, economic cooperation with the Soviet Union and China boosted the exploration and led to discovery of many fluorite deposits in central and eastern Mongolia (Kishimoto 1982).

Initial research, prospecting, and exploration results of fluorite deposits in Mongolia were reported by Marinov (1958, 1980), Kalenov and Khasin (1965), Konstantinov and Zimina (1966), Kovalenko et al. (1976), Khrapov et al. (1977), Ontoev et al. (1977), Kotov et al. (1979), Ontoev et al. (1979a, b), Volchanskaya and Korytov (1980), Shuvalov et al. (1980), Batjargal et al. (1982, 1985), Kandinov and Dobrolyubov (1984), Koshelev (1985), and Jamsran et al. (1986). Mineralogy, geochemistry, and genesis of fluorite deposits in Mongolia have been studied by Lkhamsuren and Batjargal (1975), Naumov and Ivanova (1975), Lkhamsuren (1976, 1984, 1988), Lkhamsuren et al. (1979), Korytov et al. (1980), Frikh-Khar and Volchanskaya (1982), Tumenbayar et al. (1986), Voinkov and Lkhamsuren (1986), Tumenbayar (1987), Batjargal and Lkhamsuren (1987), and Lkhamsuren and Hamasaki (1998).

J. Lkhamsuren · Y. Majigsuren (✉)
Museum of Geology and Mineral Resources, School of Geology and Mining Engineering, Mongolian University of Science and Technology, Ulaanbaatar, Mongolia
e-mail: lkhamsuren@must.edu.mn; majigsuren@mus.edu.mn

O. Gerel et al. (eds.), *Mineral Resources of Mongolia*, Modern Approaches in Solid Earth Sciences 19, https://doi.org/10.1007/978-981-15-5943-3_11

In this book, we review the classification and characteristics of fluorite deposits in Mongolia mainly based on ore genesis with basic tonnage and grade requirements. Various estimation resources and reserves of fluorite deposits and occurrences in different classification of categories were reported. A, B, and C_1 categories of resource or reserve by Russian classification are comparable to category measured resource or proven reserve by CRIRSCO classification. B, C_1, and C_2 categories of resource and reserve are comparable with indicated resource or probable reserve. C_2 and P_1 categories are regarded as a category of inferred by CRIRSCO classification.

11.2 Fluorite Mineralization in Mongolia

The geological history of the Mongolian territory is characterized by subduction and subsequent collision tectonics. Proterozoic, Lower Paleozoic, and Mesozoic fold belts of Mongolia were formed by tectonics related to the movement of the Siberian and Chinese cratons. The mantle-derived magmatisms were developed in these fold belts. Especially during the Mesozoic period, large tectonic basins and volcanism developed in Inner Baikal and Eastern Mongolia. Majority of fluorite mineralizations are clustered in the central and eastern parts of Mongolia (Fig. 11.1). The fluorite mineralization of the late Jurassic to early Cretaceous periods is important for the genesis of economic fluorite deposits (JICA MMAJ 1993).

11.3 Classification of Fluorite Deposits in Mongolia

Fluorite deposits in eastern Mongolia have been categorized by Kalenov and Khasin (1965), Konstantinov and Zimina (1966), Marinov (1980), and Filippova et al. (1984). Jamsran et al. (1986) and Lkhamsuren (1988) have classified fluorite deposits in Mongolia more precisely.

Fluorite deposits in Mongolia are mainly hydrothermal and divided into two groups both genetically related to magmatic rocks. One group is related to volcanic rocks and the other to plutonic rocks. The former are classified as epithermal fluorite deposits and carbonatite-hosted deposits. The latter are classified as bertrandite-phenacite-fluorite deposits, albitite, tin-tungsten vein, tungsten-molybdenum vein, greisen, and pegmatites depending on the main mineral assemblage. The classification is summarized in Table 11.1 Except for the epithermal type, fluorite mineralization in other types has no commercial significance and is only recoverable as by-products.

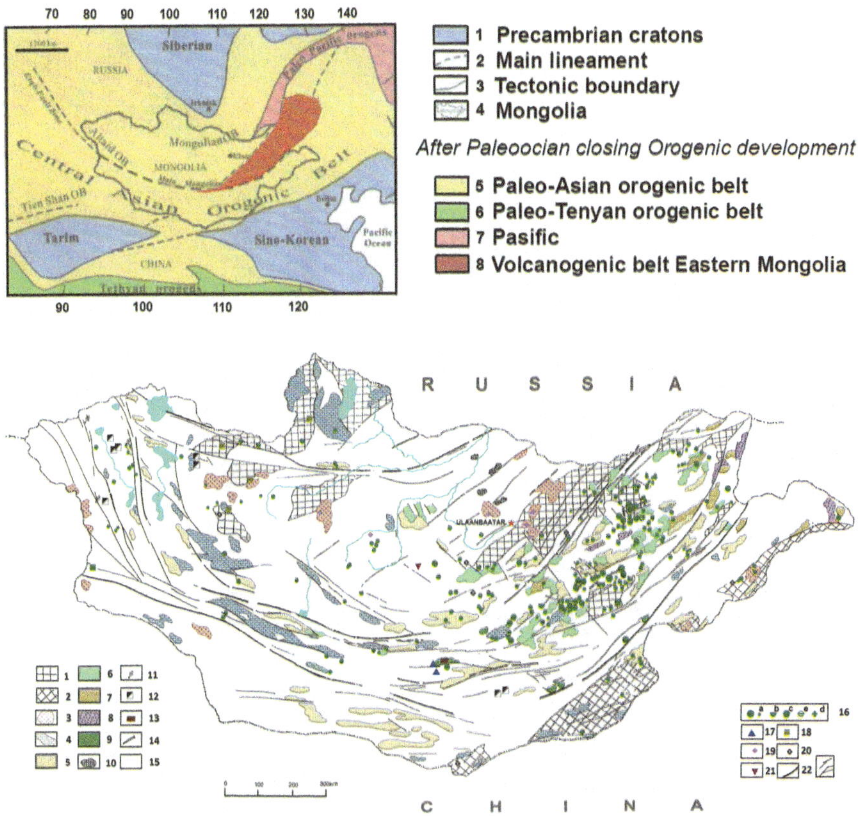

Fig. 11.1 Schematic geological map and position of fluorite deposits in Mongolia (modified after Lkhamsuren 1988). (1–4) Pre-Mesozoic rocks ((1) Proterozoic basement in Caledonian folding area, (2) Early Paleozoic basement in Hercynian folding area, (3) Late Paleozoic to early Mesozoic granite, (4) various ages of carbonate-chert and carbonate formations); (5) Mesozoic rocks (molasse and coal-bearing molasse formations, with some volcanic rocks); (6–9) volcanic rocks (6) basalt, andesitic basalt, trachybasalt, and trachyandesite, (7) dacite to rhyolite and trachyrhyolite, (8) trachyandesite, trachyrhyolite, and andesite to rhyolite, (9) potassium alkaline basalt); (10–15) intrusive rocks ((10) gabbro, (11) diabase and dolerite, (12) potassium ultramafic alkaline, (13) potassium ultramafic alkaline with carbonatites, (14) alkaline basaltoid, (15) rocks which are genetically not related with fluorite mineralization in composition and age); (16–21) fluorite mineralization types (16) epithermal fluorite deposit; main mineral assemblage is (a) quartz-fluorite (large and small deposit), (b) calcite-quartz-fluorite, (c) barite-quartz-fluorite, (d) sulfide-quartz-fluorite, and (e) adularia-quartz-fluorite; (17) fluorite deposit in carbonatite host; (18) fluorite-bearing albitite; (19) fluorite-bearing pegmatite; (20) Mo-W and W-Sn deposit; (21) bertrandite-phenacite-fluorite deposit); (22) main regional and deep fault; and (23) deep fault and fracture zone

11.3.1 Epithermal Fluorite Deposits

Epithermal fluorite deposits are the only economically viable deposit type in Mongolia at present. Major minerals of the epithermal type are quartz and fluorite. Jamsran et al. (1986) and Lkhamsuren (1988) distinguished five types of mineral

Table 11.1 Classification of fluorite deposits in Mongolia

	Deposit type	Mineralization style	Mineral assemblages: Main minerals	Mineral assemblages: Minor minerals	Representative deposit	Genetically related rock	Radiometric age (Ma)
1	Epithermal fluorite deposit	– Vein – Brecciated – Disseminated – Irregular – Metasomatic body – Lens	Qtz, Fl	Kao, Cal, Adu, Fe-Mn-oxide	Berkh, Delgerkhaan, Khar Airag, Adag, Bor Undur, Urgun	– Alkaline basalt – High potassium trachybasalt – Trachyrhyolite – High alkaline differentiated from basalt-trachyrhyolite series	131–116
			Cal, Qtz, Fl	Py, se, Adu, Chp, Sid	Zuun Tsagaan Del, Chuluut Tsagaan Del, Bujgar, Khongor		
			Bar, Qtz, Fl	Py, gin, Cal	Galshar		
			Adu, Qtz, Fl	Kao, Cal, Fe-Mn oxide	Dojir		
			Slf, Qtz, Fl	Py, gin, Sph, Tet, Cin, Dik, Chp	Idermeg Bayan khan		
2	Fluorite deposit in carbonatite host	– Vein – Disseminated – In stock	Ap, Mgt, Fl	Fd, Cal, Bar, Cel, Qtz, do	Mushugai Kudag	– Alkaline ultramafic rock	153–136
			Qtz, Cal/Cel, Fl	Bar, Ank, Bst, Phg			
3	Bertrandite-phenacite-fluorite deposit	– Disseminated – Irregular	Qtz, F, Brd, Ph, Fl	Brl, Py	Dulaan Ovoo	– High alkaline granitic intrusion	154–180
4	Albitite	– Disseminated – Irregular	Ab, Qtz, Fl, me, Nb oxide, Bst	Ta-Nb-Ti oxide, cry, Ilm, Cbt, Tt, Pyc, Zr, Arf, Rbe, Lpd	Altan boom, Buural Khangai	Alkaline granitoid	
5	Mo-W deposit	– Greisen vein – Stockwork	Qtz, W, Mo, Fl	Brl, me, Tpz, Py, gin, Sph, bi	Eguzer, Ikh Khairkhan	– Granite – Alaskite	126–210
	W-Sn deposit		Qtz, W, Cas, Fl	Tpz, Zin, Sch, Py, gin, Sph, bi	Baga Gazariin Chuluu, Modot		175–205
	Pegmatite	– Lenticular – Columnar – Vein	Me, Qtz, Fl	Mica, Ab, tin, Tpz, Brl	Gorikhi, Zuun Bayan	– Granite	205–220

Ab albite, *Adu* aduralia, *Ank* ankerite, *Ap* apatite, *Arf* arfvedsonite, *Bar* barite, *Bi* bismuthinite, *Brd* bertrandite, *Brl* beryl, *Bst* bastnaesite, *Cas* cassiterite, *Cal* calcite, *Cbt* columbite, *Cel* celestite, *Cin* cinnabar, *Chp* chalcopyrite, *Cry* cryolite, *Dik* dickite, *Do* dolomite, *Fd* feldspar, *Fl* fluorite, *Gin* galena, *Ilm* ilmenite, *Kao* kaolinite, *Lpd* lepidolite, *Mc* microcline, *Mgt* magnetite, *Mo* molybdenite, *Pyc* pyrochlore, *Ph* phenacite, *Phg* phlogopite, *Py* pyrite, *Qtz* quartz, *Rbe* riebeckite, *Sch* scheelite, *Se* sericite, *Sid* siderite, *Slf* sulfide, *Sph* sphalerite, *Tet* terahedrite, *Tpz* topaz, *Tt* tantalite, *W* wolframite, *Zin* zinnwaldite, *Zr* zircon assemblages in epithermal fluorite deposits based on major minerals in the ore body. These types are quartz-fluorite (1), calcite-quartz-fluorite (2), barite-quartz-fluorite (3), adularia-quartz-fluorite (4), and sulfide-quartz-fluorite (5) (Fig. 11.1).

Quartz-fluorite is the most dominant type in the fluorite mineralization in Mongolia. Quartz usually forms as chalcedony-like, fine-grained, or columnar crystals. Fluorite is often coarse-grained, massive, columnar, rarely fine-grained, and microcrystalline or even cryptocrystalline. Colors of fluorite are varied as violet, dark violet, green, yellow, and bluish-green. Sometimes it is pale or colorless (Fig. 11.2).

Most of the calcite-quartz-fluorite type of deposits was formed by hydrothermal metasomatism of limestone. Fluorite content in ore of this type is 36 to 64%, and sometimes exceptionally very high grade ores of 90.0–98.5% have been identified at the Berkh deposit. The fluorite deposits occur as veins, irregular, and lenticular in shape in the brecciated and disseminated zones. The vein deposits are dominant. The veins are 1–30 m thick and 30–1300 m long, with an exceptional 3000 m long deposit at Bor Undur. 60–70 m thick veins were observed at the Khongor and Bujgar deposits. Host rocks are Precambrian limestones, schists, Mesozoic basalts,

Fig. 11.2 Fluorite structure types: (1) crystal of fluorite, (2) zoned crystal of fluorite, Bujgar deposit. (3) Crystal of fluorite and aquamarine, Khavtsal deposit. (4) Massive fluorite, Berkh deposit. (5) Brecciated fluorite, Zuun Tsagaan Del deposit. (6) Growth zones of fluorite ore, Khar Airag deposit. (7) Radiated and zoned fluorite, Khongor deposit. (8) Calcite-quartz-fluorite vein, Khamar Us deposit. (9) Banded structure, Urgun deposit (photos from Museum of Geology and Mineral Resources of Mongolia)

rhyolites, basaltic and rhyolitic tuffs, and Paleozoic granites. For example, long fluorite veins are observed in granite at Berkh. Several veins in basalt, andesite at Zuun Tsagaan Del, and large veins in volcanogenic rocks at Bor Undur have also been identified. Ore bodies in volcano-tectonic structures at Khongor and metasomatic bodies and lens-like shape ore bodies have been described at the Bujgar deposit. Ratios of quartz and fluorite vary in a wide range in the deposits. The formation temperature of all types of fluorite is 270–120 °C based on the fluid inclusion study (Lkhamsuren 1988). The radiometric age of the epithermal fluorite deposits varies from 131 to 116 Ma (Kandinov and Dobrolyubov 1984). Representative deposits are Berkh, Khar Airag, Zuun Tsagaan Del, Bor-Undur, Urgun, Chuluut Tsagaan Del, Khongor, Gal-Shar, Dojir, and Tsagaan- Elgen.

11.3.2 Carbonatite-Hosted Fluorite Deposits

Carbonatite-hosted deposits were discovered in southern Mongolia (Baskina and Volchanskaya 1976). The deposits occur as veins, in disseminated zones, stocks, and dikes. Vein-type ore bodies are dominant. Based on the main mineral assemblage, they can be distinguished into two types of veins:

1. Fluorite-apatite-magnetite-quartz.
2. Fluorite-calcite or celestite-quartz which is accompanied by feldspar, barite, dolomite, ankerite, bastnaesite, or phlogopite as minor minerals. They are 5–70 m in length and 3–30 m in thickness. Fluorite content is 30–40% in the ore bodies. The veins are related to alkaline intrusive rocks or syenites.

The radiometric age of the alkaline volcano-plutonic complex is 153–135 Ma (Kovalenko et al. 1976; Baskina et al. 1978). The representative deposit is Mushgai Khudag.

11.3.3 Bertrandite-Phenacite-Fluorite Deposits

Bertrandite-phenacite-fluorite deposits are found in central Mongolia (Kostyrev et al. 1977; Korytov et al. 1978). Veins in the deposits are up to 1.5 m thick and are located in a fractured zone of few tens of meters in length and up to 10 m thick in the exocontact with the early Mesozoic subalkaline granite massifs. The major minerals are feldspar, quartz, phenacite bertrandite, and fluorite, and minor minerals are pyrite and beryl. The highest grade of CaF_2 in the ore is 30%. The fluorite is richer in feldspar-fluorite veins with phenacite, although bertrandite is a characteristic mineral for this deposit type. The radiometric age of the deposits varies from 154 to 180 Ma (Kotov et al. 1979).

11.3.4 Albitite-Type Fluorite Deposit

Metasomatic albitite is associated with syenite and alkaline granite, confined to deep faults from the late Paleozoic to early Mesozoic periods. The albitite is defined by a high content of Zr, Y, Ta, Nb, Be, Sn, and REE, which may be included in zircon, pyrochlore, columbite, and cryolite with bastnaesite and fluorite (Yashina 1975, 1982). Zonal albitite was occurred as a result of multistage post-magmatic activities. The major minerals are quartz, albite, microcline, fluorite, and bastnaesite, whereas minor minerals are lepidolite, cryolite, ilmenite, tantalite, columbite, arfvedsonite, and riebeckite.

11.3.5 Mo-W and W-Sn Greisen

Rare metal-bearing fluorite deposits associated with molybdenum-tungsten and tungsten-tin mineralization form as greisen or veins. Ivanova (1972, 1976) studied the geology, geochemistry, and mineralogy of fluorite-bearing Mo-W and W-Sn deposits in eastern Mongolia. Typical deposits are Baga Gazariin Chuluu, Eguzer, Elstei, Ikh Khairkhan, Tumen Tsogt, Buren Tsogt, and Modot. Particularly in the Eguzer Mo-W deposit, fluorite is occurred in greisen, vein, and shatter zones. In this deposit fluorite formed in two stages. In the early stage, fluorite occur at 460–220 °C in relationship with quartz, beryl, molybdenite, wolframite, and topaz. In the late stage, fluorite occur at 200–100 °C with pyrite, chlorite, calcite, and siderite. Ore-bearing greisens and greisenized granite porphyries are 400–500 m in diameter and 30–40 m in thickness. Quartz-wolframite veins in hornfels are 70–80 m in length and 0.1–0.6 m in width.

11.3.6 Pegmatite

Mesozoic granitoids containing fluorite in miarolitic pegmatite are situated in the near-margin part of the eastern Mongolian volcanic belt. Miarolitic pegmatite cavities are chiefly consist of quartz and feldspar and together with topaz, fluorite, beryl, albite, tourmaline, biotite, and magnetite. They often have a symmetric zonal structure. The formation temperature of fluorite is 250–130 °C and rarely 480–250 °C in the early stage. Representative deposits are Gorikhi, Janchivlan, Zuun Bayan, and Bayan Tsogt.

11.4 Production of Fluorite

Mongolia is one of the world's largest producers of acid- and metallurgical-grade fluorite, producing 15% of the world's fluorspar. The total reserve of fluorite in Mongolia amounts to 48.3 Mt. by 2019. This reserve ranks fifth in the world. The

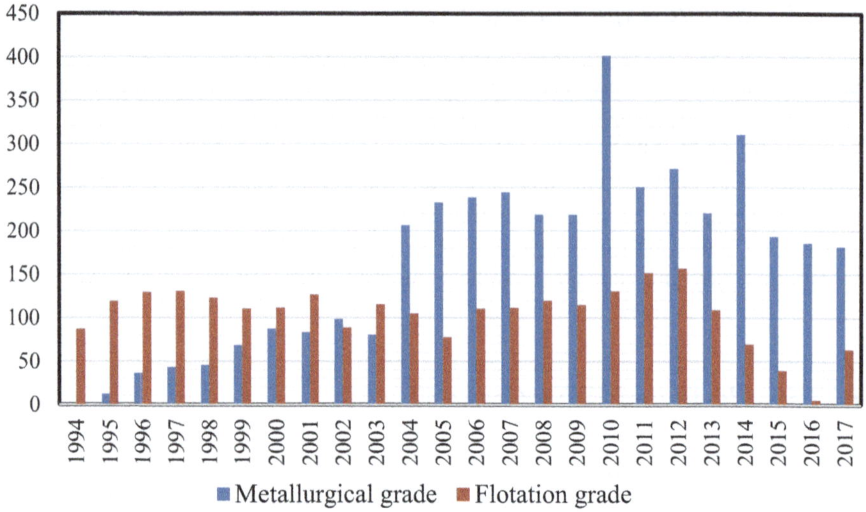

Fig. 11.3 Fluorite export of Mongolia (by thousand tonnes)

first production of fluorite in Mongolia began at the Berkh deposit in 1954. In the middle 1970s, the joint enterprise of the Soviet Union and Mongolia, Mongolrostsvetmet was established, and Mongolo-Czechoslovak followed in 1980. These enterprises account for a large proportion of fluorite production in Mongolia. The total annual production of fluorite reached 563 thousand tonnes in 2010 (Fig. 11.3).

Fluorite mines are open pit and underground types. The open pit-type mines produce 63% of the total ores. At present, Bor Undur, Adag, Bujgar, Zuun Tsagaan Del, Tsagaan Elgen, Urgun, and Delgerkhaan deposits are in commercial production. Further exploration in each mine is being carried out. Bor Undur is the biggest mining and processing complex in Mongolia. Today fluorite ore processed at the Bor Undur plant averages 30% CaF_2 with an applied cut-off grade of 26–28% CaF_2 (oral comm., 2008). Both metallurgical-grade (>75% CaF_2), flotation-grade, and acid-grade (>97% CaF_2) fluorspar concentrates are being produced and exported by railway mainly to Russia and China.

11.5 Description of Selected Fluorite Deposit Districts

Most fluorite deposits belong to the Eastern Mongolian fluorite belt (Fig. 11.4) with its subdivisions centers and districts depending on the tectonic-structural characters. Average grade and reserve information of fluorite deposits were listed in Table 11.2. Reserves are estimated by Russian ore reserve code.

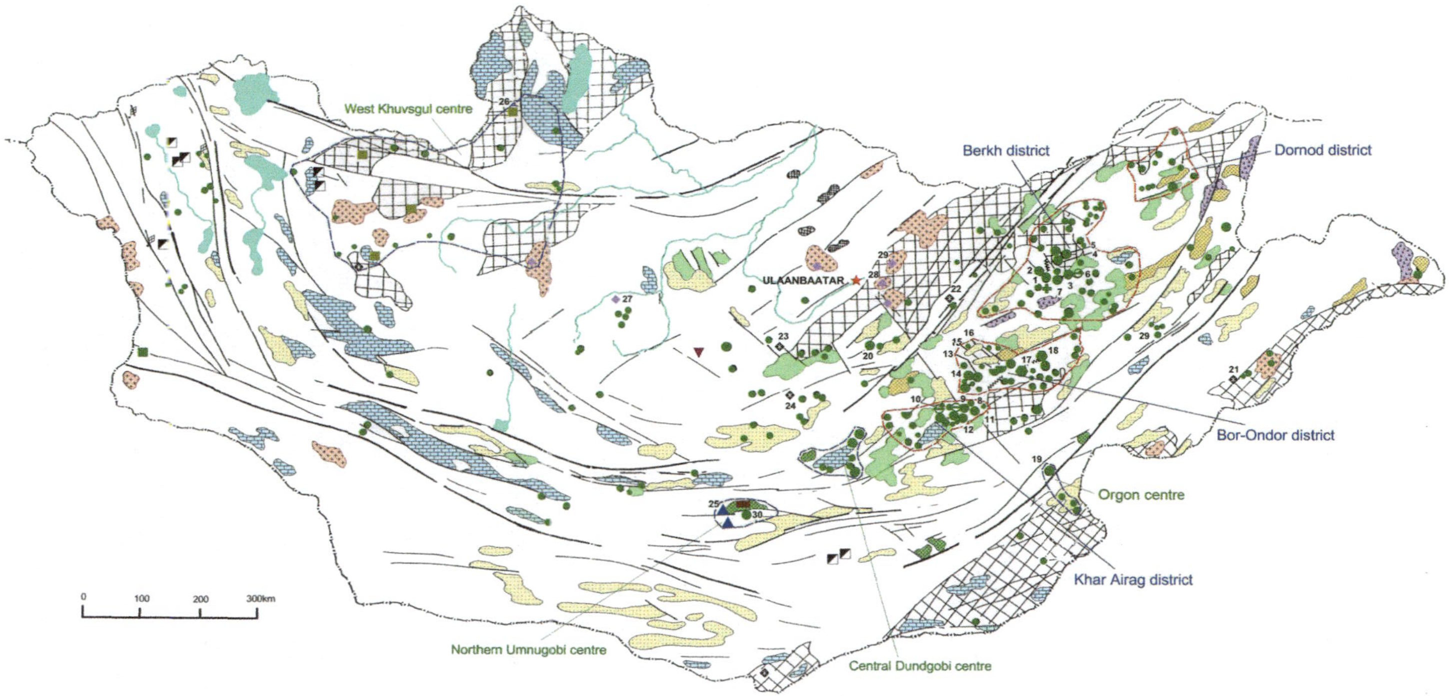

Fig. 11.4 Fluorite deposits and occurrences in Mongolia (Lkhamsuren 1988). Deposit No: (1) Berkh, (2) Chimid, (3) Delgerkhaan, (4) Yamaat, (5) Shuvuutai, (6) Dojir, (7) Idermeg Bayan Khan, (8) Khar Airag, (9) Khongor, (10) Bujgar, (11) Khairt, (12) Tsagaan Del, (13) Bor Undur, (14) Adag, (15) Khajuu Ulaan, (16) Zuun Tsagaan Del, (17) Tsagaan Elgen, (18) Galshar, (19) Urgun, (20) Chuluut Tsagaan Del, (21) Eguzer, (22) Modot, (23) Ikh-Khairkhan, (24) Baga Gazariin Chuluu, (25) Mushgai Khudag, (26) Altanboom, (27) Dulaan Ovoo, (28) Gorkhi, (29) Zuun Bayan, (30) Bayan Ovoo

Table 11.2 Selected features of fluorite deposits in the Bor Undur district

Deposit name	Locality	Mineralization	Major mineral and content	Genetic type	Reserves
Adag	10 km north-west of the Bor Undur mine	Three vein systems 3.640 m, 1.300 m length and 4.37 m, 2.39 m thick	Fluorite 43.28%	Epithermal	741,230 tonnes
Dai uul	25 km south-east of the Bor Undur mine	Five vein system 260–1080 m length, 1.68–6.70 m thick	Quartz-fluorite vein. CaF_2 31.95%	Epithermal	4,641,000 tonnes
Khajuu Ulaan	Khentii, Darkhan soum	Seven different ore bodies, made up of steeply dipping veins	Fluorite-quartz calcite 49.2–57.7% CaF_2	Epithermal	601,880 tonnes
Zuun Tsagaan Del	Dornogovi Ikhhet soum	Seven ore bodies. Three ore bodies are economic importance	Fluorite-quartz calcite and siderite vein. CaF_2 32.7%	Epithermal	7,500,000 tonnes
Galshar	Khentii, Galshar soum.	Four veins of 200–550 m long and 0.4–6.6 m thick	57.9–81.3% CaF_2, 9.27–20.51% SiO_2, and 0.35–19.68% $BaSO_4$,	Epithermal	185,300 tonnes
Khamar us	Dornogovi, Ikh het soum	Five ore bodies with complicated internal structure, 130–280 m length, 1.05–17.15 m thick.	35.63–43.38% CaF_2	Epithermal	1,589,200 tonnes
Tsagaan Chuluut	Khentii, Batnorov soum	Two mineralized zones: Zone #2 is 410 m length and 0.1–6 m thick	CaF_2 30.5–55.2%	Epithermal	320,520 tonnes
Tsagaan Elgen	Dkunkgkvo, Ikhhet soum	Four ore bodies; 30–2000 m length, 1.67–1.84 m thick	Quartz-fluorite 12.0–51.5% CaF_2	Epithermal	938,000 tonnes
Saikhan Gashuun	20 km south-west of the Bor Undur plant	Several vein bodies up to 880 m length and 0.50–4.74 m thick	30–75% CaF_2 Average 48.16%	Epithermal	424,600 tonnes
Undur Ovoo	5 km west of the Bor Undur plant	Several fluorite vein systems; average 700 m in length, and 2.1 m thick	Quartz-fluorite vein. 33.72 CaF_2, 56.12% SiO_2, 0.88% $CaCO_3$, 0.06% P_2O_5	Epithermal	546,000 tonnes
Bor Khujir	40 km north-west of the	One vein of 660 m length and 25–28 m	Calcite-quartz-fluorite	Epithermal	1,126,000 tonnes

(continued)

Table 11.2 (continued)

Deposit name	Locality	Mineralization	Major mineral and content	Genetic type	Reserves
	Bor Undur mine	thickness (on average 26.5 m)	34.9% CaF_2, 0.46% $CaCO_3$, and 0.15% S.		
Tsagaan Undur	25 km north-east of the Bor Undur mine	Two ore bodies 530 m length and 0.66–28.88 m thick, on average 6.86 m	18.12–47.55% CaF_2, 43.1–61.85% SiO_2, 0.81–1.05% $CaCO_3$, and 0.06–0.16% P_2O_5	Epithermal	47,270 tonnes
Omno Baga Chuluu	Dundgovi, Altanshiree	One steeply dipping main vein of 2250 m average length, only 0.3–0.4 m thick	1.87–92.1% CaF_2, average 49.95%, 5.89–92.8% SiO_2, 0.27–5.46% $CaCO_3$	Epithermal	135,070 tonnes

11.5.1 *Bor Undur Fluorite District*

The Bor Undur ore field district is a tectonic block bounded by faults on all sides, and it is the largest mineralized area in Mongolia. This district consists of 16 deposits, most of them in production or even mined out, and numerous occurrences. In the core of the block, there are Triassic granites crosscut by a system of dikes and dike-like bodies which are typically high-angled vein-like mineralization zones composed of host rock breccia with fluorite and quartz-fluorite cement. Permian acidic effusive occurs to the north whereas late Mesozoic andesitic basalt is developed to the east. Northwest and northeast trending faults constitute the most important ore-controlling structures. Ore body thickness ranges from 0.5 to 32 m, extending for 100–3400 m along strike and 20–350 m down dip. The average content of major components varies within the following limits: CaF_2 28–44%, SiO_2 40–55%, and $CaCO_3$ 0.8–2.2% (Harald et al. 2011).

Representative deposits are described below in detail, and selected features of some fluorite deposits and occurrences in the Bor Undur district are summarized in Table 11.2.

11.5.1.1 Bor Undur Deposit

Bor Undur I and Bor Undur II fluorite deposits were discovered as early as 1956. Both open pit and underground mining operations started up in 1982 by Mongolrostsvetmet. The Bor Undur I fluorite deposit consists of four steeply dipping quartz-fluorite veins with an average ore grade of about 42.5% CaF_2 (Fig. 11.5).

The Bor Undur II fluorite deposit contains about 20 steeply dipping veins. Grades of CaF_2 are very homogenous varying from 39.68% to 41.72%. As of the beginning

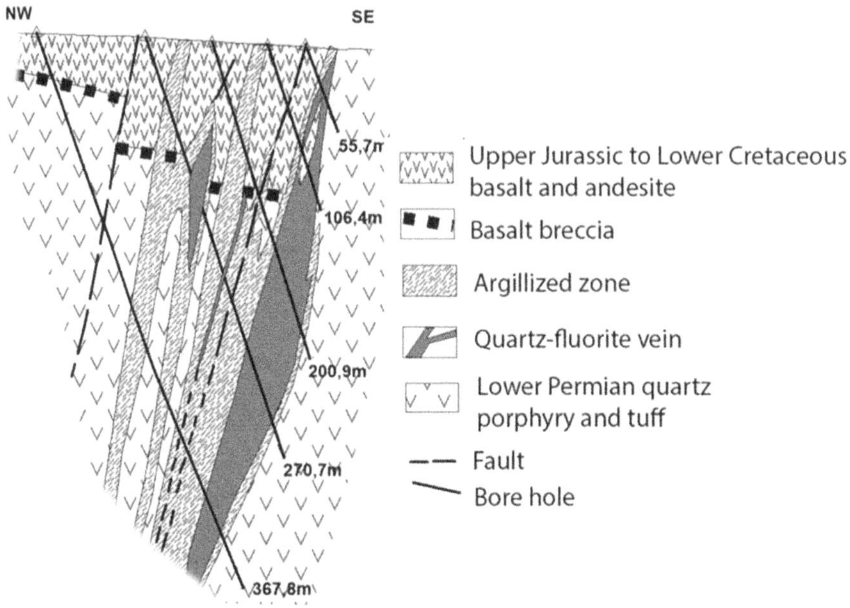

Fig. 11.5 Schematic vertical cross section of Bor Undur deposit (Lhamsuren 2011)

of 1989, the reserves were calculated at 12.5 and 4.5 Mt. for the Bor Undur I and Bor Undur II deposits, respectively.

11.5.1.2 Zuun Tsagaan Del Deposit

The Zuun Tsagaan Del fluorite deposit is 50 km north of the Bor Undur plant. Within the deposit there are seven ore bodies cut by Mesozoic andesite-basalt (Fig. 11.6). A quartz-fluorite-calcite exposed vein with north orientation is 530 m long and 32–75 m thick (14 m thick in average). Three ore bodies are lens-like, and they have economic importance. The total resources estimated in category C1 are 7.5 Mt. of ore with grade of 30.19–35.77% CaF_2.

11.5.1.3 Khavtsal Deposit

The Khavtsal fluorite deposit is located 50 km north-northwest of Tuvshishiree soum in Sukhbaatar province, 180–200 km east of the Bor Undur fluorite mine. The Khavtsal fluorite deposit is situated along the fracture zone with a NW 320–325° orientation within the Khar Nuur plutonic complex. A quartz-fluorite exposed vein with a SW 82–85° orientation is 900–950 m long and 0.4–1 m thick with the thickest vein reaching 2–4 m.

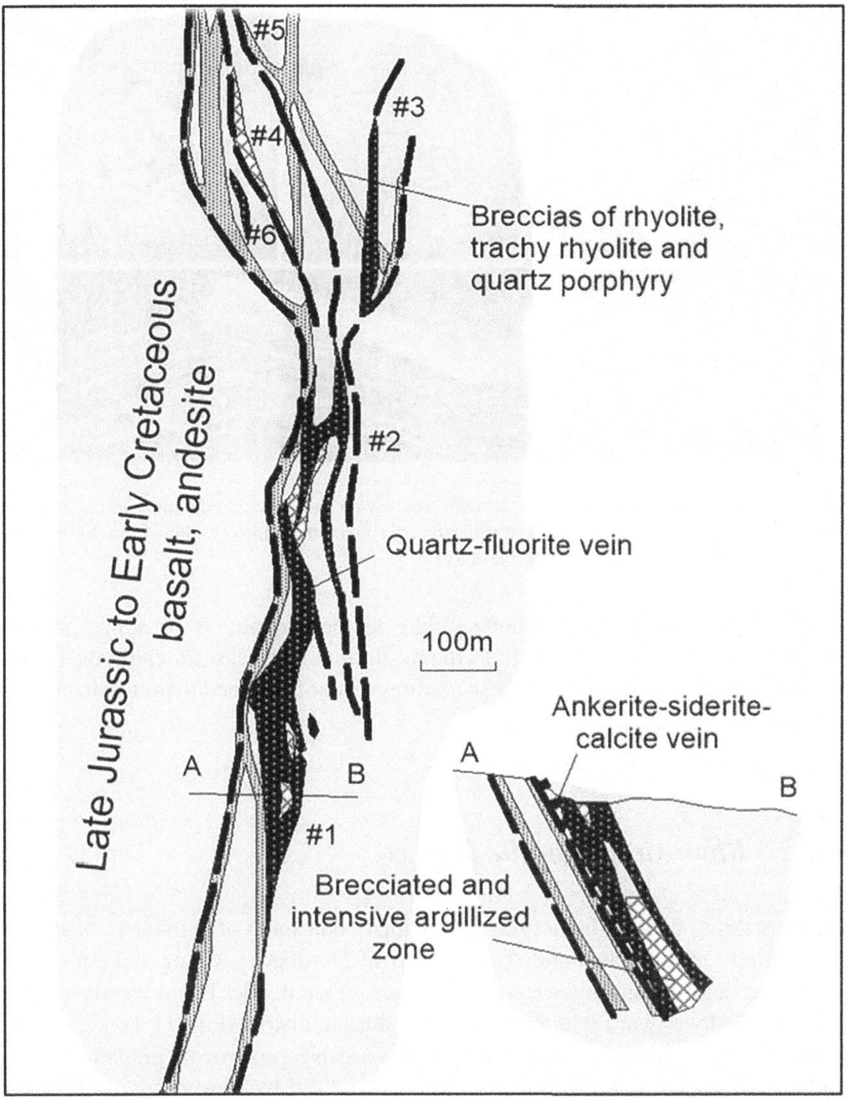

Fig. 11.6 Geological map of Zuun Tsagaan Del deposit (Lkhamsuren 1988). #1–6 Quartz-fluorite veins

Granitic rocks are strongly cataclastic and altered by chlorite, kaolinite, and secondary quartz. The ore body is characterized by breccias, breccias with cocard texture and grades of CaF_2 between 30% and 98%. Several stages of brecciation occurred which are characterized by macro- and micro-breccias and macrocockade and megacockade textures corresponding to mineralization. On the other hand, it is proven that tectonic activities occurred over a long time period. The "faience" (name

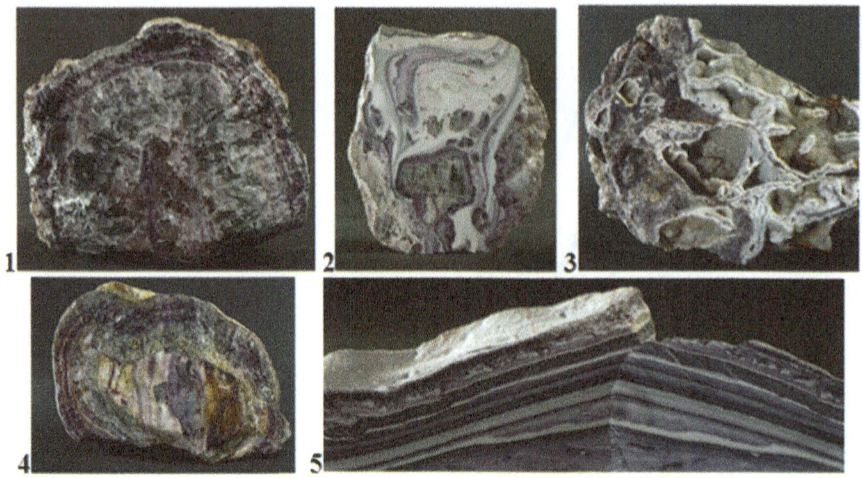

Fig. 11.7 Structure types of Khavtsal deposit. Sukhbaatar province: (1) radial fluorite, (2) fluidal fluorite, (3) vuggy fluorite, (4) cockade fluorite, (5) layered fluorite (Photos from Museum of Geology and Mineral Resources of Mongolia)

in English for fine tin-glazed pottery)-like textured quartz is of a fine-grained, granular type (Fig. 11.7). Amorphous quartz-fluorite is developed within the central and marginal parts of the veins. These textures can be explained by milonitization of granitic rock by faults.

11.5.2 Khar Airag Fluorite District

The Khar Airag fluorite district lies in Bayanjargalan soum of Dundgovi aimag and neighboring Dalanjargalan and Airag soums of Dornogovi Aimag and consists of 20 fluorite deposits and numerous occurrences. This district is not clearly defined and runs southwestward into the Bor Undur fluorite district (Fig. 11.4).

Khar Airag district is predominately composed Neoproterozoic graphite-bearing marbleized limestone, rarely schists, and quartzite cut by Lower Paleozoic granitoids. In local molds and grabens is developed low-carbonaceous sandstone-conglomerate with layers of acidic volcanics. Neoproterozoic formations are breached by small intrusions of Upper Paleozoic and Lower Mesozoic granitoids, accompanied by dike bodies of granite porphyry and quartz porphyry. In the surrounding area distributed Upper Jurassic-Lower Cretaceous basalt, andesite, and rhyolite, also Jurassic and Cretaceous sediments. Most of fluorite deposits and occurrences of Khar Airag district localized in the zones of sub-latitudinal faults. Thus, the Naidvar, Bujgar, and Khongor deposits are confined to the Khongor-Aytay fault.

Representative deposits are described below in detail, and selected features of some fluorite deposits and occurrences in the Khar Airag district are summarized in Table 11.3.

11.5.2.1 Khongor I, II, and III Deposits

The Khongor I, II, and III fluorite deposits are all relatively small and are located close to each other, belonging as they do to a common volcanic structure (Fig. 11.8).

Khongor I consists of one steeply dipping (63°–80°) vein of 680 m length, 65 m depth, and 1.63–11.33 m thickness (4.91 m on average). Grades of CaF_2 vary between 26.36 and 51.64%. The content of $CaCO_3$ is 2.01–5.52% (3.92% on average). Using cut-off grades of 15% CaF_2, category C_1 reserves were calculated at 244,600 tonnes of ore with an average grade of 33.96% CaF_2 (i.e., 83,100 tonnes of CaF_2).

Khongor II is made up of 22 ore bodies of which numbers 1, 2, 3, 4, 4a, and 5 are the biggest ones between 1977 and 1990; the Khongor II deposit was explored in more detail and successfully mined by Mongolrostsvetmet. In total, 522,800 tonnes of ore with 176,100 tonnes of net CaF_2 (average grade is 33.68% CaF_2) were processed and 52,600 tonnes of low-grade ore dumped.

Khongor III lies in the southwest of the Khongor area. It contains three steeply dipping veins, and reserves up to 120 m depth were calculated at 592,700 tonnes of ore with an average grade of 33.24% CaF_2 (i.e., containing 197,000 tonnes of CaF_2 in category C_1).

11.5.2.2 Khoit Khongor Deposit

The Khoit Khongor fluorite deposit lies very close to the Khongor I–III deposits. During prospecting work from 1986 to 1988, two ore belts, with five ore bodies/veins in total, were discovered. The total resources estimated in category P_1 are 170,235 tonnes of ore with grade of 46.26% CaF_2 (i.e., containing 78,700 tonnes of CaF_2) and 4.67% $CaCO_3$ on average.

11.5.2.3 Bujgar I and Bujgar II Deposits

The Bujgar II and Bujgar II fluorite deposits lie 11 km south and 11 km west of the Khar Airag train station. Bujgar I consists of 12 different ore bodies with an average composition of ore varying from 33.23% to 50.48% CaF_2 and 0.78% to 9.6% $CaCO_3$ (3.12% on average). The other average grades are 39.82% SiO_2 and 1.31% Fe_2O_3. Mining at Bujgar II started in 1964, with mining operations taken over by Mongolrostsvetmet in 1983. Initial resources were calculated as 517,300 tonnes of economically mineable ore in category C_1.

Table 11.3 Selected features of fluorite deposits in the Khar Airag fluorite district

Deposit name	Locality	Mineralization	Major mineral and content	Genetic type	Reserves
Khar Airag	Dornogovi, Airag soum	Two major ore bodies of 115 and 195 m average length, and 7.2 and 4.3 m average thickness	Quartz-fluorite vein 70.4% CaF_2, 29.38% SiO_2, 1.47% $CaCO_3$, and 1.08% Fe	Epithermal	
Bor Tolgoi (Umnud Khar Airag)	Dornogovi, Airag soum	Two ore bodies; 265 m, 100 m average length, and 3.0–12.67 m, 0.95 m average thickness	Quartz-fluorite Average 44% CaF_2	Epithermal	87,670 tonnes
Khuren	Dornogovi Airag soum	16 ore bodies; 20–195 m length, 0.5–4.0 m thick	42.35–72.52% CaF_2 average	Epithermal	119,950 tonnes
Suul Undur	Dundgovi Bayanjargalan	Spread over 7 distinct sites some 0.5–4.5 km apart 10–500 m length, 0.1–6.72 m thick	41.38% CaF_2 11.54% $CaCO_3$	Epithermal	402,530 tonnes
Khukh Del	Dundgovi Bayanjargalan	21 ore bodies spread over an area of 30 km^2	Carbonate-quartz-fluorite ore with grades of 29.18–53.5% CaF_2, quartz-fluorite ore with grades of 25.25–47.48% CaF_2,	Epithermal	574,700 tonnes
Baruun Tsagaan Del	Dornogovi Airag	Seven sites containing 67 small ore bodies in total	32.26–51.24% CaF_2	Epithermal	402,570 tonnes
Maikhant I	31 km west-southwest of the Dalanjargalan Soum	Lens-type of three ore bodies 1.5 km length, 2.43–9.55 m thick	30% CaF_2	Hydrothermal-metasomatic	931,900 tonnes

(continued)

Table 11.3 (continued)

Deposit name	Locality	Mineralization	Major mineral and content	Genetic type	Reserves
Maikhant II	34.5 km west-southwest of the	Four ore bodies; average depth is 50–300 m, average thickness is 0.6–12.9 m	17.4–50% CaF_2	Hydrothermal-metasomatic	197,600 tonnes
Tsagaan Takhilch	Dornogovi Dalanjargalan	Four steeply dipping fluorite-bearing veins	Quartz-fluorite 40.99% CaF_2, 48.51% SiO_2, 1.67% $CaCO_3$	Epithermal	578,600 tonnes
Khairt		Lenses are 85–310 m long and 2.6–19 m thick. In the eastern area, the lenses are 95 m and 100 m long and 12 m and 1.2 m thick	20–80%, CaF_2	Epithermal	164,800 tonnes
Naidvar	4 km northwest of the Khar Airag	Ten veins and 2 lenses 40 and 180 m length, 0.70–7.94 m thick	33.57% CaF_2 on average	Epithermal	63,800 tonnes

Bujgar II is made up of five ore bodies with main parameters as shown in Fig. 11.9. Some 52,800 tonnes of ore were mined from this low-grade deposit with estimated remaining category C_2 resources down to 50 m depth amounting to 271,500 tonnes of probably mineable ore, 105,500 tonnes of probably non-mineable ore, and 11,400 tonnes of probably non-mineable ore due to high $CaCO_3$ content.

11.5.3 Berkh Fluorite District

The Berkh fluorite district is located in the southern part of the North Kerulensky ore district and covers an area of about 7.5 thousand km^2. The territory is dominated by lower and Upper Paleozoic granitoids, and Upper Jurassic-Lower Cretaceous basalts and andesite-basalts. Neoproterozoic metamorphic rocks, Upper Paleozoic-Jurassic volcanics, Mesozoic granitoids, and Cretaceous and Cenozoic continental sediments are rarely distributed. The Berkh fluorite district encompasses some

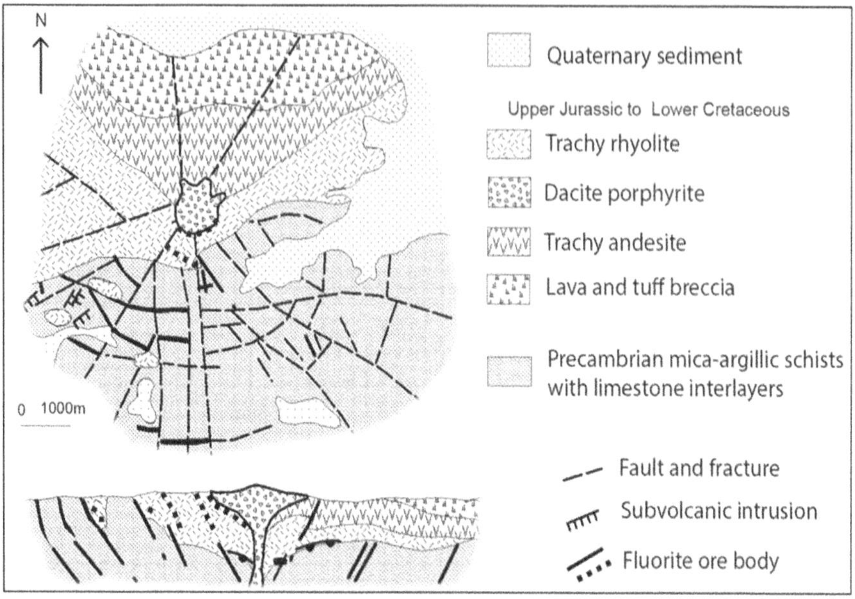

Fig. 11.8 Geological scheme of volcano-tectonic fluorite-bearing structure of Khongor deposit (Batjargal et al. 1979)

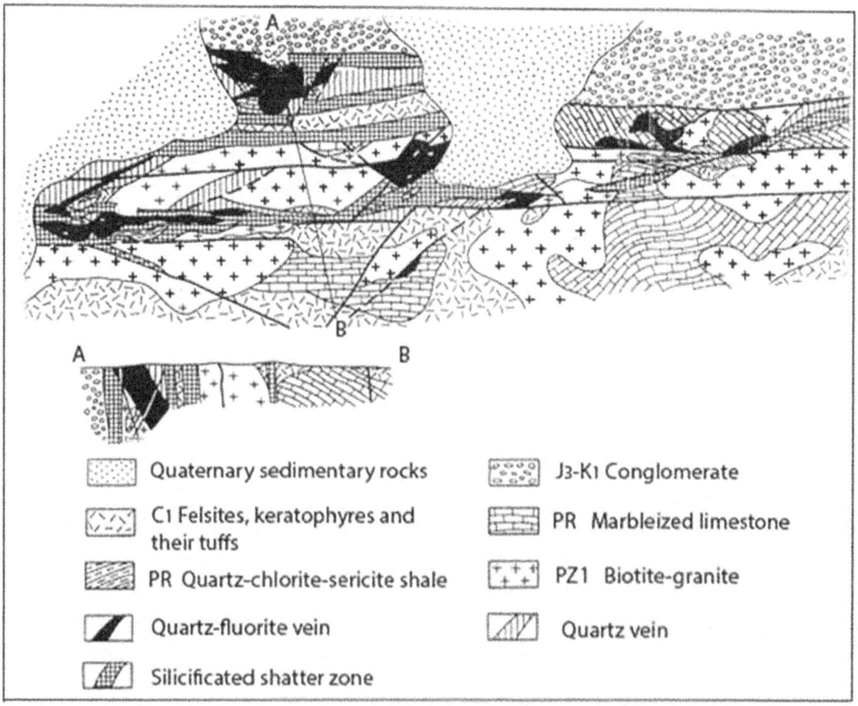

Fig. 11.9 Geological sketch map of the Bujgar fluorite deposit (Khrapov et al. 1977)

20 fluorite deposits, most of them mined out or noneconomic and about 2 dozen occurrences. The mining of fluorite in the Berkh fluorite district started in 1946.

Representative deposits are described below in detail, and selected features of some fluorite deposits and occurrences in the Berkh fluorite district are summarized in Table 11.4.

11.5.3.1 Berkh Deposit

Berkh deposit belongs to Khentii aimag, Batnorov soum, 55 km east from Chingeshaan city. Mining the Berkh fluorite deposit commenced in 1946 by Sovmongolmetall, first in an open pit only, and from 1956 to 1995 in an underground mine by Mongolrostsvetmet, or rather its predecessor company Mongolsovtsvetmet. In contrast to other deposits of the district, it is characterized by higher-quality ores suitable for metallurgical production without previous enrichment. The deposit is confined to the northern side of the Late Mesozoic Olon Nuur volcano-tectonic depression.

There were two major vein systems: one is 1100 m length along the strike and 0.55–5 m thick. Second vein is 320 m length and 1.5–5 m thick (Fig. 11.10). The veins are zonal in structure. Near the contact of vein is consist of fine and medium crystalline fluorite; it is followed by banded quartz-fluorite ore in center. The central part is composed of coarse-crystalline massive fluorite (Fig. 11.11). The major ore-forming minerals are fluorite and quartz, frequently chalcedony-like.

Initial demonstrated total reserves were 6.8 million tonnes of ore grading at 43.61–73.22% CaF_2. Close to the surface, the ore graded up to 91.7% CaF_2, with the grade decreasing to 65% CaF_2 with depth. The average processed ore graded 82–84% CaF_2. This high-grade ore was hand-sorted, transported by truck to the railroad in Choibalsan, from where it was exported to Russia.

11.5.4 Dornod (Khuv Bulag, North Choibalsan) Fluorite District

This poorly prospected and even less explored fluorite district is in northwestern Mongolia and includes only two deposits and several occurrences, all of them in Dornod Aimag. It covers the northern part of the East Mongolian volcanic belt. The territory is predominately composed of Permian and Mesozoic volcanic rocks and Upper Paleozoic granitoids. Fluorite deposit and occurrences are located to the intersection sites of variously oriented faults.

Representative deposits are described below in detail, and selected features of some fluorite deposits and occurrences in the Dornod fluorite district are summarized in Table 11.5.

Table 11.4 Selected features of fluorite deposits in the Berkh fluorite district

Deposit name	Locality	Mineralization	Major mineral and content	Genetic type	Reserves
Delgerkhaan	13 km south-east of Berkh town	8 steeply dipping veins of which four are of most economic importance 960, 880, 186 m length, 1.2, 0.4–3.4, 0.68–3.75 m thick	Quartz-fluorite veins 13.76–77%, 27.5–91.0%, 31.8% CaF_2	Epithermal	2,181,100 tonnes
Chimid	15 km north-west of Berkh town	3 major ore bodies and several small veins; 80–350 m long 1.0–4.45 m thick	Quartz-fluorite veins 20–75% CaF_2 5–7% SiO_2	Epithermal	87,990 tonnes
Kholiin Kholboo	50 km north of Berkh town	Two areas with two major steeply dipping veins; 350–1100 m length, 0.2–3.4 m thick	Up to 84.86% CaF_2 0.2% Ag, up to 12% Pb, Zn, and cu oxides.	Epithermal	123,300 tonnes
Khavtgai I	36 km northeast of Berkh town	One major steeply dipping vein system of up to 1700 m length (675 m on average) and 0.96–10.16 m thickness on average and 2.68 m at the surface	48.41% CaF_2, 45.41% SiO_2, and 1.01% $CaCO_3$ on average	Epithermal	252,000 tonnes
Khavtgai II	44 km northeast of the Berkh town	Three steeply dipping vein system; 100–250 m length and 1.31–4.1 m thickness on average	Breccious to massive ore varies 21.87–69.48% CaF_2	Epithermal	78,100 tonnes
Khavtgai III	20 km the Berkh town	Three steeply dipping ore bodies 150 m, 125 m average length, 1.06–2.30, 1.18–2.71 m thickness	26.5–61.7% on average, up to 60% CaF_2	Epithermal	176,700 tonnes
Khaidelgerkhaan	Northwestern part of the Berkh town	One major vein of 775 m average length, 200 m proven depth, and 0.32–6.94 m thickness	40.94% CaF_2 on average	Epithermal	348,200 tonnes
Dojir	*Khentii Bayan-Ovoo Soum*	Nine quartz-fluorite veins 100–800 length, 0.2–5 m thick.	77.7–82.16% CaF_2	Epithermal	555,150 tonnes
Yamaat	40 km northeast of the Berkh town	Six steeply dipping veins, with 400 m (up to 680 m) length, 200 m depth, and 1.0–4.05 m thickness (on average 1.25 m)	67.75–88.93% CaF_2 8.89–30.59% SiO_2	Epithermal	123,000 tonnes

Bulgan Uul	Khentii Kherlen	Six ore bodies, formed by steeply dipping veins with multiple apophyses and fine veins	43.77–61.51% CaF_2	Epithermal	231,850 tonnes
Serven	35 km south-east of the Berkh town	Several veins; it is over 700 m long, steeply dipping and 1.47 m thick on average	43.98% CaF_2, 45.08% SiO_2, 1.39% $CaCO_3$, 0.026%, P_2O_5, 0.01% S, and 0.1–0.04% $BaSO_4$	Epithermal	72,254 tonnes
Khailuur Jonsh	Sukhbaatar Tumentsogt	Five ore bodies	Quartz-fluorite veins, Average CaF_2 36.93%, 2.20% $CaCO_3$, 0.09% P, 1.27% Fe, and 0.01% S	Hydrothermal and metasomatic	313,880 tonnes

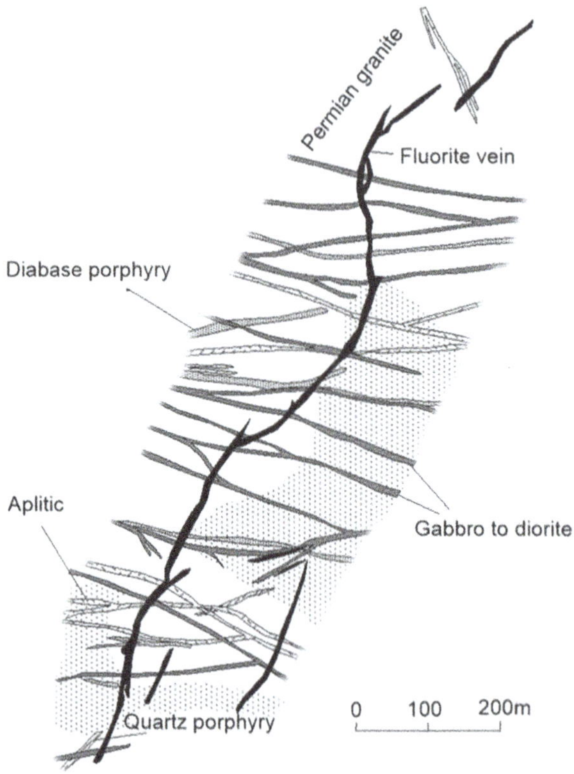

Fig. 11.10 Geological map of Berkh deposit (Khrapov et al. 1977)

Fig. 11.11 Fluorite of Berkh deposit. (1) Fluorite crystal, (2) Massive fluorite, (3) Yellow massive fluorite (photos from Museum of Geology and Mineral Resources of Mongolia)

Table 11.5 Selected features of fluorite deposits in the Dornod fluorite district

Deposit name	locality	Mineralization	Major mineral and content	Genetic type	Reserves
Baruun Suuj	50 km south-west of Dashbalbar soum	9 major and 9 minor fluorite-bearing veins up to 860 m length and 0.11–1.68 m thick	Quartz-fluorite vein Average at 43.0% CaF_2	Epithermal	1,331,000 tonnes
Khuv Bulag	50 km south-west of Gurvanzagal soum	22 vein system average length of 280 m and an average thickness of 0.80 m	Quartz-fluorite veins 64.56–96.39% CaF_2, 2.77–55.3% SiO_2, 0.13–0.16% CaO, 0.54–0.57% F_2O_3, and 0.54–0.74% other oxides	Epithermal	512,600 tonnes

11.5.5 *Chuluut Tsagaan Del Fluorite Center*

This small fluorite center, some 130 km south of Ulaanbaatar, in Bayantsagaan soum of Tuv aimag, contains one fluorite deposit and various fluorite occurrences.

11.5.5.1 Chuluut Tsagaan Del Fluorite Deposit

The Chuluut Tsagaan Del fluorite deposit consists of eight ore bodies, with 87% of all fluorspar resources contained in bodies #2 (lens of 430 m length, 10–95 m depth, 70 m thickness, and average grade of 55.65% CaF_2) and #5 (lens of 700 m length, 220 m depth, 40 m thickness, and average grade of 48% CaF_2). $CaCO_3$ contents are given as 2.3–16.0%.

Using cut-off grades of 17% CaF_2 in general and 30% CaF_2 in blocks of interest, the total category C_1 reserves in ore bodies #2, #5, and #8 are estimated at 1,107,600 tonnes of ore with a grade of 51.78% CaF_2 on average (i.e., containing 573,300 tonnes of CaF_2). Additionally, the category C_2 resources were calculated at 388,900 tonnes of ore with a grade of 47.30% CaF_2 (i.e., containing 184,000 tonnes of CaF_2). Between 1981 and 1990, ore body #2 was quarried by Mongolczechoslovakmetal with 629,000 tonnes of ore containing 334,900 tonnes of CaF_2 taken out. The mined crude ore was hand sorted to produce higher-grade lump fluorite ore, which was exported to the former Czechoslovakia. Some 20,300 tonnes of non-commercial ore were dumped.

11.5.6 Urgun Fluorite Center

This important fluorite center covering 12 km^2 lies in the far south of the Eastern Mongolian fluorite belt in Urgun Soum of Dornogovi Aimag about 60 km south of the town of Sainshand. It was discovered by the Choir Geological expedition in 1981–1983 and contains several ore bodies of economic interest.

The ore field is broken into a number of blocks by numerous faults which control the fluorite mineralization. Here there are 17 metasomatic ore zones and orebodies. The ore zones usually occur at contact between limestones and granites, limestones and schists, and less frequently within granites. Meanwhile, the total geological resources are estimated at ten million tonnes of ore. While low-carbonate ores (on average 47.0% CaF_2, 45.0% SiO_2, and 4.7% $CaCO_3$) compose the bulk (80%) of the ore bodies, high-carbonate ores (28.5% CaF_2 on average, 34.7% SiO_2, and 30.7% $CaCO_3$) are difficult to process and have to be blended with low-carbonate ores before processing (Harald et al. 2011).

Urgun (ore body #9) and also ore body #14 have been quarried by Mongolsovtsvetmet, now Mongolrostsvetmet LLC, since 1988. Ore body #9, which is of lower quality, extends for 520 m along the strike on the surface and 370 m downdip. About 90,000–120,000 tonnes of ore are being quarried every year with the majority of this tonnage transported to the Bor Undur plant for further processing. The ore bodies are of calcite-quartz-fluorite type with veins of 5–105 m thickness becoming narrower at a depth of 113 m. As of 2009, the remaining reserves and resources given by Mongolrostsvetmet LLC are 4,137,600 tonnes of ore with grades of 18.6–61.3% CaF_2 (42.98% on average), i.e., containing 1,778,200 tonnes of CaF_2. The bulk of these reserves and resources are bound to ore body #9, with ore body #14 being nearly mined out.

Chuluut (ore body #10) is currently being explored. It appears to be made up of three small mineralized lenses and one larger and less regularly shaped body of >30 m true thickness. The latter dips northwesterly and plunges at about 60° to the west. A very rough resource estimate down to 145 m depth is 170,000 tonnes of ore grading at 40.5% (30.1–53.1%) CaF_2 on average (Harald et al. 2011). Because 80% of these resources can only be mined underground, mining of the Chuluut ore body may not be feasible.

11.5.7 Central Dundgovi Fluorite Center

This newly defined fluorite center lies in the center of Dundgovi Aimag and consists of few fluorite deposits in Gurvansaikhan Soum and numerous surrounding fluorite occurrences. It has a complex mosaic block structure, due to the developed numerous sub-latitudinal, submeridional, and northwestern fault system.

11.5.7.1 Bilgekh Uul (Bilk) Deposit

The Bilgekh Uul (Bilk) fluorite deposit consists of three areas with several ore bodies/veins each. Area I contains 16 veins, of which 12 are of economic importance. Within these veins the grade of CaF_2 varies from 33.54% to 68.8% with 42.19% on average. There are 28 veins in area II, while area III has 23 veins. At a depth of 40–50 m, the thickness of the veins decreases, but the veins could continue down to 100–250 m depth. The total resources are estimated at 1,625,900 tonnes of ore, while resources in area I only (suitable for open pit mining) were re-estimated at 732,500 tonnes of ore with a grade of 42.19% on average up to 40 m depth (i.e., containing 309,000 tonnes of CaF_2).

11.5.8 Northern Umnugovi Fluorite Center

This fluorite center, which lies south of the East Mongolian fluorspar belt, is also newly defined and has just one fluorite deposit of interest discovered so far.

11.5.8.1 Bayan-Ovoo Deposit

The Bayan-Ovoo fluorite deposit is situated in the Mandalovoo-Ulziit gold zone of the South Mongolian metallogenic belt. It's located 90 km north of Dalanzadgad city. Several faults occurred in the deposit area, and fluorite-bearing veins with NE 60–80° and NW 340–350° developed along faults. Two generations of fluorite-bearing veins were formed, with the first generation of veins directed NE and the second generation directed NW by late fluid activity. The genesis of the fluorite deposit is assumed to be related to Triassic, Jurassic, and Cretaceous subvolcanic rocks, and dikes corresponding structurally with Early Permian magmatism. The mineralization type belongs to the epithermal vein type with quartz-fluorite formation. The deposit covers 5 km^2 with eight ore bodies found so far, of which one seems to have the biggest economic potential. This vein has a length of 1700 m and a thickness of 0.5–11 m. The average grade in the ore of this vein close to the surface is 76.6% CaF_2, while in boreholes, an average grade of 71.98% (44.49–88.26%) CaF_2 was determined. $CaCO_3$ ranges from 0.1% to 2.62%. The total estimated resources are 709,600 tonnes of ore in category C_2 and 2,289,800 tonnes of ore in category P_1.

11.5.9 West Khuvsgul Fluorite Center

This newly defined fluorite center lies far northwest of the Eastern Mongolian fluorite zone in Tsetserleg soum of Khuvsgul Aimag and consists of the Jandai fluorite deposit and several small occurrences.

11.5.9.1 Jandai Hydrothermal Deposit

Jandai is a typical hydrothermal fluorite-quartz vein deposit consisting of four areas with a total of ten major veins. The total resources in area III are 175,279 tonnes of ore in category C_2 and with 519,119 tonnes of ore in all areas in category P_1. Several of the fluorite deposits mentioned above were visited by a field team of BGR and MRAM geologists in 2008 and 2010.

11.6 Concluding Remarks

Mongolia is one of the leading producers of fluorite in the world. Major deposits of economic importance are of Jurassic and Cretaceous ages and are distributed in central and eastern Mongolia. Most of the fluorite deposits in Mongolia are hydrothermal or epithermal. The epithermal fluorite deposits occur as veins, irregular and lens-shaped, and in brecciated and disseminated shear zones. The vein deposits are dominant.

Major minerals of the epithermal type are quartz and fluorite. Quartz commonly occurs as a chalcedony like, fine-grained, or columnar crystals. Fluorite is frequently coarse-grained, massive, columnar, rarely fine-grained, and microcrystalline or even cryptocrystalline.

The total annual production of fluorite reached 527,000 tonnes. At present, the Bor-Undur, Adag, Zuun Tsagaan Del, Tsagaan Elgen, Urgun, Delgerkhaan, and Bayan-Ovoo deposits are in commercial production, and further exploration is being carried out. The largest mining and ore processing complex is built on the Bor Undur deposit.

References

Baskina VA, Volchanskaya I.K (1976) Novy tip mineralizatsii redkikh metallov na yuge Mongolii svyazanny s shchelochnymi vulkanicheskimi porodami. (new type of rare metal mineralization in the southern Mongolia related with alkaline volcanic rocks) Papers of Academy of Sciences of USSR. 228:670–672

Baskina VA, Volchanskaya IK, Gundsambuu Ts, Nikitina IN (1978) Celestinovaya mineralizatsiya yuga Mongolii. (Celestine mineralization of southern Mongolia) Proceedings of the USSR Academy of Sciences. Geological series. 5:96–105

Batjargal Sh, Lkhamsuren J (1987) Rudnaya struktura mestorozhdeniy epitermal'noi flyuoriteovoi rudnoi formatsii Mongolii. (Ore structure of deposits of epithermal fluorspar ore-formation in Mongolia) Sci Trans of Mineralogical Museum of the Mongolia. Poly.thec.Institute. 9:16–39

Batjargal Sh, Lkhamsuren J, Tumenbayar B, Shiiter N (1982) Geologo-strukturnyye i mineralogo-geochimicheskiye osobennosti mineralizatsii flyuorita na territorii Mongolii i ikh prakticheskoye znacheniye. (Geologic-structural and mineralogical-geochemical features of fluorite mineralization in territory of Mongolia and their practical significance) Khaiguulchin 2(39):6–14

Batjargal Sh, Shiiter N, Lkhamsuren J (1979) Fflyuoritee aguulsan volcanic-tectonic structur Fflyuoritesoderzhashchiy vulkanotektonicheskiy - «Khongor» v tsentral'noy Mongolii (Fluorite-bearing volcanic-tectonic structure “Khongor” in central Mongolia). Sci. Transaction of Mongolian State University 2(68):37-47

Batjargal Sh, Tumenbayar B, Lkhamsuren J, Shiiter N (1985) Tipomorfizm flyuorita iz endogennykh mezozoyskikh mestorozhdeniy vostochnoy Mongolii. (Typomorphism of fluorite from endogenic Mesozoic deposits in Eastern Mongolia) Problems of geology and metallogeny of eastern Mongolia. Proc. of III Sci. Conf. of Kherulen Geol. Expedition, 127–138

Filippova IB, Suetenko OD, Khasin RA (1984) Mezozoyskaya paleogeodinamicheskaya obstanovka i nekotoryye osobennosti metallogenii vostochnoy Mongolii. Mesozoic paleo-geodynamic position and some features of metallogeny in eastern Mongolia. Geology and Mineral Resources of Mongolian People's Republic: II. 27–46

Frikh-Khar DI, Volchanskaya IK (1982) Relations of the fluorite and rare-earth mineralization in Mongolia. VI symp. IAGOD. Abstracts. Tbilisi, p. 225

Harald E, Peter B, Martin S, Tamiraa A (2011) Industrial minerals and selected rare metals in Mongolia. An Investors’ guide. Prepared by the Government of Mongolia. Funded by the Federal Ministry of Economic Cooperation and Development of the Federal Republic of Germany through the Federal Institute for Geosc. Natur. Resour. (BGR), 92–142

Ivanova (1972) Geokhimicheskiye usloviya formirovaniya volframitovykh mestorozhdeniy (Geochemical conditions of formation of wolframite deposits) Moscow. Nauka p.152

Ivanova (1976) Mineralogiya i geokhimiya vol'framovykh mestorozhdeniy Mongolii. (Mineralogy and geochemistry of the tungsten deposits of Mongolia) Moscow. Nauka, p.260

Jamsran M, Lkhamsuren J, Obolensky AA, Badamgarav J, Batjargal Sh, Lbolenskaya RV, Tumenybayar B, Shiiter N (1986) Metallogeniya Mongolii. Plavikovy shpat. (Metallogeny of Mongolia. Fluorspar) Institute of Geology and Geophysics. Siberian branch, USSR Acad. of Sci: 9:1–48

JICA MMAJ (1993) Mongolia’s mineral resources and the potential. Abroad Mining Information, 268:25-37

Kalenov AD, Khasin RA (1965) Vozrast i nekotoryye prirodnyye raspredeleniya i perspektivy mineralizatsii flyuorita v vostochnoy chasti Mongolii. (age and some natural distribution and prospective of fluorite mineralization in eastern Mongolia). Soviet Geol 4:36–48

Kandinov MN Dobrolyubov VA (1984) Yestestvennoye rasprostraneniye i perspektivy flyuoritonosnost'i vostochnoy i tsentral'noy Mongolii. (Natural distribution and prospective of fluorite mineralization of eastern and central Mongolia) Geology and Mineral Resources of Mongolian People's Republic, II, Nedra, Moscow, 165–176

Khrapov AA, Zimina NA, Konstatinov NF, Kalita ED (1977) Plavikovyy shpat. Geologiya i mineral'nyye resursy Mongol'skoy Narodnoy Respubliki (Fluorspar. Geology and Mineral Resources of Mongolian People's Republic) III, Mineral deposits. Nedra, Moscow. 493–552

Kishimoto F (1982) Fluorite deposits in Mongolia. Chishitu News 329:25–34

Konstantinov NF, Zimina NA (1966) Ob ekonomiko-geneticheskikh tipakh mestorozhdeniy flyuorita Mongolii. (Economic-genetical types of fluorite deposits of Mongolia) Materials of Geology of Mongolian People’s Republic. Nedra, Moscow. 161–167

Korytov FYa, Kandinov MN, Khrapov AA and Dobrolyubov VA (1980) Poetapnoye razvitiye i zonal'nost' mineraloobrazovaniya na nekotorykh mestorozhdeniyakh flyuorita vostochnoy Mongolii. (Staged development and zonality of mineral formation on some fluorite deposits of eastern Mongolia) Geology and Mineral Deposits of Mongolia. 1. Nedra, Moscow 172–175

Korytov F.Ya, Khrapov A.A, Kandinov M.N (1978) O fenatsit-flyuoritovoy mineralizatsii Mongolii. (about phenacite-fluorite mineralization of Mongolia) Papers of Academy of Sciences of USSR, 242:190–191

Koshelev JYa (1985) Epitermal'nyye flyuoritovyye mestorozhdeniya vostochno-mongol'skogo vulkanicheskogo poyasa. (Epithermal fluorite deposits of eastern Mongolian volcanic belt) Nedra, Novosibirsk, 1–135

Kostyrev NF, Sengee D, Volkov LS (1977) Pervoye otkrytiye flyuorit-bertrandit-fenakitovogo rudoobrazovaniya na territorii Mongolii. (first discovery of fluorite-bertrandite-phenakite ore-formation in the territory of Mongolia) Khaiguulchin: 4:14–16

Kotov PA, Sambalkhundev Ts, Kotova AI (1979) O vozraste flyuoritovykh mestorozhgeniy Mongolo-Zabaykal'skoy flyuoritovoy provintsii. (About age of fluorite deposits of Mongolian-Transbaikalian fluorite province. In: Geological Setting and Natural Distribution of Important Mineral Resources in the Territory of Mongolian People's Republic) Nedra, Moscow, 142–151

Kovalenko VI, Vladykin NV, Konusova VV (1976) Proyavlenie kontsentrirovannoy redkozemel'noy mineralizatsii v yuzhnoy chasti pustyni Gobi. (an occurrence of concentrated rare earth mineralization in southern part of Gobi Desert) Papers of Academy of Sciences of USSR. 230:209–213

Lhamsuren J (2011) Khailuur jonsh. In: Metal bus ashigt maltmal VII. Mongoliin geologi, ashigt maltmal (Non-metallic mineral resources of Mongolia) 128–159

Lkhamsuren J (1976) Fluid inclusion data on the fflyuoritee occurrences in eastern Mongolia) 25th IGC. Abstracts, Canberra. 3:810–811

Lkhamsuren J (1984) Usloviya i mekhanizm obrazovaniya flyuidnykh vklyucheniy flyuorita raznogo genezisa na territorii Mongol'skoy Narodnoy Respubliki. (Conditions and mechanism of the fluid inclusions formation in fluorite of different genesis in the territory of the Mongolian People's Republic) 27th IGC. Abstracts, Moscow, 5:97–98

Lkhamsuren J (1988) Flyuoritovaya mineralizatsiya Mongolii (rudoobrazovaniye, genezis i zakonomernosti rasprostraneniya). (The fluorite mineralization of Mongolia: ore-formation, genesis and distribution regularities. Ulaanbaatar. p 54

Lkhamsuren J, Batjargal Sh (1975) Napravlennyye izmeneniya morfologii kristallov flyuorita v endogennykh otlozheniyakh Mongolii. (Directional changes of fluorite crystal morphology in endogenic deposits of Mongolia). Scientific Transaction of Mineralogical Museum of the Mongolian State Univ. 5:75–85

Lkhamsuren J, Batjargal Sh, Shitter N (1979) Temperaturnyye usloviya obrazovaniya flyuorita na mestorozhdenii Khongor po dannym izucheniya gazo-zhidkostnykh vklyucheniy v kristallakh flyuorita. (Temperature conditions of fluorite formation in Khongor deposit according to the study of gas-liquid inclusions in fluorite crystals) Scientific Papers of Mongolian State Univ. 2:87–96

Lkhamsuren J, Hamasaki S (1998) Fluorite deposits in Mongolia: an outline. Bull Geol Surv Japan 49(6):309–318

Marinov NA (1958) O vremeni formirovaniya mestorozhdeniy flyuorita v vostochnoy Mongolii (about time of forming of fluorite deposits in eastern Mongolia). Soviet Geol 9:164–167

Marinov NA (1980) Flyuoritovaya mineralizatsiya vostochnoy Mongolii. (fluorite mineralization of eastern Mongolia). Geol Ore Deposits 2:99–103

Naumov VB, Ivanova GF (1975) Termobaricheskiye usloviya obrazovaniya flyuorita v vol'framitovykh otlozheniyakh. (Thermobaric conditions of fluorite formation in wolframite deposits). Geochemistry, Academy of Science of USSR Moscow 3:387–400

Ontoev DO, Kandinov MN, Korytov FYa (1977) O temperature mineraloobrazovaniya mineralizatsii apatit-flyuorit-redkozemel'nykh elementov Yuzhnoy Mongolii. (about temperature of mineral formation of apatite-fluorite-rare earth elements mineralization of southern Mongolia) Papers of Academy of Sciences of USSR 234:1164–1166

Ontoev DO, Kandinov MN, Korytov FYa, Gundsambuu Ts (1979a) Temperaturnyye usloviya formirovaniya F-REE-Fe mestorozhdeniy (Temperatural conditions of formation of fluorine-REE-iron ore deposits. Basic Parameters of Endogenic Ore-Formation Process. Nauka, Novosibirsk. 1:199–206

Ontoev DO, Luvsandanzan B, Gundsambuu T (1979b) Geologicheskoye stroyeniye i endogennaya mineralizatsiya Mushugayskogo flyuorito-redkozemel'nogo mestorozhdeniya (Mongoliya). (geological structure and endogenic mineralization of Mushugai fluorite-rare earth deposit (Mongolia)). Geol Ore Deposit 21:27–42

Shuvalov VF, Avrov DP, Yackobson LN (1980) O vozraste flyuoritovoy mineralizatsii vostochnoy Mongolii (On the age of fluorite mineralization in eastern Mongolia) Geology and Mineral Deposits of Mongolia. 1. Nedra, Moscow, 161–171

Tumenbayar B (1987) Geokhimiya i tipomorfizm plavikovogo shpata Mongolii. (Geochemistry and typomorphism of fluorspar of Mongolia) Publication House Academy Science of Mongolian People's Republic, Ulaanbaatar, p 174

Tumenbayar B, Lkhamsuren J, Ganbat D, Bershov LV, Speransky AV (1986) Issledovaniye raspredeleniya ferromagnitnykh ionov -Mn^{2+} i Gd^{3+} vo flyuorite Mongolii (Distribution investigation of ferromagnetic ions-Mn^{2+} and Gd^{3+} in fluorite of Mongolia in: Problems of geology of Mongolia) Transaction of Geological Institute of Mongolian Acad of Sci. 98–214

Voinkov DM, Lkhamsuren J (1986) Etapy osvoyeniya i vremennogo rezhima formirovaniya mestorozhdeniy flyuorita rudnogo uzla Berkh Mongolii (The development stages and temperature regime of fluorite deposit formation of the ore-knot Berkh in Mongolia) Mineralogy of Tadjikstan. Publication House Donish 7. Dushanbe, 58–67

Volchanskaya IK, Korytov FYa (1980) Glubokiye razlomy v tsentralnoy i vostochnoy Mongolii i rasprostraneniye flyuoritovoy mineralizatsii. (Deep-seated faults in central and eastern Mongolia and distribution of fluorite mineralization) Proceedings of the USSR Academy of Sciences, Geological series. 10:123–132

Yashina RM (1975) Shchelochnye formatsii dugoobraznykh oblastey blokovkh podnyatiy i ikh metallogenicheskiye osobennosti. (Alkaline formation of arched uplift-block regions and their metallogenic peculiarities. Regularities of Mineral Deposit Distribution). XI. Nauka, Moscow 134–154

Yashina RM (1982) Shchelochnoy magmatizm skladchato-blochnykh oblastey (Alkaline magmatism of folding-block regions) Nauka, Moscow. p 274

Chapter 12
Phosphate Deposits

Baatar Munkhtsengel, Jambaa Byambaa, and Altangerel Tamiraa

12.1 Introduction

Phosphatic rocks have a high economic and scientific significance, because the element phosphorus (P) is a bioelement, vital for life, hence an integral component of organic matter. The main mineral systems that carry P are phosphorites, which, as the name implies, are P-rich (>18 wt% P_2O_5) marine chemical sedimentary rocks (Pufahl 2010). The Mid-Late Permian Phosphoria Formation in the USA is a classic representative of rich phosphorite deposits (Gulbrandsen and Krier 1980). Black shale sequences can have high P_2O_5 contents, which in many cases may have economically viable phosphorite deposits and may represent the deep-water end member of shallower chert-rich phosphorites. Phosphate ores can be magmatic or sedimentary, the latter of either marine or terrestrial origin.

Phosphorite deposits in Mongolia are divided into two types: sedimentary and phosphate nodules. Sedimentary deposits are widespread in Khuvsgul and Zavkhan areas, Northern Mongolia, originated from Ediacaran and Cambrian epiplatform cover (Khuvsgul Group) and shelf deposits (Tsagaanolom Formation) (Byamba et al. 1986; Byamba 1996).

B. Munkhtsengel (✉)
Department of Geology and Hydrogeology, School of Geology and Mining, Mongolian University of Science and Technology, Ulaanbaatar, Mongolia
e-mail: tsengel@must.edu.mn

J. Byambaa
Geological Investigation Center, Ulaanbaatar, Mongolia
e-mail: megd_byamba@yahoo.com

A. Tamiraa
Mineral Resources and Petroleum Authority of Mongolia, Ulaanbaatar, Mongolia
e-mail: tamiraa.a@mrpam.gov.mn

O. Gerel et al. (eds.), *Mineral Resources of Mongolia*, Modern Approaches in Solid Earth Sciences 19, https://doi.org/10.1007/978-981-15-5943-3_12

Deposits related with Khuvsgul Group and Tsagaanolom Formation are named as Khuvsgul phosphorite basin and Zavkhan phosphorite basin, respectively. Although these deposits are distant, they resembled carbonate-terrigenous rocks. Even though they have similar structure and composition, content of ore mineral is notably different. Phosphate nodule type is only known in Govisumber area, Southeastern Mongolia, which is considered of poor economic potential. All phosphate deposits of economic interest in Mongolia are of marine origin with phosphate ores occurring as phosphorite.

12.2 Geology of the Khuvsgul Phosphorite Basin

The Khuvsgul phosphorite basin in northern Mongolia stretches into southern Siberia and is one of the largest phosphate-bearing basins in the world. The study of the Khuvsgul phosphorite basin started in the 1970s–1980s after its discovery by a Russian-Mongolian joint expedition (Yanshin 1986). The main phosphorite deposits were prospected during this period (Osokin et al. 1973). Works by Byamba et al. (1986) and Byamba (1996) played a significant role in the investigation of the basin and assessment of the phosphorite potential of Mongolia (Ilyin 2004).

The Khuvsgul phosphorite basin has a length of some 300 km and a width of up to 170 km (Zaitsev 1981; Ilyin et al. 1986). The basin belongs to the Tuva-Mongolian microcontinent (Ilyin 1971), which, in turn, is a constituent of the Caledonian Sayan-Baikal foldbelt (Kuz'michev 2000). In the Late Riphean, the Tuva-Mongolian Massif is represented by a microcontinent spatially isolated from other continental terranes and the Siberian Platform. The microcontinent was slightly compressed as a result of the Late Baikalian orogeny. In the Vendian-Cambrian, the area accumulated carbonate sediments and transformed into a carbonate platform that was still separated from the Siberian continent by an ocean (Kuz'michev 2001). The Tuva-Mongolian microcontinent joined the Siberian Platform as a result of collision related to the Salairian orogeny (Khain and Rudakov 1995). As explained by Kuz'michev (2001), the pre-Vendian basement of Khuvsgul phosphorite-bearing basin is heterogeneous. This is reflected in both the distribution and dislocation pattern of phosphorites. Archean granulites of the Gargan Block are located north of Lake Khuvsgul and represent the oldest (Rb-Sr age $>$ 800 Ma) component of the basement. Vendian-Cambrian sediments of the carbonate platform are underlain over a large area by graywackes and volcanics of the Oka Group (Sm-Nd and U-Pb ages $>$760 Ma). The ophiolitic complex of the Shishkhid arc (Rb-Sr age 630 Ma) represents another component of the basement. In the present-day structure, ophiolites are thrusted along the Khugein Thrust over the Khuvsgul basin.

Geological setting of the basin is complex composing of various ages of rocks from Paleoproterozoic to Quaternary and various compositions of intrusive complexes (Fig. 12.1).

Paleoproterozoic rocks are represented by gneiss, granulite, amphibolites, crystalline schist, quartzite, and marble. Rocks are affected by granitization and

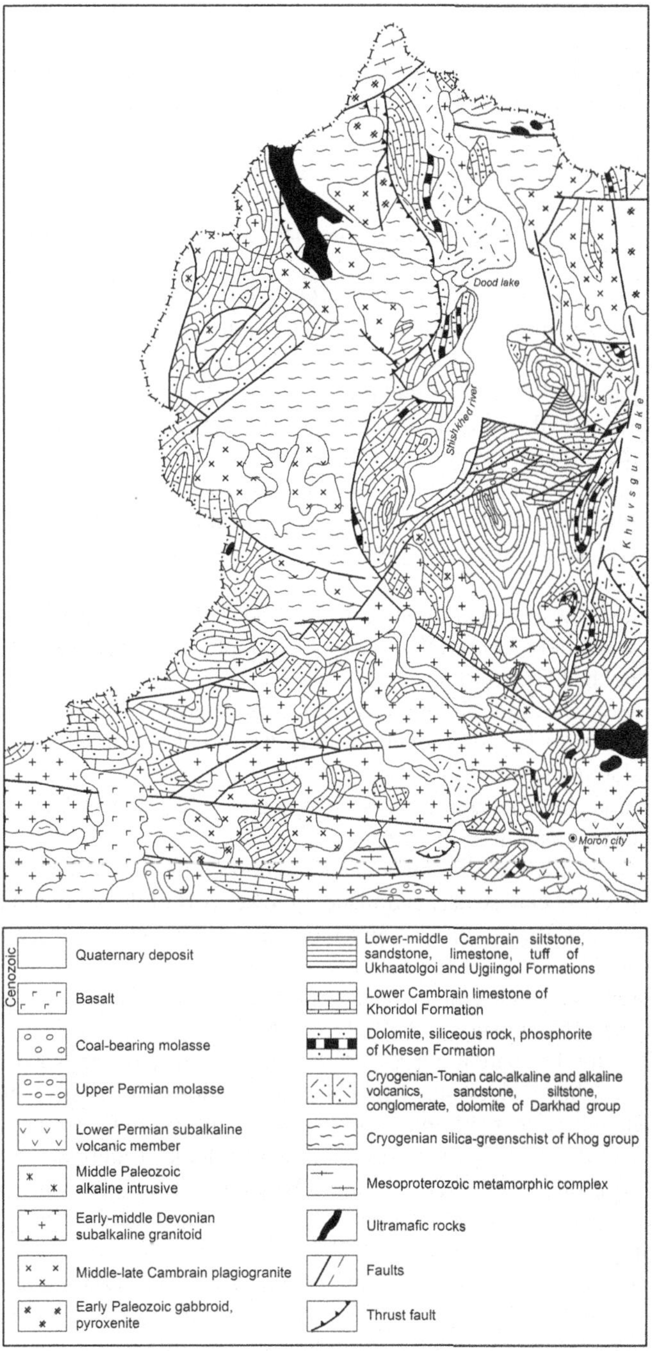

Fig. 12.1 Geological schematic map of the Khuvsgul phosphorite basin (Byamba 2012)

migmatization and form granite-gneiss mold. These metamorphosed rock outcrops compose ancient blocks of Uurgol, Kachik, Agoi, and Gargan. Thickness of the blocks varies from 3000 to 5000 m. Age of crystalline rocks older than Mesoproterozoic is on debate. Mitropanov et al. (1981) explains as Archaean, whereas Belichenko (1969) said rocks of this type belong to Early Paleozoic.

Middle and upper Mesoproterozoic schists and metamorphic rocks of Khungui and Kharsair Formations form geoanticlinal uplifting along the marginal and inner part of the phosphorite-bearing basin. Thickness of these rocks does not exceed than 3000 m.

Granite gravel in upper layers in Darkhad Group shows 823 Ma which can indicate schist is younger (Byamba 2012). Neoproterozoic terrigenous and alkaline volcanic formations are classified among the Darkhad Group. They are observed in the lower part of Ediacaran-Cambrian terrigenic-carbonate rocks of Khuvsgul Group. Quartz and arkose sandstone and siltstone play important role in terrigenous deposit of the Darkhad Group, and dolomite and porous schist occur in upper section. Calc-alkaline and alkaline volcanites are observed in marginal part of the Darkhad Group distribution. Small bodies of riebeckite-granite which show 752 Ma can be present comagmatic member of those volcanites (Byamba et al. 1986; Byamba 2012).

Thickness of the Darkhad Group rocks is about 3000 m. The Khuvsgul Group rocks are structurally conformably but stratigraphically unconformably located above the Darkhad Group rocks.

Boundary of Khuvsgul phosphorite basin is re-determined by the outcrop of the Khuvsgul Group rocks. Stratigraphy of the Khuvsgul Group is well studied in southern and eastern part of the basin.

The group is classified into Khar Khadat, Khesen, Khoridol, Ukhaa Tolgoi, and Ujgiin Gol Formations from lower to upper.

Kharkhadat and Khesen Formations are composed of dolomite, siliceous rocks, and carbonated and clay schists and make up lower part of the section. Phosphorite here is more thick (up to 100 m) and represents lowest part of the phosphorite-bearing layer. The Khoridol Formation is composed mainly of organic limestone, and gravelstone and sandstone are observed in lower part. Fragments and cements show phosphate component.

The Khoridol Formation represents the upper level of phosphorite-bearing deposit. The Ukhaa Tolgoi and Ujgiin Gol Formations are mostly composed of limestone and siliceous rocks and some sandstone, tuffite, and siltstone. Age of the Khuvsgul Group is determined as Ediacaran-Middle Cambrian by trilobite, archeociate, oncolite, and algae fossils (Byamba 2012). The Khuvsgul Group rocks are cut by Middle-Late Cambrian calc-alkaline and alkaline granitoid of the Telmen Complex.

Coarse-fragmented molasses of Upper Cambrian-Lower Ordovician Khujirtiin Gol Formation and other comparable formations cover the Khuvsgul Group rocks. In some area, fragments are wholly composed of the Telmen Complex granitoid. Devonian and Carboniferous subalkaline and alkaline intrusive massifs represent continental crust formation in the adjacent Zed, Eastern Tuva, and Lake zones. Permian Khanui Group subalkaline and calc-alkaline volcanic rocks are distributed on the southern side, whereas small outcrops of middle Jurassic coal-bearing formations of 250–300 m thickness are observed in the central and south-western part of the basin. Cenozoic deposits are represented by Quaternary and Neogenic coarse-grained reddish sediments and potassic alkaline basalts which fill up neotectonic grabens and basins between mountains.

Ediacaran-Cambrian formations that fill the Khuvsgul basin form synclinal structure along latitude. Dislocations and steep, isoclinal, and inverted folding are very common within the Khuvsgul Group, as well as older Darkhad Group rocks. Main phosphorite deposits are located along with the marginal synclinal structure. The anticlinal structure is developed in the central part of the basin, and anticlinal and synclinal structures are separated by deep faults. The lithology of the Khuvsgul Group is controlled by these faults (Byamba 2012). The marginal syncline structure is separated into several blocks by faults. Some of these blocks are down-dropped, and thickness and composition of Khuvsgul and Darkhad Group rocks within the blocks are changed (Ilyin 1990).

The most important feature in phosphorite distribution consists in the distinct confinement of all phosphorite deposits and occurrences to two submeridional (western and eastern) zones of phosphate facies (Ilyin 1981). The first zone extends along the western slope of the Darkhat Depression, while the second zone extends parallel to the western shore of Lake Khuvsgul (Fig. 12.2). The distance between these zones is approximately 80 km. The synclinal structure of the basin (Zaitsev and Ilyin 1970) promoted the deep subsidence of the basal section of the Khuvsgul Group (Khesen Formation) in the axial part, which is insufficiently studied because of the location in the not easily accessible Khoridulin Sardig Ridge. However, the base and lower part of the Khuvsgul Group are well exposed in some areas along the Arasana River. In all these areas, the Khuvsgul Group is exclusively composed of dolomites. Phosphorites and associated biogenic rocks are absent. This situation is also observed in local outliers among sediments of the Darkhat Depression. Thus, phosphate facies are presumably localized in two zones (Ilyin 2004).

The eastern zone corresponds to the inner shelf, whereas the western (Darkhat) zone corresponds to the deeper water zone (Ilyin 1998a, b).

Phosphate facies in both eastern and western zones are characterized by sharp variations in thickness and number of phosphate beds and phosphate contents therein. Each block of the formerly single basin is marked by the specific character of the pre-Vendian basement and tectonic subsidence (and corresponding, sedimentation) rates in the Vendian-Cambrian. This is reflected in the stratigraphic position of the Khesen Formation relative to the base of the Khuvsgul Group. For instance, the thickness of the subphosphate dolomite unit is not more than 300 m in the Urandush Block, which accommodates the Khuvsgul deposit (Ilyin 1990), and increases to 600 m in the neighboring block (Uliin Davaa deposit). Such variations combined with substantial differences in quantitative parameters of deposits provoked erroneous concepts of twofold (Osokin and Tyzhinov 1998) or even threefold (Zhuravleva 1974) occurrence of phosphate-bearing sequences within the Khuvsgul Group. The negligible probability of such assumptions was proved during the detailed study of these deposits (Zaitsev 1992). Phosphate deposits and occurrences in the Khuvsgul basin are listed in Table 12.1.

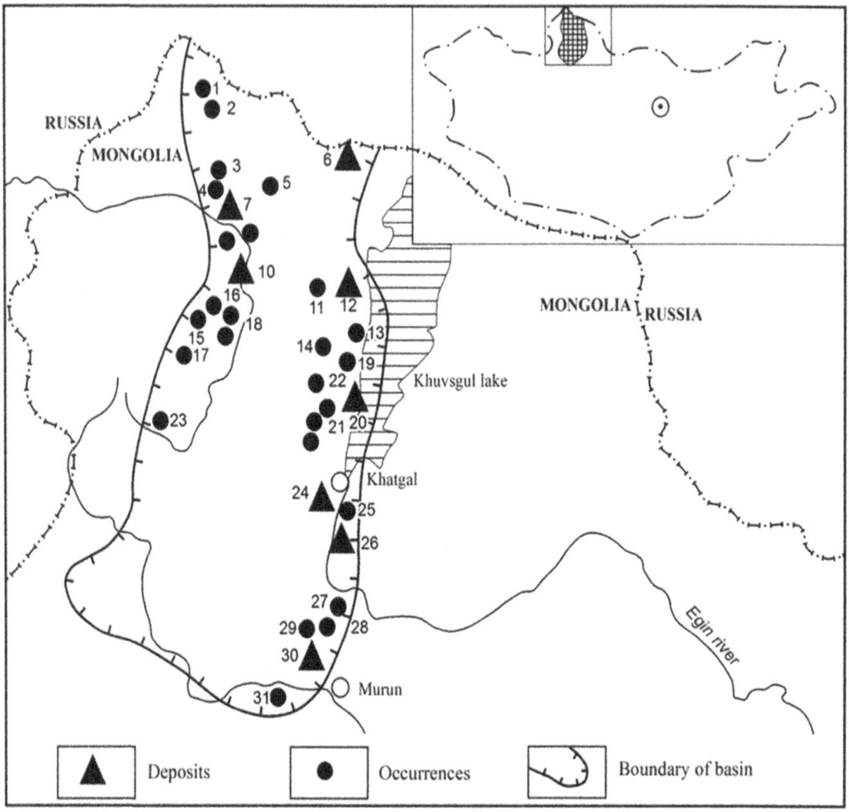

Fig. 12.2 Distribution of phosphate deposits and occurrences in the Khuvsgul phosphorite basin (Byamba 1996). Phosphate deposits and occurrences: 1. Dert; 2. Torgon nuur; 3. Middle Tengesin; 4. Usariin; 5. Khabkhain; 6. Ukhaagol; 7. Khogorgo; 8. East Dood Nuur; 9. Kharmain; 10. Tsagaan Uul (Nuur); 11. Maraat Uul; 12. Uliin Davaa; 13. Jiglig; 14. Kharaa Gol; 15. Khugiin Gol; 16. Bayangol; 17. Temeen Sulit; 18. Darkhad group occurrence; 19. Khar Us Gol; 20. Khuvsgul; 21. Arsain group occurrence; 22. Baga Tsagaan Gol group occurrence; 23. Khunkh; 24. Berkhem Uul; 25. Khuren Nuur; 26. Mankhan Uul; 27. Ikh Tsakhir; 28. Suul Tolgoi; 29. Zam; 30. Burenkhaan; 31. Tsagaan Nuur

12.2.1 Stratigraphy, Lithology of the Khuvsgul Basin

A typical stratigraphic section of the Khesen Formation begins with a lower or main bed of high-grade phosphorites with average thickness of 210 m and is capped by layers of low-grade phosphate sandstones or sandy phosphorites. The main bed is overlain by several (up to four) thinner beds separated by phosphate-bearing (eastern zone of phosphate facies) or phosphate-free (western zone) dolomites. The total thickness is 150 m, approximately 70 m of which is composed phosphate-free dolomites underlying phosphate sandstones (Ilyin 2004).

Table 12.1 Phosphate deposits and major occurrences in the Khuvsgul phosphate basin from north to south (Ilyin 1973; Byamba 1987)

Deposit	Satellite deposit	Occurrence
Ukhaa Gol = Tavkhai Nuur		
Uliin Davaa		
		Maraat Uul
Jiglig		Tashint
		Kharaa Gol
		Khar Us Gol
Khuvsgul (sensulato)		Ikh Tsagaan Gol
	Saikhan Gol	
	Khoit Uran Dush	
	Uran Dush	
		Doloon Tsagaan Khad
	Baga Tsagaan Gol	
	Bayanzurkh	
	Nuurtyn Am	
	Ulaan Khad	
	Ongolog Nuur	
	Khuvsgul	
	Jankhai	
		Asan Group
		Berkhem Ukhaa
		Khuren Nuur
Maikhanuul		
		Ikh Tsakhir
		Suul Tolgoi
		Shirenget
Burenkhaan		
		BagaTsagaanuul
		Dert
		Targaluul
		Tuv Tengis
		Utsegen
		Khabkhain
		Khogorgo
		Dood Nuur
		Kharmain
Tsagaan Nuur		
		Khugen Gol
		Bayan Gol
		Temeen Suult
		Khunkh

The Khesen Formation contains microphytolites (Zhuravleva 1974), catagraphites, acritarchs, and filamentous cyanobacteria (Zhegallo et al. 2000; Mei 2007). Based on this evidence, the formation is attributed to the Tommotian Stage, although its Late Ediacaran age cannot be ruled out (Terleev 1998).

Khuvsgul Group deposited during Ediacaran to Cambrian time in Khuvsgul basin comprises five formations: Khar Khadat, Khesen, Khoridol, Ukhaa Tolgoi, and Ujigiin Gol.

Lower part contains carbonate conglomerate with Ediacaran oncolite, sandstone, dolomite from telloids, five layers of phosphorite, siltstone, and limestone. The upper part contains intercalations of limestone with trilobite, archaeocyatha, sandstone, and siltstone. The phosphorite layer is within a 3000-m-thick dolomite-siltstone (Fig. 12.3).

Petrography: Phosphorite ore contains dark-black-colored monophosphate, layered phosphate, and fine-grained breccia and oolite. Chemical composition is defined by silica and carbonate phosphate mineral ratio. Phosphate mineral in ore consists of fluorapatite and hydroxyapatite, and ore grade fluctuates from 10% to 98%. Moreover, F, Pb, Mo, V, Ni, Sn, Co, Cu, Sr, Zn, U, TR, and Ba are detected in the ore. Highest ore grade is hosted by layered monophosphate and breccia phosphorite and granular phosphorite.

The lower horizon (lower phosphate unit, Khesen Formation, Fig. 12.4) within the lower part of Early Cambrian Khuvsgul Group contains five phosphate horizons. The lower horizon shows thickness of 25–90 m which is variable and contains separate phosphate horizons up to 15 m thick. The horizons composed of layered carbonates (15–20% dolomite, 1–30% calcite), chalcedony and quartz, chemically precipitated phosphorites, and other minerals such as limonite, sericite, various heavy minerals, hydromicas, and organic matter (0.5%, rarely up to 5%). The total rare earth element content is <300 ppm (Ilyin et al. 1986). The P_2O_5 content reaches up to 29% in the lower phosphate horizons. There are two types of phosphorites:

- Isotropic fluorapatite with finely disseminated carbonate and quartz and some organic substances. The fluorine content is about 1–2%.
- Fine crystalline low-thermal apatite.

The upper horizon represents upper phosphate unit which belongs to Erkhel Formation which is also in the lower portion of the Khuvsgul Group. This horizon is much thinner (10–15 m) and composed of siliceous and carbonaceous clastic fragments and clastic phosphorites overlying a thick dolomite bed (main dolomite) (Fig. 12.3). The average P_2O_5 content in this horizon ranges from 13.5% to 21.5%.

The chemically precipitated (primary) phosphorite of the lower horizon formed in an offshore upwelling area south of the Siberian craton in Early Cambrian time. The upper horizon was most probably formed by reworking the phosphorite from the lower horizon (Harald et al. 2011).

All phosphate deposits and occurrences at the southern margin of the basin belong to the upper horizon, whereas the phosphate deposits in the north end of the basin can be correlated with the lower phosphate horizon.

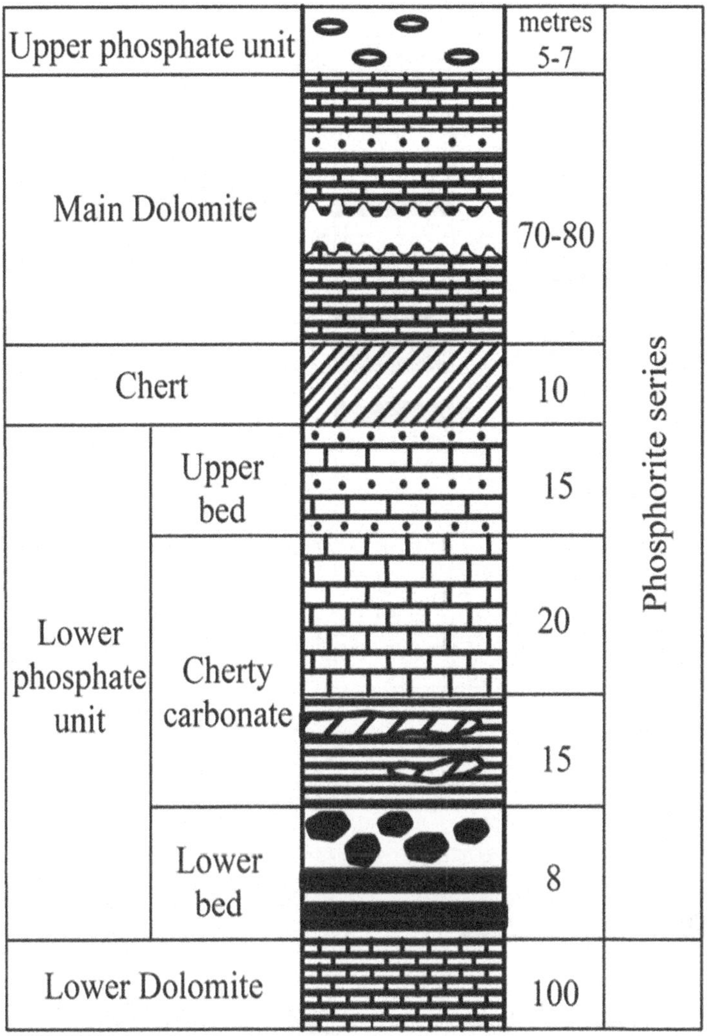

Fig. 12.3 Stratigraphy of the phosphate series of the Khuvsgul Group (Ilyin et al. 1986)

The whole Khuvsgul Group rocks were strongly folded, faulted, and fractured during the Caledonian orogeny which are now in a big synclinorium and dissected by several faults into blocks with displacement. The Khuvsgul basin was intruded by various intrusions.

The two phosphate horizons are bound to the periphery of the Khuvsgul phosphate basin, i.e., to two N-S-trending folded belts.

The eastern phosphate belt, the Khuvsgul zone, is 250 km long and 10–15 km wide which contains all of the larger phosphate deposits and occurrences. For example, the most famous within this eastern belt are Burenkhaan (in the south)

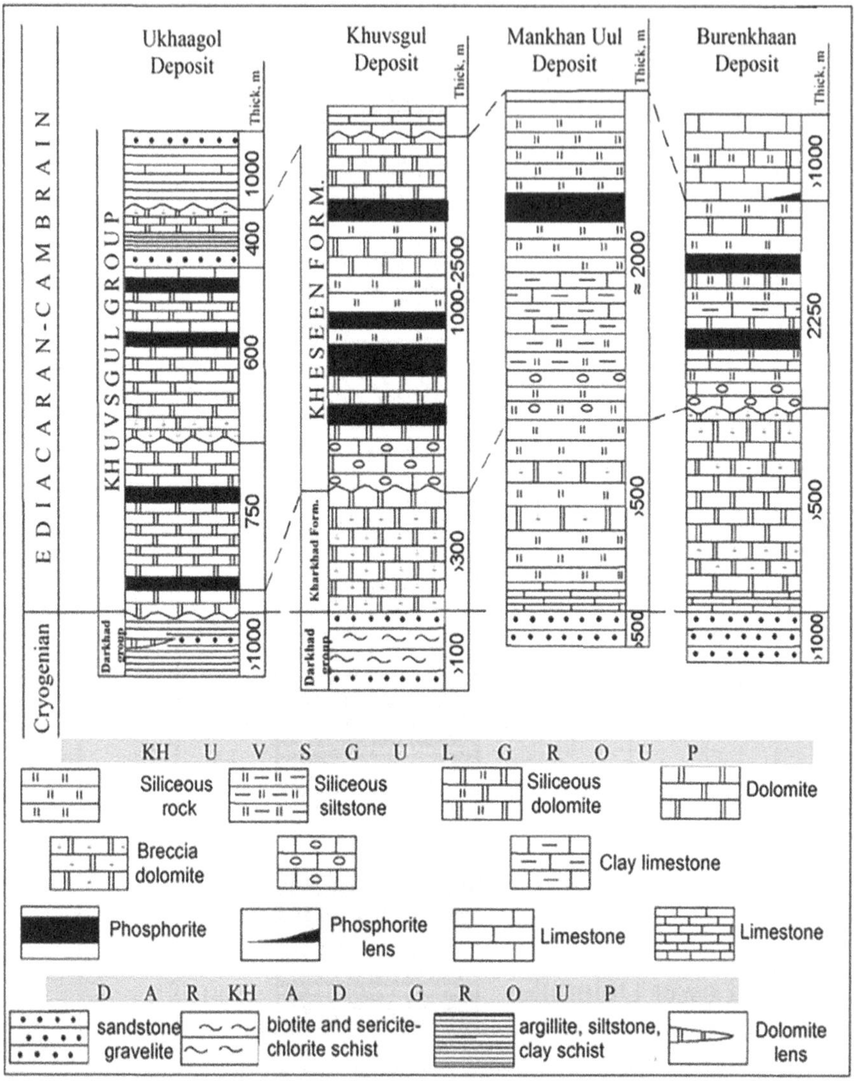

Fig. 12.4 Stratigraphic sections and lithology of the Khesen Formation, phosphorite deposits in eastern belt of the basin (Byamba and Iliyn 2012). 1–11- Khuvsgul Group: 1- siliceous rock, 2- siliceous siltstone, 3- siliceous dolomite, 4- dolomite, 5- dolomite breccias, 6- tillite and tilloid, 7- clayish limestone, 8- phosphorite horizon, 9- phosphorite lens, 10- coarse-grained limestone, 11- thinner limestone; 12–15- Darkhad Group: 12- sandstone, gravel stone, 13- biotite and sericite-chlorite schist, 14- argillite, siltstone, shale, 15- dolomite lens

and Khuvsgul (in the north), especially the latter one with several satellite deposits and occurrences in tectonically fragmented blocks. The western phosphate belt (Darkhat zone) is 200 km long and 10 km wide (Harald et al. 2011).

12.2.1.1 Composition of Phosphorite in the Khuvsgul Basin

Phosphorite in the Khuvsgul basin looks black, similar to siliceous and bituminous limestone by the naked eye. The composition of phosphorite in the basin is well studied (Ilyin and Ratnikova 1971; Ilyin 1973; Muzalevskii 1970). Researchers classify aphanitic (mono-phosphate), layered, microgranular, pelletal, brecciated, and fragmented types in terms of petrography. These types belong to lower formation level (Byamba 2012). These various types are differently observed in all deposits.

Mono-phosphate type includes massive and structureless phosphorite with a small amount of insoluble residue (3–4%) (Byamba 2012). Layered type is abundant and characterized by the intercalation of mono-phosphate and dolomite. The lower boundary of mono-phosphate layer is sharp, whereas the upper boundary is blurry and gradually changed to dolomite. The thickness of mono-phosphates ranges from a few millimeters to several centimeters. Granular phosphates comprise pelletal and oolitic type, and grain size is 0.05–0.2 mm. They are laminated, oval, and structureless. Some phosphorite varieties, e.g., characteristic oolitic-granular phosphorites with numerous phosphate films, occur in all of the basins (Ilyin 2004).

Structureless and granular phosphorites are generally similar in the composition of main chemical components, except for the following distinctions. Relative to the granular phosphorite, the structureless variety is slightly enriched in P_2O_5 and depleted in SiO_2. The Corg content decreases from 1.0 to 1.1% in the granular variety to 0.n% in the structureless variety (Ilyin 2004). Larger differences are recorded for trace elements, for instance, the granular varieties usually show notably higher concentrations of trace elements (particularly REE and Y) and depleted in Cd and V (Ilyin 2004).

The phosphorite ore reserve in the Khuvsgul basin is 3000 Mt, and the proven reserve is 2264.87 Mt (Table 12.2).

12.2.2 Descriptions of Selected Phosphorite Deposits in the Khuvsgul Basin

12.2.2.1 Phosphorite Deposits in Eastern Flank of the Basin

Phosphorites in the eastern zone, which comprise the large and well-prospected Khuvsgul, Ukhaagol, Uliin Davaa, and Berkhem Uul deposits, and Kharaa Gol, Jiglig, Khar Us Gol, Arsain, Baga Tsagaan Gol and Khuren Nuur occurrences, are substantially better studied than phosphorites in the western zone.

Table 12.2 Ore reserves in the Khuvsgul basin (Byamba 2012)

No	Phosphorite deposits, occurrences	Reserve category, average content			
		C_{1+2} (P), Mt		$P_1 + P_2 + P_3$ (Rs), Mt	
		Ore	P_2O_5 (%)	Ore	P_2O_5 (%)
East and south of the Khuvsgul basin					
1	Khuvsgul	250.0	19.6	903.0	18.7
2	Uliin Davaa			150.0	22.0
3	Ukhaa Gol			317.0	16.0
4	Berkhem Uul			17.77	12.6
5	Mankhan Uul	28.0	17.0	56.0	17.1
6	Burenkhaan	205.57	23.1	113.2	20.5
7	Erkhel Nuur			371.0	20.3
8	Suul Tolgoi			50.0	15–18.0
9	Dalan and other occurrences			19.4	21.7–27.3
West zone of the Khuvsgul basin					
1	Tsagaan Nuur			216.0	25.0
2	Khogorgo			1.5	12–21.0
	All	483.57		2264.87	

12.2.2.2 Khuvsgul Deposit

The Khuvsgul deposit is the largest among the studied deposits in the basin. It is located on western bank of the Khuvsgul Lake from Jankhai pass to Uran Dush Mountain, covering the east part of the Khoridol Saridag mountain range which continues 35 km at elevations of 2400–2600 m (asl). Structurally, it is located along Khesen syncline fold limbs (Fig. 12.5).

There are five phosphate horizons in the lower phosphate unit of the Khuvsgul deposit. Here, I phosphorite-bearing layer is the most important. The deposit can be measured for 1.2–3.5 km along strike and has a thickness of 1.0–11.7 m (average 6.2 m). The deposit is strongly folded and fractured. The facies types within the phosphorite-bearing layers are highly variable, which is also reflected in the chemical composition. The P_2O_5 content ranges from 16.84 to 37.74% (22.56% on average) in layer I. The ore contains 4.56% MgO, 2.44% R_2O_3, and 9.82% insoluble on average as an impurity. Taking into account minimum grades of 10% P_2O_5 in general and 16% P_2O_5 in ore zones of interest, the total resources for all five phosphate horizons amount to 9.9 million tonnes (Mt) of ore, grading >16% P_2O_5, i.e., 1.4 Mt of ore in categories C_1 + P (definition of categories is in Appendixes). Ilyin (1973) predicted the (probable) resources of the whole Khuvsgul phosphate deposit down to 200 m from the surface depth as 432 Mt with an average 20.7% P_2O_5 (i.e., 89 Mt of net P_2O_5). He also estimated the resources down to the Khuvsgul lake level as 517.9 Mt of ore with geological resources more than 1 billion tonne of ore. The probable resources about 900 Mt of ore grading 14.8–29.3% P_2O_5 seem to be realistic. Thus, the total P_2O_5 content is estimated around 175 Mt (Harald et al. 2011).

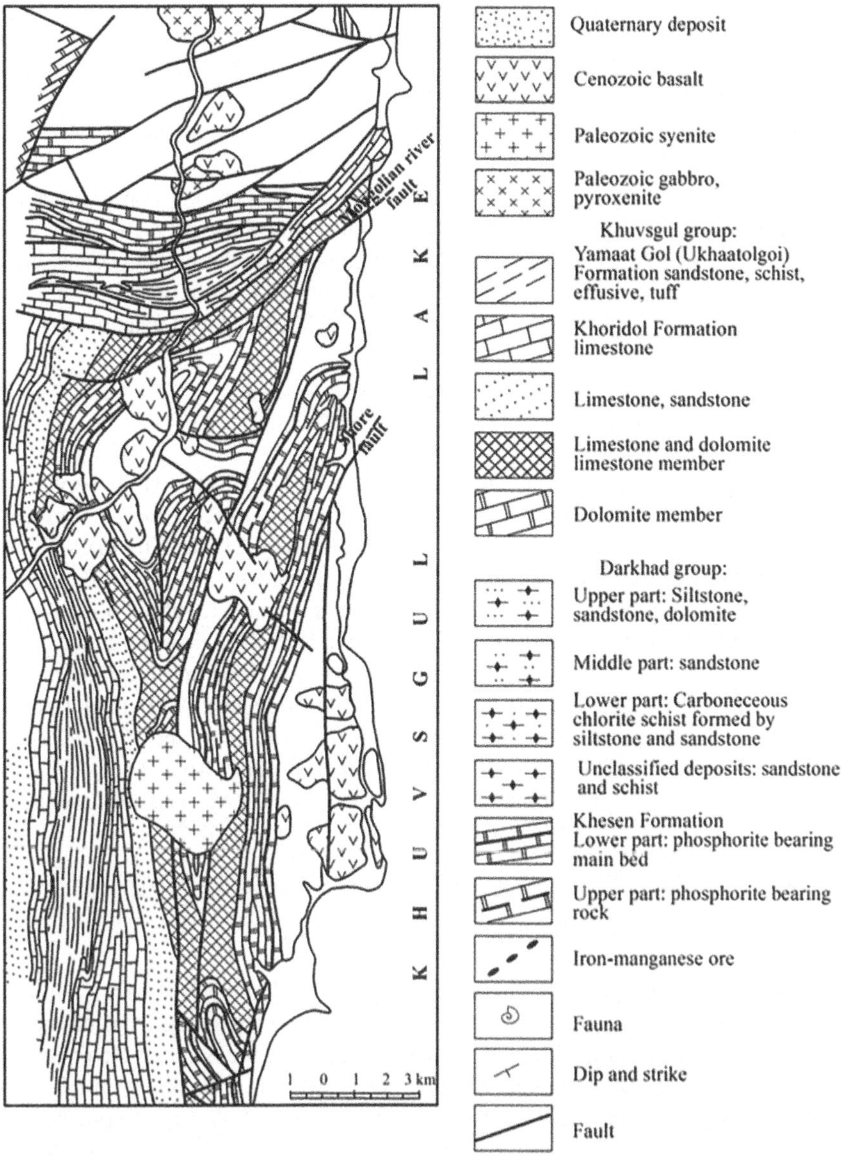

Fig. 12.5 Geological map of the Khuvsgul deposit (Ilyin 1973)

The Khuvsgul deposit consists of three parts: Ongolog, Jankhai, and Urandush. The northern end of the Ongolog deposit is about 10 km southeast from Uran Dush Mountain and extending southward for about 20 km. It comprises black phosphorite nodules of oolitic, pelitic, and breccias, found mostly in siliceous dolomite.

There are five phosphate horizons stretching 500–19,000 m (average 5000–6000 m) with thickness of 1.5–50 m (average 10–12 m) in the southern Khuvsgul area. Unwanted impurities are 0.16% Na_2O, 0.26% K_2O, and 7% MgO in ore. The total resources amount to 571.652 Mt of ore grading 18.7% P_2O_5 on average, i.e., 106.771 Mt of P_2O_5 in categories B + C_1 + C_2, and 38.108 Mt of ore grading 14.8% P_2O_5, i.e., 5.629 Mt of P_2O_5 in category P_1.

Jankhai deposit consists of five phosphate horizons where horizon I shows the biggest economic interest. The deposit is separated into three parts by faults. The average P_2O_5 grade is 27.82%, which is significantly high. Component of insoluble residue does not exceed than 20.1%. The resources are 15.8 Mt of ore with 4.4 Mt of P_2O_5 in category C_2 and 42.1 Mt of ore with 11.71 Mt of P_2O_5 in category P_2.

12.2.2.3 Urandush

The western part of the Khesen syncline connects to Urandush. One phosphorite-bearing horizon continues from the Ikh Uul granite contact zone to northeast direction to Urandush Mountain and turns to the Khuvsgul Lake. The thickness of Urandush phosphorite horizon is 15–60 m; P_2O_5 content is about 21.4% with an insoluble residue of 14.9%. The North Urandush deposit situates about 60 km north of Khatgal village, in the north of Urandush Mountain. Within dolomitic limestones there are two phosphate-rich horizons, both slightly silicified and reaching a total thickness of 5–24 m. One horizon is 2 km long, 20 m wide, and 3–8 m thick (locally up to 10 m). Another horizon shows 5 km length, 80 m wideness, and 15 m thickness (locally up to 87 m). The grade of P_2O_5 is highly variable from 2.8% to 23.5% (average of 18.05%). The estimated total resource is 99.9 Mt, and ore grading is 17.88% P_2O_5, i.e., 17.87 Mt of P_2O_5 (category P) (Harald et al. 2011).

12.2.2.4 Saikhan Gol Deposit

The Saikhan Gol phosphate deposit is located about 60 km north of Khatgal village and 10 km north of the Nuurtyn Am phosphate deposit stretching from the north-western side of Urandush Mountain to the lower part of the Saikhan River. About 6.5 km long, 55° dipping phosphorite-rich horizon of 4.6 m thickness in siltstone, dolomite, and dolomitic conglomerate is studied by trench during the prospecting work in the late 1980s. The average P_2O_5 grade in calcareous siliceous ore is 19.44% averaging 29.26%. Based on this average grade, estimated resources are 10.5 Mt of ore with 3 Mt of net P_2O_5 (by category P) (Harald et al. 2011).

12.2.2.5 Baga Tsagaan Gol Deposit

The Baga Tsagaan Gol phosphate deposit is situated in the middle part of the Baga Tsagaan River area. In the vicinity, limestone is present in 15.5 × 2.0 km area where

5.7-km-long and 6.0-m-thick one phosphate horizon was found. P_2O_5 grades of this horizon range from 11.75 to 26.89% with an average of 21.47%. The estimated resources are 20.8 Mt of ore with 4.4 Mt of net P_2O_5 in category P_2 (Harald et al. 2011).

12.2.2.6 Bayanzurkh Deposit

The Bayanzurkh deposit is located 5 km southwest of Urandush Mountain and consists of North and South ore bodies which are both highly fractured. The phosphate-rich horizon in the North ore body (average 19.78% P_2O_5) stretches for 2.7 km, with 3 m thickness. The South ore body (Umnud Bayanzurkh) (on average 17.18% P_2O_5) consists of two horizons in which the lower one stretches for 1.2 km with an average thickness of 6.3 m and the upper one stretches for 1.5 km and has a thickness of 3–5 m. Taking into account a mineable depth of 150 m, the North ore body might contain 6.4 Mt of ore or 1.2 Mt of P_2O_5, whereas the South ore body contains only 3.2 Mt of ore with 0.6 Mt of P_2O_5 (category P_1) (Harald et al. 2011).

12.2.2.7 Nuurtyn am Deposit

The Nuurtyn Am phosphate deposit is in 50 km north of Khatgal village and 8 km southwest of Urandush Mountain and is covering 72 km^2 area. There are three phosphate horizons attaining 1.2 km average length, 50 m average width, and from 13.5 to 19.4 m thickness. The average P_2O_5 grade is 23.84%. Unwanted impurities are 17.44% insoluble on average, 1.84% Fe_2O_3, and 4.43% MgO. The estimated resources by P category are 70.1 Mt of ore with 14.8 Mt of net P_2O_5.

12.2.2.8 Ulaan Khad Deposit

The Ulaan Khad phosphate deposit is located 23 km north of Khatgal village and 8 km south of Ikh Uul Mountain. Phosphorite occurs in phosphorite-rich siltstone (lower), limestone with silicified phosphorite (middle), and highly silicified limestone (top). There are two to three phosphate horizons reaching 300–3000 m length and 42–45.5 m overall thickness in the western limb and one to two phosphate horizons with 47.2 m total thickness in the eastern limb of the deposit. Phosphorite revealed down to 200 m depth by drilling. The average P_2O_5 grade in the lower, middle, and top horizons is 26.32%, 23.74%, and 21.37%, respectively. The average grade of insoluble is 18.1%. The estimated resources in category P1 are 7.56 Mt of ore grading 23.8% P_2O_5, i.e., 1.5 Mt of P_2 O_5 (Harald et al. 2011).

12.2.2.9 Ikh Tsagaan Gol Deposit

IkhTsagaan Gol phosphate occurrence is located 7 km north of the Saikhan Gol deposit and covering 18 × 4 km area. There are carbonates, quartzites, and dolomitic schists, partly enriched in phosphate (up to 30.7% P_2O_5) and highly fractured. Phosphorite is mostly ordained in 18-km-long, 4-km-wide, and 20–40-m-thick middle layer of dark limestone and dolostone dipping at 60–70°. There are three phosphorite horizons attaining 1.0–6.5 m (average 2 m) thickness. The P_2O_5 content is averaging 19.44% with grades varying from 13.37% to 21.8%. Vivianite occurs as a common secondary mineral. The estimated resources of phosphorite ore are 1.5 Mt.

12.2.2.10 Doloon Tsagaan Khad

The Doloon Tsagaan Khad phosphate occurrence is at Avgar Tsagaan Khad mountain, which can be reached via Davaany Gol Valley. Dolomitic silicified phosphorite horizon about 4 km long and 1.7 m thick was identified in dolomitic limestone, dolomite, sandstone, and siltstone. The estimated resources are 1.8 Mt of ore grading 20.69% P_2O_5 on average, i.e., containing 0.4 Mt of P_2O_5 (Harald et al. 2011).

12.2.2.11 Ukhaa Gol Deposit

The Ukhaa Gol (Tavkhai Nuur) phosphate deposit is situated in southeastern part of Sayan Mountain Range (Fig. 12.6) crossing the border of Mongolia and Russia, with the most part in the Mongolian territory. V.A. Ilyin discovered the deposit during the thematic survey in 1970 (Ilyin 1973).

The deposit is mainly composed of argillaceous micro-phosphorites comprising fine-grained quartz detritus and granular or pelletal phosphorites. Phosphorus-bearing member consists of dolomite-limestone layer with siliceous shale which can correspond with Khesen Formation (Fig. 12.6.). Total thickness is 2650 m. The main difference of Ukhaa Gol phosphorite-bearing carbonate-terrigeous section than other sections in the Khuvsgul phosphorite basin is the absence of Cambrian phosphorite. Phosphorite-bearing member is 1750 m, and totally four phosphorite layers of 4–5 m have been found. P_2O_5 content is 12.8–25%, which is significantly high. The content of insoluble residue is 23.9%. Deposit resource is 150 Mt with 22% of phosphorite.

The black phosphorites are found both in light colored carbonaceous rocks (2.5–10.0 m thick) and within an 80–100 m higher, upper layer of quartzites or cherts (3–5 m thick). The phosphate-rich (13.9–27.71% P_2O_5) horizons were found to continue far to the north and northeast, with phosphorite changing to sandy phosphorites and phosphatic carbonaceous sandstone and breccia. The resources

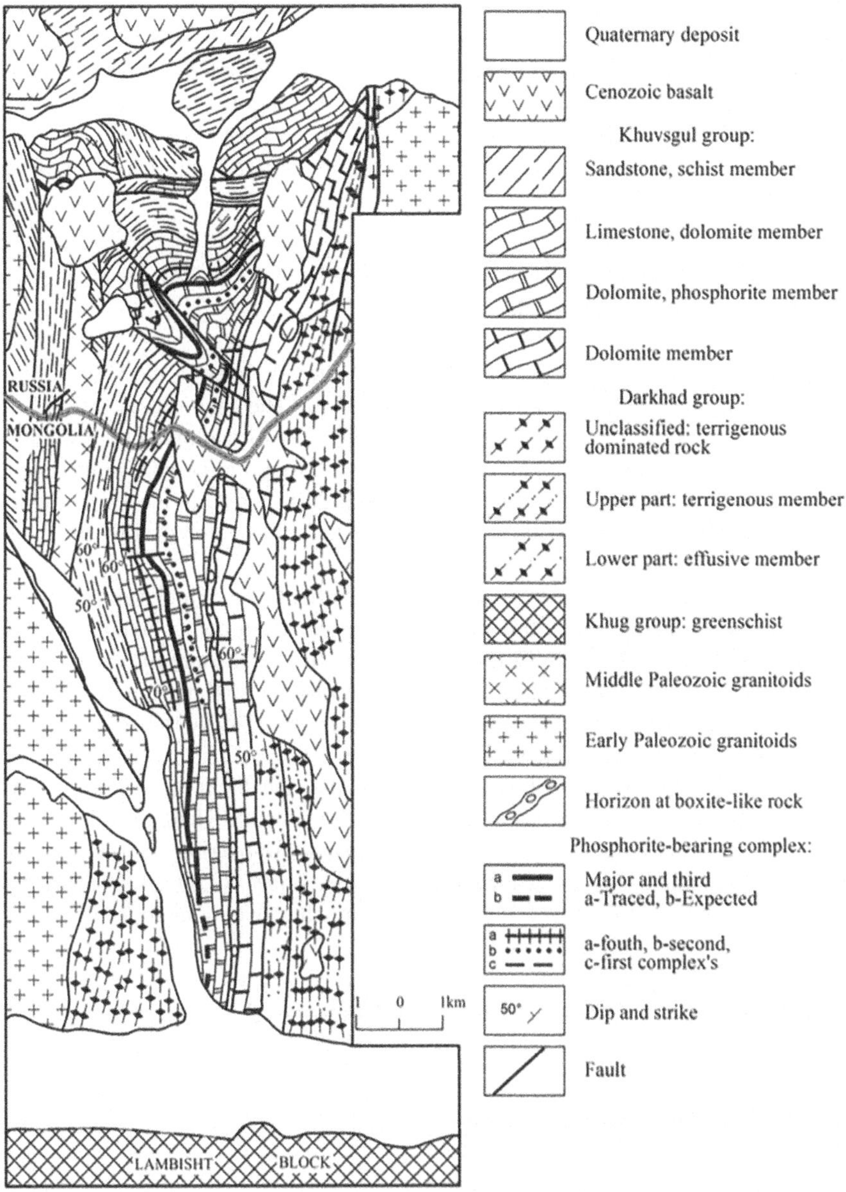

Fig. 12.6 Geological sketch map and cross section of the Ukhaa Gol phosphorite deposit (Ilyin 1973)

of these low-grade (7–9% P_2O_5) rocks are thought to be as large as 2–3 billion tonnes.

A more recent calculation of the resources in the Ukhaa Gol deposit is 10.245 Mt of ore grading 23.24% P_2O_5, i.e., 2.38 Mt of P_2O_5 in category C_2, and 20.742 Mt of ore grading 23.23% P_2O_5, i.e., 5.064 Mt of P_2O_5 in category P_1. The thickness of permafrost in the Ukhaa Gol region is 0.2–0.3 m in the north and 0.5–0.7 m in the south (Harald et al. 2011).

12.2.2.12 Uliin Davaa Deposit

Some 50 km south of the Ukhaa Gol deposit, on the east bank of the Usan Khujirt River, lies the Uliin Davaa phosphate deposit. The deposit is related to Neoproterozoic-Lower Paleozoic carbonate-terrigenous formations, in the south of the Desht uplift which is composed of the Khukh Group biotite and biotite-garnet gneiss. Below the carbonate-terrigenous formation, it is hosted in the 1000-m-thick Darkhad Group sandstone-shale member, whereas multi-colored siltstone, mudstone, and dolomite lenses are found on the top.

The lower part consists of three phosphorite-bearing layers. Thickness of this part is 2–7 m. Ore contains mainly layered or breccia type, P_2O_5 component is 24–25%, and insoluble residue is 4.3%. Cracks and fractures in the phosphorite ore are filled by malachite, azurite, hematite, and calcite. The thickness of phosphorite layer is variable and separated by dolomite and silicate layers. Thickness of second and lower parts does not exceed than 2 m.

Based on a total length of deposit as 10 km and average thickness as 11 m, 150 Mt of 22% of phosphorus anhydrite is calculated at 200 m depth. There the thickness of permafrost is also only 0.4–1.0 m. The main phosphorite-bearing layer is bound to the upper, silicified, and brecciated part of a limestone (70.09–74.08% $CaCO_3$, 3.4–5.3% $MgCO_3$, 15.9–16.9% in sol.) which stretches over a length of 2500 m, a width of 16 m, and a thickness of 5–12 m (Fig. 12.7). Estimated resources are 15.3 Mt of ore grading 20.2% P_2O_5, i.e., 3.0 Mt of P_2O_5 in category C, and 149 Mt of ore grading 3.09–16.1%, on average 15.55% P_2O_5 in category P.

12.2.2.13 Chiglel Phosphate Deposit

The Chiglel phosphate deposit is situated 15 km north of the Khuvsgul phosphate deposit on the west side of the Chiglel River. It is built up from thin layers of phosphorites 2–3 mm in diameter and phosphorite-rich limestone and sandstone. There are four phosphate horizons stretching for 2 km of 27 m attained thickness. The grade of P_2O_5 varies from 15.64% to 19.88% with estimated resources of 17.5 Mt of ore in category P_2 (Harald et al. 2011).

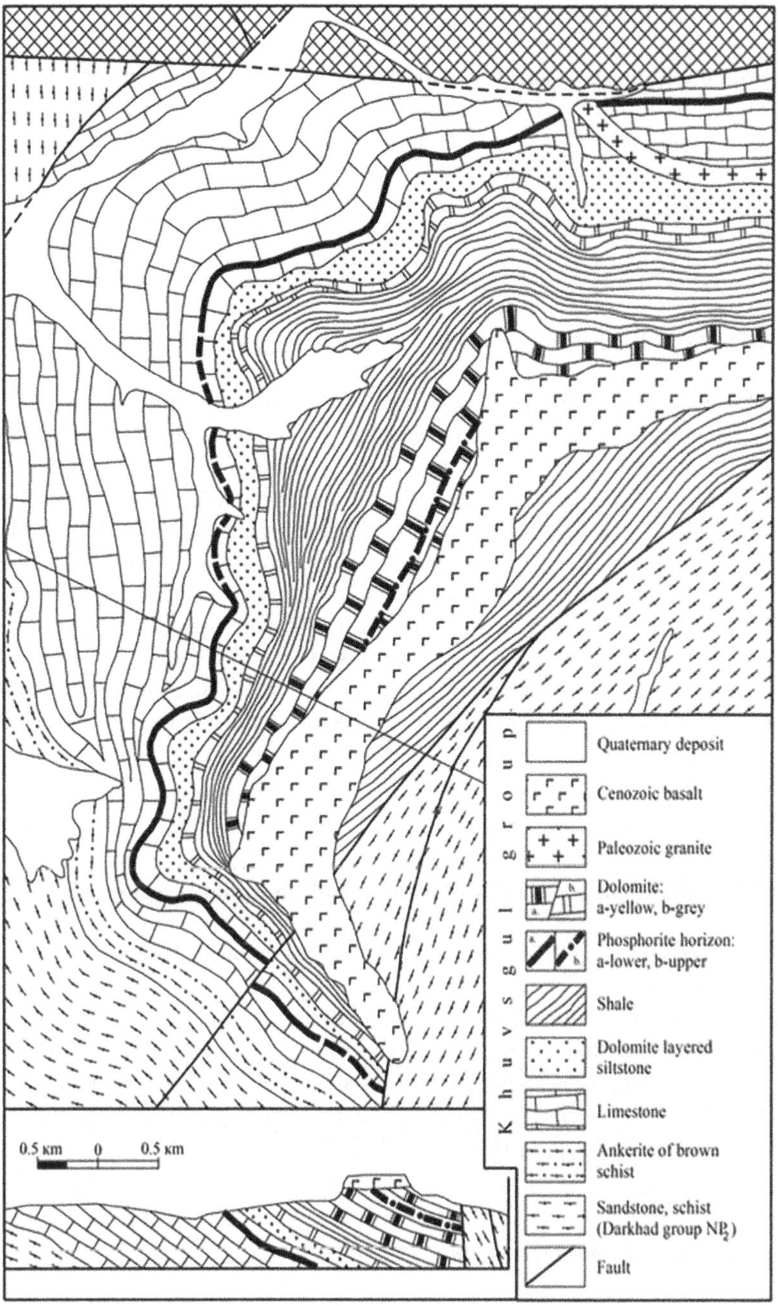

Fig. 12.7 Geological map and cross section of the Uliin Davaa phosphorite deposit (Ilyin 1973)

12.2.2.14 Tashint Phosphate Occurrence

The Tashint phosphate occurrence is located in Tashintyn Mountain surrounding its peak and covers 16 km^2. There are limestone, dolomitic limestone, and phosphate-rich dolomites in the occurrence area. One phosphate horizon of 7.5 km length and only 1.5 m thickness is found. The average grade of P_2O_5 in the occurrence is 16.59%, resulting in estimated resources of 3 Mt of ore with about 0.5 Mt of net P_2O_5.

12.2.2.15 Mankhan Uuls Deposit

The Mankhan Uuls deposit was discovered by P.V. Osokin during South Khuvsgul geological mapping and prospecting work in 1965–1966. The deposit is located about 50 km south of the Khuvsgul deposit and 20 km south of Khuvsgul Lake. The deposit is just outside of the Khuvsgul National Park. Main phosphorite-bearing rocks are silicified schists and quartzites. These rocks occur in the two limbs reaching lengths of 500–6500 m (averaging 5000 m) of an N-S-trending anticline (Ilyin 2004). The thicknesses of phosphorite-bearing layers in the eastern and western limbs are 87 m and 31 m, respectively. The ore grades are comparatively low: 15.67% (11.37–20.8%) P_2O_5 and 0.88% MgO, 2.22% Fe_2O_3, and 2.00% Al_2O_3 on average. Phosphate concentrates grading 30.61% P_2O_5 are estimated by flotation experiment. Applying a cut-off grade of 10% P_2O_5, resources amount to 9.69 Mt of ore grading 15.67% P_2O_5, i.e., 1.52 Mt of P_2O_5 in category C_1, and 41.4 Mt of ore grading 14.77% P_2O_5, i.e., 6.11 Mt of P_2O_5 in category C_2.

12.2.2.16 Berkhem Ukhaa and Berkhem Uul Phosphate Occurrences

The Berkhem Ukhaa and Berkhem Uul occurrence is located 30 km south of the Khuvsgul deposit on the border of the Khuvsgul National Park, along the western bank of the Egiin Gol River. The main feature of the occurrence is that it is hosted in a 1000-m-thick limestone member which resembles to the Khordil Saridag Formation of Khuvsgul Group. There are three low-grade (7.1–12.5% P_2O_5) phosphate horizons of 1.6–3 m thickness each. The total resources are estimated as 16.2 Mt of low-grade ore by P_1 category (Harald et al. 2011).

12.2.2.17 Suul Tolgoi Occurrence

The Suul Tolgoi phosphate occurrence lies 15 km north of the Burenkhaan deposit, at the eastern shore of Erkhel Nuur. The occurrence consists of three highly tectonically fragmented ore bodies stretching 500–3500 m (average 2000 m). Ore contains 8.0–29.8% (average 13.8%) P_2O_5 and 7.8% MgO and 6.21% Fe_2O_3 on average. Based on cut-off grade of 10% P_2O_5, total resources were estimated as 7 Mt of ore in category C_2 (Harald et al. 2011).

12.2.2.18 Shirenget Occurrence

Another rather small occurrence is Shirenget at the eastern side of the Egiin Gol valley. In this area, one 3.8-km-long horizon of highly silicified phosphorite-bearing limestone was found. The estimated resources are 2.2 Mt of ore of unknown grade (P_1).

12.2.2.19 Burenkhaan Deposit

J. Byamba studied in detail (1969–1970) structural facies aspects of phosphorite-bearing formations of Neoproterozoic-Cambrian age and phosphorite occurrences in western and eastern parts of Khusvgul basin. In this study he suggested the Murun city area is the most prospective area of phosphorite mineralization in Northern Mongolia. The Burenkhaan deposit was discovered during the prospecting work by Russian geologists, followed by more detailed exploration.

Neoproterozoic-Cambrian terrigenous-carbonate sedimentary rocks are abundant in the ore district and classified as Darkhad and Khuvsgul Groups. Upper Cambrian-Lower Ordovician molasses and Neogenic basalts are present in northern part of Murun depression. The upper part of the Darkhad Group section belongs to the Arsai Formation which is composed of greenish gray sandstone, conglomerate, and felsic and intermediate volcanic rocks and dominant terrigeneous sediments. The Arsai Formation deposit occupies the central part of the anticline structure in the Burenkhaan deposit area, and its thickness is 200 m.

Carbonate-dominated terrigenous-carbonate formation of the Khuvsgul Group covers the Darkhad Group structurally conformable but stratigraphically unconformable. The Khuvsgul Group is subdivided into Khar Khadat, Khesen, Erkhen Nuur (Khoridol), Ukhaa Tolgoi, and Ujgiin Gol Formations. The Kharkhadat Formation is the base of phosphorite-bearing Khesen Formation. Sporadic layers of carbonate conglomerate and tillitoids can be present, covered by dark gray dolomite. The Khesen Formation conformably overlies the Khar Khadat Formation. The formation is composed of upper and lower members, and the lower member (120–900 m) contains phosphorite and is divided into three beds.

The lower bed is composed of gray and dark gray limestone up to 300 m in thickness. Siliceous lenses are observed in upper part of the bed. Middle bed contains phosphorite and shows various compositions and thickness (30–260 m). There are two to three phosphorite horizons in the bed. Horizons are separated by dark colored siliceous rock, siliceous-clayish-carbonaceous schist dolomite, and limestone. The phosphorite horizon thickness varies from 5 to 100 m, and length continues 1–3 km along the strike. The upper bed which is mainly composed of dark-colored limestone overlies the phosphorite-bearing bed, and the thickness ranges from 20 to 400 m. The upper member (1400 m) is widely distributed in the deposit area and mainly composed of siliceous limestone and dolomite. The Khesen Formation thickness ranges from 800 m to 2300 m in the Burenkhaan deposit area, and the formation age is in the lower part of Cambrian Fortunian stage (Fig. 12.4).

In the Burenkhaan deposit area (Figs. 12.8 and 12.9), the Khesen Formation is conformably overlain by the Erkhel Nuur (Khoridil) Formation (1600–2400 m in thickness) of dark gray chemogenic and organic limestone dominated. The Erkhel Nuur Formation has two members. The lower member (10–40 m) contains phosphorite fragment-bearing quartz sandstone and conglomerate at the base. Phosphorite fragments make up to 30% of the sandstone. The upper member is mainly composed of black and dark-colored spotty limestone, which differs from limestone of Khuvsgul Group. Trilobites, hyolitha, and archeociates are common in the Erkhel Nuur Formation. The age of the Erkhel Nuur Formation is as the lower part of Cambrian Terreneuvian series. Southern part of the Burenkhaan deposit area belongs to Ukhaa Tolgoi Formation (1500 m in thickness) which is composed of dark gray sandstone, siltstone, and conglomerate, as well as mafic and felsic volcanic rocks and related tuffs with limestone beds. The age of the formation belongs to Stage 2 of Lower Cambrian.

The Burenkhaan phosphorite deposit is one of the 15 strategic deposits of Mongolia. Although it was well explored in 1980, the general area where the deposits occur is huge in size (140 km^2) that it in fact consists of several interesting areas with high mineralization with even larger barren areas in between. The Burenkhaan deposit consists of 25 ore bodies with lengths of 500–3200 m, widths of 20–100 m, and thicknesses of 2–60 m. Maximum thicknesses reflect strong folding and faulting including the stacking of phosphorite layers. Phosphorite-bearing layers are made up of variable amounts of coarse (1–2 mm), rounded to angular phosphate grains, apatite, dolomite, calcite, silica, and minor amounts of pyrite, limonite, sericite, other mica, clay minerals, organic matter, rutile, and garnet.

While all of them were assumed to represent typical phosphate ore, a final analysis revealed that only three are really phosphate ore (P_2O_5 = 15.5–24.26%), while the other four are more or less free of P_2O_5 (0.02–2.87%) (Harald et al. 2011). This finding corresponds very well with the reports of Russian geologists, who, after working on the phosphate ore in the Khuvsgul basin for several years, claimed that the phosphorite is extremely difficult to identify in the field. Ilyin et al. (1986) mentioned that the hand specimens of the Burenkhaan phosphorites are massive black rocks, indistinguishable from cherts. The chemical analyses prove the good quality of the Burenkhaan phosphate ore in general (Table 12.3). The reserves of the Burenkhaan deposit in total (upper horizon only) were calculated as follows:

- 33.09 Mt of ore grading 23.81% P_2O_5 on average (7.88 Mt of net P_2O_5) in category B
- 129.47 Mt of ore grading 20.36% P_2O_5 on average (26.36 Mt of net P_2O_5 in category C_1
- 12.95 Mt of ore grading 20.00% P_2O_5 on average (2.59 Mt of net P_2O_5) in category C_2, i.e., total 175.51 Mt of ore (36.83 Mt of net P_2O_5 in categories B + C.

Fig. 12.8 Simplified geological map of the Burenkhaan deposit (Byamba 2012)

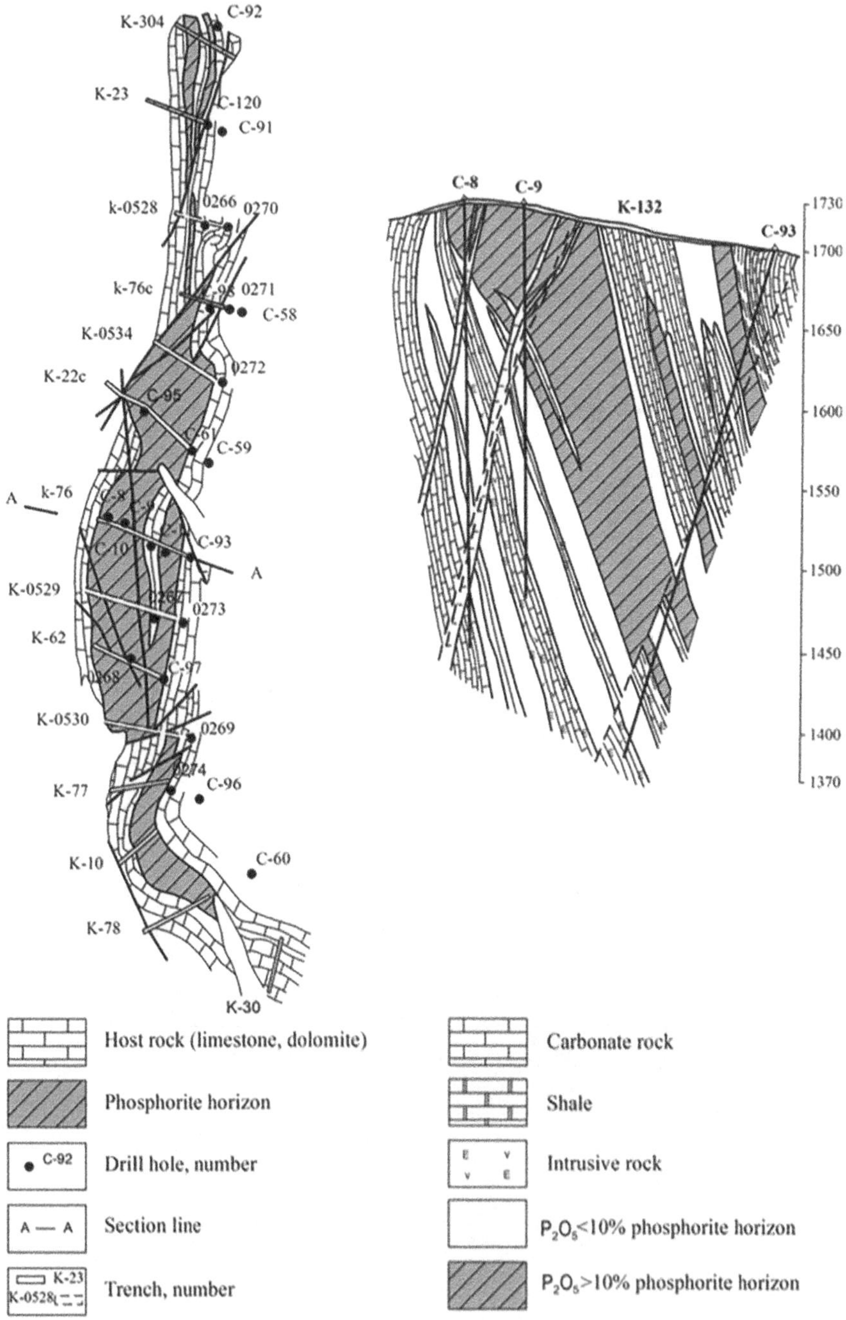

Fig. 12.9 Geological map and section (along drill core 8, 9, and 23; trench 132) of I part of the Burenkhaan deposit (Byamba 2012)

Table 12.3 Chemical composition (in wt.%) of various types of phosphate ore from the Khuvsgul and Burenkhaan phosphate deposits, after Byamba (1996)

	Burenkhaan			Khuvsgul	
Chemistry	Siliceous breccious ore	Medium-grained ore	Extremely fine-grained ore		Medium- grained ore
P_2O_5	12.56	38.20	40.05	39.26	39.60
SiO_2	51.79	2.40	0.80	0.6	1.34
TiO_2	–	–	–	–	–
Al_2O_3	4.28	0.45	0.25	1.12	0.30
Fe_2O_3	3.02	0.50	0.87	1.26	0.49
FeO	–	–	–	0.18	–
MgO	0.10	n.d.	n.d.	0.87	0.78
MnO	–	–	–	–	–
CaO	20.79	53.20	54.80	53.18	52.60
Na_2O	–	0.10	0.08	0.13	–
K_2O	–	Traces	Traces	0.05	n.d.
H_2O	–	–	–	–	–
CO_2	3.23	1.05	0.85	2.15	1.13
SO_3	0.04	Traces	Traces	–	0.50
F	–	2.75	3.04	1.75	–
Others	–	2.40	1.24	–	–

n.d. not determined

Because two phosphate horizons of the lower phosphate unit were also encountered in boreholes, their resources were additionally calculated as 28.62 Mt of ore grading 22.45% P_2O_5 on average (6.65 Mt of net P_2O_5) in category C and 108.42 Mt of ore grading 20.81% P_2O_5 on average (22.56 Mt of net P_2O_5) in category P.

12.2.2.20 Phosphorite Deposits in Western Flank of the Basin

Deposits on the western flank of the basin span a 150 km area south from Tengis River headstream to Shishhid River headstream. There are more than ten phosphorite deposits and occurrences studied in this area, and the biggest one is Tsagaan Nuur deposit. Most of the deposits and occurrences are in Dood Tsagaan Nuur area (Fig. 12.10). Ore reserves of the Tsagaan Nuur and Khogorgo deposits are calculated as about 200 Mt.

12.2.2.21 Tsagaan Nuur Deposit

Tsagaan Nuur is the only phosphate deposit on the western flank with any available resource data. It is very poorly studied and consists of three phosphate-rich areas and

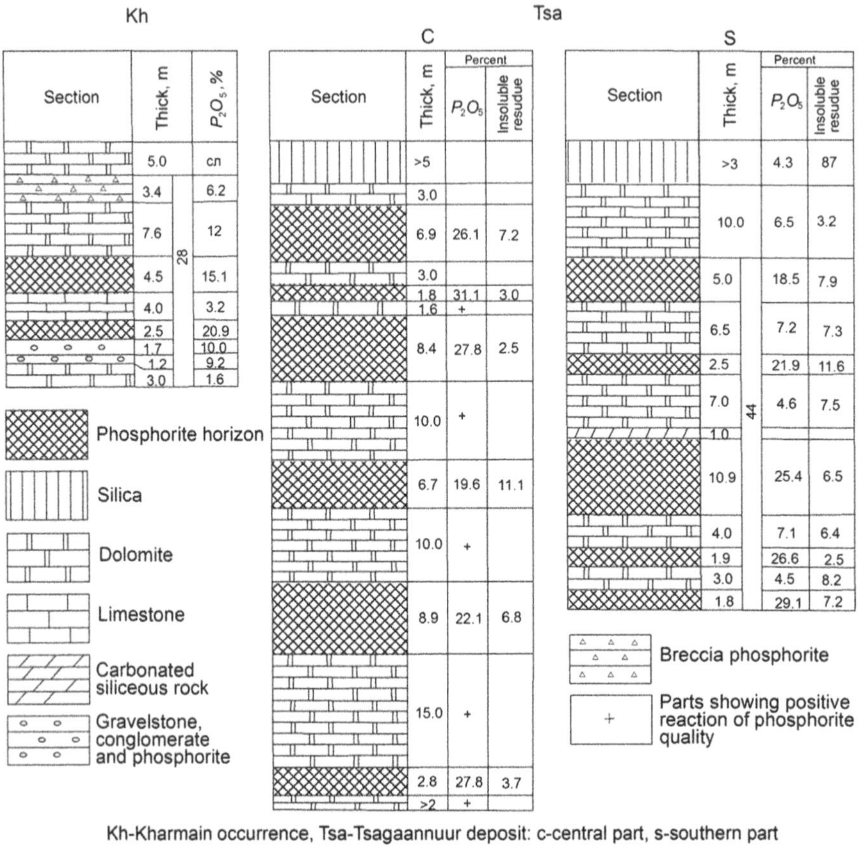

Fig. 12.10 Section of phosphorite-bearing formation in the Tsagaan Nuur group deposit and occurrences (Byamba and Iliyn 2012)

five phosphate beds. The average length of these beds is 1.5 km, with thicknesses of phosphate-rich horizons varying from 1.5 to 4.0 m.

The deposit consists of four zones separated by Central, Western, North, and South fault. The Tsagaan Nuur deposit located in the western zone is dominated by the structureless variety, which R. Sheldon proposed in 1980 to term as "primary bedded structureless phosphorite" in contrast to both granular phosphorites of Bed 5 in the Khuvsgul Basin and well-known pelletal phosphorites of the Phosphoria Formation. The bedded structureless phosphorites include rhythmically alternating phosphate and dolomite laminae. Phosphate is pigmented by organic matter that is concentrated at the base and top of interbeds. Such laminated varieties constitute the largest parts of beds, but they are frequently subordinate to dolomitic phosphorites that are typical of the eastern zone. In the west, phosphate components of dolomitic phosphorites are characterized by larger dimensions. They commonly represent variably disintegrated interlayers of the structureless phosphate. Primary bedding is retained in elongate fragments of layers. These fragments are separated by relatively large rhomboidal dolomite crystals (Ilyin 2004).

While Ilyin (1973) originally estimated resources of some 216 Mt of ore grading about 25% P_2O_5 on average, more realistic approaches end up with 25 Mt of ore. Detailed analysis proved the uranium content of this deposit to be very low (4 ppm on average) (Ilyin et al. 1986).

12.2.2.22 Khogorgo Deposit

The deposit is located along longitude within Dood Nuur Formation dolomite, in graben-like structure. Phosphorite-bearing layer is 4 km long. North and south limbs of deposit are covered by Cenozoic basalt and loose deposits. The phosphorite horizon is underlain by yellowish, brown, and brownish thin marlstone and dolomite member. Phosphorite of Khogorgo deposit comprises carbonaceous layers and coarse-grained types. Phosphorous anhydrite content of carbonate phosphorite is 27.4%. P_2O_5 content of breccia phosphorite varies from 15% to 21%. The thickness of the Dood Nuur Formation is 500–600 m within the deposit area (Byamba and Iliyn 2012).

12.2.2.23 East Occurrence of the Dood Nuur

Most of the geological features of this occurrence resemble to the Khogorgo deposit. A 100-m-thick yellowish and porous shale underlies this layer. Phosphorite-bearing bed contains two phosphorite layers each 3–5 m in thickness. Ore of this occurrence is characterized by layers, breccia, conglomerate, and sandstone (Fig. 12.11). The P_2O_5 content of ore reaches 5–20.1% (Byamba and Iliyn 2012).

12.3 Geology of the Zavkhan Phosphorite Basin

The Zavkhan basin is the one of the potential areas for phosphate mineralization in Mongolia. In this region, about 20 phosphate occurrences were discovered, mainly in limestones of Lower Cambrian age (Fig. 12.12; Table 12.4). The basin is located in western part of the Central Mongolian microcontinent comprising more than 20 thousand km^2 isometric area.

Tectonically, it is a part of the shelf zone in western Central Mongolian microcontinent. Ediacaran-Lower Cambrian terrigenous-carbonate rocks are located within graben and horst structures within the Tsagaanolom basin. South Songino uplift is the most uplifted block between Zavkhan and Khungui Rivers. This northwest-oriented, comparatively small block is surrounded largely by terrigenous deposits of Bayangol and Tsagaanolom Formations.

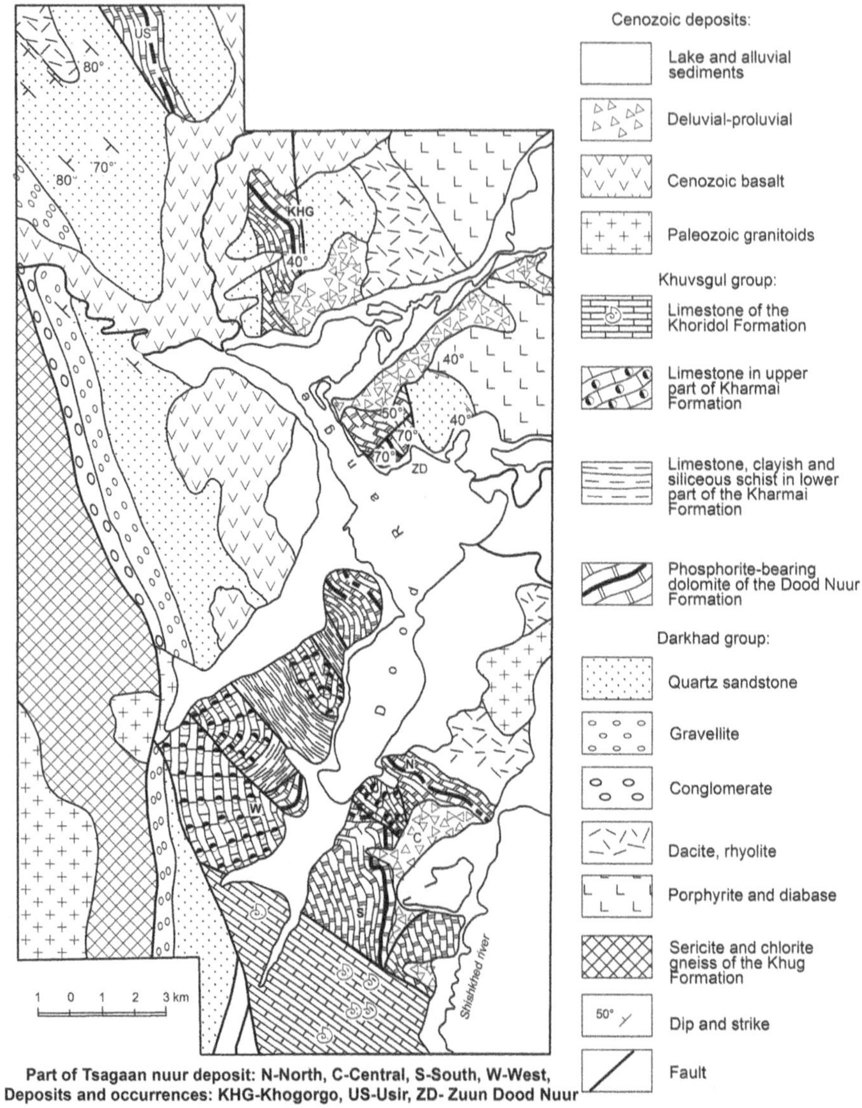

Fig. 12.11 Geological map of the Dood Nuur Group deposits and occurrences (Ilyin 1973)

Geological setting of the Zavkhan basin is complex, and most part is composed of intrusive rocks from late Proterozoic to Permian (Dorjnamjaa et al. 1977). Lower Proterozoic rocks are represented by gneiss, crystalline schist, quartzite, and marble (Fig. 12.13).

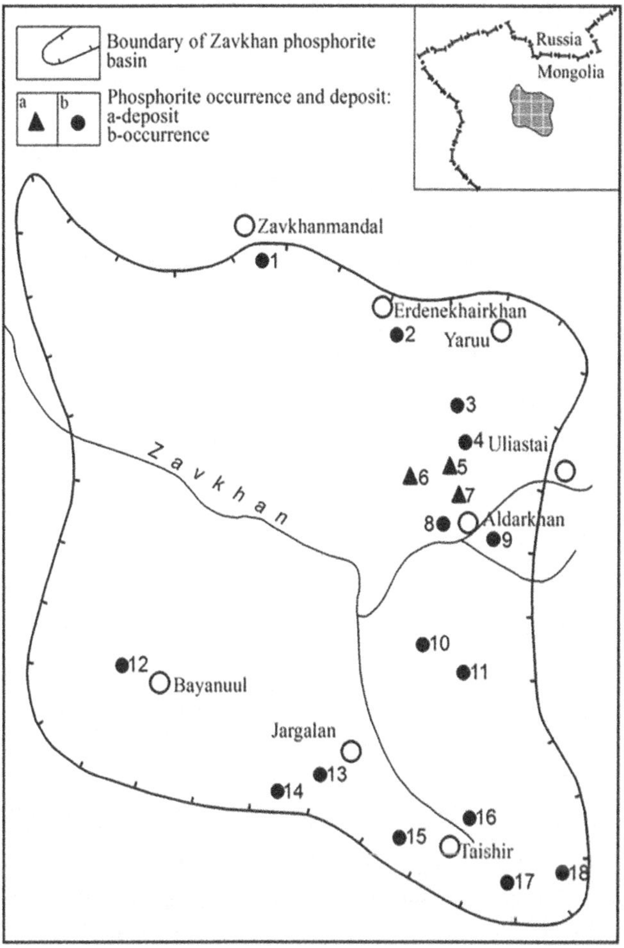

1-Argalant Uul
2-Tsagaankhairkhan
3-Tsakhirin Dorolj
4-Bunkhant
5-Ikh Khanan
6-Tsakhir Uul
7-Alagiin Davaa
8-Bogdiin Gol
9-Buural Uul
10-Baruun Arts
11-Zuun Arts
12-Urtuun Tsakhir
13-Khorgo
14-Orolchain Gol
15-Bayangol
16-Taishir
17-Khevtee Tsakhir
18-Khamarbulag

Fig. 12.12 Schematic map of the location of phosphorite deposits and occurrences in the Zavkhan basin (Byamba 2012)

12.3.1 Stratigraphy of Phosphorite-Bearing Formations in the Zavkhan Basin

12.3.1.1 Bayangol Formation

The phosphorite-bearing terrigenous-carbonate deposits of the Bayangol Formation are mainly composed of dolomite, limestone, and phosphorite, and siliceous rock, sandstone, and siltstones are also widely distributed. About 20 phosphorite occurrences are determined in the area and mostly located in north and east part of the Zavkhan basin (Fig. 12.12).

Table 12.4 Information of phosphorite deposits and occurrences in Zavkhan basin (Dorjnamjaa and Soyolmaa 2001)

No	Deposit and occurrence	P_2O_5%	Thickness of ore horizon (m)	Locality	Geological location (resources by P category)
1	Tsakhir Uul	4.6–12.71		Zavkhan aimag, Yaruu soum	Normal (30–40°), sometimes shifted by normal fault (not clear)
2	Ikh Khanan	4.8–30.0	80–100 (with Al component)	Zavkhan aimag, Aldarkhaan soum	Monoclinal (30–40°), (50 mln.t)
3	Alagiin Davaa	5.2–36.0	5–10	Zavkhan aimag, Aldarkhaan soum	Normal (30–40°); (30 Mt)
4	Khorgo	20.0–30.0	0.5–1.1	Govi-Altai aimag, Jargalant soum	Normal (25–30°) (not clear)
5	Bogdiin Gol	6.9–26.0	10–12	Zavkhan aimag, Aldarkhaan soum	Normal (35–48°) (not clear)
6	Zuun Arts	7.1–26.0	5–7	Zavkhan aimag, Tsagaan Chuluut soum	Normal (30–35°) (not clear)

Most of the phosphorite occurrences are located in upper part of the siliceous-dolomite formation, and only one phosphorite horizon is recognized in all of them, except for the Baruun Arts occurrence. There are three phosphorite horizons in the Baruun Arts occurrence; its lower horizon continues almost 15 km with various thicknesses of up to 10 m. The middle horizon occurs at 100–130 m above the lower horizon and discovered in the Baruun Arts, Bogdiin Gol, and Buural Uul occurrences. Length of middle horizon continues up to several tens of meters, and thickness ranges from 0.8 to 5 m. The upper horizon occurs in Zuun Arts and Orolchain Gol area and is composed of fragmental phosphorite 5–7 m in thickness.

Total thickness of siliceous dolomite formation is about 1500 m. This formation is gradually changed to limestone-sandstone formation, which is mainly composed of sandstone, siltstone, and mudstone. Here, limestone is rare. The lower part of the Bayangol Formation and lower and upper part of the Salaanii Gol Formation are composed of reef limestone and contain archaeocytes and algae. The rocks within the basin are not well studied, and phosphorite-bearing siliceous-dolomite formations are distributed along the basin margin and related to uplifts of syn-sedimentation (Byamba 2012).

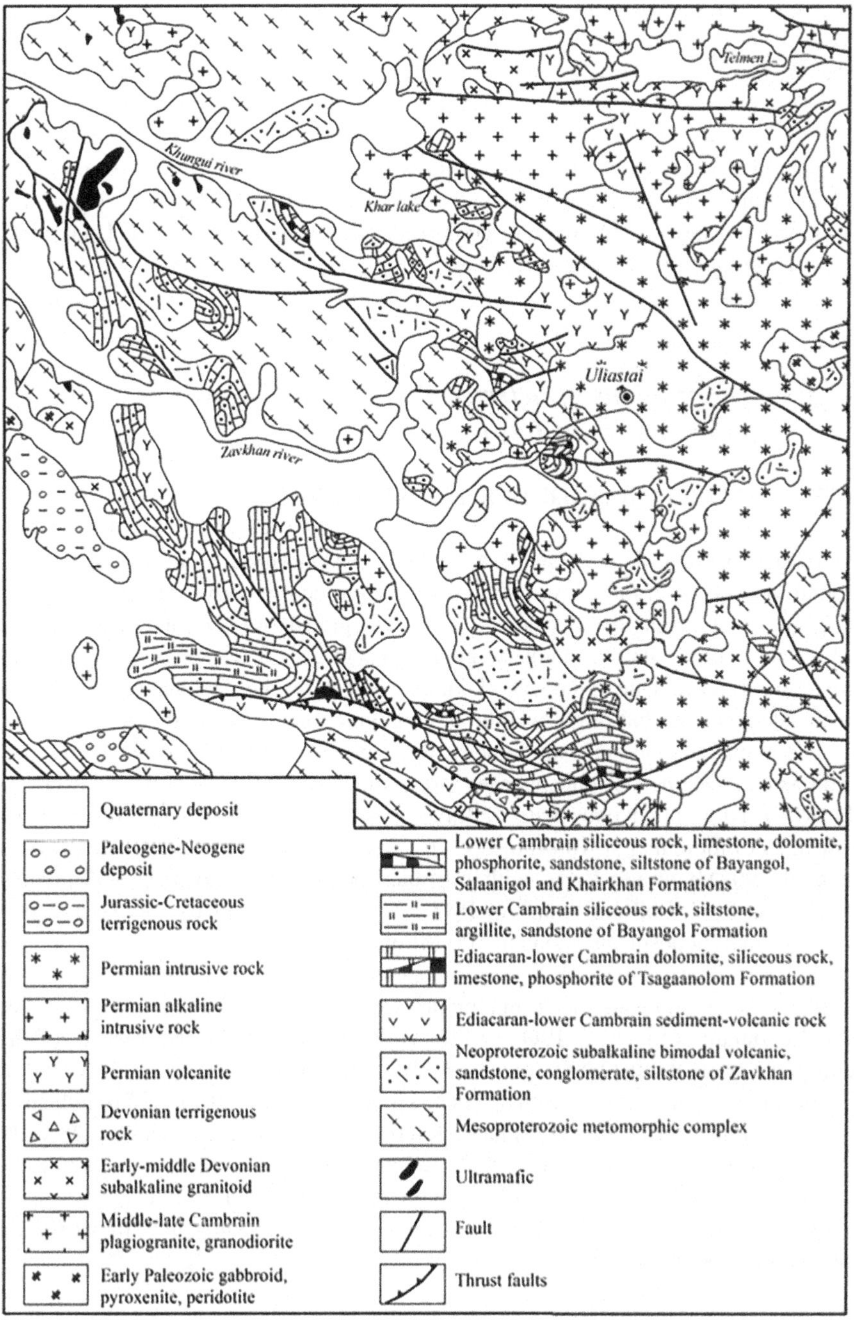

Fig. 12.13 Geological map of the Zavkhan phosphorite basin (Byamba 2012)

12.3.1.2 Tsagaanolom Formation

The Tsagaanolom Formation is located at the Zavkhan Formation volcanic rocks. The Tsagaanolom Formation rocks are widely distributed in middle part of Zavkhan and Khungui Rivers and Bayanuul soum. They are mainly composed of dark gray, black, and gray dolomite, limestone, and carbonaceous rocks. In the section along the Bogdiin Gol and Khungui River basins, siliceous rocks, siltstone, shale, and conglomerates are present. Carbonaceous rocks are strongly brecciated. Total thickness of formation is 1500 m.

12.3.2 Descriptions of Selected Phosphorite Deposits in the Zavkhan Basin

12.3.2.1 Alagiin Davaa and Tsakhir Uul Deposits

At Alagin Davaa (Alaguul) and Tsakhir Uul in Aldarkhaan Soum, phosphorite is bound to a 20–40-m-thick layer of sandstone with the ore body stretching for some 6–7 km. The tectonic displacement is weak. The grades of P_2O_5 range from 10% to 32.3% with most data plotting between 15.6% and 21.3%. The estimated resources are 31.6 Mt of ore. Due to the lack of data, resources for the Zavkhan basin have not been calculated yet but might well be of economic interest (Harald et al. 2011).

The main international requirements for deposits of phosphate ores of marine-sedimentary origin, i.e., phosphorites, are a minimum grade of 5 wt.-% P_2O_5 (= 11 wt.-% BPL, bone phosphate of lime), most often >20 wt.-% P_2O_5, and net resources of more than 5–10 Mt of P_2O_5 (= small deposits). Medium-sized deposits start at 100 Mt of net P_2O_5, with big and world-class deposits possessing net resources of >1.5 billion tonnes of P_2O_5. According to the Russian classification system, small deposits also contain 100 Mt of P_2O_5. The minimum design capacity of dressing plants is 100,000 tonnes of phosphate ore per year. For direct use as fertilizer, i.e., without any other mineral processing besides grinding, phosphate ores should contain >15 wt.-% P_2O_5 (> 33 wt.-% BPL). A total of 55 wt.-% of the phosphate content must be dissolvable in 2%-conc. Formic or citric acid. A total of 80–90 wt.-% of the crude phosphate ore has to be ground to <0.1 mm. For industrial uses, phosphate rocks, i.e., phosphate ore concentrates, need to contain >28 wt.% P_2O_5, <11 wt.% $CaCO_3$, <5 wt.% SiO_2, <3 wt.-% F, <2.5 wt.% R_2O_3, <0.25 wt.% MgO, <0.10 wt.% Cl, <1 wt.-% organic substance, <20 ppm Cd, as well as extremely low values of Pb, Hg, U, Cr, As, Se, and V. Phosphate rocks should also be fine-ground and dry.

12.4 Concluding Remarks

All deposits and occurrences in the Khuvsgul phosphorite basin constitute a major phosphate resource of Mongolia. The grades in these deposits vary from 10 to 38% of P_2O_5. The Burenkhaan and Khuvsgul deposits are the largest in the basin. Latest information by Harald et al. (2011) suggests that the Burenkhaan phosphate deposit is the best for further mining interest because of being less environmentally sensitive. All other deposits in the Khuvsgul basin are of limited interest only, as they may be of good grade and even of medium size but are lying within the environmentally highly sensitive Khuvsgul National Park or in very remote areas. The Zavkhan basin may contain some phosphate deposits of economic interest; however, a lot of exploration is still needed before any serious evaluation is possible.

Web Sources https://www.micon-international.com/mineral-resource-reporting-differences-between-cim-jorc-and-others/

Appendixes

Geological reserves of the deposit are categorized to be proven (A), measured (B), and indicated (C) reserves.

A, B, and C_1 categories by Russian State classification are equal to proven reserve and measured resources.

B, C_1, and C_2—Indicated and probable.

C_2 and P_1—Inferred.

P_2 and P_3—Exploration potential or exploration target.

References

Belichenko VU (1969) Nizhnii palleozoi Zapadnogo zabaikal'ya (Lower Paleozoic of Western Transbaikalia) Nedra, Moskva, p 207

Byamba J (1987) Drevnye fosfority Mongolii (Ancient phosphorites of Mongolia) Litologiya poleznye iskopaemye, Nedra, Moskva, 22, 1:79–88

Byamba J (1996) Drevnye struktury Mongolii i ikh fosforitonosnosti' (Ancient Structures of Mongolia and Their Phosphorite Potential) Admon, Ulaanbaatar, p 181

Byamba J (2012) Mongoliin ertnii phosphorit (ancient phosphorite of Mongolia). In: Mongoliin khojuu proterozoi-turuu paleozoin struktur ba phosphorit. Soyombo Printing, Ulaanbaatar, pp 174–212

Byamba J, Iliyn AB (2012) Phosphor (phosphate). In: Altangerel S (ed) Metal bus ashigt maltmal, Mongoliin geologi, ashigt maltmal, VII. Soyombo Printing, Ulaanbaatar, pp 21–39

Byamba J, Iliyn AB, Zaitsev NS (1986) Proterozoic and Cambrian phosphorites deposits: Khubsugul, Mongolian People's republic. In: Phosphate deposits of the world. Vol. 1, Proterozoic and Cambrian phosphorites. Cambridge University Press, pp 162–174

Dorjnamjaa D, Kepejinskas KB, Ochir L (1977) Phosphority Zabkhanskogo progiba Zapadnoi Mongolii (Phosphorites in Zavkhan basin, Western Mongolia). Geologiya i geophizika 9:29–38

Dorjnamjaa D, Soyolmaa D (2001) Zavkhanii fosforitiin sav gazar: fosfor, alt, ochir erdene (The Zavkhan phosphorite basin: phosphrous, gold, diamond), Ulaanbaatar p 166

Gulbrandsen RA, Krier DJ (1980) Large and rich phosphorus resources in the Phosphoria Formation in the Soda Springs area, Southeastern Idaho. U.S. Geol. Surv. Bul. 1494, p 25

Harald E, Peter B, Martin S, Tamiraa A (2011) Industrial minerals and selected rare metals in Mongolia. An Investors' guide. Prepared by the Government of Mongolia, Funded by the Federal Ministry of Economic Cooperation and Development of the Federal Republic of Germany through the Federal Institute for Geosc. Natur. Resour. (BGR), p 322

Ilyin AV (1971) The Tuva–Mongolia Massif, Tr. NIL Zarubezh- geologii (Proc. of NIL Zarubezhgeologiya), Nedra, Moskva, 22:27–36

Ilyin AV (1973) Khubsuguli'skii fosforitonosnyi Bassein (Khubsugul phosphate-bearing basin). In: Yanshin AL (ed) The joint Soviet-Mongolian scientific-research geological expedition, Moskva, vol 6, p 167

Ilyin AV (1981) Primary, bedded, Structureless Phosphorite of the Khubsugul Basin, Mongolia. J Sediment Petrol 51(1):1215–1222

Ilyin AV (1990) Drevnie fosfatonosnye basseiny (Ancient Phosphate-Bearing Basins) Nauka, Moskva, p174

Ilyin AV (1998a) Upper cretaceous cherty Facies in Europe. Byull Mosk O-va Ispyt Prir Otd Geol 73(5):49–60

Ilyin AV (1998b) Rare-earths geochemistry of "old" Phosphorites and probability of direct Synsedimentational precipitation and accumulation of phosphate. Chem Geol 144:243–256

Ilyin AV (2004) The Khubsugul phosphate-bearing basin: new data and concepts. Lithol Miner Resour 39(5):454–467

Ilyin AV, Ratnikova GI (1971) O fosforitakh Khubsugul'skogo basseina Mongolii (About phosphorites of the Khubsugul basin of Mongolia). Litol Polezn Iskop 1:63–75

Ilyin AV, Zaitsev NS, Byamba J (1986) Proterozoic and Cambrian phosphorite deposits: Khubsugul, Mongolian People's Republic. In: Cook PJ, Shergold JH (eds) Phosphate deposits of the world, Vol. 1: Proterozoic and Cambrian phosphorites, 162–174

Khain VE, Rudakov SG (1995) Baykal'skaya skladchatost' N.S.Shatskogo i baykal'skaya era tektogeneza (The Baikalian Folding according to NS Shatsky and the Salair Orogeny) Stratigrafiya. Geologicheskaya korrelyatsiya, 1995 3(6):25–32

Kuz'michev AB (2000) Tektonicheskoye znacheniye paleozoyskogo granitnogo magmatizma v baykalidakh Tuvino-Mongol'skogo massiva (tectonic implication of Paleozoic Granite Magmatism in the Baikalides of the Tuva–Mongolian Massif). Geotektonika 34(6):76–92

Kuz'michev AB (2001) Rannebaykal'skiye tektonicheskiye sobytiya v Tuvino-Mongol'skom massive: kolliziya Ostrovnoy dugi i kontinenta (Early Baikalian Tectonic Events in the Tuva–Mongolian Massif: Arc–Microcontinent Collision). Geotektonika 35(3):44–59

Mei MX (2007) Revised classification of microbial carbonates: complementing the classification of limestones. Earth Sci Front 14(5):222–232

Mitropanov FP, Kozakov IK, Palei IP (1981) Dokembrii Zapadnoo Mongolii i Tuvy (Precambrian of Western Mongolia and Tuva) Nauka, leningrad, p 153

Muzalevskii MM (1970) Geologicheskoye stroyeniye i osnovnye tipy fosforitov Khubsugul'skogo mestorozhdeniya MNR (Geological structure and main types of phosphorites of the Khubsugul deposit in Mongolia) Kazan', p 20

Osokin PV, Ilyin AV, and Zaitsev NS (1973) Okino-Khubsugul'skii fosforitonosnyi bassein na territorii SSSR i MNR (Okino-Khubsugul phosphorite-bearing basin in the territory of the USSR and MPR) Syr'evaya baza fosfatnoi promyshlennosti SSSR—perspektivy rasshireniya I ratsional'nogo ispol'zovaniya. Vses. Inst. Ekon. Miner. Syr'ya, Moskva: 54–56

Osokin PV, Tyzhinov AV (1998) Dokembriiskie tilloidii Okino-Khubsugul'skogo fosforitnogo basseina: Vostochnyi Sayan, Severo-zapadnaya Mongoliya (Precambrian Tilloids in the Oka–

Khubsugul Phosphorite-Bearing Basin: Eastern Sayan, Northwestern Mongolia). Litol Polezn Iskop 33(2):162–176

Pufahl PK (2010) Bioelemental sediments. In: James NP, Dalrymple RW (eds) Facies Models, 4th edn. Geological Association of Canada, pp 477–503

Terleev AA (1998) Age of the Bokson Group. Geol Geofiz 2:211–220

Yanshin AL (1986) Old Phosphorites—regional review: Asian Part of USSR and Mongolia. In: Cook PJ, Shergold JH (eds) Phosphorite deposits of the world, vol. 1: Proterozoic and Cambrian Phosphorites. Cambridge University Press, pp 63–69

Zaitsev (1981) Khubsugul'skoi fosforitonosnyi Bassein Mongol'skoi Narodnoi Respubliki (Khubsugul phosphate basin of Mongolia). Dokl AN SSSR 192(2):391–394

Zaitsev NS (1992) Drevnyaya fosforitonosnaya formatsiya Mongolii (ancient Phosphorite-bearing formation in Mongolia). Soviet Geol 1:78–79

Zaitsev NS, Ilyin AV (1970) Tektonika Prikhubsugul'skogo progiba: MNR (tectonics of the Khubsugul depression). Geotektonika 5:3–11

Zhegallo EA, Rozanov AY, Ushatinskaya GT, Hoover R, Gerasimenko L, Ragozina A (2000) Atlas of microfossils from ancient Phosphorites of Khubsugul (Mongolia). NASA, Hantsville, p 112

Zhuravleva ZA (1974) Vozrast i korrelyatsya drevnykh otlozhenii Severo-Zapadnoi Mongolii (Age and correlation of ancient phosphorites in Northwestern Mongolia). Izv Akad Nauk, SSSR Ser Geol 2:37–44

Part III
Energy and Fuel Resources

Chapter 13
Uranium Deposits

Dorjyunden Altankhuyag and Baldorj Baatartsogt

13.1 Introduction

The uranium metallogeny of Mongolia has previously been interpreted in terms of geodynamic concepts or tectonic belts (Mironov 2006). In this report, we update the tectonic classifications of Mironov (2006) and Byamba et al. (2018) by geological classification, as recommended by the IAEA (2018).

Currently, no uranium is being produced in Mongolia; however, a number of deposits are in the stage of mine development. The Government of Mongolia has approved the Agreement of mine development of Zoovch Ovoo and Dulaan Uul deposits, owned by Badrakh Energy LLC, a joint company Mon-Atom of the Government of Mongolia, Orano group of France, and Mitsubishi Corporation of Japan. The uranium will be extracted by in situ leaching (Axel et al. 2018). The first uranium geological survey and exploration was carried out in 1945 in Mongolia. During the period 1945–1960, numerous uranium occurrences were discovered in Mesozoic lignite in eastern Mongolia. Uranium exploration was conducted under a bilateral agreement between Mongolia and the former Soviet Union, throughout Mongolia in 1970. Airborne gamma-ray spectrometry surveys at a scale of 1:25,000 and 1:50,000 were conducted over 420,000 km^2, which covers 27% of Mongolian territory, in addition at a scale of 1:200,000 over 650,000 km^2, which is 28% of the territory, and at a scale of 1:1,000,000 over 224,000 km^2, comprising 14% of the Mongolian Altai, Khangai Mountains, and Gobi region.

D. Altankhuyag (✉)
Strategic Policy and Planning Department, Ministry of Mining and Heavy Industry, Ulaanbaatar, Mongolia
e-mail: altankhuyag@mmhi.gov.mn

B. Baatartsogt
Administration Department, The Nuclear Energy Commission, Ulaanbaatar, Mongolia
e-mail: b.baatartsogt@nea.gov.mn

O. Gerel et al. (eds.), *Mineral Resources of Mongolia*, Modern Approaches in Solid Earth Sciences 19, https://doi.org/10.1007/978-981-15-5943-3_13

The first uranium deposit, Dornod, was discovered in 1973 in eastern Mongolia by Russian geologists and reported by Kochkin et al. (1985), Rogov and Rubzov (1989), Rogov and Yurchenko (1987), Mironov et al. (1993), and Mironov (2003) in various publications such as Uranium Metallogeny of Mongolia under an intergovernmental agreement. This is the North Choibalsan uranium district and contains five uranium deposits (Gurvanbulag, Dornod, Mardaingol, Khavar, and Nemer), hosted in the Mesozoic volcanic depressions. Russian expeditions also found the first uranium deposit related to the Lower and Upper Mesozoic terrigenous sediments in southern Mongolia (Budunov 2006). Uranium mining commenced operation on the Dornod and Gurvanbulag deposits as underground and open-pit mines in the Mardaingol District in 1986 and produced about 535 tonnes of U with average grade of 0.12% U. The ores mined in the Mardaingol District were transported by rail 484 km to the Priargunsk mining and processing plant in Krasnokamensk, Russia. Because of political and economic changes both in Mongolia and neighboring areas of Russia, uranium production at Erdes company was closed in 1995. From this time onward, Mongolian-Russian-American Gurvansaikhan joint investment companies carried out exploration focused on the Kharaat and other south Mongolian sediment-hosted uranium deposits. The State-owned Uranium Company also explored for uranium since 1995 (Batbold 1997). The Mongolian-French Cogegobi joint company was established in 1997 and discovered Dulaan Uul and Zoovch Ovoo sediment-hosted deposits in Mesozoic basins in southern Mongolia.

13.2 Regional Tectonic and Geological Setting

Mongolia comprised part of the Central Asian Orogenic Belt (Zonenshain et al. 1990; Jahn 2004; Windley et al. 2007) or the Altaid Orogenic Collage (Sengor and Natal'in 1996) or Central Asian Orogenic Supercollage (Yakubchuk 2004). It extends from the Ural Mountains in the west to the Pacific margin in the east and comprises several generations of Neoproterozoic to Mesozoic, magmatic arcs, accretionary terranes, and back-arc basins. The belt is located between the North Asian Craton to the north and the Sino-Korean Craton to the south (Fig. 13.1). The Argun-Idermeg cratonic terrane (Fig. 13.1) has been deformed oroclinally into a tight C-shape, or the central Mongolian orocline (Yakubchuk 2004), due to subduction and closure of the Mongol-Okhotsk Ocean in Latest Jurassic to Earliest Cretaceous time (van der Voo et al. 2015). The Mongol-Okhotsk collage (Fig. 13.1) extends over 3000 km NE into Russia to the present-day Okhotsk Sea.

Southern Mongolia has been tectonically divided in a number of island arc and related terranes (Badarch et al. 2002). They are grouped by Kröner et al. (2010), as the Gobi-Altai zone or south Mongolia-Khingan collage (Nokleberg 2010), which extends as an arcuate belt through southern Mongolian for over 2000 km (Fig. 13.1). Silurian to Devonian oceanic and continental blocks formed the basement to the south Mongolia-Khingan collage (Kröner et al. 2010). During the Early

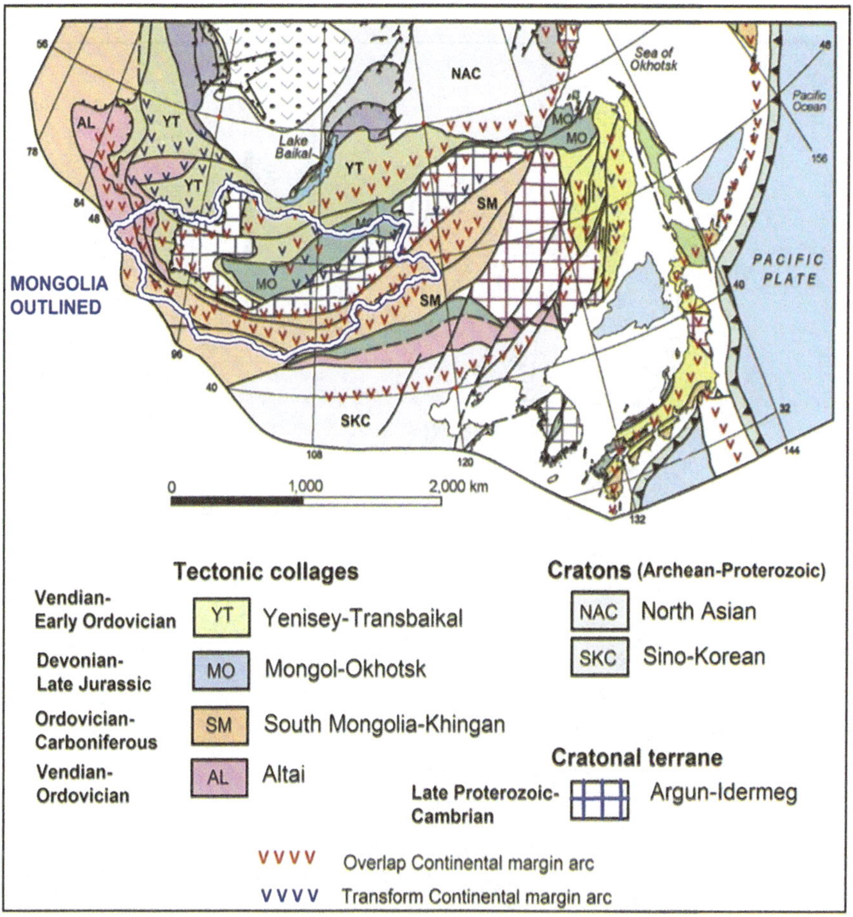

Fig. 13.1 Tectonic setting for Mongolia (Nokleberg 2010)

Carboniferous, the magmatic arcs and continental and oceanic basement in southern Mongolia were imbricated and accreted against continent margin to the north.

During the Late Jurassic to Early Cretaceous, rifting, including formation of metamorphic core complexes (Webb et al. 1999) and synrift basins with thick sediment accumulations, formed in southeastern Mongolia (Johnson 2004). The Mesozoic sedimentary basins were filled with terrigenous sediments, lignite, include basalt flows, and rhyolitic ash tuff. Extension was followed in the Cenozoic by transpressional tectonic events related to the Himalayan collision (Cunningham 2010), which dominate the present-day regional structure of southern Mongolia.

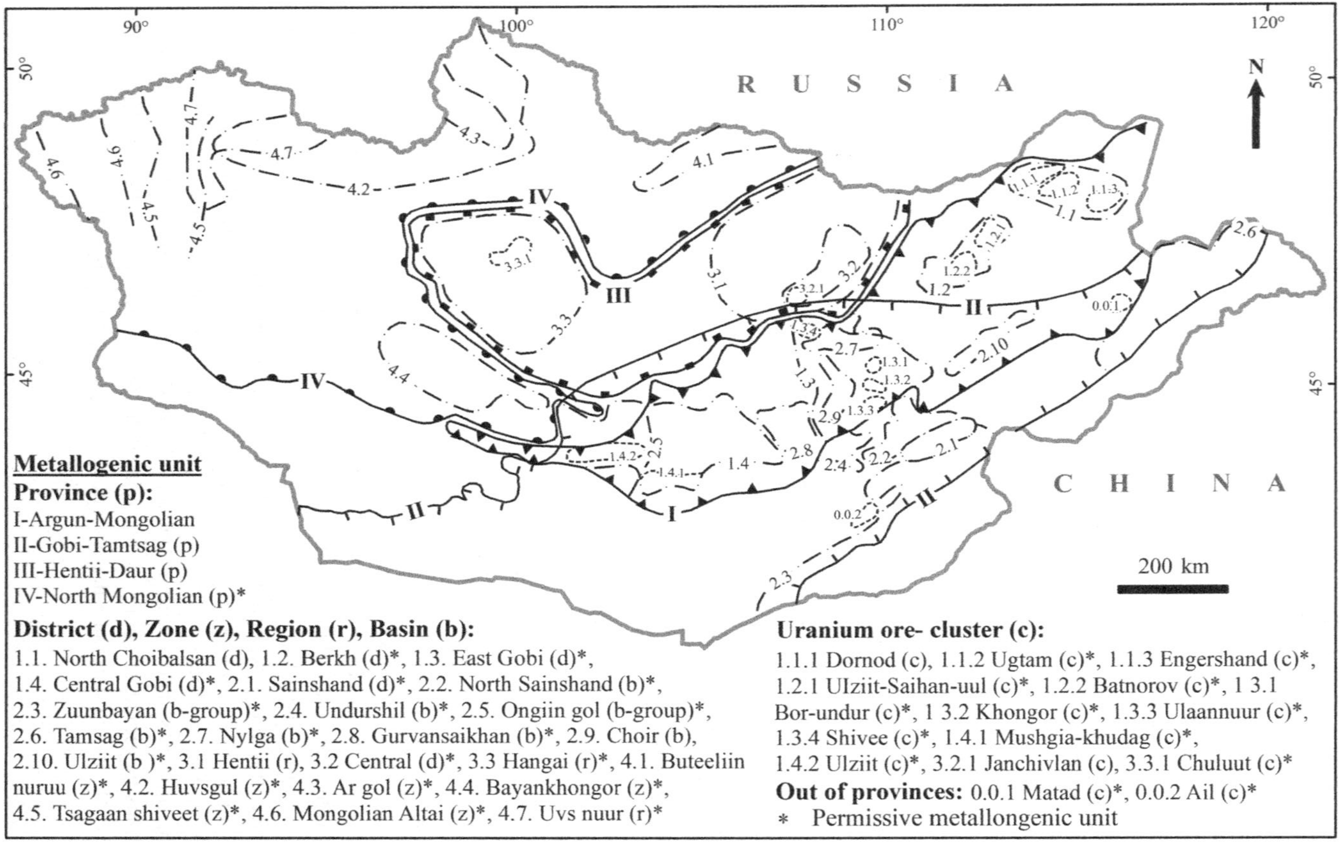

Fig. 13.2 Uranium metallogenic belts of Mongolia (Mironov 2006)

13.3 Historic Classification of Uranium Deposits

Previously, Mongolia was subdivided into four uranium ore provinces (Fig. 13.2) by Mironov (2006). These provinces are now grouped in terms of tectonic collages and cratonic terrains (Fig. 13.1).

13.3.1 Argun Mongolian Province

The Argun Mongolian province (North Choibalsan district) belongs to the Late Proterozoic-Cambrian Argun-Idermeg cratonal terrane (Figs. 13.1 and 13.2). U-Mo-F mineralization occurs as veins in Late Mesozoic sedimentary volcanics and basement rocks.

13.3.2 Gobi-Tamsag Province

The Gobi-Tamsag uranium ore province belongs to the south Mongolia-Khingan collage (Figs. 13.1 and 13.2) and is characterized by supergene uranium in block-faulted, Mesozoic terrigenous sedimentary sequences. This province is divided into Sainshand uranium ore districts (Choir, Tamsag, Gurvansaikhan, etc.).

13.3.3 Khentii-Daguur Province

The Khentii-Daguur (Daur) uranium ore province belongs to the Late Devonian-Jurassic Mongol-Okhotsk collage (Figs. 13.1 and 13.2), which contains Late Triassic-Early Jurassic leucogranitic plutons with high contents of radioactive minerals. The Janchivlin and Chuluut uranium clusters belong within this province.

13.3.4 North Mongolian Province

The North Mongolian uranium province belongs to the Ediacaran-Ordovician Yenisey-Transbaikal collage and Late Proterozoic-Cambrian Argun-Indermeg cratonal terranes (Figs. 13.1 and 13.2) and includes several uranium zones. The Buteliin Nuruu uranium zone is hosted in metasomatic rocks, migmatites, and pegmatites with brannerite, euxenite, and uraninite segregations. The Khuvsgul uranium zone is characterized by small uranium occurrences hosted in Ediacaran-Cambrian carbonaceous cherty rocks, enriched with P and V. A similar geological setting is typical of the Bayankhongor uranium zone. The Argol uranium zone is distinguished by the Late Paleozoic and Early Mesozoic plutons of subalkaline granites and syenites accompanied by impregnations of REE, U, and Th minerals.

13.4 Tectonic Classification of Uranium Deposits

Recent compilation of uranium deposits in Mongolia was carried out on the basis of the geodynamic concept of the development of the continental crust and published (Byamba et al. 2018). The territory of Mongolia was divided into the north Mongolian and south Mongolian metallogenic provinces. Each metallogenic province can be classified as a sub-province, belt, ore district, ore cluster, etc. according to Byamba et al. (2018). As regarding to their size, the province can be classified into belts. In addition, an airborne gamma-ray spectrometry survey was recently conducted over southwestern Mongolia, and uranium ore districts (Khovd, Gobi Tengeruul) and ore clusters (Khustiin Nuruu, Songint) were identified. The uranium deposits are summarized in Table 13.1 in terms of geodynamic concepts and by IAEA geological classification.

13.4.1 North Mongolian Metallogenic Province

The North Mongolian metallogenic province includes cratonic basement of Late Proterozoic to Cambrian (Argun-Indermeg) and Ediacaran to Ordovician (Yenisey-Transbaikal) and Devonian-Late Jurassic (Mongol-Okhotsk) collage (Fig. 13.2).

13.4.1.1 Tuva-Mongol Belt

Tuva-Mongol belt is characterized by uranium mineralization in Middle Paleozoic back-arc alkaline complexes, with associated REE, phosphorite, Au, Fe, and Pb-Zn. Zaanar and Kherbestei occurrences are related to dark brown carbonaceous rocks in the Tuva-Mongol basin. In the Khuvsgul ore district, uranium mineralization comprising orthite, thorite, zircon, monazite, and rare uraninite occurs at the contact of Late Ediacaran-Cambrian carbonate and alkaline intrusion, with associated quartz, chlorite, albite, K-feldspar, fluorite, sulfide, and Be mineralization. Higher-grade uranium was discovered in the Khan Khukhii ore district in Late Proterozoic carbonaceous rock located on the north edge of the Khan Khukhii block. This mineralization is related to Early-Middle Palaeozoic granosyenite, syenite, alkaline granite, and Permian granitoids.

Table 13.1 The uranium mineral systems in Mongolia (modified after Byamba et al. 2018)

Province	Sub-province, belt		Ore district	Ore cluster	Geodynamic setting	Type	Main deposit (d) and occurrence (o)
North Mongolian metallogenic province	Tuva-Mongol belt		Khuvsgul		Back-arc destruction/ extension	Phosphate Black shale Metamorphite	Zaamar (o), Erkhelnuur (o)
			Khan Khukhii		Continental arc	Granite-related	
	Nuur-Egiin Gol belt		Egiin Gol		Back-arc rift	Metasomatite Granite-related	
			Nuur		Continental arc		
	Mongolian Altai belt		Tsagaan Shuvuut	Kharkhiraa-Turgen	Continental arc	Granite-related	
				Tashant-Deluun			
				Kharus-Ulaat			
			Khovd	Khustain Nuruu			
	Central Mongolian sub-province	Buteel-Khangai belt	Buteeliin Nuruu		Intensive saturation zone	Intrusive Metasomatite	
			Chuluut	Chuluut	Plume	Surficial	Dalt (d)
			Bayankhongor	Narangiinovoo	Continental arc	Granite-related Black shale	
		Khentii-Daguur (Daur) belt	Janchivlan (central)	Janchivlan	Plume	Granite related	Tamgat (o), Elst (o)
		Dornod Mongol belt	North Choibalsan	Dornod	Marginal arc, rift	Volcanic-related	

(continued)

Table 13.1 (continued)

Province	Sub-province, belt		Ore district	Ore cluster	Geodynamic setting	Type	Main deposit (d) and occurrence (o)
		(Mongol-Arguun)				Lignite-coal Sandstone	Dornod (d), Gurvanbulag (d), Nemer (d), Mardaingol (d)
				Ugtam Turgen Engershand Suman Nuur	Marginal arc, rift	Volcanic-related Lignite-coal Sandstone	
			Berkh	Batnorov Ulziit-Saikhan Uul	Rift	Volcanic-related Sandstone	
			Choir		Post-rift taphrogenic	Sandstone	Kharaat (d)
			Shivee	Nyalga Shivee Tavansuveet	Rift, post-rift taphrogenic	Sandstone Volcanic-related	Engerar (d)
			Ulziit		Post-rift taphrogenic	Sandstone	Ulziit (d)
			Dundgovi		Post-rift taphrogenic	Metasomatite, Sandstone	
			Khairkhan	Khairkhan Gurvansaikhan	Post-rift taphrogenic Post-rift taphrogenic	Sandstone Sandstone	Khairkhan (d) Gurvansaikhan (d)
			Bor-Undur	Bor-Undur Ulaan Nuur Khongor	Rift and post-rift taphrogenic	Sandstone Volcanic-related	Ulaan Nuur (o)

South Mongolian metallogenic province	Gobi belt	Mandal-Ovoo	Uushiin Nuur, Ongigol, Mushgai Khudag	Post-rift taphrogenic	Volcanic-related Granite-related Sandstone	
		Ondorshil		Post-rift Taphrogenic	Volcanic-related Sandstone	
		Nukhetdavaa		Continental arc	Volcanic-related Sandstone	
		Sainshand	Narst	Post-rift Taphrogenic	Volcanic-related Sandstone	Narst (o), Ingen (o), Most (o), and Dorvoljin (o)
		Matad	Tamsag	Post-rift Taphrogenic	Volcanic-related Granite-related Sandstone	Zuunbulag (o)
		Zuunbayan	Unegt, *Zuunbayan*	Post-rift *Taphrogenic*	Sandstone *Sandstone*	Dulaan Uul (d), *Zuuvchovoo (d)*
		Ail	Ail	Post-rift Taphrogenic	Sandstone	Ail (o), Khad (o), Elgen (o)
	Southwest Mongolia belt	Nemegt	Songinot	Continental rift Post-rift Taphrogenic	Granite-related Sandstone	
		Baruun Khuurai		Post-rift Taphrogenic	Sandstone	
		Gobi Tengeruul		Granite-related	Granite-related	

13.4.1.2 Nuur-Egiin Gol Belt

This belt hosts two ore districts, namely, Nuur and Egiin Gol. Arigiin Gol uranium prospect was identified in east side of Egiin Gol ore district based on radioactive anomalies. Despite old basement rocks, Cretaceous and Cenozoic sediments host uranium mineralization. For example, based on airborne gamma-ray spectrometry survey, uranium anomalies were recognized in upper Jurassic-Lower Cretaceous sandstone in Khyargas Lake basin.

13.4.1.3 Mongol Altai Belt

In this belt, Lower-Middle Paleozoic terrigenous sedimentary and volcanic rocks were intruded by Middle Paleozoic-Jurassic granitoids, with associated W, Au, Ag, REE, and Fe mineralization. Uranium mineralization is found in pegmatite in the northern part of belt. Several anomalies of U, Th, and K are associated with REE mineralization. Based on the radioactive anomalies, Tashant-Deluun, Kharus Ulaat, and Kharkhiraa-Turgen ore clusters are considered prospective for uranium mineralization in the belt.

13.4.1.4 Tsagaan Shuvuut Ore District

The Tsagaan Shuvuut ore district is an extension of the Altai-Sayan uranium mineralization in Russian territory and composed of Precambrian, Paleozoic metasediments, Devonian felsic volcanoclastic, Devonian-Carboniferous rift-related syenite, and subalkaline granitoid complexes. Uranium, Th, and REE mineralization is characteristic for this district and may be related to alkaline granite complexes.

13.4.1.5 Khustain Nuruu Ore Cluster

Khustain Nuruu ore cluster is identified in the southeast end of the Khovd District by an airborne gamma-ray spectrometry survey. Middle-Upper Ordovician carbonate, shale, gneiss, siltstone, sandstone, and oil shale-bearing sequences are intruded by Late Devonian fine- to medium-grained biotite granite. Altai complex is characterized by uranium mineralization with associated Au, Be, Nb, Ta, Y in veins, and pegmatites.

In the Late Silurian Khovd complex, leucogranite, biotite granite, and plagiogranite are widespread. Au, Cu, Zn, Be, and W occur in Mesozoic terrigenous sedimentary rocks, whereas Nb-Ta, Be, REE, Sn, Cu, and Pb anomalies are found in biotite granite. The range of gamma radiation was from 50 nGy/h to 124 nGy/h and the grade is 3.5–8 g/t U, 8–25 g/t Th, whereas, the range of 60–95 nGy/h, 4.5–5.1 g/t U, 10–16 g/t Th, 1–2.6% K in massive granite (Duinharjav et al. 2018).

The Altai complex consists of granite, syenite-granite, and monzogranite enriched in K (Gerel 2012) as well as Ag. The granitoid geochemical data suggest S-type affiliation and alkaline composition. The Altai complex granitoid age was determined by K-Ar dating technique as 400–364 Ma (Gavrilova and Zaizev 1975).

13.4.2 Central Mongolian Sub-Province

Uranium mineralization in the central Mongolian sub-province is mostly related to igneous rocks. The sub-province is divided into three belts, namely, Buteel-Khangai, Khentii-Daguur (Daur), and Dornod Mongol belts.

13.4.2.1 Buteel-Khangai Belt

This belt is composed of Proterozoic and Paleozoic volcanic and sedimentary rocks and Paleozoic and Mesozoic igneous rocks with Cu, Au, rare metal, and Fe mineralization. However, there are many radioactive anomalies within the belt, including the Buteeliin Nuruu and Bayankhongor ore districts.

Buteeliin Nuruu ore district consists of Proterozoic granitoid with radioactive elements, uranium-bearing pegmatite, and metasomatite. Bayangol and Zuun Undurtseh uranium occurrences are hosted in quartz-plagioclase metasomatite within a cataclastic zone of amphibole-biotite gneiss, biotite-gneiss, and granite-gneiss. Brannerite is the main uranium mineral which is associated with euxenite, titanomagnetite, and rare elements, with grades reaching 0.44% U.

The Chuluut ore district is identified in northwest part of Dornod Khangai zone around river Chuluut, covering 400 km^2. This area includes the Dalt deposit and related occurrences. The Chuluut ore cluster trends northwest and is composed of Quaternary sediments. Uranium mineralization is located in unconsolidated sediments of paleovalleys beneath Quaternary-Neogene basaltic flows. Permian subalkaline leucogranites have been emplaced along the large transverse faults oriented northeast.

Bayankhongor ore district was identified based on the radioactive anomalies, as well as stratigraphic control in Paleogene to Neogene sediments. In addition, underlying Cretaceous sediments can also be a perspective for uranium mineralization.

13.4.2.2 Khentii-Daguur (Daur) Belt

In this belt, widespread Mesozoic granites are thought to be a significant uranium source. Radioactive anomalies and uranium occurrences within the belt have associated Au, base, and rare metal mineralization. Chikoin and Daguur uranium districts are situated on the northeast extension of this belt into the Russian territory.

Uraniferous coal-bearing Lower Cretaceous sediments in depressions occur in the margin on the southeastern border of Khentii-Daguur belt. Another prospective ore district is Janchivlan (central), in which Baruun Jargalant, North Yudeg, Elst, Tamgat, and Arshaant occurrences are located on fractures and fault zone northwest of the Janchivlan pluton. The radioactive anomalies identified at Nalaikh are related to pegmatite and greisen. Pitchblende and coffinite with associated sulfide mineralization are hosted in tectonic zones, comprising gouge, breccia, and crushed zones. The uranium mineralization is mainly controlled by veins and lenses extending tens or a few hundreds of meters along the strike and down the dip, varying in thickness from a few meters to 10–15 m.

13.4.2.3 Dornod-Mongol Belt

This belt is characterized by bimodal continental basalt-rhyolites and interbedded continental sediments. The belt is dominated by Proterozoic gneisses, schists, and anatectic granites, occurring as inliers in early Paleozoic volcanic and sedimentary rocks. Silurian and Devonian sandstones and shales are widespread. Upper Paleozoic volcanic and molasses unconformably overlie in earlier Proterozoic and Paleozoic rocks.

U-Mo mineralization is located predominantly in the volcanic depressions of the Mongol-Priargun volcano-plutonic belt such as the Tulukuev (SE Trans-Baikal area) and Dornod in Mongolia (Khomich et al. 2016). The Mongol-Priargun belt is also considered as the outer part of the large Verkhneamur marginal-continental volcano-plutonic belt, whose inner axial part is composed of the Bolshoi Khingan volcano-plutonic belt (Gordienko et al. 2000).

North Choibalsan ore district consists of the Gurvanbulag, Dornod, Mardaingol, and Nemer uranium deposits in Upper Mesozoic volcanic sediment-filled depressions or downfaulted crustal blocks (Fig. 13.3a). The stratigraphic profiles of these deposits are similar but not identical. Each has a lower succession of sedimentary rocks, basalts, intermediate tuffs, and rhyolites, capped by a thick upper succession of a typical bimodal basalt-rhyolite association. The volcanic rocks erupted during the Late Jurassic to Early Cretaceous. Imprecise K-Ar dates indicate that the time in which the volcanic rocks erupted may have spanned 5–35 million years. The Proterozoic to Late Paleozoic basement of North Choibalsan consists of gneisses, migmatites, amphiboles, granites, aplites, and pegmatites. Strong Na and K metasomatic, propylitic, and argillic alteration may be found along faults, some of which are radioactive.

The Upper Mesozoic volcanic-sedimentary Dornod complex (Bayantumen Formation) is divided into Lower, Middle, and Upper sub-complexes (Fig. 13.3b), separated by breaks and unconformities. The Lower sub-complex is an intercalation of layers of quartz-feldspar porphyries, trachydacites, and basaltic andesite with horizons of tuff sedimentary rocks. The total thickness of the sub-complex varies from 150 to 450 m. The Middle sub-complex comprises structurally and texturally varied layers of volcanic rocks, with a total thickness of 300–800 m. The Upper

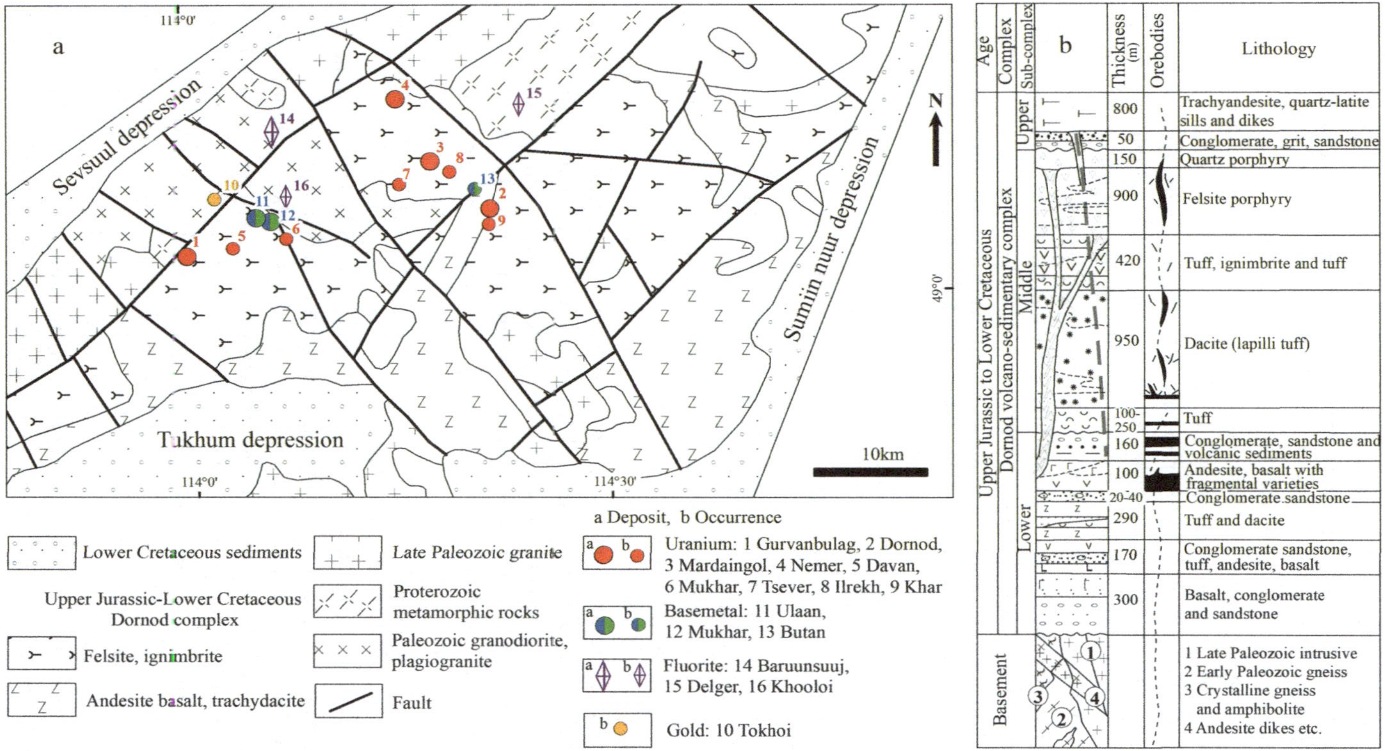

Fig. 13.3 (**a**) Simplified geology of the North Choibalsan ore district and distribution of deposits and occurrences. (Modified after Rogov and Rubzov 1989, Selzov and Vishnyakov 1990, Mironov et al. 2009). (**b**) Simplified lithostratigraphy of Dornod complex and position of U mineralization. (Modified after Rogov and Rubzov 1989, Mironov 2003, Agnerian et al. 2007)

sub-complex includes layers of trachyandesite a few hundred meters thick. The obsidian horizon is a relatively thin sheet of extensive lateral distribution. It can be traced in outcrop for 15–20 km and extends an unknown distance down dip. Mironov (2006) reports an area of 159 km^2 at Gurvanbulag level. Where known, the glass ranges from 0 to 12 m thick but is mostly 3 to 8 m.

The Gurvanbulag, Dornod, Mardaingol, and Nemer deposits occur in three tectonic blocks, respectively known as Ulaan, Erkhet, and Mardaingol. The blocks are separated from each other by prominent block-defining faults, some with measurable offsets of stratigraphic contacts. Alteration can be divided into two categories, such as high temperature in basement rocks that predates most of the volcanic cover and low temperature alteration, when uranium introduction after Late Mesozoic volcanism had ceased. The most common minerals are hydromicas, mixed layer micas, montmorillonite, carbonates, kaolinite, chlorite, feldspars, sericite, epidote, coffinite, pitchblende, uranophane, fluorite, and molybdenite. The intensity of alteration has been selective, determined by porosity, rock composition, and brittleness. Superimposition of one kind of alteration on similar but earlier alteration greatly complicates the history. The mineralization consists of cross-cutting veins (Khavar, Mardaingol) in overall stratiform sheet-like bodies but which have disconformable internal stockwork structure (Dornod) and conformable bedding plane (Gurvanbulag).

Occurrences and deposits of base metals, gold, fluorite, and coal also occur in the North Choibalsan area. Base metal and fluorite are hosted in veins and in the basement, as well as in volcanic sedimentary rocks, whereas gold occurs in the Early Mesozoic diorite and granodiorite.

Berkh ore district comprises the Ulziit-Saikhan Uul, Ikh Bulag, Nimgen, and Batnorov volcanic tectonic structures with associated fluorite and gold mineralization. The fluorite deposits are closely related to uranium. The ore district is divided into the Batnorov and Ulziit-Saikhan Uul ore clusters. The uranium content ranges from 5.8 to 11.0 g/t U in rhyolite at Ulziit-Saikhan Uul, from 4.8 to 5.7 g/t U in rhyolite at Khentii, and from 4.1 to 5.7 g/t U in terrigenous sediments at Nimgen. These ore clusters are composed of felsic lava, volcanic tuff, and ignimbrite. The Ikh Bulag and Tanai uranium occurrences and radioactive anomalies were identified in the southwest part of Batnorov ore cluster.

Choir ore district is located in the 150-km-long, 10–25-km-wide sedimentary basin filled with Cretaceous sediments enclosing highly radioactive Mesozoic granites. The sandstone-type Kharaat uranium deposit and many other occurrences were located by radioactive anomalies within sediment sequences as a result of oxidation.

Shivee ore district comprises the Nyalga, Shivee, and Tavansuveet ore clusters classified by Mironov (2006) and the Bayan Airag basin. The Engerar sandstone-type uranium deposit and Durtai, Tsaidam Nuur, Naidal, Naranbulag, and Elgen occurrences were discovered in this district by follow-up of radiometric anomalies. Th, REE, fluorite, Cu, and W occurrences were also identified. The Shivee volcanic tectonic structure is the main structure for uranium mineralization within this district. The rhyolite contains 5.3 g/t U and 33 g/t Th, while volcanic glass contains 17 g/t U and 53 g/t Th. The mafic volcanics contain 2.5 g/t U and 9 g/t Th, and the granite-gneissic massive basement contains 17 g/t U and 1.7 g/t U.

Ulziit ore district basin is filled with Lower Cretaceous-Upper Cretaceous-Lower Paleogene sediments. The basin was formed during the late-stage Mesozoic tectonic-magmatic activation as a result of vertical block movement. Ulziit uranium deposit is located in the northwestern part of the basin and composed of crystalline basement and various age granitoids, as well as Jurassic and Lower Cretaceous volcanic rocks. The surrounding crystalline basement of the deposit exhibits higher uranium grade. The Middle-Upper Jurassic granite and granosyenite were likely a significant source of uranium in the Late Cretaceous sedimentary sequences in the basin. The radioactive anomalies were identified within the basin, and uranium mineralization with grades between 0.01 and 0.05% U occurs around 180 m depth within the sequences.

Dundgovi ore district is characterized by many radioactive anomalies and rare volcanic occurrences. The basement of this district is composed dominantly of Late Proterozoic carbonate rock.

The Khairkhan ore district is hosted in an elongate basin extending approximately 150 km southwest to northeast and 10–25 km southeast to northwest and filled with Cretaceous sediments. Lens and tabular-shaped ore bodies accumulated in a paleovalleys as a result of supergene processes. The Khairkhan basin is composed of Cretaceous lignite containing gray sediment.

Bor-Undur ore district included the Bor-Undur, Ulaan Nuur, and Khongor uranium ore clusters. The Bor-Undur I, Bor-Undur II, and Zurkh uranium deposits overlap with a coexisting fluorite ore cluster, and occurrences were identified (Banzragch and Chuluundorj 1979). These occurrences are surrounded by regional faults and characterized by radiochemical anomalies. The basement of the ore clusters is composed of Lower Paleozoic granite and Lower Proterozoic metamorphic rocks and is covered by Upper Jurassic-Lower Cretaceous volcanic and Lower Cretaceous molasses. The Ulaan Nuur ore cluster occurs in an elongated 30-km-long volcano-tectonic structure and in a granite-metamorphic basement covered by volcano-sedimentary rocks. Recently, the resources have been established for the Ulaan Nuur uranium deposit. More than 20 primary and secondary uranium occurrences have been identified in the cluster.

13.4.3 South Mongolian Province

The south Mongolian province corresponds to the Ordovician to Carboniferous south Mongolia-Khingan collage (Fig. 13.1). The south Mongolian province can be divided into Gobi belt and southwest Mongolia belt, which are covered by Cretaceous sedimentary rocks in the eastern Mongolia and Cenozoic sedimentary rocks in the western part of belt, respectively.

13.4.3.1 Gobi Belt

The Gobi belt is composed of Proterozoic and Paleozoic sediments and volcanics, which were intruded by Middle-Late Paleozoic and Mesozoic intrusions hosting Cu, base, rare metal, fluorite, and REE mineralization. Cretaceous sediments are dominantly within the belt and contain coal, uranium, petroleum, and gypsum deposits and occurrences. The belt is classified into Dornogovi, Zuunbayan uranium basin, Baruunkhuurai, Nemegt, Mandal-Ovoo, Undurshil, Sainshand, Ail, Matad, and Sulinkheer perspective districts.

The Dulaan Uul and Zoovch Ovoo uranium deposits occur in Upper Cretaceous sediments. Numerous other radioactive anomalies were identified. The Narst uranium occurrence is hosted in Lower Cretaceous sediments. Sandstone-hosted uranium deposits are also important within the belt. Small-scale occurrences are located along faults and volcano-tectonic structures. Uranium could be accumulated in Cenozoic sediment as well.

The basement of the Mandal-Ovoo ore district consists of Proterozoic and Paleozoic sediments and volcanic which were intruded by Middle-Upper Paleozoic and Mesozoic intrusions with Cretaceous and Cenozoic sedimentary cover. The Ulziituul and Teeg uranium occurrences are related to volcanic rocks. The Khashaat, Gerovoo, and other occurrences occur as uranium phosphate mineralization.

The Uushiin Nuur, Ongigol, and Mushgaikhudag uranium ore clusters have also been identified. The Ondorshil ore district is classified based on the radioactive anomalies. Uranium occurrences and anomalies were identified in Cretaceous sediment sequences. The Nukhetdavaa ore district is characterized overlapping radioactive zone. The Saikhan Uul and Tsagaan Uul uranium occurrences occur within the Yuguzer tungsten deposit area.

The Sainshand ore district covers an area in the Sainshand and Dornogovi basins. Deposits include the Narst, Ingen, Most, and Dorvoljin uranium occurrences and many uranium anomalies. The basin is composed of red-colored Cretaceous-Neogene, Upper Cretaceous, and overlying unconformable Lower Cretaceous sediments. Uranium-bearing Upper Cretaceous sediment is 40–50 m thick and consists of sand clay over a length of 6 km. The mineralized zone is 800–100 m long, 100–400 m wide, and 0.3–7.0 m thick, with uranium grades between grade range 0.02 and 0.67% U.

The Matad ore district is characterized by the distribution of Cretaceous sediments within the Tamsag radioactive zone. A uranium anomaly is associated with an area of 150x400 m at the Zuunbulag lignite occurrence. Radioactive anomalies extend over 16-km-long strike which includes the Zuunurt Uul and Bayan Soum-9 occurrences. This ore district is also having potential for petroleum, and Toson Uul and Tamsag petroleum deposits were discovered at a depth approximately 2000 m. The Zuunbayan ore district and/or basin comprise Zuunbayan, Unegt subbasins, and western extension of the Sainshand basin. The Dulaan Uul sandstone-type deposit was discovered in Unegt subbasin, and the Zoovch Ovoo deposit is in the Zuunbayan subbasin. The basin is filled with Upper Cretaceous Sainshand

Formation of sandy terrigenous molasses over a basement of the Jurassic Sharil Formation. The upper member of Sainshand Formation hosts uranium accumulation in sediments and is 200–300 m thick. It is composed of fine- to coarse-grained sands deposited in paleovalleys.

The Ail ore district occurs in a volcanic tectonic structure in a radio-geochemical anomalous zone and hosted in Cretaceous sediments. The Khad, Ail, and Elgen occurrences were identified in this ore district and are considered to be a prospective for uranium mineralization.

13.4.3.2 Southwest Mongolia Belt

This belt is comparably less studied. Silurian and Devonian sediments and monzonite, gabbroic syenite, and ultramafic intrusions are widely distributed in the belt and are characterized by Au, Ag, and base metal occurrences. There is potential for uranium mineralization occurring within fault structures and also for Cretaceous or later-age sediment sandstone-hosted uranium mineralization.

The Nemegt ore district is prospective for sandstone-hosted uranium deposit based on the occurrences of radioactive anomalies and lithology structure criteria. Lower Cretaceous sediments are distributed in the northern part of the district and upper Cretaceous sediments in the south. They are rarely covered by Paleogene and Neogene sediments. The Songinot (Bayanundur) area in Nemegt ore district is composed of Late Paleozoic and Early Mesozoic leucogranite, Devonian terrigenous sediments and volcanics, and Jurassic molasse and coal-bearing Cretaceous sediments. The Sn, Mo, fluorite, Zn, Ag, and REE mineralization occurs in the Songinotuul area with mineralization grades 21.3 ppm Th, 3.4% K, and 5.2 ppm U. South of this ore district the Burkhan-Khailaast Uul area occurs. This has similar geology and mineralization and is composed of Late Permian-Early Triassic age lithium-rich intrusive rocks and pegmatites and metasamotites with quartz veins hosting Sn, Be, Li, and REE mineralization.

Based on the regional geological and airborne geophysical survey, anomalies of U and Th were identified in the Baruunkhuurai ore district. NE-SW-trending, highly anomalous radioactive rock consists of Carboniferous felsic intrusion and Mississippian-Pennsylvanian volcanics and sediments covered by Quaternary sediments. The range of gamma radiation was from 20 nGy/h to 95 nGy/h, whereas mineralization grade is 1.8–5 g/t U, 2–14 g/t Th, and 0.5–3.8 g/t K (Dagvadorj and Munkjargal 2018).

According to the airborne gamma-ray spectrometry survey in the Gobi Tengeruul ore district, the gamma radiation was 90 nGy/h, with grades of 7.0 ppm U, 12.0 ppm Th, and 2.6% K in gabbro, granodiorite, and granite of Early Carboniferous age. The Gobi-Tengeruul, Late Carboniferous Soyottolgoi complexes were emplaced in a continental margin tectonic environment, and these anomalies are overlapped by Silurian-Devonian and sometimes Cretaceous sediments. The Cu, Pb, Zn, Au, Mo, Mn, and Th anomalies are identified in these rocks in the south of Altai ridge, where gabbrodiorite, diorite, monzonite, monzodiorite, and granodiorite, granite (saturated aluminum, alkaline, mostly K-rich) occur. The age is 345 ± 2 Ma (Economos et al. 2008), determined by U-Pb method on zircons. Kröner et al. (2010) determined an age of 399 Ma. The south Mongolian complex consisting of early Permian subalkaline gabbro-monzodiorite and granite in the Khatansuudal-Taliinmeltes district located at the Mongolian-Chinese border zone is relatively high radiometrically and associated with Pb, TR, Au, Zn, Cu, Pb, Mo, and Sn geochemical anomalies. Felsic and alkaline sub-volcanics are distributed in the complex.

13.4.4 Age of Uranium Mineralization

Recent discoveries and research on uranium suggest that there are three dominant epochs of uranium ore formation and mineralization in Mongolia.

Most of the known volcanic-related uranium deposits (Dornod, Gurvanbulag, Mardaingol, and Nemer deposits) were formed during the Late Mesozoic epoch. The main stage of hydrothermal activity in the Dornod volcanic structure was dated at 140 ± 3 Ma by the K-Ar method for mica and 135–138 Ma by U-Pb for uranium minerals. Uranium deposit formation is related to the Late Mesozoic tectono-magmatic reactivation of continental structures triggered by the development of the Pacific mobile belt. The second, Cenozoic epoch, is characterized by the formation of uranium mineralization in Mesozoic and Cenozoic sediments related to the secondary groundwater transportation and stratigraphic redox processes (Kharaat, Khairkhan, Zoovch Ovoo, and Dulaan Uul deposits). The Cenozoic epoch corresponds to arching and blocks movements within the global orogenic belt. Secondary uranium deposits have been relatively recently discovered in Quaternary alluvial sediments (Uyakhan and Altankhuyag 2015).

Uranium mineralization occurs in various geological settings and ages in Mongolia. Metallogenic epochs and related mineralization in Mongolia can be classified as Late Proterozoic to Middle Paleozoic U with some Mo and W occurrences. The Late Paleozoic veins W and W-Mo; Th in pegmatites; Early Mesozoic W, Sn, and Mo; high U with Zr-REE-Th, Li, and Be and U in crush zones in granites. For example, Proterozoic granites, granite gneisses, rarely host uranium in the north Mongolian provinces, whereas the Paleozoic carbonaceous cherty rocks contain uranium which is most likely of synsedimentation origin.

13.5 Geological Classification of Uranium Deposits

Uranium deposits in Mongolia can be classified based on IAEA classification scheme (Bruneton and Cuney 2016) and described in order from deep primary magmatic deposits to sedimentary and surficial deposits following the geological cycle (Cuney 2010). Subtypes can be classified according to definition by Dahlkamp (2009). Volcanic-related, sandstone, and surficial type of uranium deposits are found, in addition to granite-related, metasomatite, lignite-coal, and phosphate types of uranium mineralization (Jamsrandorj et al. 2016; Jamsrandorj and Altankhuyag 2016).

According to the investigation of confirmed sandstone-type uranium deposits (OECD(NEA)/IAEA 2004) and oil gas, coal exploration, and excavation, 82% of the total sandstone-type uranium deposits coexist with confirmed oil-gas fields and coal fields in the same basins (Lui et al. 2007). Energy resources, such as coal, oil, and uranium, are supplied from large sedimentary basins in Mongolia. In recent years, sandstone uranium deposits closely associated with hydrocarbon-bearing basins have been identified in the Zuunbayan basin, and similar basins are found at Matad, eastern Mongolia and Ail, and southern Mongolia. Energy basins and sub-basins distributed in territory of Mongolia are shown in Fig. 13.4.

Volcano-sedimentary, sedimentary basins, and grabens cover much of southwest, south, and northeast Mongolia as well as adjacent region in China. Pirajno (2013) discussed the Junggar, Turpan-Hami, and Tarim basins in northwest, the Ordos and Qaidam basins in central, and the Bohai and Songliao basins in eastern China and, where warranted, its respective uranium mineral systems. These basins are economically important for hydrocarbon resources and sandstone-hosted U deposits in both countries.

13.5.1 Volcanic-Related Uranium Deposits

This type of uranium deposit is commonly located within volcanic calderas filled with bimodal volcanic and volcanoclastic sediments (Kochkin et al. 1985; Rogov and Yurchenko 1987; Mironov et al. 1990). Deposits are structurally controlled and strata-bound and unconformable mineralization. In Mongolia, medium-to-low temperature hydrothermal uranium mineralization is found in the Late Mesozoic volcano-tectonic structures and the underlying crystalline basement.

Volcanic-related type of deposits is characterized by fluorite-molybdenite-pitchblende and pitchblende-fluorite-apatite mineralization in Mongolia. The mineralization is closely related to the bimodal rhyolite-basalt magmatism in the Argun-Mongolian volcanic belt (Mironov et al. 1990; Chuluun and Oyunbaatar 1994).

The coincidence of the slab boundary projections and area of their influence with the spatial position of large F-Mo-U ore clusters (Streltsovska, Dornod and Guyuan-Doulung) separated from each other by hundreds of kilometers is regarded as evidence for the possible influence of deep geodynamics (Khomich and Boriskina 2018). Byamba et al. (2018) also concluded that the bimodal volcanic tectonic zone was formed by the influence of a mantle plume in Dornod area. Dornod, Gurvanbulag, Mardaingol, Khavar, and Nemer deposits and numerous occurrences in northeastern and central Mongolia are volcanic-related type. These uranium deposits are found in Upper Jurassic-Lower Cretaceous volcanic and sediments, characterized by rhyolite and basaltic andesite composition, named as the Dornod

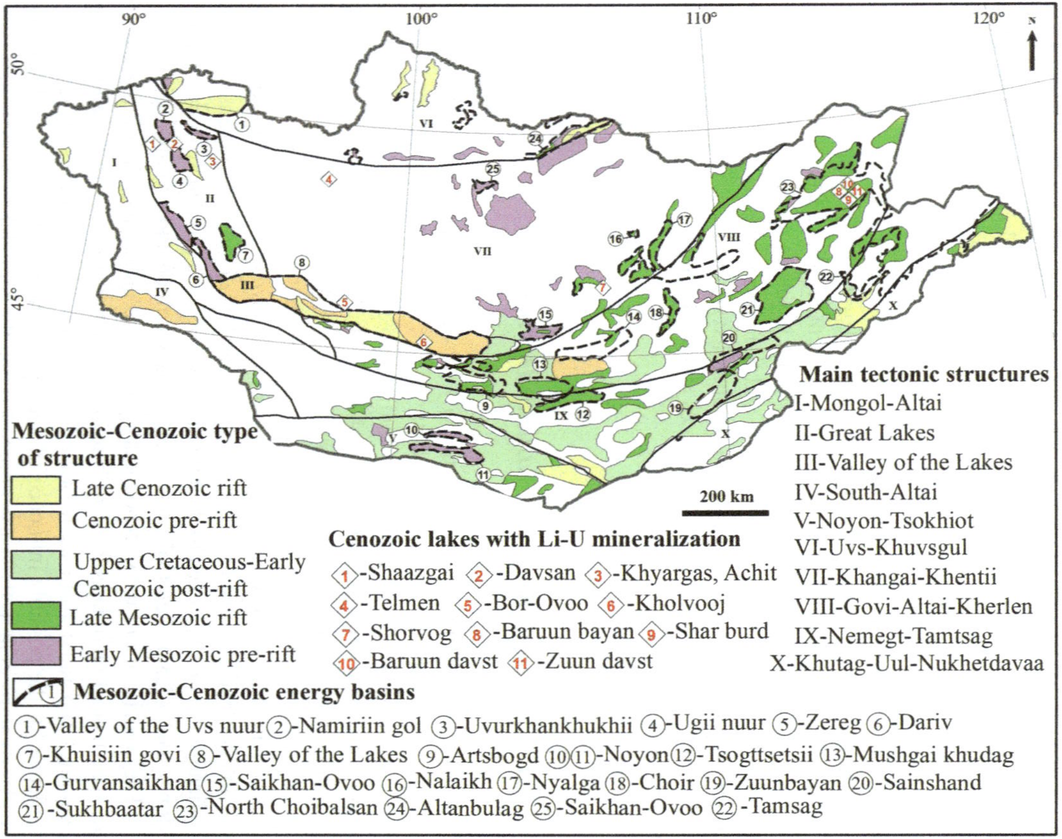

Fig. 13.4 Distribution of energy basins and main Mesozoic-Cenozoic tectonic structures. (After Badamgarav et al. 2011)

complex (Bayantumen Formation). Ore bodies of those deposits are distributed in similar geological settings and levels.

The U-F-apatite type is represented by rather widespread prospects and occurrences within volcano-tectonic structures and in superimposed depressions of southwest part of the Dornod Mongolian belt. The Ulaan Nuur deposit in the Bor-Undur ore district is characterized by mineralization hosted in the brecciated and fractured rocks within steeply dipping fault zones. The vein-like and stratiform bodies consist of collophane and francolite. The U grade is low, and nonuniform, orebodies are small in size, often as pockets and small lenses. The P_2O_5 content reaches 10%.

The most important volcanic-related uranium deposits of the North Choibalsan ore district (Dornod, Gurvanbulag) are considered in this chapter.

13.5.1.1 Dornod Deposit

The Dornod deposit is hosted in volcanic rocks, composed of amygdaloidal and brecciated basalts and andesite, within a volcanic-sedimentary sequence as much as 1500 m thick. This sequence comprises a series of subhorizontal flow lavas and pyroclastic and sedimentary rocks. The Dornod deposit comprises a number of separate uraniferous horizons spread over an area of 1000 m by 500 m. The most continuous zone is a 30–40 m thick, tabular body of high-grade uranium mineralization, that occurs at vertical depths between 410 m and 450 m below the surface. Uranium mineralization occurs as pitchblende-coffinite assemblages associated with carbonaceous partings and fragments in areas of structural preparation (Jamsrandorj et al. 2015). Within the brecciated andesite, uranium mineralization is commonly associated with thin slivers and discontinuous seams of carbonaceous material which are incorporated within the quartz-carbonate cement and as disseminations and small impregnations within highly strained and cracked volcanic blocks and fragments. A zone of high-grade uranium surrounded by a halo of lower-grade mineralization has been identified in a single thick horizon. The uranium mineralization was deposited in two stages (Mironov 2006). The first stage is evident as low-temperature metasomatically altered rocks. The alteration displays zoning, with an outer kaolinite-montmorillonite argillic assemblage, a central mica assemblage, and an inner hematite-bearing quartz-mica-chlorite assemblage. The second stage of fissure filling is subdivided into the (1) pre-uranium mineralization, (2) uranium ore mineralization, and (3) post-uranium mineralization substages. Quartz-fluorite-sulfide veinlets with galena, sphalerite, and pyrite were formed at the pre-mineralization substage. Coffinite and pitchblende in association with galena, molybdenite, and pyrite precipitated at the uranium ore mineralization substage, and t-quartz-carbonate-kaolinite stockwork with marcasite characterizes the post-mineralization stage. Uranium-bearing keritic carbonaceous material was detected in the ore in addition to coffinite and pitchblende. Molybdenum is of economic importance in the stockwork zone together with uranium. Uranophane, autunite, uranotil, and uranospinite occur in the zone of oxidation. The temperature of ore formation of the Dornod deposit was 120–170 °C. Hematite alteration is directly associated with uranium mineralization, and the intensity of alteration correlates positively with the ursolvedanium ore grade. Uranium mineralization is associated with Pb, As, and Mo geochemical anomalies which form extensive haloes around the fault-controlled mineralization.

13.5.1.2 Gurvanbulag Deposit

The geological basement of this deposit consists of Proterozoic metamorphic rocks and Paleozoic intrusions. The Proterozoic rock is composed of amphibolite, biotite-hornblende and hornblende crystalline slates and schists, and biotite-muscovite and biotite-sillimanite gneisses with intercalations of quartzite, calciphyre, and marbleized limestone. The Paleozoic rocks consist of gabbro-diorites, diorites, granodiorites, and leucocratic biotite granites with many conformable contacts with the host metamorphic rocks and characterized by retaining the original gneiss foliation and by the presence of xenoliths of unaltered substrate rocks.

The Upper Mesozoic sedimentary-volcanic deposits of the Dornod complex (Bayantumen Formation) are divided into the lower, middle, and upper sub-complexes, separated by breaks and unconformities. The lower sub-complex is an intercalation of layers of quartz-feldspar porphyries, trachydacite, and basaltic andesite with horizons of tuff sedimentary rocks. The total thickness of the sub-complex varies from 150 to 450 m. The middle sub-complex comprises structurally and texturally varied layers of volcanic rocks, with a total thickness of 300–800 m. The upper sub-complex includes layers of trachyandesite a few hundred meters thick (Rogov and Yurchenko 1987). The Gurvanbulag uranium deposit is 1.5–4 km wide, 9 km elongated to NE, and dipping down 5–15 degree to the SE and comprises of central, intermediate, and southwest ore zones. The central zone of Gurvanbulag deposit is situated at the elevation of 620–980 m and contains 86% of total tonnage of resources the deposit.

The Gurvanbulag deposit occurs in close spatial association with a volcanic glass horizon, up to 15 meters thick, within the middle sub-complex. Mineralization is primarily hosted in the vitroclastic, ashy felsite tuffs in the immediate footwall to the obsidian horizon (Fig. 13.5) but is also locally well-developed in the hanging wall felsite. The obsidian occurs at the base of the upper felsite. Thin beds of tuffs may intervene, but mostly the obsidian (150 Ma) is in contact with felsite. The obsidian layers variably mineralized and commonly barren.

The felsite is hard and brittle mainly because of high silica content (72 to 78% SiO_2) and the fine grain size. Drilling shows them to be intensely fractured resulting in breccia-like textures in drill core. Bleaching can be observed along fractures. On the open-fracture surfaces, white and beige clay minerals, ochre-colored iron hydroxides, uranophane, blue and purple fluorite, and black manganese oxides may be seen. Widespread pervasive mica alteration is developed in the volcanic rocks. They represent pre-uranium mineralization alteration, main stage uranium mineralization, and post-uranium mineralization alteration. The earliest alteration is represented by widespread micas and carbonates. The micas are mainly illite and

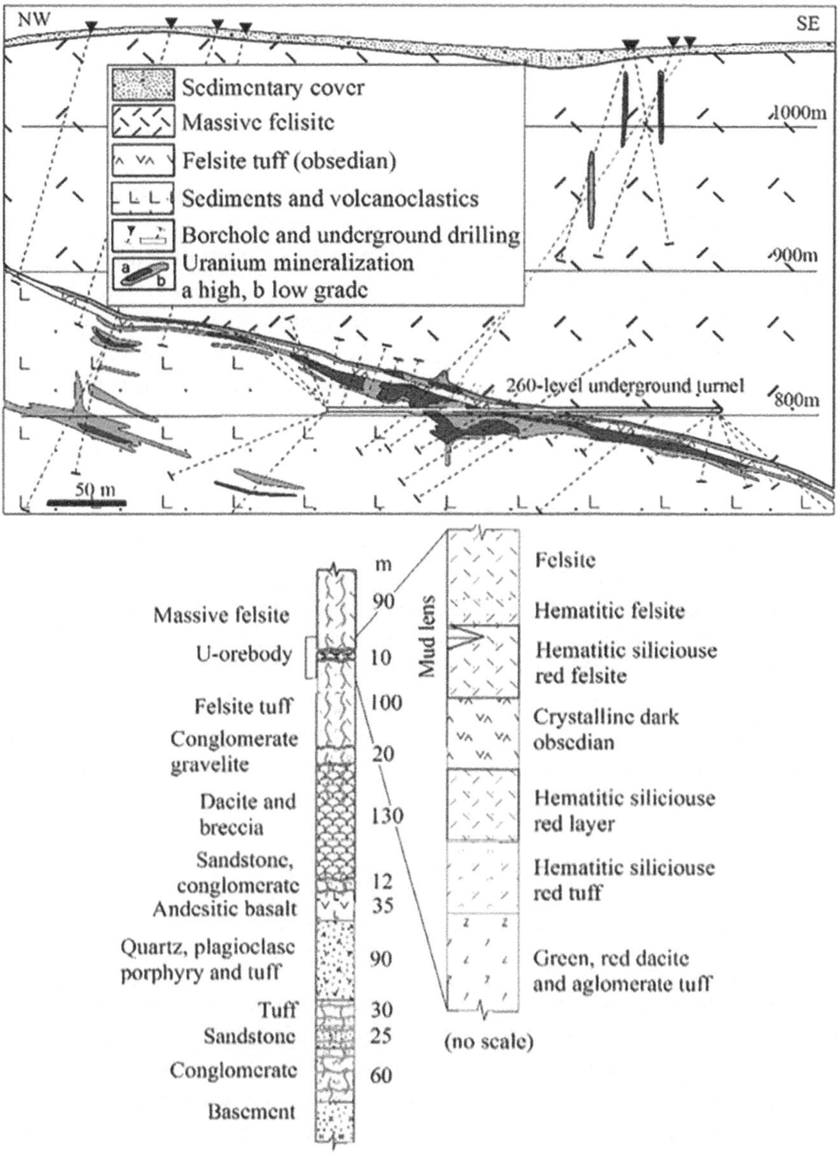

Fig. 13.5 (**a**) Simplified cross section, (**b**) stratigraphy of the Gurvanbulag U-deposit, position of the Main zone. (After Lee and Nowak 2006 and Li 2016)

sericite formed from the felsite and occurring as minute flecks and elongate wisps. K-Ar dating on micas from several rock types has yielded consistent ages in a narrow range between 143 and 141 Ma.

The main uranium minerals are coffinite, uranophane, and uraninite/pitchblende. Alteration accompanying uranium mineralization is generally weak, being limited to silicification and hematite and chlorite. Some fluorite may have been introduced at this stage. Post-uranium mineralization alteration minerals are quartz, fluorite, micas, chlorite, kaolinite (dickite), montmorillonite, zeolites (analcime), and carbonates. Alteration in the main mineralized zone at Gurvanbulag consists of white, putty-like clay minerals, silicification, deep red hematite, deep red feldspars, chlorite, carbonates, albite, analcite, and fluorite.

Aside from molybdenite, other sulfides occurring in small quantities including galena, sphalerite, and chalcopyrite. However, in a volcanic setting, they do appear to be a special case in that rhyolites are always dominant. What sets the Dornod complex apart from all other complexes is the presence of a laterally extensive volcanic obsidian horizon below massive the felsite/rhyolites at Gurvanbulag and the laterally extensive bedded conformable uranium mineralization in parts of this glass horizon (Fig. 13.5). The highest-grade uranium concentrations found to date occur in the upper parts of the lower sub-complex (Dornod deposit) and the lower part of the middle sub-complex (Gurvanbulag deposit).

Also, this mineralization correlates positively with the most complete and thickest sections of the lower sub-complex. Veins occur most commonly in volcanic rocks associated with steeply dipping sub-volcanic intrusions, feeder pipes, or fissures. The highest-grade mineralization occurs in bedded conformable and sheet-like bodies. These bodies include intraformational faulting and cooling and contraction fracturing. The grain size of the host tuff and breccia has produced increased permeability. This has resulted in increased fluid movement resulted in chemical alteration and precipitation by reductants including carbonaceous plant remains, iron sulfides, and titanium oxides.

13.5.2 Sandstone-Hosted Uranium Deposits

The sandstone uranium occurs in medium- to coarse-grained terrigenous sedimentary rocks in Late Mesozoic-Cenozoic basins in Mongolia (Fig. 13.6). This type of uranium mineralization is closely related to uranium-enriched source rocks and permeable sandstone aquifers. The Kharaat, Khairhan, Dulaan Uul, Zoovch Ovoo, Gurvansaikhan, Ulziit, and Engerar deposits and many other occurrences are situated in the Choir, Khairkhan, Zuunbayan, Unegt, and other basins of the southeast Gobi. Typically, sandstone-hosted uranium occurrences are generated by reduction processes. The highest grade of mineralization is hosted in the medium- to course-grained sandy, lacustrine sediments. Sandstone-hosted uranium deposits of Mongolia can be classified roll-front and basal channel or paleovalley subtype (Budunov 2002; Hocquet 2016).

The most important sandstone-hosted uranium deposits of the Gobi belt (Dulaan Uul, Zoovch Ovoo) are considered in this chapter.

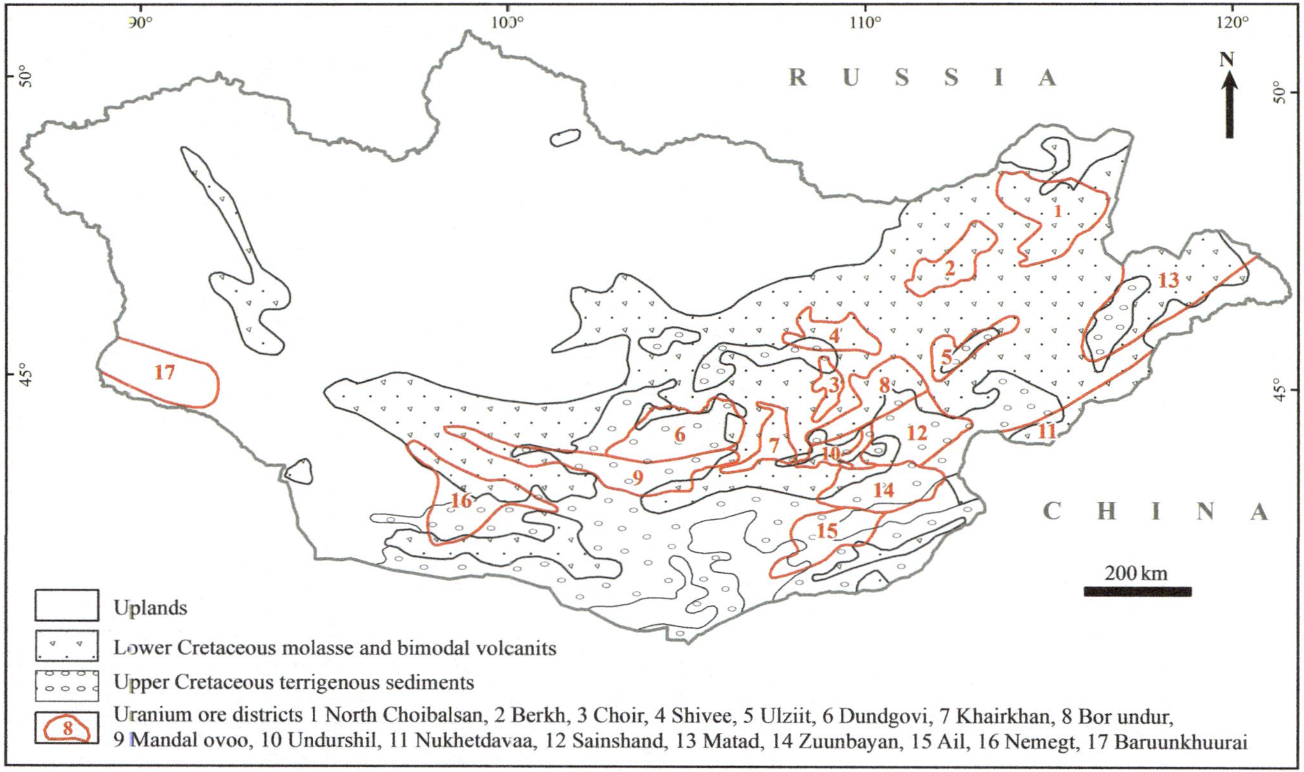

Fig. 13.6 Distribution of Cretaceous sediments and uranium ore districts and/or basins. (After Makhbadar schematic map in Byamba 2011, Byamba et al. 2018)

13.5.2.1 Dulaan Uul and Zoovch Ovoo Deposits

The Dulaan Uul and Zoovch Ovoo deposits have been well studied. A 1:200,000 scale geological mapping program was conducted throughout south Mongolia in the period of 1979–1985, and numerous radiometric anomalies were detected within the basins. An airborne geophysical study (Schmelev et al. 1982) was completed at a scale of 1:200,000. Based on the results of this survey, supergene uranium occurrences were identified within Cretaceous conglomerate, sandstone, and clay-rich sedimentary formations.

Deposits are hosted within terrigenous formations of the Upper Cretaceous which comprise the Sainshand basin, which is a post-Late Cretaceous-Early Cenozoic rift structure according to the Mesozoic-Cenozoic tectonic classification of Mongolia by Badamgarav et al. (2011). As a result of the neo-tectonic movement, the Zuunbayan depression separated into two subbasins and created the horst-graben system separated by a major fault. Due to the subsequent development stages of the rift, the valleys were covered by sediments of the Sainshand and Bayanshiree Formations.

The Unegt and Zuunbayan subbasins were identified as above. The Unegt subbasin, where the Dulaan Uul deposit is located, is isolated from the Zuunbayan subbasin by the regional deep fault (Fig. 13.7). The Dulaan Uul is the sedimentary deposit formed in relation to the Upper Cretaceous Sainshand Formation. This formation has non-permeable thin argillaceous layers dominated by sand units of uneven thickness. It can be divided into U0-U3 layers. An oxidized zone of the formation is distributed mainly in the south of the valley, whereas a redox zone is occurring mainly in the north and also in the central part of the Unegt subbasin.

The Dulaan Uul deposit comprises Myagmar, Australia, and southeastern and southwestern uranium accumulations which are lithologically different and described below (Jamsrandorj et al. 2013). Myagmar is located mostly in the U4 layer which is close to the surface, and also mineralization occurs in the U2 layer. The uranium accumulation extends from the southwest to the northeast as an oval-shaped ore body with thicknesses between 1.5 and 18.5 m. A total 2688.01 tonnes of U reserve was determined with an average grade 239.47 g/t U. The Australia deposit is hosted mainly within the U2 layer with an area of 695.7 thousand m^2. Exploration has identified ore bodies with average thicknesses of 1.5–8.5 m containing an estimated resource of 482.87 tonnes of U, with an average grade of 407.97 g/t U. The southeastern deposit is hosted mainly in the U2 layer and is laterally extensive in a north-south direction. The thickness of the ore bodies varies between 1.55 and 7.5 m. According to the results of 116 exploration wells drilled where uranium reserves were identified with an average grade of 293.67 g/t under the measured and indicated resources. A few bodies are hosted in the U3 layer. The southwestern deposit is hosted mainly in the U2 layer and has an isometric shape with an area of 7746.9 thousand m^2. The thickness of ore bodies ranges between 1.55 and 20.0 m, and a total of 2237.36 tonnes of U reserves were estimated with an average grade of 175.03 g/t U (Hocquet 2016; Batbayar et al. 2016, 2018).

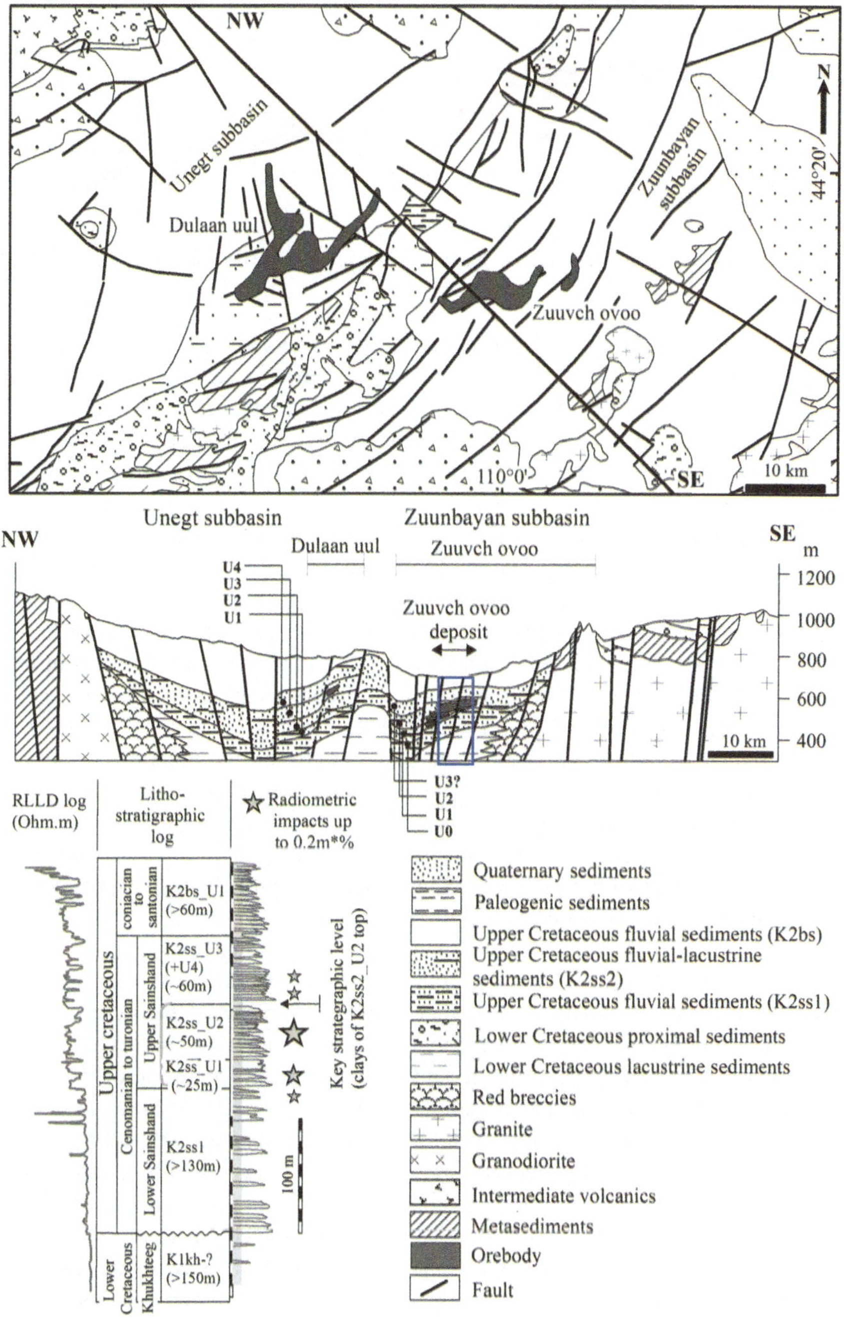

Fig. 13.7 Geology map of the Zuunbayan area and cross section of Unegt and Zuunbayan subbasins. (After Hocquet 2016)

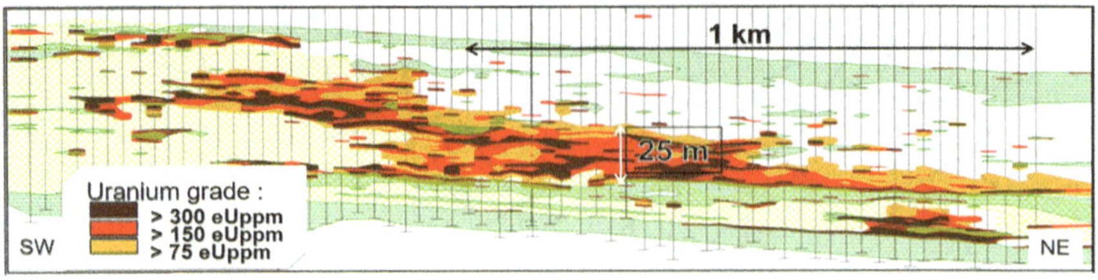

Fig. 13.8 Cross section and distribution of uranium of Zoovch Ovoo deposit. (After Hocquet 2016)

The source of the Zoovch Ovoo deposits is the granitoid surrounding the valley to the southeast and south. Granitoid magmatism occurred in several stages. Here are Devonian, Middle-Late Carboniferous, and Permian complexes as well as medium-size granite blocks of the Mesozoic. The uranium in the Zoovch Ovoo deposit in the Zuunbayan basin occurs mainly within the in U1 and U2 layers of the Sainshand Formation upper member. The thickness of the sandy layer with the main uranium ore ranges between 35 and 40 m. The host unit has a uniform composition with small amounts of organic material. The lower non-permeable layer is composed of red argillaceous sediments of the Sainshand Formation. The upper non-permeable layer is 6–16 m thick consisting of red sediment located in the upper section of the U2 layer. The upper water non-permeable layer isolates the ore-bearing sediment from the general oxidation zone. The uranium mineralization in the sediment reaches up to 20 km in extension from southwest to northeast.

The oxidation-redox boundary is the section between the unoxidized and oxidized sandstone (Fig. 13.8). In the oxidized zone, gray-yellow sandstone, pyrite, kaolinite, and rarely hematite are present at the edge of the front. The unoxidized sand is composed of gray feldspar and coal chips, and higher radioactivity levels are observed at the base of the sandstone. Limonite spots, initial hematite spots, kaolinite, and pyrite occur within the mineralized zone.

Based on the gamma-ray spectrometry mapping, radioactive anomalies were determined over the Late Paleozoic granitoid. The southeastern border of Unegt-Zuunbayan is considered to have higher potential than the northern and western edges. Therefore, it may be considered that it was the predominant source of uranium migration. The neo-tectonic structure and hydrogeological condition of this valley played important roles for the accumulation of uranium. This shows that stratigraphic factors are extremely important for mineralization in the Dulaan Uul deposit.

Petrographic thin section analysis shows that host sandstones are relatively well sorted and coarse grained. The sandstone unit is composed of quartz, feldspars muscovite, biotite, organics, and some xenoliths of rounded felsic and mafic rocks. Zircon, apatite, iron oxide, iron-titan, and rutile are present as accessory minerals, whereas tourmaline and pyrite are rarely seen. Electron microscopy analyses on the uranium-bearing samples indicated that uranium occurs in the lignite and rarely uranium-iron micro grains in the quartz grain.

13.5.2.2 Kharaat

The Kharaat uranium deposit is located in a Late Mesozoic extensional basin, which is one of the prominent geological and topographic features of central east Mongolia. The basins are interpreted as having formed in an intracontinental, back-arc tectonic setting in response to extensional faulting. These basins, likely fault-bounded grabens and half grabens, were filled with eroded sediments during the Jurassic and Cretaceous periods. The Choir basin is a linear depression about 150 km long and from 10 to 20 km wide. Basement around the Choir basin comprises Proterozoic schist, gneiss, and limestone, Paleozoic granite, Permian alkaline volcanic, Mesozoic leucogranitic rock, and associated volcanic (Budunov and Jamsrandorj 2012).

The Choir basin consists of non-lithified sediments with a total thickness of approximately 1500 m. The Lower Cretaceous sediments of the Zuunbayan Formation are divided into two facies, with the first typically variegated and the second normally gray. The variegated section is comprised of conglomerate, sandstone, and siltstone and occurs mainly on the margins of the depression. The second facies is comprised of lacustrine sediments, typically clays and argillaceous sandstone, with interbeds of brown coal and disseminated iron sulfides. The Upper Cretaceous section is comparatively thin in the Choir basin and is generally from 5 to 40 m thick. It is typically composed of variegated sand and gravel with limonite-goethite cementation. The Uranium grades vary between 1.2% and 4% in the lens-shaped ore bodies in the paleovalleys. Uranium mineralization may occur in the upper layer of the lower Cretaceous sediment as tabular, basal channel shapes with large areal extent (about 20 km long, 0.5–2 km wide, 1–30 m thick) parallel to the groundwater table. Uranium mineralization in the Choir basin is associated with a number of separate uranium geochemical anomalies and occurs in small deposits such as Kharaat. Mineralization is present in significant concentrations both above and below the water table. Typically, mineralization below the water table occurs in gray-colored rocks or at the interface of the red oxidized rocks. Mineralization below the water table occurs primarily as uraninite with smaller amounts of coffinite. The uranium minerals occur as films and diffuse haloes on and between individual rock fragments and sand grains. In the oxidizing environment above the water table, secondary uranium minerals are present and include autunite, bergenite, phosphuranylite, schroeckingerite, torbernite, and minor amounts of other uranophosphates.

In the Choir basin, greater than 70% of the known mineralization with potentially economic grade occurs above the natural water table. Full saturation of the ore-bearing section is the normal condition for in situ recovery; however, the ISR test at Haraat included leaching from both saturated and unsaturated horizons.

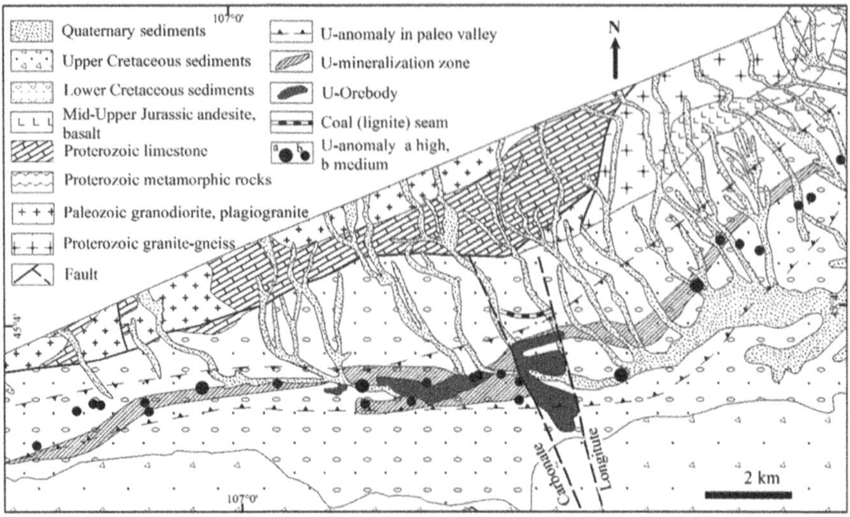

Fig. 13.9 Simplified geology of Khairkhan deposit and distribution of uranium. (After Budunov 2006)

13.5.2.3 Khairhan

The Khairhan deposit is located in the northwest part of the Khairhan basin (Fig. 13.9). It was discovered by drilling at the end of the 1996. The Khairhan basin is filled with the coal-bearing Cretaceous Zuunbayan Formation and covered by Sainshand Formation around the southern margin. The host depression is about 150 km long and 10–25 km wide. The deposit is situated along the paleovalley hosted in unconformable clastic and lacustrine sandstones and shales about 25 km elongated in an east-west direction. The host sedimentary rocks of the Khairhan deposit consist of unconformable clastic and lacustrine sandstones and shales. Precambrian granites and Paleozoic and Mesozoic rocks form the northern edge of the depression. The thickness of the paleochannel varies between 30 and 70 m, whereas at the end of channel, it varies up to 150 m. The length of the ore body varies and may extend from several hundred meters up to 1200 m, whereas the width ranges from 50 to 500 m. Mineralized lenses occur at depths from 10 m to 200 m, with an average depth in the 60 m to 80 m range. Uranium grade ranges between 0.01% and 0.1% U (Budunov and Jamsrandorj 2013).

The main ore reserves are confined to the central block bounded by the carbonate fault in the west and the longitude fault in the east. The carbonate fault extends in the north-northwest direction and is characterized by sporadic carbonation of sediments; as a result, they acquire the appearance of solid lithified rocks and occur as columns, lenses, and pockets from a few meters to a few tens of meters across. The carbonated rocks do not contain ore mineralization. The longitude fault is marked by some indirect signs, first of all, by the sharp drop of groundwater table and by the replacement of channel fill and flood-plain sediments occurring to the west of the fault by lacustrine sediments in the eastern wall. The ore-forming implication of the fault remains obscure; however, it is evident that the highest-grade and contrasting ores with a productivity of 70 kg/m^2 are localized exactly along this tectonic line that probably served as a hydrodynamic, lithological, and geochemical barrier on the way of filtration of oxygen-bearing groundwater enriched in uranium. It cannot be ruled out that this fault also was a conduit for gaseous reductants (hydrogen sulfide, hydrocarbons, etc.) supplied into the zone of surface and ground oxidation from below.

The ore bodies are nearly horizontal with a slight eastward inclination. The ore is aluminosilicate and carbonate-free. Uranium mineralization is mainly represented by pitchblende; coffinite is of the secondary importance. Autunite and uranophane occur in the oxidized rocks. Molybdenum is the main associated element; the Mo grade in ore reaches 0.05%. The hydrogeological conditions at the deposit are complex but generally favorable for uranium recovery by underground leaching. The ore mineralization is localized below the groundwater table, being largely hosted in permeable rocks. The ore formation was related to the multistage epigenetic processes. The uranium concentration is controlled by reduction, and the zone of stratal oxidation serves as a supplier of uranium to the geochemical barrier. The uranium concentration is controlled by reduction, and the oxidized zone serves as a supplier of uranium to the redox front.

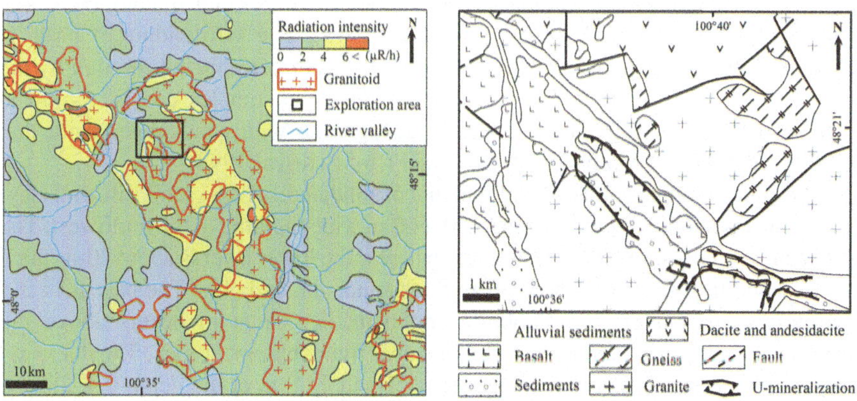

Fig. 13.10 (**a**) Distribution of radiation and Egiin Davaa granitoid complex; (**b**) simplified geology of Dalt deposit and distribution of uranium mineralization. (After Jamsrandorj et al. 2019)

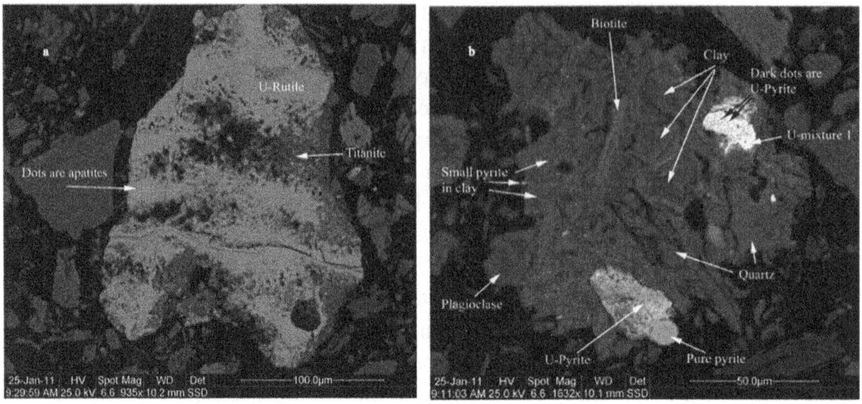

Fig. 13.11 Backscatter electron images showing U-rutile and U-pyrite. (**a**) Apatite, titanite, and U-rutile; (**b**) quartz, plagioclase, biotite, pyrite, clay, and uranium in pyrite (Jamsrandorj et al. 2019)

13.5.3 Surficial Uranium Deposits

This type of deposits is defined as young and near-surface uranium accumulations in sediments and soils. Recently, Dalt uranium deposit was discovered in Chuluut district based on the result of radioactive anomalies.

13.5.3.1 Dalt

Dalt uranium deposit was formed in the fluvial valley filled with sediments and covered by basaltic flows. Radioactive anomalies were determined along the valley and surrounding granitoids in the past (Fig. 13.10). The surrounding of basin is composed of the Paleozoic granitoids and the Permian volcanic cut through by the Late Triassic-Early Jurassic Egiin Davaa complex. The complex consists of red to pinkish gray, medium-grained porphyritic leucogranite with subalkaline granite that were dated 166–127 Ma based on K-Ar and Rb-Sr (Koval 1998).

The valley contains alluvial, deluvial unconsolidated gravels, sand, silt, and clay, and its size is about 8.8 km long and 150–600 m wide trending to the northwest. The sediment is totally more than 50 m thick, whereas the lower layer is about 20-m-thick gravels, and upper layer is about 30-m-thick sand, silt, and clay. The basement is a granite that was altered to kaolinite and hematite. Uranium occurs above the weathered or decomposed 20 m thick zone underlying basement rock. The thickness of uranium accumulation ranges between 1.1 and 15.9 m at the depth 0.4–28.7 m from the surface, and uranium grade reaches 0.12% in gray, organic component bearing gravel and small-to-medium grained sand and clay. Uraninite, meta-otenite, uranium-associated rutile, and pyrite were determined by XRD. Uranium is present in otenite, meta-otenite, rutile, and pyrite (Fig. 13.11).

13.5.4 *Other Type of Uranium Mineralization*

Intrusive igneous-related mineralization can be divided into granite-monzonite, peralkaline complex and carbonatites classes. World-class Cu-Au-Mo porphyry deposits were discovered in south Mongolia. Alkaline and peralkaline intrusive-related REE, Th, and U mineralization occur in Khalzan Burgedei and Khan Bogd, and REE-bearing carbonatite, Nb (Ta)-Zr-REE-bearing alkaline metasomatite, and pegmatite were found in Mushgai Khudag, Lugiin Gol, and Khotgor. These porphyry deposits and REE mineralization may be an unconventional resources, and uranium occurs as a by-product.

13.5.4.1 Granite-Related Uranium Mineralization

This type of mineralization is widespread in Mongolia and is related to the high-radioactive leucogranitic and alkali granitic plutons emplaced at the early stage of the Mesozoic tectono-magmatic reactivation. The granitic plutons of the high-radioactive leucogranites are distributed in the Khentii-Daguur region within the Janchivlin pluton (Elst, Tamga, Urt, Arshaan), in the Bayan-Delger pluton (Borj), in the Nukht Davaa uplift (Saikhan Uul, Tsagaan Uul), in the Argun-Mongolia volcanic belt (Tumen Tsogt, Baruun Tsogt), and in the Khangai uplift of central Mongolia. The intragranitic uranium mineralization in the Janchivlan pluton occurs as pitchblende and coffinite in association with sulfides including marcasite, galena, and sphalerite which occur with montmorillonite in veinlets. The zone of oxidation of sulfide and uranium ores is extensive at the surface and traceable down to 400 m depth. The mineralization is related to the high-radioactive leucogranite and is hosted in greisen zones with low-temperature hydrothermal alteration and subsequent supergene alteration, associated with Sn-W mineralization.

13.5.4.2 Metasomatite-Related Uranium Mineralization

Numbers of occurrences of Th-U mineralization are hosted in various metasomatic rocks, migmatites, and pegmatites localized in Precambrian metamorphic rocks in the Buteeliin Nuruu uranium ore district. They are characterized by Samovich (1984). The best studied occurrences are situated in the eastern Buteeliin Nuruu inlier in the framework of a granite-gneiss dome, where a system of conjugated northeast and northwest trending faults is clearly deciphered on the satellite images. The quartz-feldspar metasomatic rocks replace biotite, biotite-amphibole, and granite gneisses close to the cataclastic zones. The northwest trending metasomatic zones extend as far as 200 m having a thickness of 1.0–1.5 m; radioactivity/radiation varies from 250 to 1500 μR/h. The U-bearing metasomatic rocks consist of equant quartz grains and coarse tabular microcline replaced by albite laths, biotite flakes, and isolated almandine crystals. The ore disseminations are composed of euxenite and brannerite intergrown with titanomagnetite and REE minerals. The occurrences of U- and Th-bearing minerals are typical of metasomatic rocks related to subalkaline igneous rocks.

13.5.4.3 Uraniferous Lignite

Uranium mineralization occurs in Cretaceous lignite-bearing sediments of Choir, Choybalsan, Ulziit, and other basins. At the Ulziit coal deposits, the uranium is related to the oxidized strata of the coal seam that extends for 2 km; the width of the U-bearing zone is 300–1500 m. The U content at the seam surface is overlapped by yellow sand, a product oxidation. Grades reaches 0.115% U over a thickness of 1 m. The U content declines to 0.035% with depth. Only low grades were recorded in boreholes within a depth interval of 31.2–44.8 m. The zones of ground or strata oxidation are the most spectacular at the Dorbolj occurrence localized in the coal-bearing sedimentary rocks of the lower Cretaceous Zuunbayan Formation. Uranium is concentrated here over a $4 \times 2\ \mathrm{km}^2$ area at depths of 100 m in the dark gray coaly siltstone close to the lower redox contact within a sand unit. The thickness of the orebodies does not exceed 0.5–1.5 m; the U grade is 0.01–0.015%. In the Choir basin, the uranium mineralization of Shivee Ovoo, Ereen, and Jargalant Nuur is related to the disintegrated coal-bearing sediments in contact with oxidized sandstones and within iron-rich and bleached sandstones.

13.5.4.4 Phosphorite

Uranium in sediments with continental-shelf origin was found at Khuvsgul in north Mongolia. Radioactive anomalies were identified among the phosphorite deposits, and this type of deposit contains uranium as unconventional resources but of very low grade.

13.6 Uranium Exploration Activities and Resources

Mongolia has adopted the State policy on exploitation of radioactive minerals and peaceful uses of nuclear energy as well as Nuclear Energy Law in 2009. This regulates the exploration, development, and mining of uranium and gives the state a greater degree of ownership and control of uranium resources. The Law provides the government the right to take ownership without payment of not less than 51% of the shares of a project or joint venture if the uranium mineralization was discovered by state-funded exploration and not less than 34% if state funding was not used to find the mineralization. Licenses to conduct uranium exploration and production of any radioactive minerals must be obtained under this law.

Table 13.2 Main uranium deposits and resources in Mongolia

Deposit	Province	Lat N/Long E	Deposit type	Resource (t)	Average grade (% U)
Mardaingol	Dornod	114°15′13″/ 49°09′41″	Volcanic	1948	0.05
Nemer	Dornod	114°15′13″/ 49°09′41″	Volcanic	5937	0.11
Dornot	Dornod	114°21′40″/ 49°05′30″	Volcanic	27,221	0.15
Ulaan	Dornod	114°05′00″/ 49°09′41″	Volcanic	270	0.14
Gurvanbulag	Dornod	113°56′47″/ 49°02′30″	Volcanic	13,058	0.15
Kharaat	Dundgobi	108°24′20″/ 45°41′10″	Sandstone	7288	0.026
Khairkhan	Dundgobi	107°04′10″/ 45°02′05″	Sandstone	8406	0.071
Gurvansaikhan	Dundgobi	107°07′47″/ 45°30′04″	Sandstone	4250	0.034
Ulziit	Sukhbaatar	110°45′48″/ 46°29′58″	Sandstone	3075	0.036
Dulaan uul	Dornogobi	109°44′77″/ 44°12′00″	Sandstone	7054	232.74 (g/t)
Dulaan uul-SW	Dornogobi		Sandstone	4841	188.83 (g/t)
Zuuvch ovoo	Dornogobi	110°02′01″/ 44°10′01″	Sandstone	67,706	0.023
Engerar	Tuv	108°00′00″/ 46°56′00″	Sandstone	80	0.05
Dalt	Arkhangai	100°35′57″/ 48°20′04″	Surficial	15,518	0.032

As for uranium, the major preparatory work for the recovery is underway due to the completion of deposit development agreements between the government and potential stakeholders. Mongolia is focusing on projects which can be developed to provide uranium to the global market in the near future.

The history of uranium exploration in Mongolia could be divided into three phases. The first phase started in the middle of the 1940s, whereas the second phase covers the period of 1970–1990 where specialized geological and geophysical surveys were conducted with the result of 1600 radiometric anomalies and occurrences identified by the joint expedition with the Ministry of Geology of the Soviet Union under a bilateral agreement between the People's Republic of Mongolia and the Soviet Union. The third phase started in the 1990s with private stakeholders' engagement including local and foreign entities. There is a prevalence of well studied regions in the areas of east, southeast, center Mongolia. It is estimated that there are about 1.47 and 1.0 Mt of U resources (Mironov et al. 1993; Batbold 1997; Byamba et al. 2018). According to the statistics of Mineral Resources and Petroleum

Authority of Mongolia (2019), uranium resources are assessed at 166,657 tonnes of U in total (Table 13.2).

In recent years, the government has conducted airborne and ground geophysical surveys covering southwest Mongolia. Based on the result of gamma-ray spectrometry surveys, the uranium potential zones at Baruunkhuurai, Gobi-Tengeruul, Khustiin Nuruu, and Songint were described. Most of these anomalies could be related to alkaline granites with REE mineralization.

Global exploration strategy has changed to identify low-cost uranium, especially sandstone-type deposits. Consequently, uranium exploration in Mongolia has focused and targeted on Mesozoic and Cenozoic basins in southeast Mongolia.

There are currently eight deposits on which feasibility studies have been completed so far. The Dornod open-pit mine and underground orebody was mined between 1988 and 1995. The ore was railed 480 km to Krasnokamensk in Siberia for treatment by Priargunsk. Only about 535 tonnes of U was produced, all from open-pit mining. Gurvanbulag, about 30 km west of Dornod, had extensive underground development down to 560 meters and was readied for production in the Soviet era. A potential development project is at Zoovch Ovoo and includes the Dulaan Uul uranium project in Dornogovi province in east-south region of Mongolia. The deposit will be mined by in situ leaching methods using sulfuric acid. The pilot test work on the Dulaan Uul deposit was carried out in 2011. Test work is currently underway at the Zoovch Ovoo.

13.7 Concluding Remarks

There are several types of uranium deposits in Mongolia, and the main types are volcanic and sandstone deposits. Well-studied volcanic-related uranium mineralization was identified in an area of 1300 km by 300 km in southeastern Mongolia, and the belt is dominated by Proterozoic gneisses, schists, and anatectic granites, occurring as inliers in early Paleozoic volcanic and sedimentary rocks. Volcanic deposits occur in eastern Mongolia and are associated with Late Jurassic-Early Cretaceous bimodal volcanic rocks. Volcanic-related type of deposits is characterized by F-Mo-U and U-F-apatite mineralization in southeastern Mongolia.

Sandstone-hosted uranium-type deposits occur in Mesozoic basins in southeastern Mongolia. The uranium mineralization is accumulated in the medium- to course-grained sandy, lacustrine sediments. Sandstone-hosted uranium deposits of Mongolia can be classified roll-front and basal channel or paleovalley subtype. Recently, sandstone uranium-type deposits closely associated with hydrocarbon-bearing basins have been identified in eastern and south Mongolia.

References

Agnerian H, Bocking KA, Cox JC, Heymann LH (2007) Report on the Dornod Uranium Property between 2004–2007, Mongolia: Scot Wilson Technical report for Khan resources Inc.

Axel A, Paul S, Patrier P, Beaufort D, Amiard G, Michael D (2018) Use of Digital Autoradiography to characterize the effect of acidic leaching on mobilisation of 238U series radionuclides in ISR context—Case study of the Dulaan Uul experimental site (Mongolia)

Badamgarav J, Badamgarav D, Orolmaa D, Tumurkhuu D, Narantsetseg Ts, Dembereldagva Ts, Idermunkh T (2011) Map of the Mesozoic-cenozoic tectonic structures of Mongolia. Scale 1:1000000

Badarch G, Cunningham WD, Windley BF (2002) A new terrane subdivision for Mongolia: implications for the Phanerozoic crustal growth of Central Asia. J Asian Earth Sci 21:87–110

Banzragch J, Chuluundorj Ch (1979) Regional pattern of localization of mineral deposits in Mongolia People's Republic. In Geological structure and localization of mineral resources in the territory of Mongolia People's Republic, Nedra, Moscow pp 38–40 (in Russian)

Batbayar N, Hocquet S, Batbayar Ts (2018) Resources and reserves of the Zoovch Ovoo uranium deposit. Presentation for Mongolia mineral exploration roundup-2018

Batbayar Ts, Hocquet S, Batbayar N (2016) Development stages of the Zoovch Ovoo uranium deposit. The result of the detailed exploration of Zoovch Ovoo uranium deposit. Presentation for Mongolia mineral exploration roundup-2016

Batbold T (1997) Uranium favorability and evaluation in Mongolia (Phase II), resent events in Uranium resources and production in Mongolia. The Assessment of uranium deposit types and resources—worldwide perspective. TECDOC 1258, IAEA, Vienna, pp 93–100

Bruneton P, Cuney M (2016) Geology of uranium deposits. In: Hore-Lacy I (ed) Uranium for nuclear power, resources, mining and transformation to fuel, Woodhead, London, pp 11–53

Budunov AA (2002) Gidrogennnyye uranovyye mestorozhdeniya Mongolii (Hydrogenic uranium deposits of Mongolia) Materials on the geology of uranium deposits Inform sb KNTS 144 Moscow VIMS 84–89

Budunov AA (2006) Summary report of results of prospecting work in the Gobi Region of Mongolia for 1994–1996. Ulaanbaatar

Budunov AA, Jamsrandorj G (2012) Kharaat uranium deposit. PRO 47:76–81. (in Mongolian)

Budunov AA, Jamsrandorj G (2013) Khairkhan uranium deposit. PRO 48:205–209. (in Mongolian)

Byamba (ed.) (2011) Mongolian geology and mineral resources, Stratigraphy (Mesozoic), 1: p 547

Byamba J, Mironov YB, Jamsrandorj G (2018) Geodynamics and uranium metallogeny of Mongolia. Ulaanbaatar p 302

Chuluun O, Oyunbaatar U (1994) Hydrogenic and metasomatic type of uranium ore mineralization in Mongolia. Report of Mongolian Delegation to Technical committee IAEA, Ottawa, pp 171–182

Cuney M (2010) Evolution of uranium fractionation processes through time: driving the secular variation of uranium deposit types. Econ Geol 105:449–465

Cunningham D (2010) Tectonic setting and structural evolution of the late Cenozoic Gobi Altai orogen: Geological Society Special Publication, London, 338:361–387

Dagvadorj D, Munkjargal B (2018) Result of the Airborne geophysical survey on Altai-1 area. Prospector 59:176–189. (in Mongolian)

Dahlkamp FJ (2009) Uranium deposits of the world, vol 4. Springer, Berlin

Duinharjav M, Battulga B, Dashbat A, Gombosuren B, Ochirkhuyag Ch (2018) Report of Airborne geophysical survey on Altai-2 area between 2014–2017 at scale 1: 200,000, Ulaanbaatar (in Mongolian)

Economos RC, Hanzl P, Hrdlickova K, Burianek D, Said L, Gerdes A, Paterson SR (2008) Geochemical and structural constraints on the magmatic history of the Chandman Massif of the eastern Mongolian Altay Range, SW Mongolia. J Geosci 53:335–352

Gavrilova NS, Zaizev (1975) Granitoids and alkaline formation in Western and Northern Mongolian structures Nauka, Moscow, pp 288

Gerel O (ed) (2012) Intrusive rock. Soyombo Printing, Ulaanbaatar, p 460
Gordienko I, Klimuk VS, Heng Q (2000) Verkhneamurskiy vulkan-plutonicheskiy poyas Vostochnoy Azii: struktura, sostav i geodinamicheskaya obstanovka (the upper Amur volcano plutonic belt in Eastern Asia: structure, composition, and geodynamic setting). Geol Geofiz 41 (12):1655–1669
Hocquet S (2016) Zoovch Ovoo and Dulaan Uul uranium deposits. Cogegobi LLC, IAEA-UNECE Workshop, 16/08/2016, Ulaanbaatar
IAEA (2018) Geological Classification of Uranium Deposits and Description of Selected Examples. Vienna IAEA-TECDOC-1842
Jahn BM (2004) The Central Asian Orogenic Belt and growth of the continental crust in the Phanerozoic. In: Malpas J, Fletcher CJN, Ali JR, Aitchison JC (eds) Aspects of the tectonic evolution of China: Geological Society, London, Special Publication, vol 226, pp 73–100
Jamsrandorj G, Altankhuyag D (2016) Uranium metallogenic problems of Mongolia. PRO 53:163–176. (in Mongolian)
Jamsrandorj G, Baatartsogt B, Altankhuyag D (2016) Resent discoveries and resources of uranium in Mongolia. J East China Univ Technol 39:25–31
Jamsrandorj G, Dejidmaa G, Ukhnaa G (2015) Uranium: the geology, world overview of mineral resources and mining. Ulaanbaatar (in Mongolian)
Jamsrandorj G, Enkhamgalan S, Nolph JL, Baldorj B, Bayarmaa A, Altankhuyag D (2013) Roll-front type Dulaan Uul uranium deposit. PRO 48:269–276. (in Mongolian)
Jamsrandorj G, Khaliun S, Amar O, Davaademid D, Purev-Ochir M (2019) Geology and uranium mineralization of the Dalt deposit. PRO 60:118-126. (in Mongolian)
Johnson CL (2004) Polyphase evolution of the East Gobi basin: sedimentary and structural records of Mesozoic-Cenozoic intraplate deformation in Mongolia. Basin Res 16:79–99
Khomich VG, Boriskina NG (2018) Geodynamic, tectonic, and magmatic indicators of large uranium clusters of the Transbaikal-Mongolia-North China province. Russ J Pac Geol 12 (6):487–499
Khomich VG, Boriskina NG, Santosh M (2016) Geodynamic framework of large unique uranium ore belts in Southeast Russia and East Mongolia. J Asian Earth Sci 119:145–166
Kochkin GB, Smislov AA, Terentiev BM (1985) Uranium Metallogeny of Mongolia. Ministry of Geology of the USSR. Leningrad pp 330
Koval PV (1998) Regional'ny geochemicheskiy analiz granitoidov (Regional geochemical analysis of granitoids). Siberian Branch RAS SPC UIGGM, Novosibirsk 489
Kröner AJ, Lehmann K, Schulmann A, Demoux O, Lexa D, Tomurhuu P, Štípská D, Otgonbator D, Liu Y, Wingate MTD (2010) Lithostratigraphic and geochronological constraints on the evolution of the Central Asian Orogenic Belt in SW Mongolia: early Paleozoic rifting followed by late Paleozoic accretion. Am J Sci 310:523–574
Lee C, Nowak M (2006) Independent Technical Report and Resource Estimate for the Gurvanbulag Deposit; Saddle Hills Uranium Project, Mongolia, SRK consulting, p 129
Li S (2016) Application of the United Nations Framework Classification for Fossil Energy and Mineral Reserves and Resources 2009 to the Uranium Resources of the Gurvanbulag Uranium Deposit, Mongolia, UN Economic and Social Council, ECE/ENERGY/GE.3/2016/11
Lui CY, Qiu XW, Wu BL, Zhao HG (2007) Subdivisions of the Central-East Asia multi-energy minerals metallogenetic domain and types of those basins. Energy Explor Exploit 27 (3):153–166
Mineral Resources and Petroleum Authority of Mongolia (2019) Data source from National Geoscience Database at Mineral Resources Information Technology Center
Mironov YB (2003) Uranium of Mongolia. Sankt Petersburg, p 326. (in Russian)
Mironov YB (2006) Uranium of Mongolia. Center for Russian and Central EurAsian Mineral Studies (CERCAMS), London, p 230
Mironov YB, Afanasev AM, Bulichev AB et al. (2009) Uranium deposits in Mongolia. Sankt-Petersburg (in Russian)

Mironov YB, Naumov SS, Jamsrandorj G, Chuluun O (1993) The uranium resource of Mongolia. In Resent development in uranium resources and supply. TECDOC 823, IAEA, pp 177–192
Mironov YB, Solovev NS, Shleider VA, Maximov EA (1990) Uranium deposits of Mongolia. In: Chebotirev VN, Baras Z (eds) Economic mineral deposits of Mongolia, Ulaanbaatar, pp 91–121
Nokleberg WJ (ed.) (2010) Metallogenesis and Tectonics of Northeast Asia: U.S. Geol Surv Prof Paper 1765
OECD(NEA)/IAEA (2004) Uranium 2003: resources, production and demand. OECD, Paris, p 186
Pirajno F (2013) The geology and tectonic settings of China's mineral deposits, Chap. 8. In: Volcano-sedimentary and sedimentary basins in China. Springer, Netherlands, p 682
Rogov YG, Rubzov G.B (1989) Geology and mineralization of North Choibalsan district, Eastern Mongolia. Dornot-Leningrad-Irkutsk
Rogov YG, Yurchenko IA (1987) Report on the results of detailed surveys in 1982–87 with calculation of reserves as of July 1, 1987. Ministry of Geology of the former Union of Soviet Socialist Republics (USSR)—All-Union Geological Prospecting Association "Sosnovskoye" Order of Lenin PGO—Mongolian Geological Mapping Expedition. Irkutsk. Mongolian Geo-information Centre Report # 5251
Samovich DA (1984) Otchet o vozdushno-geofizicheskom i geologicheskom kartografirovanii territorii Altaya v masshtabe 1: 1 000 000 (Report on airborne geophysical and geological mapping of Altai area at a scale of 1:1,000,000) Ministry of Geology USSR Internal report 2456
Schmelev YS, Panev VI, Gavrilov YM (1982) Report of Airborne gamma-ray spectrometry survey result in Undurshil area during 1981–1982 at scale 1: 200,000
Selzov BM, Vishnyakov BE (1990) Genetic Model for Uranium Deposit in Continental volcanism. (In case Za-Baikal, Mongolia, and Middle Asia) Moscow
Sengor AMC, Natal'in BA (1996) Palaeotectonics of Asia: fragments of a synthesis. In: Yin A, Harrison M (eds) Tectonic evolution of Asia. Cambridge University Press, Cambridge, pp 486–640
Uyakhan Z, Altankhuyag D (2015) Structure and stratigraphic-lithological features on sedimentary uranium mineralization of Mongolia. PRO 53:177–182. (in Mongolian)
Voo van der R, Hinsbergen van DJJ, Domeier M, Spakman W, Torsvik TH (2015) Latest Jurassic-earliest Cretaceous closure of the Mongol-Okhotsk Ocean: a paleomagnetic and seismological-tomographic analysis: Geological Society of America, Special Paper 513:539–606
Webb LE, Graham SA, Johnson CL, Badarch G, Hendrix MS (1999) Occurrence, age, and implications of the Yagan-Onch Hayrhan metamorphic core complex, southern Mongolia. Geology 27:143–146
Windley BF, Alexeiev D, Xiao W, Kröner A, Badarch G (2007) Tectonic models for accretion of the central Asian Orogenic Belt. J Geol Soc Lond 164:31–47
Yakubchuk AS (2004) Architecture and mineral deposit settings of the Altaid orogenic collage: a revised model. J Asian Earth Sci 23:761–779
Zonenshain LP, Kuzmin MI, Natapov LM (1990) Geology of the USSR: a platetectonic synthesis. In: Page BT (ed) Geodynamics series, geodynamic series, 21. American Geophysical Union, Washington, DC, p 242

Chapter 14
Coal Deposits

Bat-Orshikh Erdenetsogt and Luvsanchultem Jargal

14.1 Introduction

The first information of Mongolian coal was published in 1884, and mining commenced at Nalaikh in 1912. Systematic geological surveys for coal were started in 1920s, and more than 300 coal deposits and occurrences have been catalogued so far. In addition, open-pit mines were opened in all aimags (aimag – administrative subdivision of Mongolia) and produce coal for local power stations and household fuels.

Coal reserves are calculated at 36 Gt, of which 66% are lignite, located mostly in Eastern Mongolia. Currently, 8 Mt of lignite is mined annually and used for electricity and heat generation. The majority of bituminous coal reserves are in Southern and Western Mongolia, from where 25.8 Mt of coal was exported in 2016, and in that year, Mongolia was considered as the ninth major coal exporter in the world.

The purpose of this chapter is to introduce Mongolian coal, their occurrences, and characteristics. Coal quality data included here are a guide only and could not represent seam properties of a given deposit. Furthermore, in the herein chapter, the terms, "coal-bearing basin" and "coal-bearing area," are used. A coal-bearing basin is defined as a large region, where economic coal seams are distributed extensively. Basin origin is linked with the development of a regional tectonic structure, and coal seams were accumulated in a certain geological period. In contrast, a coal-bearing area is a region, where the coal potential or the genetic relation of coal-hosting formations is not clear due to limited geological information.

B.-O. Erdenetsogt (✉) · L. Jargal
Department of Geology and Geophysics, National University of Mongolia, Ulaanbaatar, Mongolia
e-mail: tsogo@num.edu.mn; jargal_L@num.edu.mn

O. Gerel et al. (eds.), *Mineral Resources of Mongolia*, Modern Approaches in Solid Earth Sciences 19, https://doi.org/10.1007/978-981-15-5943-3_14

However, few coal deposits or occurrences have been discovered within the area (Erdenetsogt et al. 2009).

14.2 Geology of Mongolian Coal-Bearing Provinces and Basins

Mongolia is a landlocked country with an area of 1.56 million km^2. Much of this area is covered by grassy steppe, with mountains to the north and west and the Gobi Desert to the south (Fig. 14.1).

Since the 1880s, more than 300 coal deposits and occurrences have been discovered, so far. Based on origin, age, tectonic setting, coal characteristics, and coal-hosting formations, the deposits can be classified into 2 coal-bearing provinces, 12 basins, and 3 areas (Bat-Erdene 1992) (Fig. 14.2). This section summarizes geology of these provinces, basins and areas.

14.2.1 Overview of Depositional History of Coal

The deposition of coal commenced with Carboniferous sequences and continued with Permian, Jurassic, and finally with Cretaceous sequences. Carboniferous and Permian coal seams were accumulated mainly in Western and Southern Mongolia,

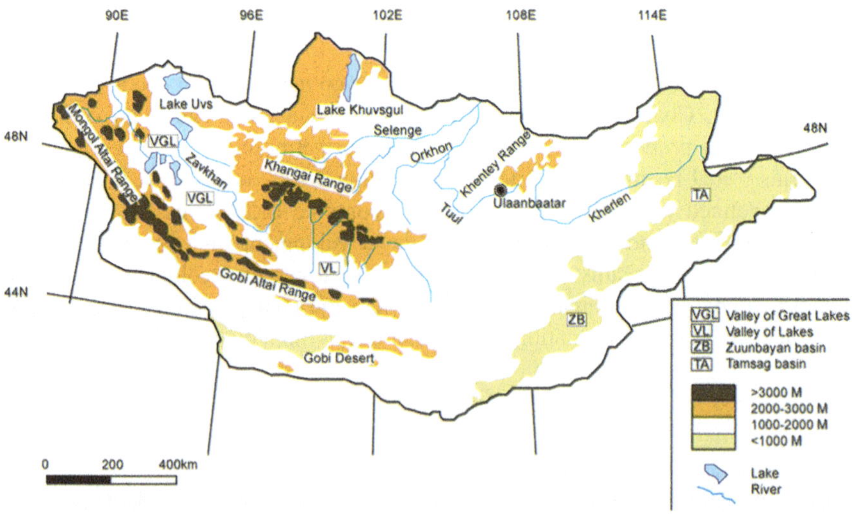

Fig. 14.1 A simplified topographic map of Mongolia (modified after Badarch et al. 2002, and published with permission of Elsevier)

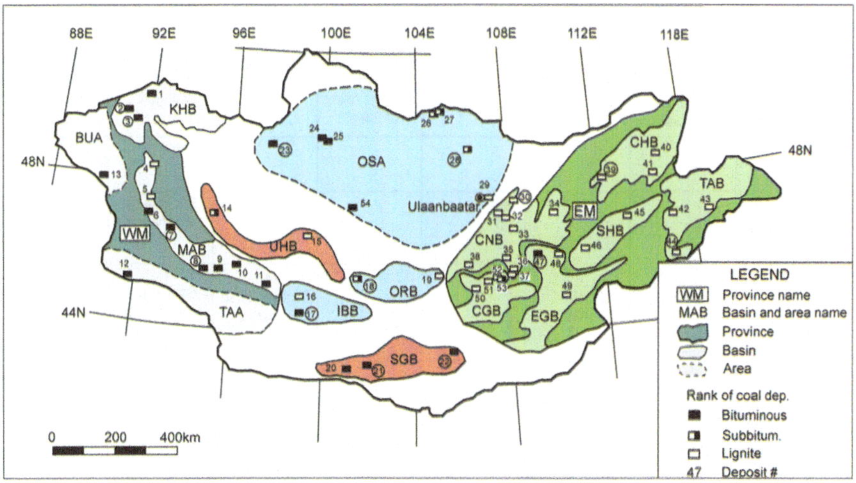

Fig. 14.2 Distribution map of Mongolian coal deposits in Mongolia (modified after Bat-Erdene 1992). The age of major coal coal-bearing sequence is depicted by colors: slate gray for Carboniferous in Western Mongolia; brown for Permian in Southern Mongolia; blue for Jurassic in Central and Northern Mongolia; green for Cretaceous in Eastern Mongolia. Provinces: WM, Western Mongolia; EM, Eastern Mongolia; Basins: KHB, Kharkhiraa; MAB, Mongol-Altai; SGB, South Gobi; IBB, Ikh Bogd; UHB, South Khangai; ORB, Ongi river; CNB, Choir-Nyalga, CHB, Choibalsan; SHB, Sukhbaatar; TAB, Tamsag; CGB, Central Gobi; EGB, East Gobi; Areas, TAA, Trans-Altai; BUA, Bayan-Ulgii; OSA, Orkhon-Selenge; Coal deposits: 1, Khuden; 2, Nuurst Khotgor; 3, Khartarvagatai; 4, Mantag, 5, Jargalant,; 6, Gantsmod; 7, Khushuut; 8, Maanit; 9, Khuren Gol; 10, Tsagaan Gol; 11, Zeegt; 12, Olonbulag; 13, Rashaant; 14, Tsakhiurt Urt; 15, Uvurchuluut; 16, Alagtsakhir; 17, Khotgor; 18, Bayanteeg; 19, Tsagaan-Ovoo; 20, Nariinsukhait; 21, Gurvantes; 22, Tavantolgoi; 23, Ovoot; 24, Nuurstei; 25, Jilchigbulag; 26, Ulaan-Ovoo; 27, Khujirt; 28, Shariin Gol; 29, Nalaikh; 30, Baganuur; 31, Maint; 32, Tsaidam; 33, Tugrug; 34, Tsaidam Nuur; 35, Shivee-Ovoo; 36, Ikh Ulaannuur; 37, Uvdugkhudag; 38, Tevshiin Gobi; 39, Khulstnuur; 40, Utaat Minjuur; 41, Aduunchuluun; 42, Zuunbulag; 43, Bulangiin Khooloi; 44, Bayantsogt; 45, Talbulag; 46, Ulziit; 47, Alagtogoo; 48, Olongiin Ukhaa; 49, Khamriin Khural; 50, Unst Khudag; 51, Taliin Khudag; 52, Nukhet; 53, Khootiin Khonkor; 54, Bayantsagaan. Note: The deposits with circled numbers are shown in Figs. 14.4–14.6

respectively. Jurassic coals were deposited in Western, Central, Northern, and Eastern Mongolia. In the Cretaceous, extensive coal and oil shale were formed mostly in Eastern Mongolia (Fig. 14.3). Coal rank is controlled by the age of the coal-bearing sequences. Carboniferous and Permian coals are high-rank bituminous (up to semianthracite) and mid-rank bituminous, respectively. Jurassic coal is low-rank bituminous and subbituminous, and Cretaceous coal is lignite (Erdenetsogt et al. 2009).

According to Erdenetsogt et al. (2009), coal-bearing basins in Mongolia can be broadly classified as compression tectonic-related and extension tectonic-related basins. Carboniferous and Permian basins in Western and Southern Mongolia are foreland basins formed by compression tectonic. In contrast, Cretaceous basins in

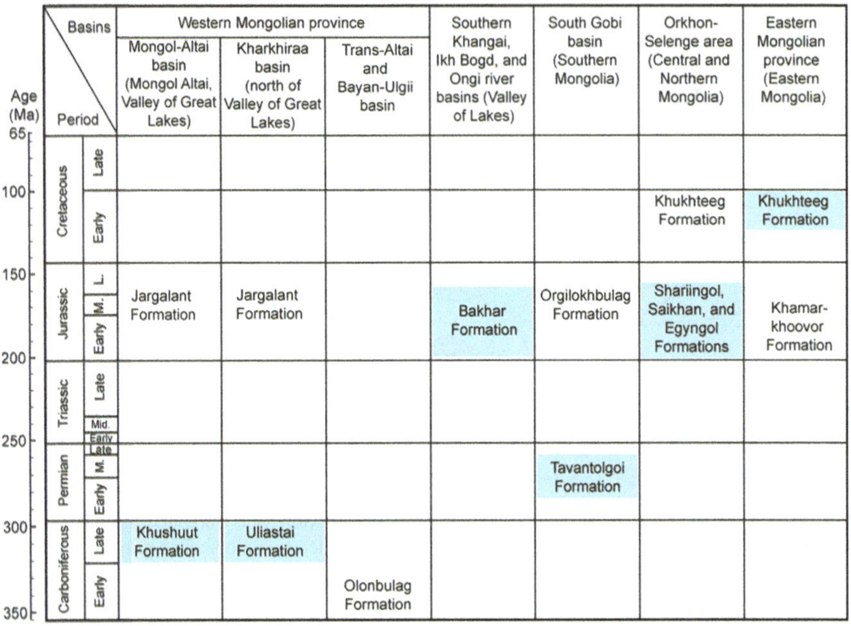

Fig. 14.3 Stratigraphic distribution of coal-bearing Formations in Mongolia. Note: Important Formations are colored

Eastern Mongolia are rift basins formed through extension tectonics. Tectonic settings of Jurassic basins are unclear due to lack of geological documentation. Jurassic structures in the north of Orkhon-Selenge Area appear to be formed by contractile deformation due to closure of Mongol-Okhotskh ocean, whereas coal-bearing Jurassic sedimentary sequences in Central Mongolia (e.g., Ongi River Basin) were probably accumulated in rift setting (Erdenetsogt et al. 2017a; Hasegawa et al. 2018). However, north-south-directed contractile deformation has been well-documented in many parts of North and Northwest China and South Mongolia (e.g., Lamb et al. 2008)

14.2.2 Geology of Western Mongolian Coal-Bearing Province

The Western Mongolian coal-bearing Province, which has a northeast-southwest orientation, extends approximately 280,000 km^2 area in Western Mongolia. The Province includes Kharkhiraa and Mongol-Altai Basins and Trans-Altai and Bayan-Ulgii Areas (Fig. 14.2). The Carboniferous volcano-sedimentary rock is the most prolific coal-bearing sequence, with some additional seams in Jurassic sequences (Erdenetsogt et al. 2009). Within the Province, more than 50 coal deposits and

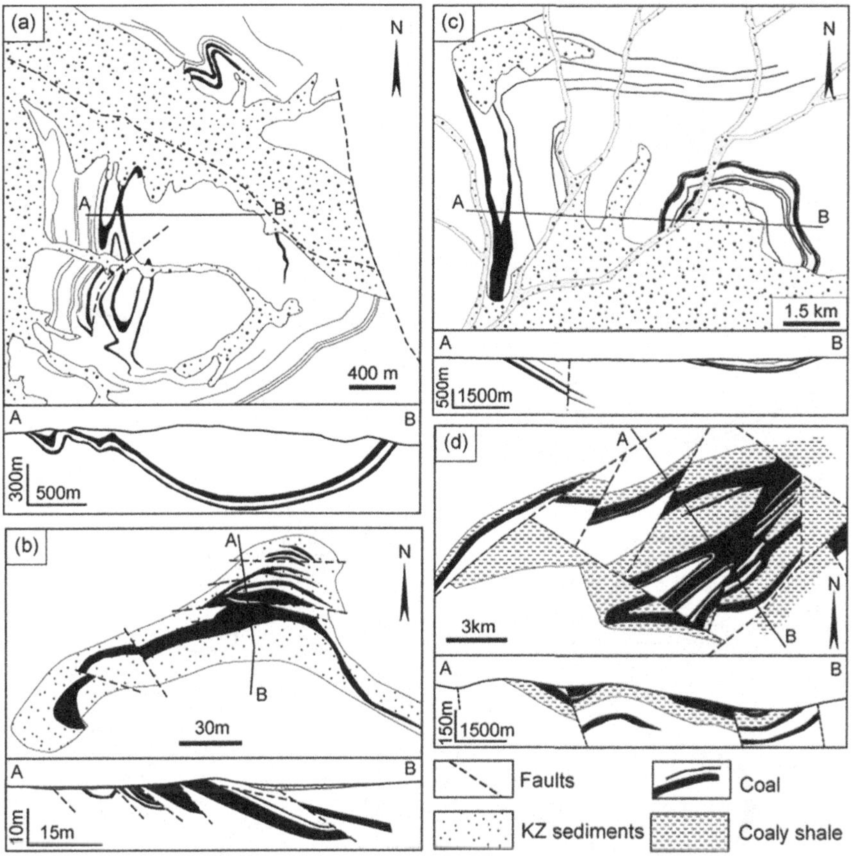

Fig. 14.4 Coal seam maps with cross sections of selected deposits in Western Mongolian Province (published with permission of the Elsevier). (**a**) Khushuut, (**b**) Maanit, (**c**) Nuurst Khotgor, (**d**) Khartarvagatai; Note: The locations of deposits are shown in Fig. 14.2

occurrences are discovered, and several open-pit mines (Khushuut, Khartarvagatai, Nuurst Khotgor, Maanit, Zeegt, etc.) are operational. The biggest mine is Khushuut, which produced 1.3 Mt of bituminous coal in 2017 and exported to China (MEC 2018).

The basement of the Province consists of a collage of island arc, backarc, foreark, and accretionary wedge terranes. During the early Paleozoic, the terranes were collided with the central Mongolian Precambrian Block (Traynor and Sladen 1995; Ruzhentsev and Burashnikov 1996; Badarch et al. 2002) and were tectonically reactivated in the late Cenozoic (Cunningham 2005; Vassallo et al. 2007). Coal seams hosted in Carboniferous volcano-sedimentary rocks were preserved in open thrusts, ramps, and graben-like structures that are separated by basement uplifts and controlled by thrust and normal faults (Erdenetsogt et al. 2009). The dip of Carboniferous rocks varies from 15 to 60 degree, and fault displacement reaches up to

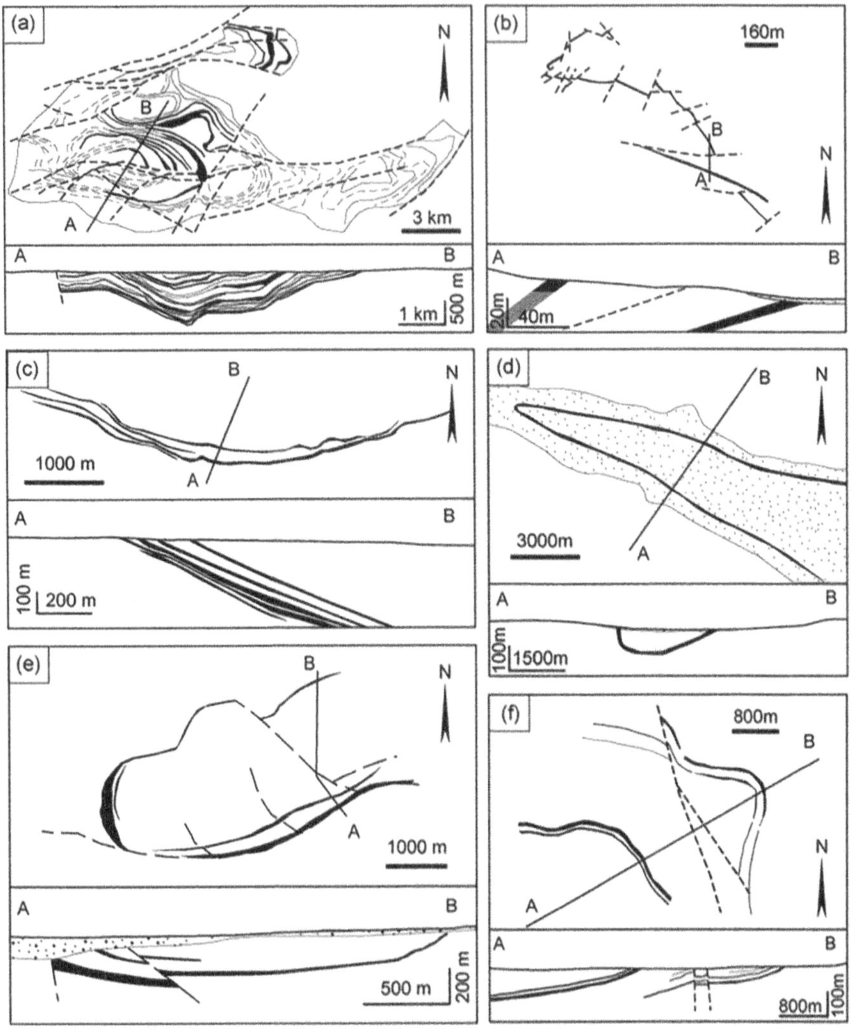

Fig. 14.5 Coal seam maps with cross sections of selected deposits in the Basins and Area in South, Central, and North Mongolia. (**a**) Tavantolgoi, (**b**) Gurvantes, (**c**) Khotgor, (**d**) Bayanteeg, (**e**) Ovoot, (**f**) Shariin Gol. For legend, see Fig. 14.4. Note: The locations of deposits are shown in Fig. 14.2

350 m. Due to this intense deformation, coal seams are disrupted highly in all deposit (Fig. 14.4).

The first extensive work on stratigraphy of Western Mongolian Carboniferous coal-bearing sequences, Pennsylvanian volcano-sedimentary Altai Group consisting of three Formations, was compiled in 1992 (Bat-Erdene 1992). This stratigraphic nomenclature has changed considerably, because of the age revision of some

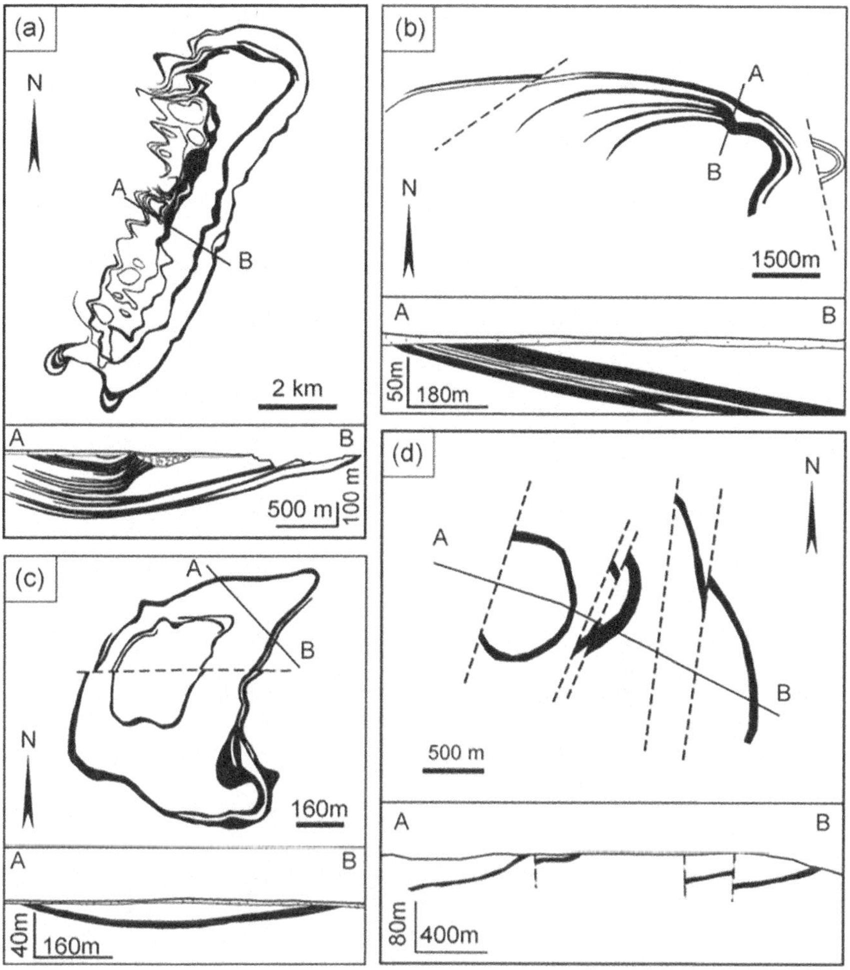

Fig. 14.6 Coal seam maps with cross sections of selected deposits in Eastern Mongolian Province. (**a**) Baganuur, (**b**) Khulstnuur, (**c**) Bayantsogt, and (**d**) Alagtogoo. For legend, see Fig. 14.4. Note: The locations of deposits are shown in Fig. 14.2

Formations of Altai Group from Pennsylvanian to Mississippian (Ariunchimeg 2011).

Coal measures in Trans-Altai and Bayan-Ulgii Areas are hosted in Mississippian Olonbulag Formation, whereas those in Mongol-Altai and Kharkhiraa Basins are hosted in Pennsylvanian Khushuut and Uliastai Formations, respectively (Fig. 14.3). Olonbulag Formation is composed of conglomerate, sandstone, and siltstone with the thin layers of limestone, coaly and muddy shale, and coal seams. The thickness of the Formation ranges from 500 to 600 m. Khushuut Formation consists of sandstone, siltstone, coaly shale, and coal seams. The Formation rarely hosts thin

Table 14.1 Number and thickness of coal seams in the Western Mongolian coal-bearing province (for location, see Fig. 14.2). Data from Erdenetsogt et al. (2009) and published with permission of Elsevier

Basin	Deposit and age	Number of coal seam	Total net thickness, up to, (m)	Basin/area	Deposit and age	Number of coal seam	Total net thickness, up to, (m)
Mongol-Altai Basin	Khushuut (C)	6	84	Mongol-Altai Basin	Mantag (J)	2	30
	Maanit (C)	1	75		Jargalant (J)	2	13
	Khurengol (C)	12	205	Kharkhiraa Basin	Khuden (C)	6	26
	Tsagaangol (C)	3	12		Nuurstkhotgor (C)	13	199
	Zeegt (C)	2	28		Khartagvagatai (C)	1	85
	Gantsmod (C)	5	17	Trans-Altai area	Olonbulag (C)	3	54

layers of conglomerate and volcanogenic rocks. The thickness of the Formation varies from 300 m to 1200 m. Uliastai Formation is characterized by conglomerate, coarse-grained sandstone, siltstone, and coal seams. Thickness of the Formation ranges from 550 to 1100 m.

Mesozoic and Cenozoic sedimentary sequences are distributed widely in the Valley of Great Lakes, where Jurassic coals have been catalogued, as well (see Fig. 14.1). A coal-bearing middle Jurassic Jargalant Formation, which unconformably rests on the Ediacaran-Lower Cambrian rocks, is characterized by conglomerate, interbedded medium to coarse sandstone, siltstone, and mudstone, with coal seam. The thickness of the Formation ranges from 600 m to 1400 m (Khosbayar 1972).

Table 14.1 shows the number and total net thickness of coal seams in Carboniferous and Jurassic coal-bearing Formations of the deposits in the Province. Carboniferous Formations contain a total of 22 seams with complicated seam splitting, whereas Jurassic Formation contains five seams. In general, Jurassic coals are relatively stable and do not have complex seam splitting.

The ages of the Carboniferous and Jurassic coal-bearing sequences are postulated on the basis of fauna and flora. Fossils that have been found from the Olonbulag Formation are Tolmatchoffia, Piloricilla, Kazakhstania sp., Michiganites sp., Lepidodendropsis concinna Radcz. etc.; from the Khushuut Formation are Angaropteridium cardiopteroides (Schm.) Zal., *Angaropteridium* sp., chakassicum Radcz., *Angaropteridium* sp. Abaeanum Zal., *Angaropteridium* sp., Paracalamites sp., *Cordaites* sp., Carpolithes sp., etc.; and from the Uliastai Formation are Neoggerathiopsis cf. theodori Zal. et. Tschern., N. tomiensis Radcz., Sphenophyllum sp., Crassinerva sp., Gaussia cristata Neub., Samaropsis sp. etc. (Ariunchimeg 2011; Ariunchimeg et al. 2014). A summary of previously studied flora, insects, and mollusks from the Jargalant Formation can be found in Sjostrom et al. (2001).

14.2.2.1 Mongol-Altai Basin

The Mongol-Altai Basin lies on the eastern flank of Mongol-Altai Mountain range to the west and the western part of the Valley of Great Lakes to the east. The Basin has a northwestern-southeastern orientation, and the total area is approximately 60,000 km^2 (Fig. 14.2).

Within the Basin, the Pennsylvanian Khushuut Formation contains coal seams. The Formation rests on Devonian marine sedimentary and volcanic and Mississippian sedimentary rocks and is unconformably overlain by Permian volcano-sedimentary rocks. The thickness of the Formation varies from 330 m to 1200 m but thins toward the Basin margins. The Formation hosts more than ten coal seams, of which six seams are widely spread. Average thickness of each seam ranges from 4 to 18 m. At the center of the Basin, the thickness of the thickest coal seam ranges from 25 to 35 m but decreases from 1.9 to 5.5 m at the north and east Basin margins (Erdenetsogt et al. 2009).

Table 14.2 Coal characteristics of Western Mongolian coal-bearing province (for location, see Fig. 14.2)

Basin/area	Deposit and coal seam number/name	Mean thickness, in (m)	Moist. (dry, wt.%)	V.M. (wt.%) d.a.f.	Ash (dry, wt. %)	C.V. (MJ/kg) d.a.f.	S. (dry, wt.%)	R_{max} (%)
Mongol-Altai Basin	Khushuut, “C”	34	1.2	19.3	15.5	29.9[a]	0.5	1.85
	Maanit, I	17	1.3	28.3	11.6	31.9	0.6	0.92
	Khurengol, IV	22	0.6	19.5	44.9	34.9	0.4	1.59
	Tsagaangol, VI	6	2.5	31.6	33.2	–	–	3.66
	Zeegt, “upper”	16	4.7	34.4	16.6[b]	28.2	0.4	0.86
	Gantsmod, III	4	2.7	37.6	17.8	31.1	0.5	–
	Mantag, “lower”	12	6.9	45.5	20.6	28.3	1.0	–
	Jargalant, I	5	7	47.6	14.0	22.0	0.6	0.4
Kharkhiraa Basin	Khuden, VI	9	3.7	32.1	20.6	29.5	0.5	–
	Nuurst khotgor, I	53	5.1[b]	27.1[b]	18.5[b]	30.4	0.4[b]	1.2
	Khartarvagatai, “Avraga”	32	4.0	29.9	16.4	31.2	0.5	1.03
Trans-Altai area	Olonbulag, III	37	2.7	16.6	46.0	27.6	0.6	1.66
Bayan-Ulgii area	Rashaant, I	3	3.0	14.4	7.5	29.8	1.8	2.83

Data from Erdenetsogt et al. (2009), Bat-Erdene (2009a), Bat-Erdene (2014), Bat-Erdene (2015), and Sjostrom et al. (2001)

Moist. Moisture content, *V.M.* volatile matter, *C.V.* Calorific value, *S.* Total sulfur, R_{max} maximum vitrinite reflectance in oil, *d.a.f.* dry, ash-free basis

[a]Expressed on as received basis

[b]Expressed on air-dried basis

Table 14.3 Petrographical and elemental composition of coal deposits in the Western Mongolian coal-bearing province (for location, see Fig. 14.2)

			Petrographic characteristics (maceral abundance, vol.%, m.m.f.)			Ultimate analysis (wt.%)	
Basin	Deposit and coal seam number/name	Number of sample studied	Vitrinite	Inertinite	Liptinite	Carbon	Hydrogen
Mongol-Altai Basin	Khushuut, "C"	22	70.8	28	1.2	88.4	4.2
Kharkhiraa Basin	Nuurstkhotgor, VII	22	44.9	53.3	1.8	82.5	4.7
	Khartarvagatai, "Avraga"	20	47.7	48.4	3.9	77.4	4.5

Data from Erdenetsogt et al. (2009) and published with permission of Elsevier
m.m.f. mineral matter-free basis

The coal characteristics of selected deposits in the Mongol-Altai Basin are shown in Tables 14.2 and 14.3. Coal deposits of Mongol-Altai have moisture contents varying from 0.6% to 4.7 wt.% (average 2.2 wt.%) and volatile matter ranging from 19.3 wt.% to 37.6 wt.% (average 28.5 wt.%). Calorific value and ash contents range from 28.2 to 34.9 MJ/kg (average 31.5 MJ/kg) and 11.6 wt.% to 44.9 wt.% (average 23.3 wt. %), respectively. The maximum vitrinite reflectance values vary from 0.86% to 1.85%, with an average of 1.40%. Tsagaangol coal (see deposit 10 in Fig. 14.2) has relatively high vitrinite reflectance (R_{max} – 3.66%) due to the local extrusion of volcanic rocks (Bat-Erdene and Jargal 1995). Sulfur content is low, averaging 0.6 wt.%, and does not exceed more than 1 wt.% (Erdenetsogt et al. 2009).

Previous petrographic studies have been rather scarce. Petrographic works have only been performed for the Khushuut deposit. According to that study, the Khushuut coals are dominated by vitrinertite lithotypes and are rich in vitrinite maceral groups (70.8 vol.%) and to a lesser extent by inertinite (28 vol.%) and liptinite (1.2 vol.%) maceral groups.

The carbon and hydrogen content of Khushuut coal are 88.4 wt.% and 4.2 wt.%, respectively (Table 14.3). The coal seams in the Basin are characterized by high- to low-volatile bituminous coal (Erdenetsogt et al. 2009).

One of the main deposits in the Basin is Khushuut deposit, which has been mined since the 1970s (deposit 7 in Fig. 14.2).

In 2017, 1.3 Mt of run-of-mine coal was produced, and 0.6 Mt of bituminous coal was exported to China (MEC 2018). The coal-bearing Pennsylvanian Khushuut Formation in the deposit is 1200 m thick. The lower part of the Formation is composed mainly of conglomerate with the thin layers of sandstone, siltstone, mudstone, and coal. The upper part is represented by sandstone, siltstone, mudstone, and thick coal seams. Deformation of coal seams is very intense. Two anticlines and three synclines are revealed, and the dip of coal seams ranges from 35° to 70°. In total, 22 faults with displacement usually less than 45 m, but up to 300 m, are determined (Fig. 14.4a).

Six seams are minable, and the average thicknesses of seams vary between 3.2 m and 33.9 m. The contents of moisture and volatile matter range from 0.8 to 1.2 wt.% (air-dried basis –a.d.b.) and from 19.3 to 10.8 wt.% (a.d.b), respectively. The ash content and net calorific value of the coals are between 15.5 and 30.1 wt.% (dry basis –d.b.) and 24.4 MJ/kg and 30.1 MJ/kg, respectively. Total sulfur contents range from 0.4% to 0.55% (d.b.). Only the uppermost two seams, called C and B seams, are suitable for coke manufacture indicated by caking index (G index) ranging from 44 to 86. The coal rank is low-volatile bituminous, and reserve is ~260 Mt.

14.2.2.2 Kharkhiraa Basin

The Kharkhiraa Basin, about 45,000 km^2, is situated in the north of the Province (Fig. 14.2). The Basin lies by the Kharkhiraa Mountain and the drainage basin of Uvs and Khyargas lakes. There are several disconnected synclinal structures that are filled with the sedimentary rocks of Pennsylvanian Uliastai Formation, hosting very thick coal seams.

The Uliastai Formation contains more than ten coal seams, with a total net thickness of 200 m. The thickness of the Formation is 1100 m but thins to the southwest. Similar to the Mongol-Altai Basin, the coal seam thickness also decreases from the center of the basin to the edges (from 200 m to 9 m). The Uliastai Formation, resting on Mississippian volcano-sedimentary rocks, is unconformably overlain by Jurassic sedimentary rocks.

Coals in the Basin contain 29.9 wt.% to 32.1 wt.% volatile matter, 3.7 wt.% to 4.0 wt.% moisture, and 16.4 wt.% to 20.6 wt.% ash (Table 14.2). Calorific values of the coals range from 29.5 to 31.2 MJ/kg. Maximum vitrinite reflectance values vary from 0.89% to 1.03%. The carbon and hydrogen contents range from 77.4 wt.% to 82.5 wt.% and 4.5 wt.% to 4.7 wt.%, respectively. Petrographic studies of the two largest deposits show that the Kharkhiraa coal are dominated by inertinite (48.4 vol.% to 53.5 vol.%) and vitrinite group macerals (44.9 vol.% to 47.7 vol.%). Liptinite group varies from 1.8 vol.% to 3.9 vol.% (Table 14.3). Inertinite content is high compared to that of the Mongol-Altai Basin. The coal of the Kharkhiraa Basin is high-volatile bituminous coal (Erdenetsogt et al. 2009).

The largest coal deposit in the Basin is Nuurstkhotgor (deposit 2 in Fig. 14.2). The deposit lies in a plunging synclinal structure that is filled with Devonian and Carboniferous sedimentary rocks. Coal-bearing Pennsylvanian Uliastai Formation consists of 200 m thick conglomerate, 250 m thick sandstone with coal seam and siltstone, 160 m thick siltstone with sandstone and coal seam, and 550 m thick sandstone with siltstone and coal seams. The dip of sedimentary rock of Uliastai Formation is 30°–60° at the structure margin but 10°–15° at the center (Fig. 14.4c). The Uliastai Formation contains 13 seams, with thicknesses ranging from 2 to 64 m. The thickness of Seam I, the lowermost one, reaches up to 200 m at the west of the deposit. The moisture and volatile matter contents range from 1.3 to 5.1 wt.% (a.d.b.) and from 21.3 to 27.1 wt.% (a.d.b), respectively. The ash content and gross calorific value of the coal are between 18.5 and 29.7 wt.% (a.d.b.) and 26.9 and 33.1 MJ/kg, respectively. Total sulfur contents range from 0.3 to 0.6% (a.d.b.). Seams I and II have high G index with an average of 85 and 70, respectively. Average G index values of the seams deteriorate from 85 for Seam I, the lowermost seam, to ~0 for Seam XIII, the uppermost seams. The coal rank is high- to medium-volatile bituminous, and coal reserve is calculated at 1.0 Gt (Bat-Erdene 2014).

14.2.2.3 Bayan-Ulgii and Trans-Altai Areas

Coal deposits in Bayan-Ulgii and Trans-Altai Areas are still largely unexplored, because of their remoteness, even though geological surveys have been investigated since the 1920s. There are few occurrences and only two known deposits (Fig. 14.2).

The Bayan-Ulgii Area is located in the westernmost part of the Province and occupies 35,000 km^2. The Trans-Altai Area, about 72,000 km^2, lies in the southernmost part of the Province. Within both Areas, three coal seams, with a thickness ranging from 0.7 to 45.8 m, are hosted in the Mississippian Olonbulag Formation. At Olonbulag deposit, in the Trans-Altai Area, coal measures were determined to be 42 m (average net thickness). At Rashaant, in the Bayan-Ulgii Area, the thickness of coal seam is less than 3 m.

According to the data from the two studied deposits, coal in the Bayan-Ulgii and Trans-Altai Area have 2.7 wt.% to 3.0 wt.% moisture, 14.4 wt.% to 16.6 wt.% volatile matter, and 7.5 wt.% to 46 wt.% ash content. Sulfur contents and calorific values are from 0.6 wt.% to 1.8 wt.% and 27.6 to 29.8 MJ/Kg calorific values, respectively (Table 14.2). Maximum vitrinite reflectance values are 1.66% in the Olonbulag and 2.83% in the Rashaant. It suggests that the rank of coal is low-volatile bituminous, possibly semianthracite (Erdenetsogt et al. 2009).

14.2.2.4 Jurassic Coal

Two small deposits, Mantag and Jargalant, and few occurrences have been discovered within the Valley of Great Lakes (see deposits 4 and 5 in Fig. 14.2). Coal potential is predicted to be huge, because 1000 m thick Middle Jurassic Jargalant Formation, hosting more than five coal seams, is extensively distributed in the Valley of Great Lakes (Erdenetsogt et al. 2009). The Jurassic coal contains 45.5–47.6 wt.% volatile matter and 14.0–20.6 wt.% ash and have 22.0–28.3 MJ/kg calorific value (Table 14.2). Moisture content is 7.0 wt.%. Sulfur content ranges from 0.6 wt.% to 1.0 wt.%. Maximum vitrinite reflectance of coal at Jargalant deposit is 0.4%.

14.2.3 Geology of Coal-Bearing Basins and Area in South, Central, and North Mongolia

14.2.3.1 South Gobi Basin

Coal deposits in Southern Mongolia occur in the South Gobi Basin, which covers 40,000 km^2 with an extension of 600 km running from east to west along its axis (Fig. 14.2). Main coal resource is hosted in Permian sedimentary rocks; however coal presence also to be found in Jurassic sequence (e.g., Nariinsukhait, see deposit 20 in Fig. 14.2). The geology of south Mongolia is subdivided into several terranes, including island arc, cratonal terranes, and forearc/back arc basins. The Devonian and Carboniferous periods were a time of growth and accretion within long-lived volcanic arc systems. Amalgamation was completed by the Permian period, except for the Sulinkheer terrane, located in the southeasternmost part of Mongolia (Lamb and Badarch 2001; Badarch et al. 2002). The terrains are overlain by Permian, Triassic, Jurassic, and Cretaceous volcanic and sedimentary rocks (Badarch et al. 2002; Hendrix et al. 2001).

Coal-bearing Middle Permian Tavantolgoi Formation fills several sub-basins and is folded and faulted by thrust and normal faults. Thickness of the formations varies from 950 to 1150 m (Fig. 14.3). The lower section of the Formation is characterized by medium- to fine-grained sandstone, siltstone, shale, and coal seams. It also contains thin layers of conglomerate and siltstone with siderite. The middle section is represented by conglomerate, sandstone with siltstone, and coal seams. The upper section of the Formation consists of conglomerate with shale and medium- to fine-grained sandstone and coal measures. The Formation hosts 16 individual coal seams. The age of the coal-bearing Formation is established from the fossil flora, which are identical to that of the central region of Angaraland (Durante 1976; Orolmaa and Ariunchimeg 2011).

The South Gobi Basin contains a predominant portion of the known coking coal reserves of Mongolia; however, significant amounts of coking coals exist in the Western Mongolian Province (Erdenetsogt et al. 2009). Two big deposits in the Basin are Middle Permian Tavantolgoi with reserves of 6.4 Gt and Middle Jurassic Nariinsukhait with reserves of 724.9 Mt. In 2016, three mines at Tavantolgoi deposit produced 14.7 Mt of coal and exported 13.7 Mt of coking coal to China. In the same year, four mines at Nariinsukhait deposit exported 11.9 Mt of semisoft coking and thermal coal to China (MRPAM 2018).

The Tavantolgoi is the largest coal deposit of Mongolia located in 50 km long and 15 km wide structure, called Ulaannuur sub-basin. This east-west trending structure is divided into several synclines (Tsankhi, Ukhaahudag, Borteeg, Onchkharaaat, Oortsog, Bortolgoi, etc.), which are separated by basement uplifts and controlled by east-west trending thrust and normal faults (Fig. 14.5a). Faulting is relatively intense, and displacement reaches 200–300 m. In general, Permian sedimentary rock dip up to 45°, but dips of less than 10° at the center of syncline are more common. Middle Permian Tavantolgoi Formation, 960 m thick, hosts 16 coal seams that are numbered from Seam 0 (the lowermost) through to Seam XV (the uppermost). The thickness ranges from 0.6 to 88 m (Table 14.4). Main seams are Seam 0, III, IV, V, and VIII, and all have very complicated seam splitting. For instance, the Seam IV has 29 partings, of which thickness ranges from 0.09 to 60.0 m.

Table 14.5 shows the coal characteristics of the economic seams of Tavantolgoi deposit. Moisture and volatile matter contents range from 0.5 wt.% to 2.1 wt.% and from 23.1 wt.% to 34.2 wt.%, respectively. Ash yield varies from 19.9 wt.% to 27.0 wt.%, and calorific values are from 32.6 to 34.4 MJ/kg. Total sulfur contents of coal seams are similar, 0.7 wt.%. The value of maximum vitrinite reflectance ranges between 0.74% and 1.28%. Free swelling index (FSI) of the Seams III, IV, and V ranges from 5 to 7.5. The rest of seams have a FSI varying from 1 to 3.5. The rank of Tavantolgoi coal decreases from low-volatile bituminous (Seam 0) to high-volatile bituminous (Seam VIII).

Petrographic studies indicate that the coal is characterized by vitrinertite and duroclarite. The coal seams developed in the lower part of the Tavantolgoi Formation have the highest inertinite content, which gradually decreases toward the upper section of the Formation (Table 14.6). Vitrinite contents increase from 55.4 vol.% at the bottom seams to 72 vol.% at the top seams. It is most likely that the lower seams

Table 14.4 Number and net thickness of coal seams in the Southern Khangai, Ikh Bogd, Ongi River, and South Gobi Basins (for location, see Fig. 14.2)

Basin	Deposit and age	Number of coal seam	Total net thickness, up to, (m)
Southern Khangai Basin	Tsakhiurt Urt (P)	2	6.2
	Uvurchuluut (K)	2	17
	Sharig Nuruu (P)	5	25
Ikh Bogd Basin	Alagtsakhir (J)	1	28
	Khotgor (J)	5	45
Ongi River Basin	Bayanteeg (J)	1	36
	Tsagaan-Ovoo (J)	1	30
South Gobi Basin	Tavantolgoi (P)	16	179
	Nariin sukhait (J)	12	117
	Gurvantes (P)	4	15

Data from Erdenetsogt et al. (2009) and published with permission of Elsevier

of the Tavantolgoi coal were formed under low water level conditions. With increasing subsidence, water levels increased, and the upper seams, rich in vitrinite, developed under more anaerobic conditions, which impeded oxidation. The increase in the ash content of coals from the base to the top of the Tavantolgoi Formation suggests that the amount of clastic influx into the basin increased over time due to the increased influence of running water (Erdenetsogt et al. 2009).

14.2.3.2 Southern Khangai, Ikh Bogd, and Ongi River Basins

The three Basins, covering an area ranging from 4800 to 35,000 km^2, are located in the Valley of Lakes (Fig. 14.2). This east-west-oriented narrow valley is bounded by the Gobi-Altai ranges to the south and southwest and the Khangai Mountains to the north (Fig. 14.1). Middle Permian, Lower-Middle Jurassic, and Lower Cretaceous nonmarine sedimentary sequences contain coal seams in the Basins (Erdenetsogt et al. 2009) (Table 14.4). Coal-bearing middle Permian sequence is characterized by siltstone, shale, coaly shale, and coal seams. At Tsakhiurt Urt deposit, the thickness of Permian sequence is 210 m. Lower-middle Jurassic Bakhar Formation consists of conglomerates, sandstone, shale, and coal, all of which intercalate with volcanic rocks. Formation thickness reaches 2700 m (Fig. 14.3). Lower Cretaceous Andkhudag Formation is composed of sandstone, thick oil shale, coal-bearing mudstone with thin limestone, and marls. The Basins, especially the Southern Khangai Basin, are intensively folded and faulted due to Cenozoic uplift and the Altai and Khangai Mountain building process (Erdenetsogt et al. 2009).

Table 14.5 Coal quality of Southern Khangai, Ikh Bogd, Ongi River, and South Gobi Basins (for location, see Fig. 14.2)

Basin	Deposit and coal seam number/name	Mean thickness, in (m)	Moist. (dry, wt.%)	V.M. (wt. %) d. a.f.	Ash (dry, wt. %)	C.V. (MJ/kg) d.a.f.	S. (dry, wt. %)	R_{max} (%)
Southern Khangai Basin	Tsakhiurt Urt, I	3	6.5	41.0	22.0	33.2	–	–
	Uvurchuluut, II	5	14.0	37.7	15.6	25.8	1.2	–
	Sharig Nuruu, III	3	4.8	28.3	33.6	29.8	0.3	1.18
Ikh Bogd Basin	Alagtsakhir, I	14	19.1	44.8	13.4	21.1	1.3	–
	Khotgor, V	7	2.2	41.4	22.0	33.8	0.5	0.79
Ongi river basin	Bayanteeg, I	25	2.2	51.9	22.6	30.2	–	0.51
	Tsagaan-Ovoo, I	12	6.1	41.1	14.4	21.3	1.6	–
South Gobi Basin	Tavantolgoi, XIII	7	2.1	34.2	27.0	32.6	0.7	0.74
	Tavantolgoi, IX	20	0.5	30.1	24.6	33.7	0.7	0.95
	Tavantolgoi, VIII	23	0.6	29.8	24.0	33.7	0.7	0.96
	Tavantolgoi, IV	12	0.5	26.3	21.5	33.7	0.7	1.05
	Tavantolgoi, III	6	0.6	25.9	19.9	34.4	0.7	1.09
	Tavantolgoi, 0 + I	19	0.5	23.1	25.1	33.5	0.7	1.28
	Nariin sukhait, V	60	1.2[a]	31.7[a]	10.1[a]	33.8	1.2	0.78
	Gurvantes, I	9	2.3	34.3	20.7	32.5	0.9	–

Data from Erdenetsogt et al. (2009), Bat-Erdene (2009a), Baigalmaa et al. (2012), Ulambadrakh and Baigalmaa (2014), and Erdenetsogt et al. (2014)
[a]Expressed on air-dried basis

Thickness of coal seams ranges from 3.0 to 25 m. Coal ranks are from bituminous coal (Permian and Jurassic) to lignite (Cretaceous). The average volatile matter and moisture contents vary from 28.3 wt.% to 51.9 wt.% and from 2.2 wt.% to 19.1 wt. %, respectively. Calorific value and ash contents of the coal are 21.1 to 33.8 MJ/kg and 13.4 wt.% to 33.6 wt.%, respectively. Total sulfur content varies from 0.3 wt.% to 1.6 wt.% (average 1.0 wt.%). In Tsagaan-Ovoo coal (see deposit 19 in Fig. 14.2), average carbon and hydrogen contents are 75.9 wt.% and 5.7 wt.%, respectively. Maximum vitrinite reflectance value of Permian coal is 1.18%, whereas that of Jurassic coal ranges from 0.51 to 0.79 (Table 14.5). The petrographical studies suggest that coal in the Basins is dominated by vitrinite group macerals (76.8 vol.

Table 14.6 Elemental and petrographical compositions of Southern Khangai, Ongi River, Ikh Bogd, and South Gobi Basins (for location, see Fig. 14.2)

Basin/area	Deposit and coal seam number/name	Petrographic characteristics (maceral abundance, vol.%, m. m.f.)			Ultimate analysis (wt.%)	
		Vitrinite	Inertinite	Liptinite	Carbon	Hydrogen
Southern Khangai Basin	Sharig Nuruu, III	85.0	14.0	1.0	–	–
Ikh Bogd	Khotgor	76.8	20.6	2.6	88.3	5.6
Ongi River Basin	Bayanteeg, I	96.6	2.0	1.4	–	–
	Tsagaan-Ovoo, I	87.3	1.0	11.7	75.9	5.7
South Gobi Basin	Tavantolgoi, XIII	72.0	21.0	7.0	–	–
	Tavantolgoi, IX	78.0	19.0	3.0	–	–
	Tavantolgoi, VIII	75.5	22.5	2.0	–	–
	Tavantolgoi, IV	73.5	25.0	1.5	–	–
	Tavantolgoi, III	63.0	36.0	1.0	–	–
	Tavantolgoi, 0 + I	55.4	43.6	1.0	–	–
	Nariin sukhait, V	86.9	11.5	1.6	84.7	5.3

Data from Erdenetsogt et al. (2009), Ulambadrakh and Baigalmaa (2014)

% to 96.6 vol.%). Inertinite group varies from 1 vol.% to 20.6 vol.%. Tsagaan-Ovoo coal has up to 11.7 vol.% liptinite group, which is notably high when compared to the other Mongolian coal (Table 14.6).

Khotgor deposit is located in Ikh Bogd Basin (see deposit 17 in Fig. 14.2). Five coal seams are hosted in the 1100 m thick lower-middle Jurassic Bakhar Formation, which forms northwest-southeast trending monocline structure and dips 30° in a northeasterly direction. The thickness of the coal seams ranges from 2 to 38 m (Fig. 14.5c). The uppermost seam or Seam V is economically important and average thickness is 7 m.

The ash yield of coal ranges from 14.8 to 25.1 wt.%. Moisture and volatile matter contents vary from 1.6 to 2.2 wt.% and 33.4 to 41.4 wt.%, respectively. Total sulfur content is low, ranging from 0.2 to 0.7 wt.%. Gross calorific value ranges from 33.8 to 34.7 MJ/kg, and the maximum vitrinite reflectance value is from 0.74% to 0.81%. FSI and G index of washed coal are from 5 to 7 and from 59 to 72, respectively. The coal rank is high-volatile bituminous and reserve is calculated at 160 Mt (Ulambadrakh and Baigalmaa 2014).

14.2.3.3 Orkhon-Selenge Area

The Orkhon-Selenge Area, about 240,000 km^2, comprises the Khangai and Khuvsgul mountains (Figs. 14.1 and 14.2). The basement of the Area is composed of Precambrian metamorphic complex, which is overlain by Paleozoic marine turbidites and volcanic-plutonic rocks (Badarch et al. 2002). Within the Area,

Table 14.7 Number and net thickness of coal seams in the Orkhon-Selenge Area (for location, see Fig. 14.2)

Area	Deposit and age	Number of coal seam	Total net thickness, up to (m)
Orkhon-Selenge Area	Ovoot (J)	3	133
	Ulaan-Ovoo (J)	6	69
	Shariingol (J)	11	50
	Nalaikh (K)	9	54
	Khujirt (J)	5	39
	Jilchigbulag (J)	2	16
	Nuurstei(J)	13	48
	Bayantsagaan (J)	6	7

coal-bearing Jurassic and Lower Cretaceous sedimentary sequences are preserved in few scattered small grabens (Figs. 14.5e, f). Jurassic rocks host up to 13 seams of subbituminous and bituminous coal in the central and north of the Area, whereas lower Cretaceous rocks contain 10 seams coal in the east, e.g., Nalaikh (Erdenetsogt et al. 2009) (Table 14.7). In the northwest of the Area, Lower-Middle Jurassic Egyngol Formation hosts coal seams. The lower part of the Formation is characterized by coaly shale, sandstone, and coal measures with conglomerate, whereas upper part consists mainly of conglomerates. The average thickness is 1100 m. In the central part of the Area, Middle Jurassic Saikhan Formation hosts coal seams. The Formation is 670 m thick and is composed of conglomerate, sandstone, siltstone, mudstone, and coal seams. In the northeast of the Area, Middle Jurassic Shariingol Formation contains coal seam, which consists of three members (Fig. 14.3). The bottom member is represented by conglomerate with thin layers of sandstone. Coal-bearing middle member is composed of sandstone, siltstone, and coal seams. Upper member consists of dark gray shales with sand, stone, and siltstone. The average thickness of the Formation is 950 m. The ages of the Jurassic Formations are postulated from plant fossils (Makhbadar 2011). In the east of the Area, Lower Cretaceous Khukhteeg Formation of the Zuunbayan Group hosts coal. At Nalaikh deposit, the Formation is characterized by sandstone, clay shale, siltstone, and shale interbedded with coal seams. The thickness of Formation is 180 m (see deposit 29 in Fig. 14.2).

Moisture content of the coals in the Orkhon-Selenge Area ranges from 0.6 wt.% to 10.1 wt.% and volatile matter content ranges 28.7 wt.% to 46.0 wt.%. Ash content and calorific values are from 11.2 wt.% to 34.7 wt.% and 29.6 to 34.2 MJ/kg, respectively. Total sulfur contents vary between 0.2 and 1.4 wt.%. Coals in the northwest have high vitrinite reflectance, ranging from 1.3% to 1.2%, but it decreases to 0.5% in the northeast (Table 14.8). The results of petrographic studies show that vitrinite is dominant maceral group, ranging from from 81.5 vol.% to 98.9 vol.%. Inertinite and liptinite maceral groups range from 0.6 vol.% to 15.2 vol.% and from 0.5 vol.% to 3.5 vol.%, respectively. It suggests that the coals belong to the vitrinertite and vitrite microlithotypes. The contents of carbon and hydrogen vary

Table 14.8 Coal quality of Orkhon-Selenge Area (for location, see Fig. 14.2)

Basin	Deposit and coal seam number/name	Mean thickness, in (m)	Moist. (dry, wt.%)	V.M. (wt. %) d. a.f.	Ash (dry, wt. %)	C.V. (MJ/kg) d.a.f.	S (dry, wt. %)	R_{max}, (%)
Orkhon-Selenge Area	Ovoot, "upper"	26	0.6	33.5	16.3	34.2	1.4	1.2
	Ulaan-Ovoo, 'upper"	48	7.3	46.0	11.2	31.1	0.3	0.52
	Shariingol, "Belikan"	50	6.9	41.4	20.5	29.8	0.9	0.55
	Nalaikh (mean)	–	10.1	45.7	17.9	27.4	0.2	–
	Khujirt, II-1	6	4.8[a]	43.4	11.7[a]	29.6	1.0	0.50
	Jilchigbulag (mean)	–	2.5	41.9	20.1	30.0	0.6	–
	Nuurstei, mean	–	0.7	28.7	34.6	35.4	0.8	1.34
	Bayantsagaan, IV	2	6.7	36.6	23.7	29.7	1.2	–

Data from Erdenetsogt et al. (2009), Bat-Erdene (2009a), Erdenetsogt and Barsbold (2014), Erdenetsogt et al. (2014), and Jargal et al. (2017)
[a]Expressed on air-dried basis

Table 14.9 Elemental and petrographical compositions of Orkhon-Selenge Area (for location, see Fig. 14.2)

Basin/area	Deposit and coal seam number/name	Petrographic characteristics (maceral abundance, vol.%, m. m.f.)			Ultimate analysis (wt.%)	
		Vitrinite	Inertinite	Liptinite	Carbon	Hydrogen
Orkhon-Selenge area	Ovoot, upper	98.9	0.6	0.5	88.1	5.2
	Ulaan-Ovoo, "upper"	91.7	4.8	3.5	74.7	3.9
	Shariingol, "Belikan"	91.6	6.0	2.4	75.6	5.1
	Khujirt, II-1	81.5	15.2	3.3	–	–

Data from Erdenetsogt et al. (2009), Erdenetsogt and Barsbold (2012)

between 74.7 wt.% and 88.1 wt.% and from 3.9 wt.% to 5.2 wt.%, respectively (Table 14.9).

Ovoot deposit is located in west of the Area (see deposit 23 in Fig. 14.2). The deposit is the second largest coking coal resource after Middle Permian Tavantolgoi in Southern Mongolia. Coal seams are hosted in Lower-Middle Jurassic Egyngol formation, consisting of two members. The total thickness is 1100 m. Lower coal-bearing member is mainly composed of intercalated of multi-grained sandstone, siltstone, coal measures and coaly shale layers. Upper member consists of only conglomerates at the margin of Basin. However, at the center of Basin, the member is characterized by sandstone and conglomerates. The age of coal-bearing sequence is postulated based on plant remains (Erdenetsogt and Barsbold 2014).

The Jurassic coal-bearing sequence in Ovoot is gently folded into a northeast-southwest trending syncline. Seams generally dip 6° toward the fold axis (Fig. 14.5e). There are three main seams called upper, lower, and OVB. The seams have multiple splitting. The upper seam thickness increases from north (2 m) to south (50 m). Thickness of the Lower seam varies between 4 m and 62 m. A third and lowermost coal seam, called as OVB seam, is locally developed in a restricted basement low, possibly the result of a large sink hole in the limestone basement. Average net coal thickness of the OVB seam is 21 m.

Coal quality of the upper seam is shown in Table 14.8. Ovoot coal has high content of vitrinite with an average of 98.9 vol. %. Maximum vitrinite reflectance ranges from 1.15% to 1.25%. The average FSI and G index of washed coal are 9 and 88, respectively. In addition, the coal has very high Gray-King coke types, G11. Ovoot coal is high-volatile bituminous coal, and its reserve is calculated at 281 Mt (Erdenetsogt and Barsbold 2014).

14.2.4 Geology of Eastern Mongolian Coal-Bearing Province

The eastern Mongolian Province has major northeast extension and encompasses about 450,000 km^2 area in Eastern Mongolia. The Province includes six basins – Choir-Nyalga, Choibalsan, Sukhbaatar, Tamsag, East Gobi, and Central Gobi (Fig. 14.2). Cretaceous nonmarine sedimentary rocks host main coal reserves in the Province. However, Jurassic and Permian sequences also include minor coals. For rank, Cretaceous coals are lignite, whereas Jurassic and Permian coal are subbituminous to bituminous coal (Erdenetsogt et al. 2009). Within the Province, more than 100 deposits/occurrences have been discovered, and several open-pit mines, including two large mines – Baganuur and Shivee-Ovoo, are in operation. In addition to coal, the basins also host thick oil shale seams that are accounted as main petroleum source rock (e.g., Traynor and Sladen 1995; Yamamoto et al. 1998; Johnson et al. 2003; Hasegawa et al. 2018).

The Province has structurally complex pre-Mesozoic basement that is composed of late Neoproterozoic metamorphic complex, Paleozoic volcanogenic, and carbonate sedimentary rocks. The main amalgamation of the terrains took place during the period from Permian to Triassic (Badarch et al. 2002). Jurassic and Lower Cretaceous nonmarine volcano-sedimentary sequences fill many rift grabens, formed through the late Jurassic-early Cretaceous extension. Deformation of the coal-bearing sedimentary rocks is relatively gentle compared to that of Paleozoic coal-bearing rocks; dip is mostly around 10°. However, it can be exceeded at some deposits due to local geological condition (Fig. 14.6).

Coal-bearing Lower-Middle Jurassic Khamarkhuuvur Formation is characterized by conglomerate, sandstone, coaly siltstone, coaly shale, and coal measures. Thickness ranges from 500 to 800 m. Age of the formation is postulated from fauna and flora (Makhbadar 2011). Coal and oil shale-bearing lower Cretaceous volcano-sedimentary Zuunbayan Group is subdivided into three formations: at the bottom, oil shale-bearing Shinekhudag Formation, coal-bearing Khukhteeg Formation, and Baruunbayan Formation. The thickness of the group is approximately 1800 m. The oil shale-bearing Shinekhudag Formation consists mainly of mudstone with sandstone, limestone, dolomite, marl, and siltstone. The formation hosts more than 100 m thick oil shale seams with up to 21.3 wt. % of organic carbon contents. The coal-bearing Khukhteeg Formation is represented by conglomerate, sandstone, siltstone, coaly shale, and coal seams. There are very thick 20 coal seams (up to 110 m) with complicated seam splitting. The thickness of the formation varies from 730 to 770 m. The Baruunbayan Formation is preserved only in large grabens and characterized mainly by sandstone with conglomerate. Thickness of the formation is highly variable, ranging from 30 to 250 m (Erdenetsogt et al. 2009).

Table 14.10 shows the total net thickness and number of coal seams in Khamarkhuuvur and Khukhteeg Formation of selected deposits in the Province. At the deposit margins, the most of thick coal seams are split.

The fossil assemblages of the Khamarkhuuvur Formation indicate Lower-Middle Jurassic age. The age of Shinekhudag Formation is Hauterivian-Barremian, whereas that of Khukhteeg and Baruunbayan Formations is Aptian-Albian. However, a recent study suggests an early Aptian age for Shinekhudag (Hasegawa et al. 2018). Detailed information on the fossil assemblage is given in Jerzykiewicz and Russell (1991), Khand et al. (2000), and Makhbadar (2011).

14.2.4.1 Choir-Nyalga Basin

This Basin, approximately 50,000 km^2, is in the northwest of the Province and is divided into several fault bounded sub-basins that are filled with Mesozoic sequences (Fig. 14.2). The Shinekhudag Formation is characterized by sandstone, siltstone, and mudstone with oil shale seams. Thickness of the Formation is 900 m. The Khukhteeg Formation is composed of conglomerate, sandstone, siltstone, and thick lignite. The Formation thickness is 450 m. The Baruunbayan Formation is composed mostly of conglomerate and sandstone with rare volcanogenic layers (Erdenetsogt et al. 2009). In the Basin, two major open-pit mines, Baganuur and Shivee-Ovoo, are producing thermal coal for domestic electricity and heat generation, and several smaller mines are exporting thermal coal to China (deposits 30 and 35 in Fig. 14.2).

Coals in the Choir-Nyalga Basin have moisture contents ranging from 7.0 wt.% to 13.3 wt.% with an average of 9.3 wt.% and volatile matter content varying from 41.1 to 50.6 wt.% with an average of 44.8 wt.%. The calorific values and ash contents of the coal are from 25.9 to 29.6 MJ/kg (average 27.3 MJ/kg) and from 10.2 wt.% to 20.9 wt.% (average 14.8 wt.%), respectively. According to data from three deposits, maximum vitrinite reflectance ranges from 0.32% to 0.41% with an average of 0.36%. Sulfur contents vary between 0.2 wt.% and 2.8 wt.% (Table 14.11). Average carbon and hydrogen contents of the coal are 65.2 wt.% and 4.3 wt.%, respectively. The results of petrographic studies indicate that coals are mainly composed of vitrinertite lithotype, and average contents of huminite, inertinite, and liptinite

Table 14.10 Number and net thickness of coal seams in the Eastern Mongolian coal- and oil shale-bearing province (for location see, Fig. 14.2). Data from Erdenetsogt et al. (2009) and Bat-Erdene (2009a)

Basin	Deposit and age	Number of coal seam	Total net thickness, up to, (m)	Basin	Deposit name	Number of coal seam	Total net thickness, up to, (m)
Choir-Nyalga Basin	Tsaidam nuur (K)	5	53	Sukhbaatar basin	Talbulag (K)	7	50
	Uvdugkhudag (K)	5	81		Ulziit (K)	1	17
	Ikh Ulaannuur (K)	4	99	Tamsag basin	Bulangiin khooloi (K)	2	5
	Shivee-Ovoo (K)	10	39		Zuunbulag (K)	3	20
	Baganuur (K)	20	180		Bayantsogt (K)	14	4
	Tsaidam (K)	3	120	Central Gobi basin	Khootiin khonkhor (J)	3	23
	Tugrug (K)	7	17		Nukhet (K)	2	15
	Tevshiin gobi (K)	5	94		Taliin khudag (K)	3	2
	Maint (K)	1	10	East Gobi basin	Khamriin khural (K)	1	15
Choibalsan Basin	Aduunchuluun (K)	2	68		Olongiin Ukhaa (K)	4	27
	Khulstnuur (K)	15	31		Alagtogoo (J)	1	7
	Utaat minjuur (K)	2	7				

Table 14.11 Coal characteristics of the Eastern Mongolian coal- and oil shale-bearing province (for location, see Fig. 14.2). Data from Erdenetsogt et al. (2009) and Bat-Erdene (2009a)

Basin	Deposit and coal seam number/name	Mean thickness, in (m)	Moist. (dry, wt.%)	V.M. (wt.%) d.a.f.	Ash (dry, wt. %)	C.V. (MJ/kg) d.a.f.	S. (dry, wt.%)	R_{max}, (%)
Choir-Nyalga Basin	Tsaidam nuur, II	40	13.3	41.4	10.2	25.9	0.9	0.32
	Baganuur, IIa	27	11.4	44.1	14.2	28.4	0.3	0.41
	Tsaidam		10.0	43.5	15.0	29.6	0.6	–
	Tugrug, "upper"	7	7.3	50.6	14.9	26.1	0.8	–
	Uvdugkhudag, IV	53	7.0	43.4	15.6	26.9	2.4	0.35
	Ikh Ulaannuur, "upper"	71	7.3	44.2	13.4	26.1	2.8	–
	Shivee-Ovoo, V	22	8.2	45.7	17.8	28.0	0.2	–
	Tevshiin gobi, II	35	11.2	45.5	20.9	26.9	0.7	–
	Maint, "lower"	20	8.2	45.2	12.0	27.6	0.5	–
Choibalsan Basin	Aduunchuluun, "lower"	34	9.4	48.1	16.7	27.1	1.1	0.32
	Khulstnuur, "deed"	26	10.2	47.5	12.7	27.3	0.7	–
	Utaat minjuur (mean)	–	7.2	43.0	17.7	19.5	1.2	–
Sukhbaatar Basin	Talbulag, "T3"	27	9.5	47.0	14.0	25.1	0.8	0.25
	Ulziit, I	13.2	8.0	54.4	10.0	26.0	0.8	–
Tamsag Basin	Bulangiin khooloi, "lower"	3	8.4	48.6	15.6	26.0	0.6	–
	Bayantsogt, "lower"	8	13.7	44.6	28.5	21.0	1.5	–
	Zuunbulag, "upper"	6.3	9.0	52.6	26.3	19.3	1.2	–
Central Gobi Basin	Khootiin khonkhor, III	6	8.8	46.6	15.6	28.6	1.5	0.54
	Unst Khudag (mean)	–	7.5	43.9	24.9	23.1	0.8	–
East Gobi Basin	Khamriin khural, I	12	8.7	40.5	24.6	23.0	1.1	–
	Olongiin Ukhaa (mean)	–	13.3	50.6	18.1	27.4	0.4	–
	Alagtogoo, I	4	3.5	33.1	27.8	30.4	1.3	–

Table 14.12 Elemental and petrographical composition of coals in the Eastern Mongolian coal- and oil shale-bearing province (For location see Fig. 14.2)

Basin	Deposit and coal seam number/name	Number of samples studied	Petrographic characteristics (maceral abundance, vol.%, m.m.f.)			Ultimate analysis (wt.%)	
			Huminite	Inertinite	Liptinite	Carbon	Hydrogen
Choir-Nyalga Basin	Tsaidamnuur, II	9	81.7	15.4	2.9	65.4	4.4
	Baganuur, IIa	10	73.8	20.1	6.1	65.7	4.2
	Tsaidam, "upper"	13	80.0	16.6	3.4	–	–
	Tugrug, "upper"	3	65.9	31.2	2.9	65.5	4.3
	Shivee-Ovoo, V	–	81.6	16.7	1.7	61.5	3.9
	Tevshiin gobi (mean)	3	71.5	24.4	4.2	71.7	3.9
	Ikh Ulaannuur, "upper"	21	79.2	18.3	1.5	61.5	5.2
Choibalsan Basin	Aduunchuluun, "lower"	20	54.9	44.0	1.1	–	–
Sukhbaatar Basin	Talbulag, T3	28	76.9	21.9	1.2	–	–
Tamsag Basin	Bulangiin hooloi, "Dood"	7	82.9	15.0	2.1	–	–
Central Gobi Basin	Khootiin khonkhor (mean)	–	96.0	0.7	3.3	60.8	4.1
East Gobi Basin	Alagtogoo, I	5	93.5	3.9	2.6	83.3	4.8

Data from Erdenetsogt et al. (2009), Bat-Erdene (2009a), and Erdenetsogt et al. (2017b) maceral groups are 76.2 vol.%, 20.4 vol.%, and 3.2 vol.%, respectively (Table 14.12).

Baganuur is a relatively well-studied deposit in the Basin. Coal exploration was completed during the period from 1974 to 1975 and mining started in 1978. As of 2015, a total of 99.4 Mt thermal coal was produced for domestic power stations (Erdenetsogt et al. 2017b).

The deposit lies in northeast-southwest trending 3 km wide and 12 km long graben syncline structure, located on the northern edge of the Choir-Nyalga Basin. At the deposit, 280 m thick lower Cretaceous Khukhteeg Formation of Zuunbayan Group contains 23 coal seams including the three main seams – Seam 2, Seam 2A, and Seam 3. Seam 2 is the lowermost seam and has average thickness of 10 m. At the center of the syncline, seam thickness increases to 36 m. Seam 2A, the middle seam, is located 45 m above underlying Seam 2, and it has average thickness of 18 m. However, its maximum thickness reaches 52 m in the center of the structure. Seam 3, the uppermost, is separated from the Seam 2A by a sequence of 85 m thick sedimentary rocks. Seam 3 has complicate seam splitting, and its thickness reaches 98 m. However, its average thickness is 15 m (Erdenetsogt et al. 2010) (Fig. 14.6a).

Baganuur coal has 10.9 to 11.4 wt.% moisture content, 42.4 to 44.1 wt.% volatile matter, 27.7 to 28.6 MJ/kg calorific value, and 14.8 to 15.8 wt.% ash content. Sulfur content varies between 0.2 wt.% and 0.9 wt.%. Average value of maximum vitrinite reflectance of seam 2A is 0.4%. Coal reserves of Baganuur deposit is 809 Mt (Erdenetsogt et al. 2017b).

14.2.4.2 Choibalsan Basin

This Basin, covering approximately 45,000 km^2 area, lies in the northeast of the province (Fig. 14.2). The thickness of the Zuunbayan group reaches 1200 m in the major depocenters. The oil shale-bearing Shinekhudag Formation is composed of conglomerates, sandstones, and siltstones. The Formation thickness is 600 m. The coal-bearing Khukhteeg Formation is 500 m thick and characterized by sandstone, coaly shale, shale, and coal seams. The Baruunbayan Formation consists mostly of volcanogenic rocks. The thickness of the Formation is 100 m (Erdenetsogt et al. 2009).

From the three relatively well-studied deposits, the coals have 8.9 wt.% moisture, 46.2 wt.% volatile matter, 15.7 wt.% ash yield, and 24.6 MJ/kg calorific value (Table 14.11). The maximum vitrinite reflectance value of Aduunchuluun coal and the carbon and hydrogen content of Khulstnuur coal are 0.32%, 70.4 wt.%, and 4.9 wt.%, respectively (see deposits 39 and 41 in Fig. 14.2). The average total sulfur content is 1.0 wt.%. According to the results of previous petrographical studies, Aduunchuluun coal consists of 54.9 vol.% huminite, 44.0 vol.% inertinite, and 1.1 vol.% liptinite (Table 14.12).

Khulstnuur deposit is located in plunging synclinal structure that is 4 km wide and 10.5 km long, filled with Cretaceous sedimentary rocks, dipping at 8° to 15° to the south. Normal faults with displacements usually less than 30 m are identified

(Fig. 14.6b). Coal-bearing Lower Cretaceous Khukhteeg Formation of Zuunbayan Group is 250 m thick and hosts 15 coal seams including three economic seams—lower, middle, and upper seam. The thickness of lower and middle seams ranges from 2 to 10 and 0.4 m to 2 m, respectively. Upper seam is the thickest seam, ranging from 20 m to 31 m with an average of 26 m. The coal is lignite, and its quality is shown in Table 14.12.

14.2.4.3 Tamsag and Sukhbaatar Basins

The Basins, Tamsag (40,000 km^2) and Sukhbaatar (32,000 km^2), are located in the easternmost part of the Province and have a major northeast-southwest orientation (Fig. 14.2). Structurally, the Tamsag Basin consists of one trough, whereas the Sukhbaatar Basin is composed of several half grabens and grabens. The Tamsag Basin is thought to have high petroleum potential compared with any other basins of Mongolia. In the basins, the thickness of the Zuunbayan group ranges from 1000 to 1450 m. The oil shale-bearing Shinekhudag Formation is characterized mainly by conglomerate, sandstone, and siltstone. The coal-bearing Khukhteeg Formation is represented by sandstone, siltstone, shale, and coal seams. Baruunbayan Formation is associated with conglomerates, sandstone, siltstone, and shale, with thin layers of volcanogenic rocks (Erdenetsogt et al. 2009).

Coals of the basin contain 10.0 to 28.5 wt.% ash, 8.4 to 13.7 wt.% moisture, and 44.6 to 54.4 wt.% volatile matter (Table 14.11). Calorific value and total sulfur contents of the coals vary from 0.6 to 1.5 wt.% and from 19.3 to 26.0 MJ/kg, respectively. Coals are vitrinite coal, suggested by 76.9 to 82.9 vol.% huminite and 15 to 21.9 vol.% inertinite maceral groups. Liptinite group macerals are typically less than 2.1 vol.% (Table 14.12). The maximum vitrinite reflectance valuc was measured only in Talbulag coal and that is 0.25%.

At Bayantsogt deposit in Tamsag Basin, 110 m thick Lower Cretaceous Khukhteeg Formation, consisting of sandstone, siltstone, and coaly shale, hosts 2 seams (see 44 deposit in Fig. 14.2). Dip of coal seam at structure margin is 5°–15°, and fault displacement is low up to 18 m. Lower seam is main seam with an average of 8.3 m. Thickness of upper seam varies from 1.1 to 5.0 m (Fig. 14.6c). Coal characteristics are shown in Table 14.12.

14.2.4.4 East Gobi and Central Gobi Basins

The East Gobi Basin comprises approximately 60,000 km^2 area in the southeast part of Mongolia. The Central Gobi Basin, covering an area of approximately 25,000 km^2, lies in Central Mongolia (Fig. 14.2). Both Basins are divided into several small sub-basins with a general northeast-southwest orientation. Lower Cretaceous rocks contain predominant portion of coal. However, two coal deposits and several occurrences are hosted in Jurassic sedimentary strata, as well

(Erdenetsogt et al. 2009). East Gobi Basin is accounted as one of the basins that have high petroleum potential in Mongolia.

The coal in the Central Gobi and East Gobi Basins has 3.5 wt.% to 13.3 wt.% moisture, 15.6 wt.% to 24.9 wt.% ash, and 33.1 wt.% to 46.6 wt.% volatile matter. Calorific value and sulfur content are ranging from 22.7 to 30.4 MJ/kg and from 0.4 wt.% to 1.5 wt.%, respectively (Table 14.11). Petrographical studies of two Jurassic coal confirmed that the coal is composed of 94.8 vol.% vitrinite, 42.3 vol.% inertinite, and 3.0 vol.% liptinite (Table 14.12). Maximum vitrinite reflectance value of Khuutiin Khonkhor coal is 0.54% (see deposit 53 in Fig. 14.2).

Alagtogoo deposit in East Gobi Basin is hosted in lower-middle Jurassic Khamarkhuuvur Formation (see deposit 47 in Fig. 14.2). The deposit has one seam with a thickness ranging from 0.3 to 7.3 m. Dip is 6°–18° (Fig. 14.6d). The ash yield of coal is 27.8 wt.%. Moisture and volatile matter contents are 3.5 wt.% and 33.1 wt.%, respectively. Total sulfur content is 1.3 wt.%. Calorific value is 30.4 MJ/kg. Similar to other Jurassic coal, Alagtogoo coal is characterized by very high content of vitrinite (93.5 vol.%) and low inertinite (3.9 vol.%) (Tables 14.11 and 14.12). Alagtogoo coal is high-volatile bituminous rank. Coal reserve was calculated to be 133 Mt, but significant portion of the reserve has been mined since 2000.

14.3 Coal Characteristics and Production

14.3.1 Coal Rank and Characteristics

Mongolian coal tends to become younger from west to east. Carboniferous coals are in the Western Mongolian Province, and Permian coals are in the South Gobi and Southern Khangai basins. Coals in the Ikh Bogd, Ongi River Basins, and the Orkhon-Selenge Area are mainly Jurassic, whereas predominant portion of the coals in the Eastern Mongolian Province are Lower Cretaceous. Concomitant with the westward aging of coal-bearing sedimentary sequences, the coal displays an increase in rank. High- to low-volatile bituminous coals are hosted in Carboniferous rocks. High- to medium-volatile bituminous coal occurs within Permian (especially in the South Gobi Basin). High-volatile to subbituminous coal are hosted in Jurassic rocks, whereas lignite is found in Cretaceous (Erdenetsogt et al. 2009). This general tendency of coalification can be seen in Fig. 14.7.

Carboniferous, Jurassic, and Cretaceous coal have 82.7 wt.%, 78.9 wt.%, and 65.2 wt.% carbon contents, respectively (Fig. 14.7a). The average volatile matters of Carboniferous and Permian coal are 26.4 wt.% and 30.3 wt.%, whereas those of Jurassic and Cretaceous coals are 42.5 wt.% and 45.8 wt.%, respectively. The average calorific values are 30.5 MJ/kg (Carboniferous), 33.0 MJ/kg (Permian), 29.2 MJ/kg (Jurassic), and 25.5 MJ/kg (Cretaceous), respectively. The average moisture contents of Carboniferous and Permian coals 2.6 wt.% and 2.0 wt.%, respectively, and those of Jurassic and Cretaceous are 5.7 wt.% and 9.7 wt.%, respectively (Fig. 14.7b, c). The total sulfur contents of coal are relatively low,

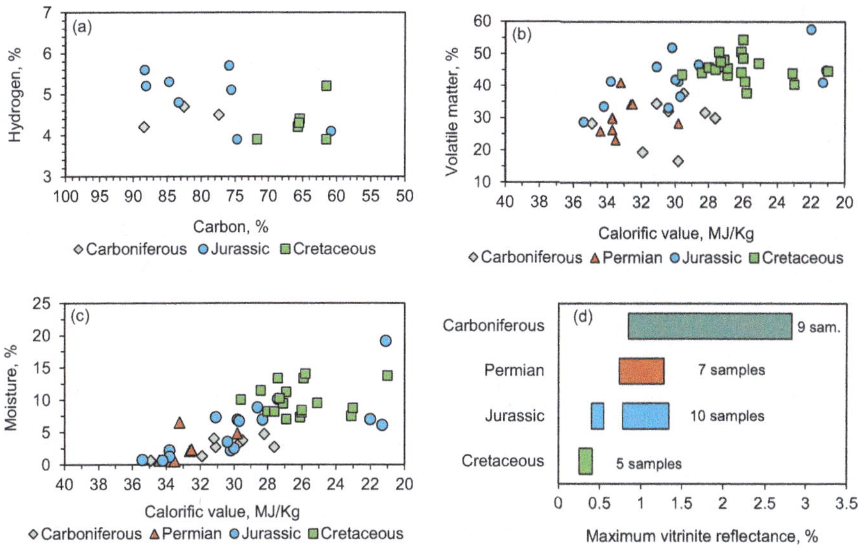

Fig. 14.7 The characteristics of Mongolian coal (modified after Erdenetsogt et al. 2009, and published with permission of Elsevier). (**a**) Hydrogen content versus carbon content, (**b**) volatile matter content versus calorific value, (**c**) moisture content versus calorific value, (**d**) maximum vitrinite reflectance

ranging from 0.2 wt.% to 2.8 wt.%, indicating a nonmarine influence (Erdenetsogt et al. 2009).

According to the results of 31 samples, the mean vitrinite reflectance values of Carboniferous coal in Western Mongolia, Permian coal in Southern Mongolia, Jurassic coal in Central and Northern Mongolia, and Cretaceous coal in Eastern Mongolia are 1.69%, 1.04%, 0.71%, and 0.33%, respectively (Fig. 14.7d). Interestingly, some Jurassic coal are almost mid-volatile bituminous, whereas others are subbituminous, even lignite. However, Mongolian Jurassic coals share common characteristics, including similar petrographic composition.

Mongolian coal is humic in type (Erdenetsogt et al. 2009). The concentrations of vitrinite and inertinite maceral groups vary from 44.9 vol.% to 96.6 vol.% and from 53.3 vol.% and 2 vol.% (on mineral matter-free basis), respectively. Liptinite maceral group contents are low, less than 11.7 vol.% (Fig. 14.8). Jurassic coal is petrographically distinct from others and represented by elevated contents of vitrinite (ranging from 87.3 vol.% to 96.6 vol.%) and liptinite (up to 11.7 vol.%). Jurassic coal is distributed throughout Mongolia. These high and relatively constant vitrinite contents in such an enormous area might be explained by paleoclimatic conditions. Hence, vitrinite in the coal seams is formed as a result of anaerobic preservation source material in swamps. It is thought that the coal with high vitrinite content was formed under more humid and raised water table conditions due to a higher rainfall than that in the Permian and Cretaceous (Erdenetsogt et al. 2009). All

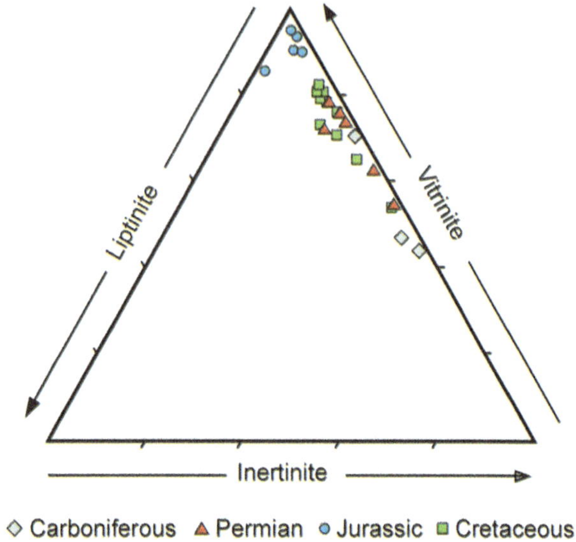

Fig. 14.8 Maceral composition of Mongolian coals based on mineral matter free basis (adopted from Erdenetsogt et al. 2009, and published with permission of the Elsevier)

Table 14.13 Mongolian coal resources classified by age (modified after Erdenetsogt et al. 2009) and published with permission of the Elsevier

Age	Coal resources (Mt)			% of reserves in total resources
	Reserves	Hypothetical resources	Total	
Carboniferous	1,763	18,400	20,163	11.3
Permian	7,634	11,300	18,934	10.6
Jurassic	2,631	10,800	13,431	7.5
Cretaceous	24,080	101,700	125,780	70.5
Total	36,108	142,200	178,308	100

other coals have similar maceral group distributions. The vitrinite/huminite and inertinite groups are varying from 44.9 vol.% to 85.0 vol.% and from 14.0 vol.% to 53.3 vol.%, respectively. The liptinite maceral group varies between 1.0 vol.% and 7.0 vol.% (Fig. 14.8).

14.3.2 Coal Resources and Production

Total geological resources of Mongolian coal are estimated to be 173 Gt. Identified reserves are calculated at 36.1 Gt. The greater portion of coal reserves are lignite in the Eastern Mongolian Province (Table 14.13). The Carboniferous and Permian coal may be suitable for metallurgical purposes, but some Jurassic and Lower Cretaceous coal can be classified as thermal coal. In addition to existing reserves, a significant tonnage of coking coal is expected to be discovered in the Western Mongolian

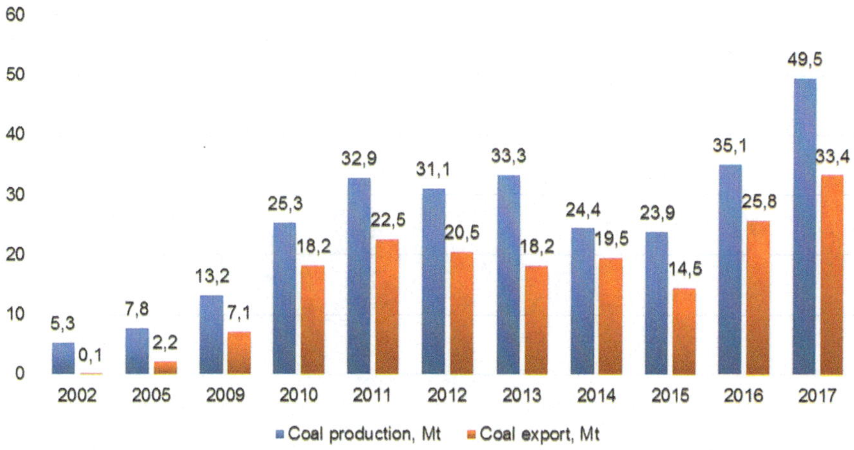

Fig. 14.9 Coal production and export of Mongolia, from 2002 to 2017 (data from MRPAM 2017)

Province and South Gobi Basin. During the last two decades, coal exploration has increased dramatically, and coal reserves of Mongolia are tripled since 2008, when it was 10.1 Gt (Erdenetsogt et al. 2009).

Coal production in Mongolia has been constantly growing over the last years and is projected to increase for the next decade, because of demand for energy from China, which is southern neighbor of Mongolia (Fig. 14.9). In 2017, domestic consumption was approximately 8 Mt, primarily for electricity and heat generation. Bituminous coal export was 33.4 Mt, rising by 29.5% from 25.8 Mt in 2016. Mongolia was the ninth major coal exporter in the world in 2016 (IEA 2017).

14.4 Oil Shale and Petroleum

Oil shale in Mongolia was hosted in Middle Jurassic sequences and Lower Cretaceous Shinekhudag Formation (Bat-Erdene 2009b; Hasegawa et al. 2018). Currently, only few Jurassic oil shales in Central and Northern Mongolia, closely related to previously known Jurassic coal seams, have been catalogued so far (Erdenetsogt and Jargal 2014). However, it is quite certain that more oil shale can be discovered in Jurassic sedimentary rocks at Western and Eastern Mongolia, where relatively bigger basins compared with Central and Northern Mongolia exist. Lower Cretaceous lacustrine oil shale is distributed widely in Central and Eastern Mongolia (Fig. 14.10).

Both Jurassic and Cretaceous oil shale-bearing sequences are composed of alternating beds of oil shale, dolomitic marlstone, dolomite, calcite, and silty mudstones. Thickness of oil shale is 300 m at Uvurjargalant (see location 5 in Fig. 14.10), 127 m at Bayanjargalan (see location 6 in Fig. 14.10), 130 m at Bayan-Erkhet (see location 7 in Fig. 14.10), and 100 m at Shavart-Ovoo (see location 8 in Fig. 14.10).

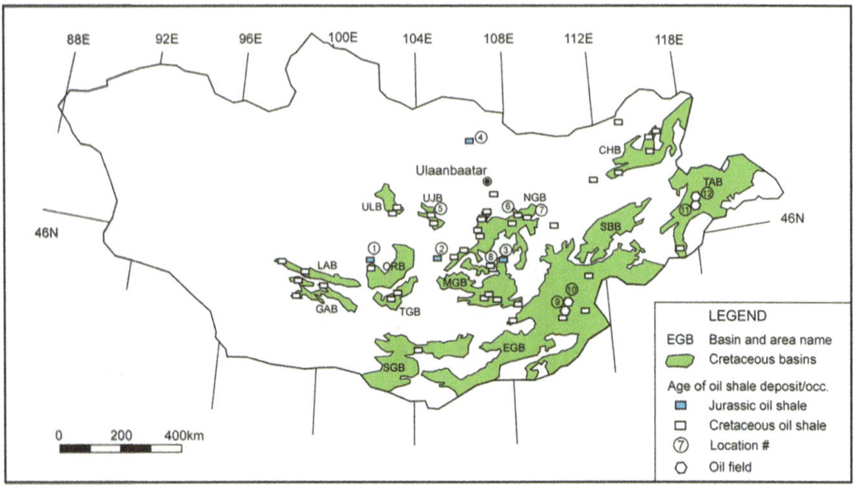

Fig. 14.10 Distribution map of Mongolian oil shale deposits/occurrences and oil shale-bearing basins with crude oil fields (modified after Bat-Erdene 2009b and published with permission of Erdenet Mine). Basins: GAB, Gobi-Altai; LAB, Nuuruudiin Khotgor; TGB, Tugrug, ORB, Ongi River; ULB, Ugii Lake, UJB;, Uvurjargalant, MGB, Middle Gobi; NGB, Nyalga; SGB, South Gobi; EGB, East Gobi; CHB, Choibalsan; SBB, Sukhbaatar; TAB, Tamsag; Oil shale deposits/ occurrences: 1, Bayanteeg; 2, Tsagaan Ovoo; 3, Eedemt; 4, Shariingol; 5, Uvurjargalant; 6, Bayanjargalan,; 7, Bayan-Erkhet; 8, Shavart Ovoo; Oil fields: 9, Tsagaan-Els; 10, Zuunbayan; 11, Toson Uul; 12, Tamsag

The shale oil yield ranges from less than 1.0 wt.% to more than 20 wt.%, e.g., 16 wt. % at Bayanjargalan, 14 wt.% at Bayan-Erkhet, and 21 wt.% at Shavart-Ovoo. Sulfur content is small, always less than 1.0 wt.% (Bat-Erdene 2009b).

Lower Cretaceous oil shales distributed in Central Mongolia have highly oil-prone Type I kerogen, emphasized by high total organic carbon (TOC) content (average of 10.1 wt.%), S2 (average of 68.7 mgHC/g rock), and HI (average of 619 mgHC/g TOC) (Erdenetsogt 2017) (Fig. 14.10). Organic matter (OM) in oil shales is accumulated in stratified lakes with anoxic bottom water, reflected by low Pr/Ph ratios (<0.28) and highly negative δ^{13}Corg (average of −30.6‰) and highly positive δ^{15}Nt (average of +10.5‰) values. The salinity of lakes was different, suggested by variable gammacerane index ranging from 0.03 to 0.44. Due to small size and instability of paleolakes, depositional environment and organic matter input type were variable that was reflected in n-alkane distributions, carbon preference index, and terrigenous/aquatic ratio of hydrocarbons of each oil shale (Erdenetsogt et al. 2018). Lower Cretaceous oil shales in eastern and southeastern Mongolia are identical to the oil shales in Central Mongolia and characterized by high TOC content, Type I kerogen, and similar depositional environment (Yamamoto et al. 1998; Johnson et al. 2003).

Middle Jurassic Tsagaan-Ovoo oil shale at Tsagaan-Ovoo coal deposit (see location 2 in Fig. 14.10) has high TOC content with an average of 15.4 wt.% and

S2 with an average of 60.5. The oil shale contains II/III kerogen. OM in the oil shale was accumulated in oxic, freshwater conditions with increased contribution from coaly materials, suggested by high Pr/Ph (>1.5) and atomic C/N (avg. 31.2 ± 0.6). This finding is also supported by stable carbon (average of −25.9‰) and nitrogen isotopic (+4.3‰) compositions as well as n-alkane distribution (Erdenetsogt et al. 2017a). Similar oil shales containing coaly OM are catalogued at Bayanteeg and Shariingol coal mines (see locations 1 and 4 in Fig. 14.10). Middle Jurassic Eedemt oil shale at Khootiin khonkhor coal deposit (see location 3 in Fig. 14.10) is different from abovementioned Middle Jurassic oil shales. Eedemt oil shale has quite high HI with an average of 714 mg HC/gTOC, which suggests Type I kerogen (Yamamoto et al. 1993). These data indicate that some Jurassic oil shale was accumulated in shallow and toxic swamps, whereas others were deposited in much deeper lakes.

In the eastern and southeastern Mongolia, four small oil fields, Toson-Uul, Tamsag, Zuunbayan, and Tsagaan-Els, have been discovered (Fig. 14.10). The oil fields are hosted in Lower Cretaceous sequences. In addition, several tar sand and oil seepages were catalogued. The proved reserves of the four oil fields are calculated at 2.1 billion barrels with 271.6 million barrels of recoverable reserves. API gravity and sulfur contents of Tsagaan-Els and Tamsag fields are 30° and 36° and 0.14% and 0.01%, respectively (Erdenetsogt and Jargal 2014). Crude oil production in 2017 was 7.6 million barrels, and all extracted oil was exported to China (MRPAM 2018).

References

Ariunchimeg Y (2011) Carbonii sistem (carboniferous system). In: Byamba J (ed) Stratigrapy (stratigraphy). Mongoliin Geologi Ba Ashigt Maltmal (Mongolian geology and mineral resources), vol 01. Soyombo, Ulaanbaatar, pp 271–334

Ariunchimeg Y, Otgonbaatar D, Buyantegsh B, Gansukh S (2014) Khushuutiin nuursnii ordiin nasnii asuudald (Khushuut deposit's age revisited). Mongoliin geosudlaach 41:42–55

Badarch G, Cunningham WD, Windley BF (2002) A new terrain subdivision for Mongolia: implications for the Phanerozoic crustal growth of Central Asia. J Asian Earth Sci 21:87–110

Baigalmaa N, Jargal C, Baatarkhuyag A, Altantsetseg D (2012) Nariinsukhaitiin chuluun nuursnii ordain petrograpiin sudalgaanaas (preliminary results of petrographical study of Nariin sukhait). Geologiin asuudluud 339(12):72–73

Bat-Erdene B (2014) Nuurstkhotgoriin ordiin nuursnii chanariin uurchlultiin zui togtol (study on the coal quality changes of Nuurstkhotgor). Geologiin asuudluud 416(13):169–180

Bat-Erdene B (2015) Khurengoliin ordiin nuursjilt, nuursnii chanariin sudalgaa (Study on coal quality and coal seam characteristics of Khurengol deposit). Master thesis, National University of Mongolia

Bat-Erdene D (1992) Nature of distribution and formational condition of coal basins in the Mongolian orogenic belt. Dissertation, State University of Geology and Exploration

Bat-Erdene D (2009a) Mongoliin nuursnii ord, savuud (Mongolian coal basins and deposits). In: Bat-Erdene D (ed) Fossil fuel. Mongoliin Geologi Ba Ashigt Maltmal (Mongolian geology and mineral resources), vol 05. Soyombo, Ulaanbaatar, pp 27–174

Bat-Erdene D (2009b) Mongoliin shatdag zanariin Ord savuud (Mongolian oil shale basins and deposits). In: Bat-Erdene D (ed) Fossil fuel. Mongoliin Geologi Ba Ashigt Maltmal (Mongolian geology and mineral resources), vol 05. Soyombo, Ulaanbaatar, pp 176–268

Bat-Erdene D, Jargal L (1995) Mongol-altain nuursnii sav gazar (Mongol-Altai coal-bearing basin). Mongol Altain geologiin asuudluud 03:36–38

Cunningham WD (2005) Active intracontinental transpressional mountain building in the Mongolian Altai: defining a new class of orogen. Earth Planet Sci Lett 240:436–444

Durante MV (1976) Paeobotanicheskoe obosnovanie stratigrapii karbona I permi Mongolii (the carboniferous and Permian stratigraphy of Mongolia on the basis of palaeobotanical data). In: Meyen SV (ed) The joint Soviet–Mongolian scientific-research geological expedition, transactions, vol 19, pp 1–280

Erdenetsogt B (2017) Preliminary results of petroleum source rock evaluation of Mongolian Mesozoic oil shales. Geologiin asuudluud 473(15):46–57

Erdenetsogt B, Barsbold T (2012) Orkhon-Selengiin talbain nuursjiltiin ontslogiin asuudald (coal characteristics of Orkhon-Selenge area). Geologiin asuudluud 339(12):51–60

Erdenetsogt B, Barsbold T (2014) Ovootiin ordiin geologiin togtots, nuursnii chanar (geological setting and coal quality of Ovoo deposit). Geologiin asuudluud 416(13):121–132

Erdenetsogt B, Hong SK, Choi J, Munkhnasan B, Mungunchimeg B, Erdene N, Baigalmaa N, Nansalmaa D (2018) Geochemistry of Lower Cretaceous lacustrine oil shales in Central Mongolia. In: Abstracts of EGU General Assembly 2018, Vienna, 8–13 April 2018

Erdenetsogt B, Hong SK, Choi J, Odgerel N, Lee I, Ichinnorov N, Tsolmon G, Munkhnasan N (2017a) Tectonic setting of Jurassic basins in Central Mongolia: Insights from the geochemistry of Tsagaan-Ovoo oil shale. In: Abstracts of 2017 AGU Fall Meeting, New Orleans, LA, 11–15 December 2017

Erdenetsogt B, Jargal L (2014) Fossil fuels hosted in Mesozoic sequences of Mongolia. In: Abstracts of the 2nd international symposium of IGCP 608, Waseda, Tokyo, 4–10 September 2014

Erdenetsogt B, Lee I, Bat-Erdene D, Jargal L (2009) Mongolian coal-bearing basins: geological settings, coal characteristics, distribution, and resources. Int J Coal Geol 80:87–104

Erdenetsogt B, Lee I, Ko YJ (2017b) Carbon isotope analysis and a solid state 13C NMR study of Mongolian lignite: changes in stable carbon isotopic composition during diagenesis. Org Geochem 113:293–302

Erdenetsogt B, Lee I, Lee SK, Ko YJ, Bat-Erdene D (2010) Solid-state 13C CP/MAS NMR study of Baganuur coal, Mongolia: oxygen-loss during coalification from lignite to subbituminous rank. Int J Coal Geol 82:37–44

Erdenetsogt B, Ts B, Gantulga B (2014) Tavantolgoi bolon Ovootiin nuursiig kholij, koksjuulah sudalgaa. I heseg: Orduudiin geologiin togtots, nuursnii chinar (study on Tavantolgoi coal blend with Ovoot. Part I: geological setting of the deposits and coal quality). Mongoliin geosudlaach 41:73–84

Hasegawa H, Ando H, Hasebe N, Ichinnorov N, Ohta T, Hasegawa T, Yamamoto M, Li G, Erdenetsogt B, Heimhofer U, Murata T, Shinya H, Enerel G, Oyunjargal G, Munkhtsetseg O, Buyantegsh B, Enkhbat D, Suzuki N, Irino T, Yamamoto K, Kouchi Y, Orihashi Y (2018) Chronostratigraphy of the Jurassic–cretaceous lacustrine deposits in Southeast Mongolia. Island Arc 27:e12243

Hendrix MS, Beck MA, Badarch G, Graham SA (2001) Triassic synorogenic sedimentation in southern Mongolia; early effects of intracontinental deformation. In: Hendrix MS, Davis GA (eds) Paleozoic and Mesozoic tectonic evolution of Central Asia — from continental assembly to Intracontinental deformation, vol 194. Geological Society of America Memoir, pp 389–412

International Energy Agency (IEA) (2017). Coal Information 2017. http://www.iea.org/ Accessed 28 July 2018

Jargal L, Erdene A, Dashkhorol A, Nansalmaa D (2017) Khujirtiin nuursnii ordain petrograpiin nairlaga, chanariin sudalgaanii ur dungees (Petrographical composition and coal quality of Khujirt deposit). Geologiin asuudluud 473(15):36–45

Jerzykiewicz T, Russell DA (1991) Late Mesozoic stratigraphy and vertebrates of the Gobi basin. Cretac Res 12:345–377

Johnson CL, Greene TJ, Zinniker DA, Moldowan JM, Hendrix MS, Carroll AR (2003) Geochemical characteristics and correlation of oil and nonmarine source rocks from Mongolia. AAPG Bull 87:817–846

Khand Y, Badagarav D, Ariunchimeg Y, Barsbold R (2000) Cretaceous system in Mongolia and its depositional environment. In: Okada H, Mateer NJ (eds) Cretaceous environments of Asia. Elsevier, Amsterdam, pp 49–81

Khosbayar P (1972) Stratigrpiya Mesozoya Zapadnii Mongolii I istoriya ee geologicheskovo razvitiya za eto vremya (Stratigraphy of the continental Mesozoic of Western Mongolia and its depositional history). Dissertation, Moscow State University

Lamb MA, Badarch G (2001) Paleozoic sedimentary basins and volcanic arc systems of southern Mongolia: new geochemical and petrographic constraints. In: Hendrix MS, Davis GA (eds) Paleozoic and Mesozoic tectonic evolution of Central Asia—from continental assembly to intracontinental deformation, vol 194. Geological Society of America Memoir, pp 117–150

Lamb MA, Badarch G, Navratil T, Poier R (2008) Structural and geochronologic data from the shin Jinst area, eastern Gobi Altai, Mongolia: implications for Phanerozoic intracontinental deformation in Asia. Tectonophys 451:312–330

Makhbadar T (2011) Mezozoi (Mesozoic). In: Byamba J (ed) Stratigrapy (stratigraphy). Mongoliin Geologi Ba Ashigt Maltmal (Mongolian geology and mineral resources), vol 01. Soyombo, Ulaanbaatar, pp 408–497

Mongolia Energy Corporation (MEC) (2018) Annual Report 2018. http://www.mongolia-energy.com. Accessed 28 July 2018

Mineral Resources and Petroleum Authority of Mongolia (MRPAM) (2017) Annual report 2017. https://mrpam.gov.mn/ Accessed 28 July 2018

Mineral Resources and Petroleum Authority of Mongolia (MRPAM) (2018) Statistical data. https://mrpam.gov.mn/ Accessed 28 July 2018

Orolmaa D, Ariunchimeg Y (2011) Permiin sistem (Permian system). In: Byamba J (ed) Stratigrapy (Stratigraphy). Mongoliin Geologi Ba Ashigt Maltmal (Mongolian geology and mineral resources), vol 01. Soyombo, Ulaanbaatar, pp 340–404

Ruzhentsev SV, Burashnikov VV (1996) Tectonics of western Mongolian Salairides. Geoteconics (English translation) 29:379–394

Sjostrom DJ, Hendrix MS, Badamgarav D, Graham SA, Nelson BK (2001) Sedimentology and provenance of Mesozoic nonmarine strata in western Mongolia; a record of intracontinental deformation. In: Hendrix MS, Davis GA (eds) Paleozoic and Mesozoic tectonic evolution of Central Asia—from continental assembly to intracontinental deformation, vol 194. Geological Society of America Memoir, pp 361–388

Traynor JJ, Sladen C (1995) Tectonic and stratigraphic evolution of the Mongolian People's Republic and its influence on hydrocarbon geology and potential. Mar Pet Geol 12:35–52

Ulambadrakh K, Baigalmaa N (2014) Ikh Bogdiin nuursnii saviin Juriin hurdsiin nuursjilt (Jurassic coals in Ikh Bogd basin). Mongoliin geosudlaach 41:67–72

Vassallo R, Jolivet M, Ritz JF, Braucher R, Larroque C, Sue C, Todbileg M, Javkhlanbold D (2007) Uplift age and rates of the Gurvan Bogd system (Gobi-Altay) by apatite fission track analysis. Earth Planet Sci Lett 259:333–346

Yamamoto M, Bat-Erdene D, Ulziihutag P, Enomoto M, Kajiwara Y, Takeda N, Suzuki Y, Watanabe Y, Nakajima T (1993) Preliminary report on geochemistry of lower cretaceous Dsunbayan oil shales, eastern Mongolia. Geol Surv Japan Bull 44:685–691

Yamamoto M, Bat-Erdene D, Ulziihutag P, Watanabe Y, Imai N, Kajiwara Y, Takeda N, Nakajima T (1998) Organic geochemistry and palynology of lower cretaceous Zuunbayan oil shales, Mongolia. Geol Surv Japan Bull 49:257–274

The manufacturer's authorised representative in the EU is Springer Nature Customer Service Centre GmbH, Europaplatz 3, 69115 Heidelberg, Germany. If you have any concerns regarding our products, please contact ProductSafety@springernature.com

Printed and bound by CPI Group (UK) Ltd, Croydon, CR0 4YY
30/09/2026
02225289-0013